Lecture Notes in Business Information Processing 527

LNBIP reports state-of-the-art results in areas related to business information systems and industrial application software development – timely, at a high level, and in both printed and electronic form.

The type of material published includes

- Proceedings (published in time for the respective event)
- Postproceedings (consisting of thoroughly revised and/or extended final papers)
- Other edited monographs (such as, for example, project reports or invited volumes)
- Tutorials (coherently integrated collections of lectures given at advanced courses, seminars, schools, etc.)
- Award-winning or exceptional theses

LNBIP is abstracted/indexed in DBLP, EI and Scopus. LNBIP volumes are also submitted for the inclusion in ISI Proceedings.

Claudio Di Ciccio · Walid Fdhila ·
Simone Agostinelli · Daniel Amyot ·
Henrik Leopold · Michal Krčál ·
Monika Malinova Mandelburger ·
Gregor Polančič · Katarina Tomičić-Pupek ·
Katarzyna Gdowska · Thomas Grisold ·
Piotr Sliż · Iris Beerepoot · Renata Gabryelczyk ·
Ralf Plattfaut
Editors

Business Process Management

Blockchain, Robotic Process Automation, Central and Eastern European, Educators and Industry Forum

BPM 2024 Blockchain, RPA, CEE, Educators and Industry Forum
Krakow, Poland, September 1–6, 2024
Proceedings

Springer

Editors
Claudio Di Ciccio
Utrecht University
Utrecht, The Netherlands

Walid Fdhila
University of Vienna
Vienna, Austria

Simone Agostinelli
Sapienza University of Rome
Rome, Italy

Daniel Amyot
University of Ottawa
Ottawa, ON, Canada

Henrik Leopold
Kühne Logistics University
Hamburg, Germany

Michal Krčál
Masaryk University
Brno, Czech Republic

Monika Malinova Mandelburger
TU Wien
Vienna, Austria

Gregor Polančič
University of Maribor
Maribor, Slovenia

Katarina Tomičić-Pupek
University of Zagreb
Varaždin, Croatia

Katarzyna Gdowska
AGH University of Krakow
Kraków, Poland

Thomas Grisold
University of St. Gallen
St. Gallen, Switzerland

Piotr Sliż
University of Gdańsk
Gdańsk, Poland

Iris Beerepoot
Utrecht University
Utrecht, The Netherlands

Renata Gabryelczyk
University of Warsaw
Warsaw, Poland

Ralf Plattfaut
University of Duisburg-Essen
Essen, Germany

ISSN 1865-1348 ISSN 1865-1356 (electronic)
Lecture Notes in Business Information Processing
ISBN 978-3-031-70444-4 ISBN 978-3-031-70445-1 (eBook)
https://doi.org/10.1007/978-3-031-70445-1

This Springer imprint is published by the registered company Springer Nature Switzerland AG
The registered company address is: Gewerbestrasse 11, 6330 Cham, Switzerland

Preface

The International Conference on Business Process Management (BPM) was established about 22 years ago as the conference where people from academia and industry meet and discuss the latest developments in the area of business process management. In 2024, the conference was organized in Krakow, Poland. This year's BPM also featured five specialized forums. This volume contains the proceedings of the Blockchain Forum, the Robotic Process Automation (RPA) Forum, the Central & Eastern European (CEE) Forum, the Educators Forum and the Industry Forum, which took place during September 3–5, 2024.

The Blockchain Forum aimed to provide a platform for the discussion of ongoing research and success stories on using blockchain for collaborative information systems. Conceptual, technical and application-oriented contributions were within the scope of this theme. The papers selected for the Blockchain Forum showcase fresh ideas from exciting and emerging topics in the area of blockchain technologies with a special focus on, yet not limited to, business process management. This was the sixth edition of the Blockchain Forum.

The concept of RPA has gained relevant attention in both industry and academia. RPA provides a way of automating mundane and repetitive human tasks requiring a low level of intrusiveness with the IT infrastructure. The RPA Forum aimed to bring together researchers from various communities and disciplines to discuss challenges, opportunities and new ideas related to RPA and its application to business processes in the private and public sectors. The forum solicited contributions related to three main RPA areas: management, technology and applications. The RPA Forum took place for the fifth time, after its previous appearances at BPM 2020, in Seville (Spain), at BPM 2021, in Rome (Italy), at BPM 2022, in Münster (Germany), and at BPM 2023, in Utrecht (The Netherlands).

The CEE Forum aims to promote and present BPM research conducted in CEE countries contributing to the theory and practice of BPM. It allows testing of existing BPM methods and techniques against the specific challenges due to cultural, political, regional or organizational differences and the various nature of the business processes of CEE countries. The CEE Forum also provides a platform for discussing innovative ideas on BPM methods and approaches applied in CEE countries. Insights on BPM in the CEE region would enable practitioners to avoid the issues faced in these countries in the process of BPM adoption and can be seen as a valuable contribution to the general BPM field. Therefore, the CEE Forum invited practitioners applying BPM methods and researchers conducting BPM research in CEE countries to submit their work to ignite discussions and collaborations to advance BPM theories and practices focused on issues important to CEE countries.

The goal of the Educators Forum is to create a space dedicated to BPM teaching, where educators can share teaching experiences, good practices, course design, examples of implementation of educational methods and tools, and much more. The large number and variety of articles presented at the Educators Forum testifies to the importance and involvement of the authors in BPM teaching and their openness to sharing and inspiring each other, as well as the strong need for substantive networking in the international community of people involved in BPM teaching.

The Industry Forum serves as a platform connecting academia and industry professionals to exchange real-world experiences and insights on leveraging Business Process Management. In 2024, the forum invited practitioners to present case studies showcasing the application of cutting-edge technologies, such as RPA, Process Mining, AI and others, which drive digital innovation and provide insights into real-world experiences in leveraging BPM. These case studies, presented during the Industry Forum, facilitated a deeper understanding of how process innovations are implemented in business practice, how strategic value is created and how change management and cross-organizational collaboration are effectively executed. The forum provided a unique opportunity for attendees to gain valuable insights into the practical applications of BPM and its potential to transform organizations in the rapidly evolving digital landscape.

The Blockchain Forum received 13 submissions, of which 7 papers were selected for presentation at the venue and for inclusion in these proceedings. The RPA Forum received 9 submissions, which led to the acceptance of the top 5 as full papers. The CEE Forum received 12 submissions, and the top 6 high-quality papers were selected for presentation and publication. The Educators Forum had 24 submissions, of which 12 were selected for publication and presentation. Lastly, the Industry Forum received 11 submissions, of which 5 were accepted for publication and presentation. For the five forums, each submission was single-blindly reviewed by at least three members of the respective Program Committee.

We hope that the reader of these proceedings will enjoy the papers presented at the forums. We would like to express our gratitude to the authors who submitted their contributions, even though some of the presented papers could not be accepted despite their quality. We also thank our colleagues who acted as reviewers in the selection process and provided the authors with meaningful and constructive comments. Finally, special thanks go to Edyta Brzychczy and Krzysztof Kluza (general chairs of BPM 2024) for organizing such an outstanding conference.

September 2024

Claudio Di Ciccio
Walid Fdhila
Simone Agostinelli
Daniel Amyot
Henrik Leopold
Michal Krčál
Monika Malinova Mandelburger
Gregor Polančič
Katarina Tomičić-Pupek
Katarzyna Gdowska
Thomas Grisold
Piotr Sliż
Iris Beerepoot
Renata Gabryelczyk
Ralf Plattfaut

Organization

Blockchain Forum Program Chairs

Claudio Di Ciccio	University of Utrecht, The Netherlands
Walid Fdhila	University of Vienna, Austria

RPA Forum Program Chairs

Simone Agostinelli	Sapienza University of Rome, Italy
Daniel Amyot	University of Ottawa, Canada
Henrik Leopold	Kühne Logistics University, Germany

CEE Forum Program Chairs

Michal Krčál	Masaryk University, Czech Republic
Monika Malinova Mandelburger	TU Wien, Austria
Gregor Polančič	University of Maribor, Slovenia
Katarina Tomičić-Pupek	University of Zagreb, Croatia

Educators Forum Program Chairs

Katarzyna Gdowska	AGH University of Krakow, Poland
Thomas Grisold	University of St. Gallen, Switzerland
Piotr Sliż	University of Gdańsk, Poland

Industry Forum Program Chairs

Iris Beerepoot	Utrecht University, The Netherlands
Renata Gabryelczyk	University of Warsaw, Poland
Ralf Plattfaut	University of Duisburg-Essen, Germany

Combined Program Committee

Aleksandre Asatiani	University of Gothenburg, Sweden
Bernhard Axmann	Technische Hochschule Ingolstadt, Germany

Saimir Bala	Humboldt Universität zu Berlin, Germany
Iris Beerepoot	Utrecht University, The Netherlands
Matthijs Berkhout	Hogeschool Utrecht, The Netherlands
Grzegorz Bełz	Wrocław University of Economics and Business, Poland
Paolo Bottoni	Sapienza University of Rome, Italy
Sarah Bouraga	University of Namur, Belgium
Saša Brdnik	University of Maribor, Slovenia
Mateja Bule	University of Maribor, Slovenia
Marco Comuzzi	Ulsan National Institute of Science and Technology, South Korea
André Coners	South Westphalia University of Applied Sciences, Germany
Szymon Cyfert	Poznań University of Economics and Business, Poland
Adela del Río Ortega	University of Seville, Spain
Andrea Delgado	Universidad de la República, Uruguay
Damian Dziembek	Czestochowa University of Technology, Poland
Mahendrawathi Er	Institut Teknologi Sepuluh Nopember, Indonesia
Peter Fettke	German Research Center for Artificial Inteilligence (DFKI) and Saarland University, Germany
Hans-Georg Fill	University of Fribourg, Switzerland
Walid Gaaloul	Télécom SudParis, France
Daniela Garbin Praničević	University of Split, Croatia
José María García	University of Seville, Spain
Luciano García-Bañuelos	Tecnológico de Monterrey, Mexico
Piotr Gawrysiak	WUT, Poland
Katarzyna Gdowska	AGH University of Krakow, Poland
José González Enríquez	University of Seville, Spain
Daniela Grigori	Paris Dauphine University, France
Marzena Grzesiak	Gdańsk University of Technology, Poland
Sergio Guerreiro	University of Lisbon, Portugal
Tomasz Gzik	Military University of Technology, Poland
Jennifer Haase	Humboldt University, Germany
Felix Härer	University of Fribourg, Switzerland
Marko Hölbl	University of Maribor, Slovenia
Hannu Jaakkola	University of Tampere, Finland
Christian Janiesch	TU Dortmund University, Germany
Slinger Jansen	Utrecht University, The Netherlands
Urszula Jessen	Eindhoven University of Technology, The Netherlands
Andrés Jiménez Ramírez	University of Seville, Spain

Marite Kirikova	Riga Technical University, Latvia
Kais Klai	Université Sorbonne Paris Nord, France
Christopher Klinkmüller	BPMotion, Australia
Krzysztof Kluza	AGH University of Krakow, Poland
Jelmer Jan Koorn	MLC, The Netherlands
Katja Kous	University of Maribor, Slovenia
Ingo Kregel	University of Applied Sciences and Arts Dortmund, Germany
Julius Köpke	University of Klagenfurt, Austria
Abderrahmane Leshob	UQAM, Canada
Orlenys López Pintado	University of Tartu, Estonia
Martin Macák	Masaryk University, Czech Republic
Fabrizio Maria Maggi	Free University of Bozen-Bolzano, Italy
Andrea Marrella	Sapienza University of Rome, Italy
Niels Martin	Hasselt University, Belgium
Benjamin Matthies	Münster University of Applied Sciences, Germany
Raimundas Matulevicius	University of Tartu, Estonia
Massimo Mecella	Sapienza University of Rome, Italy
Jan Mendling	Humboldt-Universität zu Berlin, Germany
Giovanni Meroni	Technical University of Denmark, Denmark
Fredrik Milani	University of Tartu, Estonia
Ljubica Milanović Glavan	University of Zagreb, Croatia
Andrea Morichetta	University of Camerino, Italy
Jorge Munoz-Gama	Pontificia Universidad Católica de Chile, Chile
Jannis Nacke	Enervie Gruppe, Germany
Eugenio Nerio Nemmi	Sapienza University of Rome, Italy
Andrzej Niesler	Wrocław University of Economics and Business, Poland
Alex Norta	Tallinn University of Technology, Estonia
Dan O'Leary	University of Southern California, USA
Malgorzata Pankowska	University of Economics in Katowice, Poland
Teijo Peltoniemi	University of Turku, Finland
Günther Pernul	Universität Regensburg, Germany
Igor Pihir	University of Zagreb, Croatia
Ralf Plattfaut	University of Duisburg-Essen, Germany
Pierluigi Plebani	Politecnico di Milano, Italy
Marta Podobińska-Staniec	AGH University of Krakow, Poland
Natalia Potoczek	Institute of Economics, Polish Academy of Sciences, Poland
Petr Průcha	Technical University of Liberec, Czech Republic
Luise Pufahl	TU Munich, Germany

Jana-Rebecca Rehse	University of Mannheim, Germany
Manuel Resinas	University of Seville, Spain
Yara Rizk	IBM Research Cambridge, USA
Michael Rosemann	Queensland University of Technology, Australia
Stefan Schulte	TU Hamburg, Germany
Stefan Schönig	University of Regensburg, Germany
Tijs Slaats	University of Copenhagen, Denmark
Piotr Sliż	University of Gdańsk, Poland
Koen Smit	Utrecht University, The Netherlands
Mark Staples	CSIRO, Australia
Dragana Stojanović	University of Belgrade, Serbia
Dalia Suša Vugec	University of Zagreb, Croatia
Rehan Syed	Queensland University of Technology, Australia
Stefan Tai	Technical University of Berlin, Germany
Alfred Taudes	Vienna University of Economics and Business, Austria
Lucinéia Heloisa Thom	Federal University of Rio Grande do Sul, Brasil
Martina Tomicic Furjan	University of Zagreb Faculty of Organization and Informatics, Croatia
Oktay Turetken	Eindhoven University of Technology, The Netherlands
Inge van de Weerd	Utrecht University, The Netherlands
Jan Martijn van der Werf	Utrecht University, The Netherlands
Amy Van Looy	Ghent University, Belgium
Gergana Vladova	University of Potsdam, Germany
Maximilian Völker	University of Potsdam, Germany
Moe Thandar Wynn	Queensland University of Technology, Australia
Ahad Zareravasan	Masaryk University, Czech Republic
Francesca Zerbato	Eindhoven University of Technology, The Netherlands
Kaiwen Zhang	École de technologie supérieure, Canada

Additional Reviewers

Berniak-Woźny, Justyna
Brahem, Amina
Bühlmann, Marcel
Curty, Simon
D'Ambra, John
Damwerth, Philipp
Dannecker, Achim
Ebrahimi, Elmira
Gau, Michael
Grill, Johannes
Grzybowska, Katarzyna
Henry, Tiphaine
Hewing, Michael
Homoncik, Katarzyna
Israel Sion, Sion
Kerschbaum, Clemens

Kipping, Gregor
Książek, Roger
König, Maximilian
Laue, Ralf
Lösser, Benedict
Mahringer, Christian
Marcelletti, Alessandro
Nagel, Sabine
Pirta, Rūta
Scheruhn, Hans-Jürgen
Siciński, Jedrzej
Suchenia, Anna
Szelagowski, Marek
Tönnissen, Stefan
Urban, Isabella
Vidgof, Maxim
Vollenberg, Carolin
Wuttke, Tobias
Łyp-Wrońska, Katarzyna

Contents

Educators Forum

Blockchain Forum

Preface of the Blockchain Forum

A blockchain is a distributed ledger technology that guarantees immutability and integrity, providing a practical solution to challenging problems in coordination and traceability thanks to its transparent and decentralized nature. By eliminating the need for central authorities and mutual trust between parties, blockchain-based systems open up diverse opportunities in the context of business process management. Thus, it spurs a rethink of how business collaborations are designed and implemented, impacting the whole BPM lifecycle in a wide range of fields, including healthcare, supply chain, logistics, and finance. However, these opportunities may come with challenges from security and privacy to scalability and performance, which must be addressed for successful integration and adoption in real-world applications.

Blockchain technology has already seen both academic interest and practical adoption. Various approaches exist that integrate BPM and blockchain, including executing DMN decisions, facilitating collaborative business processes, and enabling monitoring and auditing of process data through blockchain technology. These approaches also include secrecy-preserving sharing of data among actors in a multi-party process setting, utilizing process mining for smart contract analysis and validation, and much more.

The sixth BPM Blockchain Forum provided a platform for discussing and introducing new ideas related to research directions within techniques for and applications of blockchain and distributed ledger technologies in the context of BPM and beyond. It attracted 13 international submissions on a diverse range of topics within the scope of the forum. All submissions were reviewed and discussed by at least three Program Committee members. The examination of the manuscripts and the referees' comments led to the acceptance of 7 papers.

In "A Comprehensive Review of Multi-Chain Architecture for Blockchain Integration in Organizations", Sion et al. provide a summary of current literature on architectures operating across multiple blockchain networks for organizations. They analyze key architectural approaches and design principles proposed in the literature for multi-chain architectures, offering a synthesized classification of published solutions.

In their paper "Transparent Transaction Ordering in Blockchain-based Collaborative Processes", Atwi et al. tackle the key challenge of a possible divergence between the actual collation of transactions in the ledger and the intended order as perceived by the process participants. To this end, they resort to the information indicated by process choreography models to guide the buffering, ordering, and execution.

"Interpreted and Confidential Execution of Process Choreographies on a Blockchain", by Petto et al., introduces an approach that leverages process choreographies too, with the aim of enabling validation of process executions. The authors employ public blockchains for attesting to the occurrence of interactions and use zero-knowledge proofs with zk-SNARKs to hide the details that should remain confidential to the involved parties.

Zero-knowledge proofs are also central to the work of Henry and Tucci-Piergiovanni, entitled "Secure Proof Verification Blockchain Patterns", with the objective to assist practitioners in choosing the preferable non-interactive zero-knowledge proof design for blockchain applications. The authors formalize a general pattern and three related sub-patterns to this end.

Malik et al. focus on the potential contrast between the need for processes to evolve over time to adapt to new contexts, emerging norms, and contingencies, and the immutable nature of smart contracts that may automate them. In "Smart Contracts' Upgradability for Flexible Business Processes", they present an approach to take advantage of the most appropriate upgradability patterns of smart contracts to address process change patterns.

In "The Cost of Executing Business Processes on Next-Generation Blockchains: The Case of Algorand", Stiehle and Weber describe their approach of encoding process models as choreography diagrams into TEAL, the smart contract language used on the Algorand blockchain platform. Thereafter, they compare execution costs of process automated thereby with those of Ethereum- and cloud-based solutions.

Loebbecke et al. in "BPMS Blockchain Technology Soft Integration for Non-tamperable Logging" describes a novel approach that utilizes blockchain for secure reporting of process operations. Their solution is grounded in soft integration as it is based on an endowment of process management systems rather than on their redesign, thus reducing disruptiveness and implementation costs.

To conclude, we express our gratitude to the authors for their precious contribution to the Forum, the Program Committee members for their timely and constructive work, and the organizing team of BPM 2024 for their indefatigable support in the preparation of and throughout the conference.

Organization

Program Chairs

Claudio Di Ciccio	Utrecht University, The Netherlands
Walid Fdhila	University of Vienna, Austria

Program Committee

Paolo Bottoni	Sapienza University of Rome, Italy
Sarah Bouraga	University of Namur, Belgium
Marco Comuzzi	Ulsan National Institute of Science and Technology, South Korea
Hans-Georg Fill	University of Fribourg, Switzerland
Walid Gaaloul	Computer Science Department Télécom SudParis, France
José María García	University of Seville, Spain
Luciano García-Bañuelos	Tecnologico de Monterrey, Mexico
Sergio Guerreiro	University of Lisbon, Portugal
Felix Härer	University of Fribourg, Switzerland
Marko Hölbl	University of Maribor, Slovenia
Slinger Jansen	Utrecht University, The Netherlands
Kais Klai	Université Sorbonne Paris Nord, France
Julius Köpke	University of Klagenfurt, Austria
Orlenys López Pintado	University of Tartu, Estonia
Raimundas Matulevicius	University of Tartu, Estonia
Giovanni Meroni	Technical University of Denmark, Denmark
Andrea Morichetta	University of Camerino, Italy
Eugenio Nerio Nemmi	Sapienza University of Rome, Italy
Alex Norta	Tallinn University of Technology, Estonia
Günther Pernul	Universität Regensburg, Germany
Pierluigi Plebani	Politecnico di Milano, Italy
Michael Rosemann	Queensland University of Technology, Australia
Stefan Schulte	TU Hamburg, Germany
Mark Staples	CSIRO, Australia
Stefan Tai	Technical University of Berlin, Germany
Alfred Taudes	Vienna University of Economics and Business, Austria
Kaiwen Zhang	École de technologie supérieure ÉTS, Canada

A Comprehensive Review of Multi-chain Architecture for Blockchain Integration in Organizations

Sion Israel Sion[1](✉), Kaiwen Zhang[1], Alain April[1], Thomas Maketa Lutete[1], and Charlaine Bouchard[2]

[1] École de technologie supérieure, Montréal, Canada
sion-israel.sion.1@ens.etsmtl.ca,
{kaiwen.zhang,alain.april,thomas.maketa}@etsmtl.ca
[2] Université Laval, Québec, Canada
charlaine.bouchard@fd.ulaval.ca

Abstract. Blockchain technology has evolved very quickly, yielding heterogeneous blockchain platforms often featuring specific characteristics and functionalities. However, the lack of interoperability and scalability among these platforms poses significant challenges to the widespread adoption of blockchain solutions in enterprise settings. The multi-chain architecture, which entails the use of interlinked chains for heterogeneous use-cases, has recently sparked interest as an approach for addressing these issues, enabling communication and collaboration among different blockchain networks. This review aims to provide an overview of existing research on multi-chain architectures for blockchain integration in organizations, with the goal of categorizing the existing literature based on their architectural characteristics.

We conducted a systematic search and screening process, which yielded 44 relevant papers out of an initial pool of 223. The selected studies were categorized into five main themes: layered architectures, service-oriented architectures, enterprise architecture patterns, interoperability frameworks and protocols, and organizational models and governance mechanisms. Each category was analyzed through the lens of four key architectural perspectives-business, information, application, and technology-to meet our objectives. The findings of this review add to the growing body of knowledge on blockchain interoperability and enterprise integration, and provide valuable guidance to researchers, practitioners, and organizations as they explore multi-chain architectures integration.

Keywords: blockchain · interoperability · cross-chain · multi-chain · architecture · enterprise integration

1 Introduction

The rapid growth of blockchain technology has produced a decentralized, transparent, and secure approach to storing and exchanging data, with implications

C. Di Ciccio et al. (Eds.): BPM 2024, LNBIP 527, pp. 5–24, 2024.
https://doi.org/10.1007/978-3-031-70445-1_1

for a variety of industries. However, this wave of unique blockchain platforms with different features, consensus mechanisms, and scope of application - and a community that is similarly fragmented - has led to a fragmented ecosystem that hinders the widespread adoption and scalability of blockchain solutions [5,22,37]. As organizations increasingly investigate the potential of blockchain technology to streamline their operations, enhance efficiency, and develop novel business models, the necessity for interoperability and seamless integration across disparate blockchain networks has emerged as a pivotal challenge.

In order to address this challenge, the concept of multi-chain architectures has emerged as a promising approach to enable the coexistence, collaboration, and exchange of assets and data across heterogeneous blockchain networks [18,22,37]. Multi-chain architectures aim to provide a framework for the development of interoperable and scalable blockchain solutions that can support a wide range of enterprise applications and use cases [5,22,37]. Nevertheless, the design and implementation of multi-chain architectures necessitate a deep comprehension of the technical, organizational, and governance elements inherent to integrating blockchain solutions with existing enterprise systems and processes [3,38,52].

In the context of enterprise architecture practices, it is important to distinguish two fundamental concepts: Architectural models and Architectural layers. In this paper, the latter will be referred to as Architectural Perspectives, and the terms "organization" and "enterprise" will be used interchangeably to refer to any entity considering or implementing blockchain technology in its operations. Architectural models, such as microservices, service-oriented architecture (SOA), event-driven architecture (EDA), serverless architecture and monolithic architecture, serve as the backbone of systems design. They are core to the optimization of business processes and organizational workflows, for maximum efficiency and scalability [32]. Architectural perpectives, on the other hand, provide a framework for understanding and designing blockchain-based systems that align with the overall enterprise architecture. Architectural practices provide abstractions to manage complexity, enable separation of concerns, and improve lifecycle management.

As organizations increasingly explore blockchain technologies, the need for a robust architectural approach becomes paramount. Despite the potential of blockchain to revolutionize business models and operations, many enterprises face significant challenges in integrating these technologies within their existing systems such as compliance and regulatory requirements, integration of data repositories, interoperability with legacy enterprise technologies, automated process orchestration, business process management lifecycle and governance [36]. This integration is not merely a technical challenge but also an organizational one, requiring a nuanced consideration of both the technological landscape and the business context [13].

In addition, considering a multi-chain architecture raises an important aspect: the ability to seamlessly execute processes across multiple blockchain networks. This aspect, which can be referred to as process interoperability, acts as a unifying force, bringing together considerations from various organizational perspectives. It requires the alignment of cross-organizational processes, ensures consistent interpretation of data across chains, facilitates the integration of diverse blockchain applications with existing systems, and demands compatible protocols and consensus mechanisms.

Therefore, to effectively address the challenges and opportunities of multi-chain integration within organizations, there are four distinct organizational architectural perspectives to consider: Business Architecture, focusing on business motivations, value exchanges, and cross-organizational processes; Information Architecture, dealing with semantic models, shared ontologies, and data interoperability; Application Architecture, covering legacy systems, interfaces, and smart contract integration; and Technology Architecture, including protocols, platforms, infrastructure, and consensus models [3,38,52].

This holistic lens highlights the intricate, interconnected challenges that multi-chain blockchain integration in organizations presents and emphasizes the necessity of comprehensive solutions that tackle all facets of organizational architecture.

Recent years have seen a rising trend in research on blockchain interoperability and multi-chain architectures [5,22,37,41]. The state-of-the-art approaches, challenges and future directions from an organizational architecture perspective are, however, still to be comprehensively examined. Enterprise-focused models and architectures that align with the strategic priorities, business processes and information systems of blockchain-adopting organizations are also to be developed [3,38,52].

This paper reviews existing research on multi-chain architectures for blockchain systems. This review seeks to classify and categorize the existing literature based on their architectural characteristics, target domains, and organizational relevance.

To meet these objectives, this review answers the following questions:

- What are the key architectural approaches and design principles proposed in the literature for multi-chain architectures in organizations ?
- How can the existing literature on multi-chain architectures be categorized or classified based on their architectural characteristics, target domains, or organizational relevance ?

The remainder of this paper is structured as follows: Sect. 2 provides a background on blockchain technology, interoperability challenges, and the emergence of multichain architectures. Section 3 describes the methodology for the literature review - including the search strategy, inclusion and exclusion critieria, and data synthesis approach. Section 4 presents the results - categorizing and analyzing the various multi-chain architectures, frameworks, and models identified in the literature. Finally, Sect. 5 discusses the key findings and concludes the paper by summarizing the main contributions and implications of the review for researchers and practitioners in the blockchain and enterprise architecture field.

2 Background

In this section, we present the relevant background information to enhance the reader's comprehension of the current review:

2.1 Introduction to Blockchain Technology

Blockchain technology is a revolutionary innovation that provides a secure, transparent way of digitally recording transactions on a tamper-proof ledger without the need for a central intermediary [37]. At its heart is a decentralised, distributed ledger that continuously records and grows a list of ordered records called "blocks", with each block containing a timestamp and a link to the previous block. Supporting this are a number of key underpinning technologies, including cryptographic hash functions, digital signatures, and consensus mechanisms, that together provide the integrity, immutability, and security of the recorded transactions [14,37].

There are three main types of Blockchain networks distributed based on both their accessibility and governance models - Public (permissionless), Private (permissioned) and Consortium (semi-permissioned) [14,37]. Public blockchains like those behind Bitcoin and Ethereum are open to anyone and allow participants to join the network without prior permission. Private blockchains, however, are restricted to a particular group of participants and are typically used within a single organisation or enterprise. Consortium blockchains sit somewhere in the middle, with a set of pre-selected participants responsible for the consensus and validation of each block transactions [14,37].

2.2 Blockchain Interoperability and Multi-Chain Architecture

Despite the rapid growth and adoption of blockchain technology across various sectors, the proliferation of isolated blockchain networks has led to several limitations that hinder their widespread adoption [5]. These limitations include scalability issues, as individual blockchain networks may struggle to handle high transaction volumes; liquidity constraints, as assets are often locked within specific chains; and restricted functionality, as the features and capabilities of a single blockchain platform may not be sufficient to address complex, real-world use cases [5,22].

Surmounting these challenges has been made possible with the rise of blockchain interoperability [5,22]. This term describes the ability of various blockchain networks to communicate and transact with one another without interruption, while sharing information seamlessly. The literature on blockchain interoperability has explored various designs, each with its advantages and limitations. Early approaches, such as notary schemes and trusted relays [29], relied on trusted third-parties to validate or relay transaction data across chains, compromising the decentralized nature of blockchains. Sidechains and pegged sidechains [27,29] offered an alternative by creating application-specific chains connected to a main chain, but they typically employed centralized validators, raising security concerns. Decentralized approaches using atomic swaps based on hash-time locked contracts (HTLCs) [29] have also been proposed, enabling trustless cross-chain asset transfers without intermediaries.

Subsequently, as the limitations of isolated blockchain networks became apparent, the concept of multi-chain architectures gained traction as a potential solution [18,22,37,45]. Multi-chain architectures facilitate the coexistence and collaboration of multiple blockchain networks, each optimized for specific purposes, such as scalability, privacy, or functionality. By allowing different blockchains to interoperate and share resources, multi-chain architectures can help to overcome the challenges of siloed networks and unlock new business opportunities.

Multi-chain architectures in blockchain systems, particularly focusing on interoperability challenges and solutions, has garnered significant attention in recent research literature [5,8,43]. However, there is a lack of comprehensive reviews that specifically address multi-chain architectures for integrating blockchain into organizations, particularly from an architectural perspective in the context of organizational adoption. This paper aims to bridge this gap by providing a targeted analysis of multi-chain architectures and their implications for blockchain integration in organizational contexts.

2.3 Enterprise Architecture and Organizational Perspectives

Enterprise Architecture (EA), as defined by [23] can be described as a comprehensive system of principles, techniques, and models that are applied in the planning and implementation of an organization's structure, business operations, information technology, and infrastructure. It describes the enterprise from a holistic perspective, and is used to enable better-informed decision-making, improved efficiencies, and for enabling greater agility in the face of changing business needs. It offers several important insights and benefits which are applicable to the context of blockchain integration as well, where it can offer insights into making sure that the technical solutions that are implemented are the right ones, and that they are connected to the rest of the architecture of the organization and the priorities that it identifies [46].

Several prominent EA frameworks have been developed to guide organizations in their EA efforts, including the Zachman Framework [49], the "4+1" View Model [21], and TOGAF [17]. These frameworks provide a structured way to describe and analyze an enterprise, from different perspectives, in order for a holistic view to be developed of the enterprise's structure, processes, and technology.

The four main architecture domains identified in TOGAF that are closely aligned with the organizational architectural perspectives, we considered in this paper are Business Architecture, Data Architecture, Application Architecture, and Technology Architecture [17].

- Business Architecture: This perspective focuses on the business motivations, value-exchanges and cross-organizational processes. It outlines the alignment

of the enterprise architecture with the organization's goals and objectives, including its business strategy, governance, organization, and key business processes;
- Information Architecture: This perspective (Data Architecture in TOGAF terms) deals with semantic models, shared ontologies and data interoperability. It defines the structure of an organization's logical and physical data assets and data management resources, ensuring that the data is well-governed, secure, and accessible.
- Application Architecture: This perspective details the applications and their interactions, their interfaces and their integration with legacy systems. It defines the individual applications, as well as their connections and connections to the core business processes of the organizations;
- Technology Architecture: This perspective encompasses the middleware, protocols, platforms, infrastructure and network that support the deployment and operation of a solution. It encompasses the necessary logical software and hardware functionalities needed for the successful implementation of business, data, and application services.

By following principles and best practices associated with these four architectural perspectives, organizations can ensure that their blockchain implementations are technically-sound and are aligned with their business strategy and information management practices, their application landscape, and their technology infrastructure and operations [3]. This holistic approach is essential to realizing the full potential of this transformative technology and driving successful integration in enterprise contexts.

3 Methodology

This research utilized a systematic literature review approach to map the primary concepts, important sources, and types of evidence accessible in the multi-chain architecture research area. We relied on Google Scholar as our primary data source, given its extensive index of significant digital libraries and proceedings, such as the ACM Digital Library, IEEE Xplore, Springer Nature, and Lecture Notes in Computer Science. Additionally, Google Scholar indexes grey literature libraries like arXiv and Cryptology ePrint Archive, as well as other resources like books, theses, or online repositories.

The literature search was conducted using Google Scholar's keyword search, focusing on papers published since 2015 to ensure the relevance and timeliness of the reviewed literature. The search query that was utilized to locate papers within the scope of our research is provided below: (multi-chain OR cross-chain) AND (architecture OR framework OR model) AND (blockchain OR distributed ledger). The search produced 4,930 outcomes. We ceased our search at the 250th citation, as no pertinent studies were discovered past the 200th.

In the first analysis, we filtered the results based on the language (English) and the relevance of the title and abstract to our research objectives. To guarantee the pertinence of the articles integrated, every resource was subjected to evaluation by at least two of the authors and contributors to the paper.

Our final tally for studies was 27. To supplement the 27 studies we initially retrieved, we applied forward reference techniques, leading us to an additional 17 studies. These 17 studies, coupled with the original 27, provide a comprehensive overview of the relevant research for our study, as presented in Fig. 1.

Data extraction was performed on the selected papers to capture key information about each study, including architectural approach, design principles (Components and services, interoperability techniques), organizational adoption benefits, integration challenges addressed, applicability to organizations and architecture. This information was summarized in a data extraction table to facilitate the analysis and synthesis of the reviewed literature.

The data synthesis process involved a qualitative approach, focusing on identifying the architectural organizational perspective, key themes, concepts, and challenges related to multi-chain architectures for blockchain integration in organizations. The reviewed papers were categorized based on their main contributions and the types of architectures or frameworks proposed.

We acknowledge that our methodology has certain limitations. Firstly, our investigation may not embrace the entire research domain since we primarily depend on Google Scholar as our search engine. Secondly, the constraints in available documentation may have limited the depth and breadth of our analysis. Some papers may not have provided sufficient technical details or may have lacked clarity in their descriptions of the proposed architectures or frameworks, making it challenging to fully assess their contributions and potential impact.

Despite these limitations, we believe that our systematic literature review provides a valuable contribution to the understanding of multi-chain architectures for blockchain integration in organizations. Future research could expand upon this work by exploring additional databases.

4 Multi-chain Architectures for Blockchain Integration in Organizations

After analyzing the 44 relevant papers out of the initial screened 223 papers, a pattern emerges in the context of multi-chain architectures for blockchain integration in organizations. The papers can be categorized into the following groups based on their focus and approach: Layered Architectures Sect. 4.1, Service-Oriented Architectures Section Sect. 4.2, Enterprise Architecture Patterns Sect. 4.3, Interoperability Frameworks and Protocols Sect. 4.4 and Organizational Models and Governance Sect. 4.5.

To address the organizational architectural perspective, we considered the four perspectives that an architecture should match: Business Architecture (business motivations, value exchanges, cross-organizational processes), Information Architecture (semantic models, shared ontologies, data interoperability), Application Architecture (legacy systems, interfaces, smart contract integration) and Technology Architecture (protocols, platforms, infrastructure, consensus models).

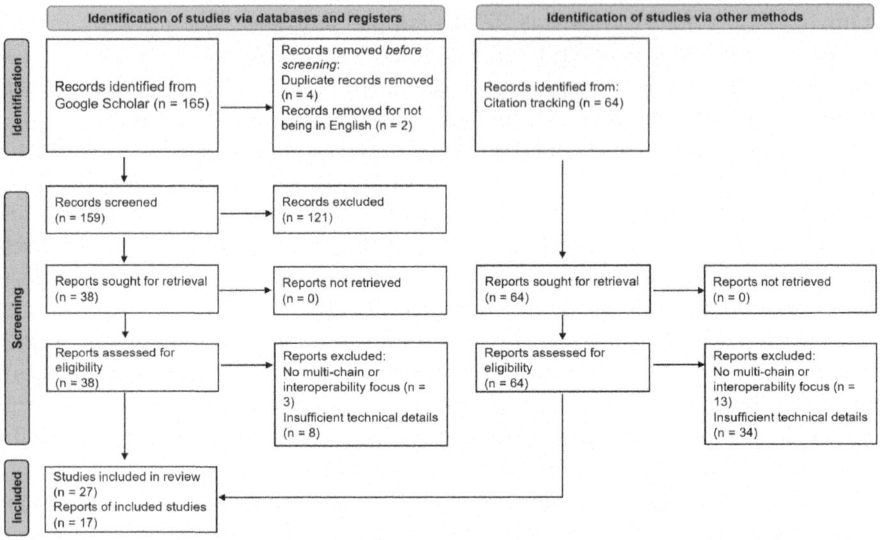

Fig. 1. PRISMA flow diagram for systematic review.

The Table 1 provides a summary the categorization of selected papers across different architectural perspectives. Each category includes references to relevant papers, and the breakdown highlights their contributions to Business Architecture, Information Architecture, Application Architecture, and Technology Architecture.

The papers reviewed in this study cover various aspects of these architectural perspectives, providing insights into how multi-chain architectures are designed to align with the needs of organizations.

4.1 Layered Architectures

Layered architectures have emerged as a popular approach to design the multi-chain systems for blockchain integration in organizations [3,18]. By separating the functionality of the blockchain network into distinct layers, these architectures aim to improve modularity, scalability, and interoperability [18]. The analyzed papers propose different layered architectures, each with their unique characteristics and benefits. A common theme among such layered architectures is to explain separation of the data layer from the consensus and application layers. For example, [18] introduces a multi-layer architecture that comprises of a data layer, network layer, consensus layer, and application layer. This separation allows the data storage and management functions to be optimized, while keeping the consensus mechanism and application logic independent and flexible. This is in line with the Information Architecture perspective, enabling efficient data management and interoperability. Similarly, [3] proposes a standardized multi-layered architecture that spans data, network, consensus, contract,

Table 1. Categorization of papers

Category	Business Architecture	Information Architecture	Application Architecture	Technology Architecture
Layered Architecture	[3, 14]	[3, 18, 24, 27]	[38]	[14, 28]
Service-Oriented Architecture	[12, 41]	[41]	[48]	[32]
Enterprise Architecture Patterns	[20, 42]	[20, 44]	[40]	[20, 40, 44]
Interoperability Frameworks and Protocols	[2, 33, 47],	[6, 7, 10, 11, 22]	[4, 9, 19, 26, 29, 31, 39, 43]	[1, 5, 16, 25, 29, 37, 39, 45, 50, 51]
Organizational Models and Governance	[30, 35]	[52]	[34]	[30, 34, 52]

trust, application, and enterprise layers. This architecture addresses the Business Architecture perspective, by enabling cross-organizational processes and value exchanges. Scalability is another key benefit of layered architectures. [14] presents a multi-layer model with a consortium blockchain as the control layer, and multiple side-chain blockchains managing specific IoT device clusters. In the architecture, parallel transaction processing and decentralized data storage is achieved using IPFS, resulting in improved performance and scalability in IoT environments. This is a reflection of the Technology Architecture perspective, addressing scalability and performance requirements.

Layered architectures enhance the functionality and performance of blockchain networks by integrating off-chain components, such as storage systems and oracles, as demonstrated in [38]. This architecture addresses the Application Architecture perspective by facilitating the integration of legacy systems and external data sources. Interoperability is a key aspect of multi-chain architectures, and layered designs support communication and data exchange among distinct blockchain networks. [27] and [24] discuss Layer 2 protocols and dual-chain designs for secure off-chain operations and seamless asset transfer, aligning with the Information Architecture perspective. The modularity of layered architectures allows for the development of application-specific solutions, such as the dual-chain architecture presented in [15], which enables secure data sharing and traceability for industry collaboration, addressing the Business Architecture perspective.

4.2 Service-Oriented Architectures

Service-Oriented Architectures (SOAs) have received significant attention in the context of multi-chain environments for blockchain integration in organizations. SOA and microservice principles are used to create modular, interoperable, and scalable blockchain-based systems. Several papers reviewed in this section look into various aspects of SOA and microservices in the blockchain space, applications, and benefits.

[12] introduces the Blockchain Service Bus (BSB) architecture, an architecture that extends the Service-Oriented Architecture (SOA) and Enterprise Service Bus (ESB) concepts using blockchain-based governance and messaging to securely and compliantly integrate microservices across enterprise boundaries. The BSB architecture combines the message bus with blockchain-based governance using smart contracts to secure and compliantly connect the microservices. This approach aligns with the Business Architecture perspective by enabling cross-organizational processes and value exchanges.

In [41], the authors propose the concept of the Internet of Blockchains (IoB), that models blockchain interactions as a dependency graph. This dependency graph can be further utilized for discovery, orchestration, oracles, scalability, security, cross-chain transactions, and data integration. This paper aligns with the information architecture perspective, as it enables data interoperability and communication between the different blockchains.

The proposed B-CSTP (Blockchain-based construction Service Trading Platform) in [48] introduces a novel construction service trading platform built upon consortium blockchain with smart contracts for Task Decomposition and Service Composition (TDSC). This architecture aligns with the Application Architecture perspective by combining smart contracts that automatically enforce the agreements for service composition and are natively integrated into the blockchain, allowing other participants to collaborate seamlessly.

[32] introduces a framework that combines blockchain technology with algebraic service composition to create a decentralized service marketplace. This approach ensures decentralized control, enhances security, and facilitates transparent and automated service composition and execution. This approach addresses the technology architecture perspective by leveraging blockchain technology to create a secure and decentralized service marketplace.

The use of SOA and blockchain technologies together enhances security and trust in multi-chain environments. Smart contracts can be used to enforce policy and automate processes, ensuring compliance and reducing reliance on intermediaries, and promoting transparency. Blockchain's immutability and distributed nature provides further security and integrity in the system and creates a tamper-proof history of service interactions and transactions. However, the integration of SOA and microservices in organizations leveraging blockchain also presents challenges. These can be standardization and interoperability protocols, performance, due to the overhead introduced by blockchain operations, and a more complex management and orchestration of services across many blockchain networks.

4.3 Enterprise Architecture Patterns

Enterprise Architecture Patterns are playing a critical role in designing and implementing multi-chain architectures for blockchain organisations. These patterns are critical for addressing enterprise blockchain integration challenges such as scalability, privacy, security and governance. The discussed papers in this section are exploring various enterprise architecture patterns and approaches to address these challenges and to unlock their benefits in the context of blockchain based organisations.

[40] presents an IoT enabled multi-chain decentralized blockchain architecture for petroleum supply chains. The provided architecture consists of a private blockchain network and it leverages IPFS for extended storage. The proposed architecture ensures data integrity, security and privacy while allowing for autonomous decision making and operations energy efficient suitable for resource constrained IoT devices. It falls into the category of Technology Architecture due to the specific issues of IoT environments and resource constrained devices that it addresses in the Data, Network, Consensus and Application layers.

[44] presents a Virtual Chain Storage Model for cross-blockchain transactions. Virtual chain is a basic model based on target and source chain in the architecture in store. Virtual Chain architecture allows Secure transaction processing and data processing can be extended to the blockchain. It falls into the category of Data layer and Information Architecture perspective, as it allows seamless data sharing and transaction processing across different blockchain networks.

A double-blockchain architecture to connect heterogeneous blockchain networks is presented in [20]. This architecture is composed of a public chain for transactions and a private chain for asset backing. These two chains are linked by color blockchains for logically structured networks. This approach aligns with the Business Architecture perspective by enabling cross-organizational processes and value exchanges between public and private blockchain networks.

In [42], a master-slave cross-chain architecture with hierarchical management structure was proposed to supervise and govern power systems. The architecture is composed of a master chain that provides governance, supervision gateways, and multiple business slave chains. It uses smart contract-based governance rules to automatically supervise the system. This architecture addresses the Business Architecture perspective by enabling the efficient governance and supervision of power systems across multiple blockchain networks.

The integration of enterprise architecture patterns for organizations leveraging blockchain also comes with some challenges. The integration of legacy systems with blockchain networks may be complex, requiring close consideration of existing infrastructure and careful system design. Additionally, the overhead associated with cross-chain communication and consensus mechanisms could lead to performance and latency issues. Furthermore, ensuring the security and integrity of data across multiple chains will require robust authentication, access control, and data validation mechanisms.

4.4 Interoperability Frameworks and Protocols

Interoperability is a critical necessity for organizations in a multi-chain setting. To enable seamless communication, data exchange, and value transfer among blockchain networks with heterogeneous designs, several interoperability protocols and frameworks have been proposed in the literature. Papers discussed in this section aim at proposing standards, protocols, and solutions that enable cross-blockchain interoperability to face the challenges that arise in unifying diverse blockchain systems.

[37] presents a multi-chain architecture that integrates blockchain technology with supply chain management systems, enabling interoperability among various blockchain platforms and traditional systems through standardized interfaces. This aligns with the Information Architecture perspective, facilitating data exchange and interoperability between disparate systems. [45] describes a multi-blockchain scenario that enables cross-blockchain communication between relay and coordination nodes, using a flat, decentralized system structure. This supports the Technology Architecture perspective by providing a decentralized and viable solution for cross-chain communication.

[22] provides an overview of cross-chain architectures, classifying them as blockchain ecosystem-based, relay-based, sidechain-based, and peer-based systems, and emphasizes the need for standards and supporting technologies to enable interoperability among diverse blockchain platforms, aligning with the Information Architecture perspective. [47] proposes a distributed fusion cross-chain model and architecture for financial transactions, employing cross-agent nodes with standardized interfaces to connect heterogeneous blockchains, representing the Business Architecture perspective and enabling cross-organizational processes and value exchanges in the financial domain.

[1,31], and [6] focus on generic interoperability frameworks and protocols. [31] presents a user-driven, transaction-based cross-chain communication model leveraging smart contracts for information exchange and state updates among multi-blockchain networks, aligning with the Application Architecture perspective by providing a framework for secure and verifiable cross-chain transactions. [1] proposes Chain-Net, a generic interoperability framework with cross-chain logic modules, protocols, storage, and cross-chain state, enabling communication and transaction execution across blockchains and addressing the Technology Architecture perspective as a comprehensive solution for cross-chain interoperability. [6] characterizes interoperability systems as blockchain ecosystems, relay-based, sidechain-based and peer-based systems and analyses the architecture, benefits and challenges of each. This paper aligns with the Information Architecture perspective. It provides a taxonomy of interoperability solutions and their characteristics.

[2,10], and [4] explore specific interoperability solutions and mechanisms. [10] presents an oracle-based architecture for cross-chain data migration, employing a three-phase migration mechanism and consensus algorithms to ensure data confidentiality, integrity, and security, aligning with the Information Architecture perspective. [2] proposes a cross-chain interoperability architecture for smart city environments, enabling peer-to-peer data exchange among blockchain networks using mixed nodes and a main chain, fostering cross-organizational processes and data sharing under the Business Architecture perspective. [4] discusses the

importance of distributed ledger technology interoperability solutions, considering technical, semantic, organizational, and legal aspects, and proposes evaluation frameworks for assessing the maturity and compatibility of interoperability solutions.

Additional research in this area offers novel technical solutions for enhancing blockchain interoperability. [9] suggest a performance-optimized system for blockchain-enabled distributed network function virtualization, utilizing multi-access edge computing to improve efficiency. [16] introduce a novel architecture that promotes interoperability through a three-layer design and a cross-chain layer with dedicated protocols. [19] develop an automated benchmarking framework for interoperability platforms, using standardized metrics for cross-chain communication efficiency.

[26] present a dual-chain framework for cross-border e-commerce, which combines a consortium blockchain for transactions and a public chain for traceability. [29] provide a comprehensive review of cross-chain mechanisms, including notary schemes, sidechains, and hash-locking protocols. [39] propose a framework that uses partitioned on-chain and off-chain storage to enhance privacy and interoperability in healthcare data management. [50] design a cross-chain architecture system that incorporates smart contracts and homogeneous consensus mechanisms to facilitate asset transfers between heterogeneous blockchains.

The remaining papers in this section address key aspects of interoperability and multi-chain systems: [5] surveys blockchain interoperability approaches, categorizing them into public connectors, blockchain of blockchains, and hybrid connectors, [7] proposes a cross-chain model enabling communication between private and public blockchains, [11] introduces a scheme for secure enterprise value chain networks with cross-enterprise authentication, [25] explores cross-chain consensus models in complex trust environments, [33] discusses a supervisory mechanism for power master-slave chain architecture, and [43] provides a comprehensive review of blockchain interoperability progress, exploring general principles and procedures for interoperable blockchain systems. All of these papers stress the importance of developing secure, robust and efficient protocols and frameworks that can enable seamless collaboration and data sharing across blockchain based organizations.

For blockchain-based organizations that operate in multi-chain environments, several benefits arise from leveraging interoperability frameworks and protocols. These mechanisms enable different blockchain networks to communicate and exchange data with ease, allowing organizations to collaborate more effectively and share vital information across organizational boundaries. Furthermore, organizations that utilize interoperability solutions will be better positioned to create new business activities and services that span multiple chains, thereby creating new business models and value exchanges.

The development and implementation of interoperability frameworks and protocols also hold challenges. A critical concern is ensuring that cross-chain transactions are secure and private, as the mechanisms themselves may expand the attack surface and create new vulnerabilities. A primary design goal is to execute cross-chain transactions without the need for trusted third parties, but in practice, establishing trust between blockchain networks and achieving consensus on the validity of cross-chain transactions requires strong governance models and consensus mechanisms.

4.5 Organizational Models and Governance

Organizational models and governance mechanisms play a crucial role in the successful integration of blockchain in a multi-chain environments. As blockchain technology enables new forms of decentralized collaboration and decision-making, traditional organizational structures and governance approaches need to be adapted to leverage the benefits of blockchain while ensuring trust, accountability, and efficient coordination among participants. The papers reviewed in this section propose novel organizational models and governance mechanisms for multi-chain environments, addressing challenges such as trust establishment, decision-making processes, and collaboration incentives.

[30] presents a cross-chain model, that employs parallel blockchains fused with smart contracts to simplify the complexity of the rice supply chain. The proposed model is designed to serve as an organizational model that illustrates how multi-chain architectures can simplify complex supply chain processes and increase transparency and trust amongst participants. This aligns with our Business Architecture perspective.

[35] presents a generic organizational model for blockchain systems, based on the Agent/Group/Role paradigm. The model specifies a set of generic roles, groups, and their interactions that capture the core functionalities in blockchains. It provides a unified modeling framework for various blockchains, enabling the rapid prototyping and development of new blockchain solutions by reusing model components. The model conforms to the Business Architecture perspective, serving as a framework for understanding and designing blockchain.

The cross-chain smart contract deployment model in [52] provides a new approach for the verification of Decentralized Identifiers (DIDs) across multiple blockchains. It depends on gateway nodes to decide verification of claims, as well as a credit evaluation system for deciding the credibility of such claims. This cross-chain DID verification model greatly strengthens security, privacy, and efficiency, and is similar to those security and privacy supporting aspects in the Information Architecture perspective.

[34] introduces a heterogeneous multi-chain blockchain network of credit banks and online courses in each region based on the Polkadot framework. The architecture is based on a parallel chain for credit banks and for online courses, a relay chain for cross-chain consensus and cross-chain transaction, and a decentralized storage system via IPFS. The organizational model enables trustless and efficient integration among credit systems and MOOC platforms, cross-chain transactions and sharing of data. The benefits of this approach include enhanced scalability through regional multi-chain structure, interoperability with other blockchain types using bridges, and efficient storage optimization by combining blockchain and IPFS. This model aligns with the Application Architecture

perspective, enabling the integration of different blockchain-based systems and applications.

An important aspect to consider is that adopting new organizational and governance structures for blockchain in organizations is not without challenges. Ensuring that data and transactions are secure and private across multiple blockchain networks requires strong cryptographic mechanisms and access control policies, whereas building trust and consensus among participants with potentially divergent interests and incentives can be complex and may require the creation of new governance frameworks and dispute resolution mechanisms. Furthermore, integrating new blockchain-based organizational models and governance processes with existing legal and regulatory institutions raises challenges of ensuring compliance with relevant laws and regulations, such as data protection and privacy regulations, as well as necessitating adaptations in organizational structures and governance mechanisms.

5 Discussion and Conclusion

The objective of our systematic literature review was to address two key research questions regarding multi-chain architectures for blockchain integration in organizations. To this end, we identified the primary architectural approaches and design principles proposed in the literature. These include layered architectures, service-oriented architectures, enterprise architecture patterns, interoperability frameworks and protocols, and organizational models and governance. Secondly, the existing literature was classified based on architectural characteristics, target domains, and organizational relevance. Each category was analyzed through the lens of four key architectural perspectives: Business Architecture (BA), Information Architecture (IA), Application Architecture (AA), and Technology Architecture (TA).

Our analysis, as shown in Table 1, reveals that while these architectural models present various multi-chain architecture approaches, each solution often addresses only a subset of the enterprise architecture facets. There is a noticeable lack of comprehensive models and architectures that holistically integrate all four facets, which is necessary for the full consideration of blockchain integration in enterprises.

Papers focused on layered architectures tend to address predominantly the IA and TA facets, with limited focus on BA and AA. Service-oriented architectures, on the other hand, predominantly delve into BA and AA, while enterprise architecture patterns are more IA and TA centric. Interoperability frameworks and protocols encompass various aspects of BA, IA, AA, and TA, but none of them provide a comprehensive treatment of the four facets. Also, papers on organizational models and governance predominantly talk about BA, IA, with scarce mention of AA and TA.

In a more practical sense, it can be observed that while significant progress has been made in addressing technical and semantic interoperability, the field has yet to fully grapple with the challenges of process interoperability across all four architectural perspectives. This gap becomes particularly evident when we consider real-world enterprise scenarios. For instance, in supply chain management, a process might span multiple blockchain networks - from sourcing raw materials to manufacturing and distribution. Each step may involve different blockchain platforms, smart contracts, and data models. The seamless execution of such a process requires more than just technical connections between chains; it demands a holistic approach that considers business logic, data semantics, application integration, and technological compatibility simultaneously.

This fragmented approach to addressing enterprise architecture facets highlights the necessity for further research and development of holistic, organization-centric blockchain architectures that align with the principles of the organizational architectural perspectives. Future research should focus on creating comprehensive models and architectures that integrate all facets to enable the full potential of blockchain technology in organizational contexts.

To address these gaps, we propose and discuss the research challenges and opportunities for developing a holistic, enterprise-centric approach to the design and implementation of multi-chain architectures and applications. We suggest a research agenda focused on five key areas:

- Formalizing process interoperability in multi-chain environments using modern mathematical approaches and techniques as a foundation for system interoperability to represent the constantly evolving connections between different blockchain entities, each with its own data structures and ontologies and establish a comprehensive, unified framework that addresses all four architectural perspectives.
- Developing standards and protocols for blockchain interoperability that capture all four enterprise architecture facets (BA, IA, AA, and TA) to guarantee the "automatic and seamless" integration of blockchain applications and architectures with an organization's business processes and goals in mind.
- Integrating blockchain technology with established enterprise architecture frameworks (e.g. TOGAF, Zachman Framework) to define comprehensive and enterprise-centric blockchain architectures that account for the challenges and requirements of deploying blockchain technology within organizational contexts.
- Designing middleware solutions, APIs, integration frameworks, and trust-enabling components that may be employed to seamlessly integrate blockchain technology with an organizations existing systems, applications, and databases as seen from the BA, IA, AA, and TA perspectives.

Conducting case studies and action research to validate the proposed enterprise-centric blockchain architectures and uncover the best practices, challenges, and success factors that present during the design and implementation of multi-chain architectures and applications in real-world enterprise blockchain implementations. Sufficient and timely attention to these research areas will not only contribute to the emergence of a new multi-chain reality but also significantly expand the scope of information systems research and practice in the domains of enterprise blockchain architectures, applications, and governance.

However, this paper does not conduct an in-depth vertical analysis of organizational perspectives within each horizontal category of multi-chain architectures, as it was beyond the current scope. Future research could build on our findings by examining these architectural perspectives more granularly across the identified categories.

The insights gained from the review provide an empirical foundation for future research and development efforts in the nascent area of multi-chain architectures for blockchain integration in organizations. By addressing the challenges and gaps identified and by considering the distinct roles of architectural models and perspectives, researchers and practitioners can work towards unlocking the full potential of blockchain technology in enterprise contexts. This will ultimately lead to more efficient, secure, and collaborative business processes and organizational structures.

References

1. Abdullah, S., Arshad, J., Alsadi, M.: Chain-net: an Internet-inspired framework for interoperable blockchains. Distrib. Ledger Technol. **1**(2), Article no. 7 (2022). https://doi.org/10.1145/3554761
2. Amet, M., Darshan, M., Srivastava, G., Crichigno, J.: A cross-chain interoperability architecture for smart city environments. In: 2022 IEEE Globecom Workshops (GC Workshops), pp. 263–268 (2022). https://doi.org/10.1109/GCWkshps56602.2022.10008645
3. Anthony Jnr, B.: Enhancing blockchain interoperability and intraoperability capabilities in collaborative enterprise-a standardized architecture perspective. Enterp. Inf. Syst. **18**(3) (2024). https://doi.org/10.1080/17517575.2023.2296647
4. Belchior, R., Riley, L., Hardjono, T., Vasconcelos, A., Correia, M.: Do you need a distributed ledger technology interoperability solution? Distrib. Ledger Technol. **2**(1), Article no. 1 (2023). https://doi.org/10.1145/3564532
5. Belchior, R., Vasconcelos, A., Guerreiro, S., Correia, M.: A survey on blockchain interoperability: past, present, and future trends. ACM Comput. Surv. **54**(8), Article no. 168 (2021). https://doi.org/10.1145/3471140
6. Chervinski, J.O., Yu, J., Xu, X.: Characterizing blockchain interoperability systems from an architecture perspective. In: Paiva, S., et al. (eds.) SmartCity 360 2021, vol. 442, pp. 504–520. Springer, Cham (2022). https://doi.org/10.1007/978-3-031-06371-8_33
7. Darshan, M., Amet, M., Srivastava, G., Crichigno, J.: An architecture that enables cross-chain interoperability for next-gen blockchain systems. IEEE Internet Things J. **10**(20), 18282–18291 (2023). https://doi.org/10.1109/JIOT.2023.3279693
8. Falazi, G., Breitenbücher, U., Leymann, F., Schulte, S.: Cross-chain smart contract invocations: a systematic multi-vocal literature review. ACM Comput. Surv. **56**(6), 1–38, Article no. 142 (2024). https://doi.org/10.1145/3638045
9. Fu, X., Yu, F.R., Wang, J., Qi, Q., Liao, J.: Performance optimization for blockchain-enabled distributed network function virtualization management and orchestration. IEEE Trans. Veh. Technol. **69**(6), 6670–6679 (2020). https://doi.org/10.1109/TVT.2020.2985581
10. Gao, Z., Li, H., Xiao, K., Wang, Q.: Cross-chain oracle based data migration mechanism in heterogeneous blockchains. In: 2020 IEEE 40th International Conference

on Distributed Computing Systems (ICDCS), pp. 1263–1268 (2020). https://doi.org/10.1109/ICDCS47774.2020.00162
11. Hao, K., Xin, J., Wang, Z., Yao, Z., Wang, G.: Efficient and secure data sharing scheme on interoperable blockchain database. IEEE Trans. Big Data **9**(4), 1171–1185 (2023). https://doi.org/10.1109/TBDATA.2023.3265178
12. He, S., Ning, Y., Xiong, D., Ma, J.: A blockchain-based micro-services architecture for distributed business. In: Wang, Q., Feng, J., Zhang, L.J. (eds.) ICBC 2023. LNCS, vol. 14206, pp. 38–53. Springer, Cham (2023). https://doi.org/10.1007/978-3-031-44920-8_3
13. Janssen, M., Weerakkody, V., Ismagilova, E., Sivarajah, U., Irani, Z.: A framework for analysing blockchain technology adoption: integrating institutional, market and technical factors. Int. J. Inf. Manag. **50**, 302–309 (2020). https://doi.org/10.1016/j.ijinfomgt.2019.08.012
14. Jiang, Y., Wang, C., Wang, Y., Gao, L.: A cross-chain solution to integrating multiple blockchains for IoT data management. Sensors **19**(9), 2042 (2019)
15. Jiang, Z., Zha, C., Li, X., Xu, Z., Zhang, X., Yin, H.: A cross-chain framework for industry collaboration and transaction. In: 2022 IEEE Smartworld, Ubiquitous Intelligence & Computing, Scalable Computing & Communications, Digital Twin, Privacy Computing, Metaverse, Autonomous & Trusted Vehicles (SmartWorld/UIC/ScalCom/DigitalTwin/PriComp/Meta), pp. 2436–2443 (2022). https://doi.org/10.1109/SmartWorld-UIC-ATC-ScalCom-DigitalTwin-PriComp-Metaverse56740.2022.00341
16. Jin, H., Dai, X., Xiao, J.: Towards a novel architecture for enabling interoperability amongst multiple blockchains. In: 2018 IEEE 38th International Conference on Distributed Computing Systems (ICDCS), pp. 1203–1211 (2018). https://doi.org/10.1109/ICDCS.2018.00120
17. Josey, A.: TOGAF® Version 9.1 - A Pocket Guide. Van Haren Publishing (2016)
18. Kan, L., et al.: A multiple blockchains architecture on inter-blockchain communication. In: 2018 IEEE International Conference on Software Quality, Reliability and Security Companion (QRS-C), pp. 139–145 (2018). https://doi.org/10.1109/QRS-C.2018.00037
19. Kazemi, M., Yazdinejad, A.: Towards automated benchmark support for multi-blockchain interoperability-facilitating platforms. arXiv:2103.03866 [cs] (2021)
20. Košt'ál, K.: Multi-chain architecture for blockchain networks. Inf. Sci. Technol. **12**(2), 8–14 (2020)
21. Kruchten, P.: The 4+1 view model of architecture. IEEE Softw. **12**, 45–50 (1995). https://doi.org/10.1109/52.469759
22. Ladleif, J., Friedow, C., Weske, M.: An architecture for multi-chain business process choreographies. In: Abramowicz, W., Klein, G. (eds.) BIS 2020. LNBIP, vol. 389, pp. 184–196. Springer, Cham (2020). https://doi.org/10.1007/978-3-030-53337-3_14
23. Lankhorst, M.: Beyond enterprise architecture. In: Lankhorst, M., et al. (eds.) Enterprise Architecture at Work. TEES, pp. 303–308. Springer, Heidelberg (2013). https://doi.org/10.1007/978-3-642-29651-2_12
24. Li, C., Zhang, G., Mao, X., Zhang, J., Xing, C.: Multi-chain model and cross-chain communication protocol for financial transactions. In: 2022 IEEE 22nd International Conference on Software Quality, Reliability, and Security Companion (QRS-C), pp. 547–551 (2022). https://doi.org/10.1109/QRS-C57518.2022.00087
25. Liu, C., et al.: Research on cross-chain consensus models in complex trust environments. In: International Conference on Networking and Network Applications (NaNA), pp. 350–358 (2023). https://doi.org/10.1109/NaNA60121.2023.00065

26. Liu, Z., Li, Z.: A blockchain-based framework of cross-border e-commerce supply chain. Int. J. Inf. Manag. **52**, 102059 (2020). https://doi.org/10.1016/j.ijinfomgt.2019.102059
27. Mohanty, D., Anand, D., Aljahdali, H.M., Villar, S.G.: Blockchain interoperability: towards a sustainable payment system. Sustainability **14**(2), 913 (2022)
28. Niu, X., Guan, J., Gao, X., Jiang, S.: Scalable and reliable SDN multi-controller system based on trusted multi-chain. In: IEEE 33rd Annual International Symposium on Personal. Indoor and Mobile Radio Communications (PIMRC), pp. 758–763 (2022). https://doi.org/10.1109/PIMRC54779.2022.9977820
29. Ou, W., et al.: An overview on cross-chain: mechanism, platforms, challenges and advances. Comput. Netw. **218**, 109378 (2022). https://doi.org/10.1016/j.comnet.2022.109378
30. Peng, X., et al.: Research on the cross-chain model of rice supply chain supervision based on parallel blockchain and smart contracts. Foods **11**(9), 1269 (2022)
31. Pillai, B., Biswas, K., Muthukkumarasamy, V.: Cross-chain interoperability among blockchain-based systems using transactions. Knowl. Eng. Rev. **35**, e23 (2020). https://doi.org/10.1017/S0269888920000314
32. Qian, C., Zhu, W.: Toward a decentralized service marketplace: the interplay between blockchain and algebraic service composition. In: Zhang, Q., Wang, Y., Zhang, L.-J. (eds.) CLOUD 2020. LNCS, vol. 12403, pp. 181–198. Springer, Cham (2020). https://doi.org/10.1007/978-3-030-59635-4_13
33. Qin, C., Guo, B., Zhao, Y., Shen, Y.: A secure and reliable construction scheme for enterprise value chain networks. In: 2022 IEEE 24th International Conference on High Performance Computing and Communications, 8th International Conference on Data Science and Systems, 20th International Conference on Smart City, 8th International Conference on Dependability in Sensor, Cloud and Big Data Systems and Application (HPCC/DSS/SmartCity/DependSys), pp. 250–258 (2022). https://doi.org/10.1109/HPCC-DSS-SmartCity-DependSys57074.2022.00065
34. Qu, Y., Chong, Z.: A blockchain system for MOOCs and credit bank. In: 2021 International Conference on Education, Information Management and Service Science (EIMSS), pp. 355–358 (2021). https://doi.org/10.1109/EIMSS53851.2021.00083
35. Roussille, H., Gürcan, Ö., Michel, F.: AGR4BS: a generic multi-agent organizational model for blockchain systems. Big Data Cogn. Comput. **6**(1), 1 (2022)
36. Saheb, T., Mamaghani, F.H.: Exploring the barriers and organizational values of blockchain adoption in the banking industry. J. High Technol. Manag. Res. **32**(2), 100417 (2021). https://doi.org/10.1016/j.hitech.2021.100417
37. Schulz, K.F., Freund, D.: A multichain architecture for distributed supply chain design in industry 4.0. In: Abramowicz, W., Paschke, A. (eds.) BIS 2018. LNBIP, vol. 339, pp. 277–288. Springer, Cham (2019). https://doi.org/10.1007/978-3-030-04849-5_25
38. Shi, X., Li, G., Chen, S., Wu, Z.: Design of electronic contract architecture based on blockchain technology. In: 2020 IEEE 3rd International Conference of Safe Production and Informatization (IICSPI), pp. 458–461 (2020). https://doi.org/10.1109/IICSPI51290.2020.9332420
39. Sonkamble, R.G., Phansalkar, S.P., Potdar, V.M., Bongale, A.M.: Survey of interoperability in electronic health records management and proposed blockchain based framework: MyBlockEHR. IEEE Access **9**, 158367–158401 (2021). https://doi.org/10.1109/ACCESS.2021.3129284
40. Umran, S.M., Lu, S., Abduljabbar, Z.A., Nyangaresi, V.O.: Multi-chain blockchain based secure data-sharing framework for industrial IoTs smart devices in petroleum

industry. Internet Things **24**, 100969 (2023). https://doi.org/10.1016/j.iot.2023.100969

41. Vo, H.T., Wang, Z., Karunamoorthy, D., Wagner, J., Abebe, E., Mohania, M.: Internet of blockchains: techniques and challenges ahead. In: 2018 IEEE International Conference on Internet of Things (iThings) and IEEE Green Computing and Communications (GreenCom) and IEEE Cyber, Physical and Social Computing (CPSCom) and IEEE Smart Data (SmartData), pp. 1574–1581 (2018). https://doi.org/10.1109/Cybermatics_2018.2018.00264
42. Wang, X., Qiu, W., Zeng, L., Wang, H., Yao, Y., He, D.: A supervisory and governance mechanism for power master-slave chain architecture. In: 2021 IEEE International Conference on Electronic Technology, Communication and Information (ICETCI), pp. 172–175 (2021). https://doi.org/10.1109/ICETCI53161.2021.9563419
43. Wang, Y., Tie, J., Hei, X., Zhao, L., Zhang, X.: A trusted privacy-preserving model for cross-chain transactions based on zk_SNARKs. In: International Conference on Networking and Network Applications (NaNA), pp. 187–192 (2023). https://doi.org/10.1109/NaNA60121.2023.00039
44. Wang, Y., Yang, B., Liu, J., Zeng, H., Xia, C.: Virtual chain: a storage model supporting cross-blockchain transaction. Concurr. Comput.: Pract. Experience **34**(12), e5899 (2022). https://doi.org/10.1002/cpe.5899
45. Xiao, X., Yu, Z., Xie, K., Guo, S., Xiong, A., Yan, Y.: A multi-blockchain architecture supporting cross-blockchain communication. In: Sun, X., Wang, J., Bertino, E. (eds.) ICAIS 2020. CCIS, vol. 1253, pp. 592–603. Springer, Singapore (2020). https://doi.org/10.1007/978-981-15-8086-4_56
46. Xu, X., Weber, I., Staples, M.: Architecture for Blockchain Applications. Springer, Heidelberg (2019)
47. Yang, G., Zang, C., Chen, J., Guo, D., Zhang, J.: Distributed fusion cross-chain model and architecture. IET Blockchain **2**(2), 29–43 (2022). https://doi.org/10.1049/blc2.12013
48. Yang, J., Zhong, B., Gao, H., Wang, Y.: Task decomposition and service composition on an innovative blockchain-based construction service trading platform to select construction services. J. Constr. Eng. Manag. **149**(12), 04023130 (2023). https://doi.org/10.1061/JCEMD4.COENG-12956
49. Zachman, J.A.: A framework for information systems architecture. IBM Syst. J. **26**(3), 276–292 (1987). https://doi.org/10.1147/sj.263.0276
50. Zhang, J., Liu, Y., Zhang, Z.: Research on cross-chain technology architecture system based on blockchain. In: Liang, Q., Wang, W., Liu, X., Na, Z., Jia, M., Zhang, B. (eds.) CSPS 2019. LNEE, vol. 571, pp. 2609–2617. Springer, Singapore (2020). https://doi.org/10.1007/978-981-13-9409-6_318
51. Zhang, L., Hang, L., Jin, W., Kim, D.: Interoperable multi-blockchain platform based on integrated REST APIs for reliable tourism management. Electronics **10**(23), 2990 (2021)
52. Zhong, T., Shi, P., Chang, J.: JointCloud cross-chain verification model of decentralized identifiers. In: 2021 IEEE International Performance, Computing, and Communications Conference (IPCCC), pp. 1–8 (2021). https://doi.org/10.1109/IPCCC51483.2021.9679363

Transparent Transaction Ordering in Blockchain-Based Collaborative Processes

Hassan Atwi[1(✉)], Tom Lichtenstein[2], Cesare Pautasso[1], and Mathias Weske[2]

[1] Software Institute, Università della Svizzera italiana, Lugano, Switzerland
{Hassan.Atwi,Cesare.Pautasso}@usi.ch

[2] Hasso Plattner Institute, University of Potsdam, Potsdam, Germany
{Tom.Lichtenstein,Mathias.Weske}@hpi.de

Abstract. Blockchain technology offers a promising solution for overcoming trust-related challenges, such as transparency in interorganizational collaborative processes. However, since the ordering of transactions in a block can affect the outcome of a process instance, fairness and transparency concerns regarding block selection may arise. This paper presents a consensus-agnostic approach to promote transparency in transaction ordering in blockchain-based business processes. The approach allows participants to align the execution order of their transactions with their business objectives, effectively mitigating the lack of transparency associated with the arbitrary selection and ordering of transactions during block selection. Our experiments reveal that the increased transparency mitigates the risks of suppression and displacement attacks at the cost of introducing additional latency and cost.

Keywords: Blockchain · Choreography · Block selection · Transaction ordering · Transparency

1 Introduction

Coordinating the execution of processes across organizational boundaries typically relies on a trusted, centralized party. Recently, blockchain technology has been studied as a potential infrastructure for cross-organizational collaboration allowing us to overcome this trust issue [2,9,17], by providing a decentralized, tamper-proof ledger of transactions. A blockchain is organized in blocks, each of which contains an ordered list of transactions, with new blocks being linked to the previous block using a cryptographic hash function [11]. New transactions are propagated through the network as pending transactions before being added to a block at the end of the blockchain.

While blockchain technology allows for transparency, ensuring fair and reliable ordering of transactions within blocks during block selection remains a challenge. In most blockchains, one leader node, which may be reassigned over

C. Di Ciccio et al. (Eds.): BPM 2024, LNBIP 527, pp. 25–39, 2024.
https://doi.org/10.1007/978-3-031-70445-1_2

time, is responsible for determining the inclusion and order of pending transactions for the next block [20]. Since transactions may represent interactions among collaborating parties, the order of transactions can affect the outcome of a blockchain-based collaborative process or the profitability of the interaction [4,7,16]. Therefore, malicious actors taking over the role of the leader node could exploit the freedom to order transactions to gain an advantage over their competitors. Given that in larger networks the pools of known pending transactions and the order in which they are received may differ for each node due to network latency, detecting fraudulent transaction orders remains a challenge [4,15].

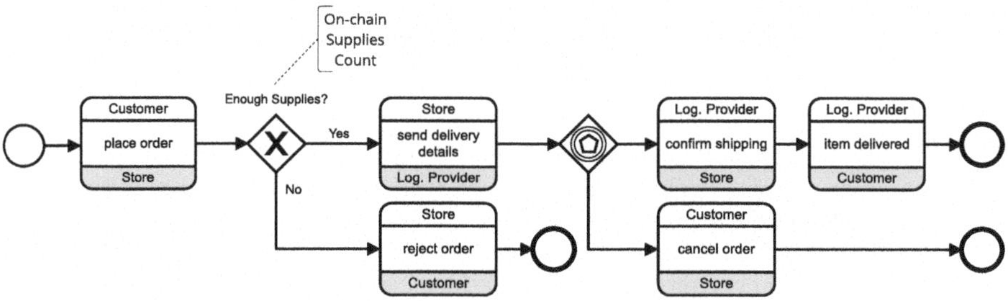

Fig. 1. Choreography diagram representing a simple supply-chain choreography.

To illustrate the impact of transaction order manipulation, we refer to a simple supply-chain choreography depicted in Fig. 1. This choreography orchestrates the interaction between a customer, a store, and a logistic provider. It begins with the customer placing an order at the store, triggering an inventory check. If supplies are insufficient, the customer is notified and the choreography ends. Otherwise, the store forwards delivery details to the logistic provider for shipping. In the latter case, the customer retains the option to cancel the order until the logistics provider confirms the shipping. Upon confirmation, the order is delivered to the customer.

The choreography is implemented on-chain using a smart contract, similar to other blockchain-based process execution approaches [17]. In such a decentralized setting where transaction orders are not centrally managed, the arbitrary nature of transaction ordering can significantly impact the fairness of the choreography. For instance, multiple customers compete for limited supplies from the same store, potentially leading to conflicts across instances of the choreography. Given an on-chain inventory check, the ordering of the transactions determines who acquires the limited supplies. In another scenario, a race condition may occur within a single instance at the event-based gateway, where the logistic provider and customer compete over who makes their choice first. In this case, the ordering of transactions decides if the order will be confirmed for shipping or canceled. While there is a natural race between multiple participants competing for shared and limited resources, malicious participants could perform front-running or censorship attacks [16], so that their transaction requests are prioritized, compromising the fairness in the choreography.

Existing solutions [1,8,12,14] typically employ a random selection and ordering of transactions to maintain fairness. However, this can result in transaction orders not aligning with the participants' intentions. To address this challenge, this paper contributes a novel consensus-agnostic, process-aware transaction ordering approach. This approach allows participants to impact the order of associated transactions directly, ensuring fairness and transparency in block creation that aligns with the participants' intentions.

The paper is organized as follows: Sect. 2 describes the underlying concepts of fairness in block selection and blockchain-based collaborative processes and outlines related work. Section 3 introduces the transaction ordering approach as the main contribution of this paper. Section 4 evaluates the protocol's resilience to transaction suppression and displacement and discusses the approach in detail. Finally, Sect. 5 summarizes the paper and points out possible future research directions.

2 Preliminaries and Related Work

This section provides an overview of blockchain-based collaborative processes and fairness in block selection.

2.1 Blockchain-Based Collaborative Processes

Blockchain technology offers a promising solution to enable collaboration between mutually distrustful organizations. Recently, model-driven approaches have been investigated to facilitate the design and execution of processes that span multiple organizations [17]. While various modeling languages are used to design blockchain-based collaborative processes, in this paper we refer to Business Process Model and Notation (BPMN) *choreography diagrams* [13].

Choreography diagrams (Fig. 1) represent interactions between two participants with *choreography tasks*, hereafter referred to as *tasks*, with participants with a white band being the *initiators*. Sequence flow arcs specify the order of tasks, while gateways express exclusive and concurrent behavior. The constraints on the order of execution are referred to as *control flow*. Choreography diagrams support two-way tasks with direct responses. We restrict this to one-way tasks, as two-way tasks can be represented by two consecutive one-way tasks without loss of generality. According to the lifecycle depicted in Fig. 2, an *initialized* task must first be *enabled* with respect to the control flow constraints before it can be executed, resulting in the *completed* state. In addition, a task may be *skipped*, for example, in the case of exclusive behavior [18].

In a blockchain environment, the execution of a task is typically reflected by a blockchain transaction, hereafter referred to as *transaction*. In second-generation blockchains, the business logic defined by a choreography diagram can be enforced on-chain via *smart contracts* [17]. In the following, we refer to smart contracts dedicated to workflow logic enforcement as *workflow contracts*.

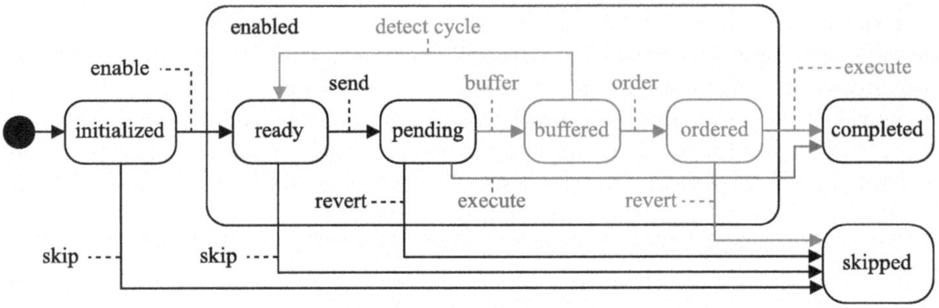

Fig. 2. State transition diagram representing the lifecycle of a choreography task in a blockchain environment using a traditional (red) and ordering contract-based (blue) implementation. (Color figure online)

As a blockchain-based task depends upon the successful completion of a transaction, the lifecycle of a blockchain-based task includes additional states between the initial enablement and the completion, as illustrated in Fig. 2. Each enabled task is initially in a *ready* state, signaling that a transaction can be sent to execute the task. Once a transaction is sent to the transaction pools of other blockchain nodes, the task is *pending*. In the following, we assume that each transaction sent will eventually be included in a block. In a traditional implementation (highlighted in red), once a transaction is included in a block, it is either *executed* or *reverted* according to the constraints of the workflow contract. The latter may occur if earlier transactions change the state of the workflow contract so that the current transaction conflicts with the workflow constraints, causing the task to be *skipped*. Thus, the course of a choreography can be affected by the execution order of the corresponding transactions [7]. A successful execution of the transaction, on the other hand, results in the task being *completed*.

2.2 Fairness in Block Selection

In current blockchain systems, the responsibility for *block selection*, i.e., selecting and ordering pending transactions to form a new block [1], is typically delegated to an elected *leader node*. Prominent protocols for leader election in public blockchains include: *proof of work* (PoW), as used by Bitcoin [11], which requires leader candidates to solve a computationally complex puzzle, and *proof of stake* (PoS), adopted by Ethereum in 2022 [19], which bases leader election on the respective balances of the nodes. Conversely, in private blockchain environments, leader election typically relies on voting, pseudorandom functions, or round-robin strategies among a predefined set of authority nodes. In the following, we will refer to the leader node as the *block proposer*.

Most common blockchain implementations grant the block proposer full control over the selection and ordering of transactions for the next block. In public blockchains, block proposers often optimize their block selection based on transaction fees to maximize their profit. This incentive can be exploited to perform front-running attacks [16], i.e., paying higher fees to place transactions first to gain an advantage over the competition. Since private blockchains generally do not charge transaction fees, front-running is only possible when block proposers use their privilege to arrange transactions to their advantage by displacing transactions or suppressing undesirable transactions, which raises the question of fairness in block selection in private networks [1]. Since latency or transmission errors can cause two nodes' transaction pools to diverge in terms of included transactions and transaction ordering, detecting malicious block proposers is particularly challenging in decentralized environments [4].

2.3 Related Work

Fairness in block selection received attention in recent blockchain literature. Kelkar et al. [4] identified *receive-order-fairness* as a relevant property for block selection, implying that the order of transactions in a block should reflect the transaction order observed by the majority of nodes. Due to Condorcet's paradox [3], receive-order-fairness is impossible to achieve in a decentralized setting. The authors propose *block-receive-order-fairness*, which ensures that transactions are not assigned to separate blocks in an order that conflicts with the observations of the majority. Wendy et al. [5] propose blockchain-agnostic preprotocols executed in parallel to the blockchain to address order fairness by having validator nodes agree on a block before proposing it to the blockchain. Similar to [4], the authors focus on ensuring order fairness between blocks. In [8], the authors introduce FairLedger, a permissioned protocol in which each node of a predefined committee has an equal opportunity to add its transactions to the block, while all nodes must be aware of all transactions to be added to the next block. The order of the transactions within the block remains undetermined.

Furthermore, Asayag et al. [1] propose Helix, a consensus protocol that uses encrypted transactions to complicate transaction suppression in block selection. In addition, the authors propose a hash-based pseudorandom ordering of transactions to ensure fair block selection. The protocol is extended in [14] with a joint block validation procedure that refines the assessment of the leader node's honesty by comparing the included transactions with those known to a committee of validator nodes. The extension also includes the concept of pending transaction declarations, which allows nodes to periodically query other nodes for their local transaction pool, further complicating transaction suppression. Another extension to Helix is presented in [12], which introduces a reputation system for leader and validator nodes. Finally, Sokolik et al. [15] propose *age-aware fairness*, which aims to limit transaction latency by reserving a predefined space in each block for senior transactions.

This work complements existing work by introducing a blockchain-agnostic approach that achieves fair block selection by 1) providing resilience to suppression, 2) making transaction ordering collaborative and transparent, 3) detecting conflicts in ordering decisions, and 4) allowing nodes to resolve them.

3 Transparent Transaction Ordering

To ensure transparency in the ordering of transactions within blockchain-based collaborative processes, we propose an approach to achieve on-chain transaction ordering based on process information derived from the choreography model. The derived information helps in enabling process-aware transaction ordering. This involves selecting orderers based on the ongoing interactions within the process model.

The approach mainly relies on the introduction of an ordering smart contract, which serves as a preprocessor to organize transaction orders before execution. This approach is consensus-agnostic: it does not depend on the specific consensus protocol adopted by the blockchain network. However, it is assumed that the consensus protocol fairly redistributes the role of the block proposer among the nodes after each added block. As illustrated in Fig. 3, there are three main phases to the approach: buffering, ordering, and execution. We refer to a run through all three phases as an epoch. In the following, we provide a detailed description of the steps within an epoch. In addition, the impact of the approach on the task lifecycle compared to traditional implementations is discussed.

3.1 Buffering Phase

In typical blockchain-based process engine architectures [9], transactions update a workflow contract's state when included in a block, allowing the block proposer to control the execution order. We propose an alternative approach using an ordering contract as a relayer between participants and the workflow contract. Participants issue meta-transactions[1] that are buffered in the ordering contract rather than executed immediately. This buffering phase delays choreography tasks state transitions by introducing an intermediate *buffered* state (Fig. 2). If a block is compromised by a malicious actor, buffered transactions can be resubmitted by its original sender in subsequent blocks before execution. This provides participants with an extended window for submitting transactions.

The buffering phase needs to be bounded so that all meta-transactions are executed within a certain time limit. However, due to the decentralized execution of smart contracts, deterministic time measurements are impossible. Oracles [10] could address this limitation by providing access to external sources of precise timing information. Depending on oracles for such essential information poses a risk of undermining the core decentralized foundations of blockchain, as the trust is placed in the hands of the timekeeping party [6]. A reliable alternative to the timestamp is the block number. Thanks to the consensus protocol used to maintain a strict block ordering along the chain, any attempt to manipulate the number of blocks is certain to fail. In our approach, the ordering contract uses the block number to introduce temporal awareness into the transaction ordering process. The block number helps establish a specific duration of the window during which the ordering contract buffers meta-transactions. Participants can

[1] https://docs.openzeppelin.com/learn/sending-gasless-transactions.

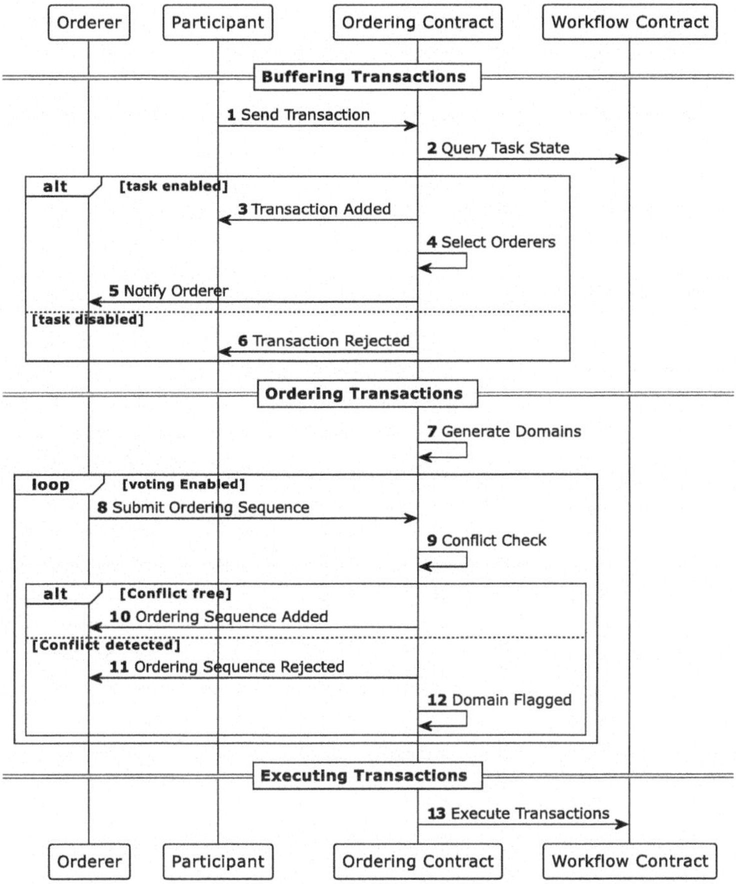

Fig. 3. An epoch consisting of the buffering, ordering, and execution phases, represented as a sequence diagram.

determine the number of blocks dedicated to this buffering period. By extending the buffering phase, participants maximize their opportunities to submit meta-transactions, which can reduce the risk of malicious attacks.

3.2 Ordering Phase

In each epoch, a set of orderers is selected from the choreography participants. The role of an orderer grants the ability to cast votes on the desired sequence for arranging the buffered transactions of an epoch. Notably, each orderer is limited to ordering only the buffered transactions corresponding to the task where they serve as either the sender or the receiver. During the ordering phase of the epoch, the ordering contract stops accepting meta-transactions and proceeds with accepting ordering sequences from the chosen orderers. Once all ordering sequences are collected, the ordering contract combines each ordering sequence, creating a merged list that reflects the consensus on the global ordering among all participating orderers.

To facilitate the explanation of the orderer selection process, we represent the buffered transactions as follows (Fig. 4). Each node in the graph corresponds to a participant. In a choreography instance, each participant assumes a specific role, which may vary from one choreography instance to another. Nodes are connected by edges if a buffered transaction represents a task involving both participants. The set of orderers for each epoch is the subset of participants involved in two or more buffered transactions. Hence, let V be the set of nodes in the graph where each $v \in V$ represents a participant, and N_v is the set of neighbors of the node v. The set of orderers can be expressed as: $O = \{v \mid v \in V \wedge |N_v| > 1\}$. This simple strategy employed for selecting orderers will result in a group of intermediaries who act as *mediators* between two or more participants, thereby establishing them as neutral parties responsible for enforcing the most suitable order. In a plain implementation, there is a risk that if two transactions race against each other, one of the senders might be the block proposer determining the order. Our strategy ensures instead that the mediator chooses the order. For instance, as shown in Fig. 4, in $Epoch_{n+1}$, $store_1$ serves as a neutral party ordering transactions for both customers. Transactions occurring in isolation, such as tx_5 in domain 2, lack any form of competition, eliminating the need for an orderer. Consequently, isolated transactions are inserted into the block regardless of their ordering.

As each orderer individually submits their vote for a preferred sequence of transactions, cycles may result from their combined preferences. The existence of cycles in collective preferences introduces an additional layer of complexity to the ordering process, which could potentially affect other instances by halting the entire network. To mitigate this impact, the ordering contract isolates each interaction graph during an epoch into separate domains. This isolation ensures that if one domain faces cycles in voting, the other domains can proceed with execution independently, thereby deterring any broader impact on the entire network.

Each domain is isolated from the others, as they do not share any common transactions. Orderers exclusively order the transactions within their respective domains. Consequently, the global order between domains is not a concern, since they are not interconnected within an epoch. For instance, in Fig. 4, $Epoch_n$ is depicted with two distinct domains, namely 'Domain 1' and 'Domain 2'. Each domain independently establishes its order of transactions. The final ordering sequence of 'Domain 1' comprises transactions tx_1, tx_2, tx_3, and tx_4, arranged according to the specific orders within 'Domain 1'. 'Domain 2' encompasses only one transaction, tx_5, thus no ordering is required. The global order of $Epoch_n$ results from combining the orderings of both domains. The ordering contract disregards the order between the two domains in the global order since they do not share any common transactions.

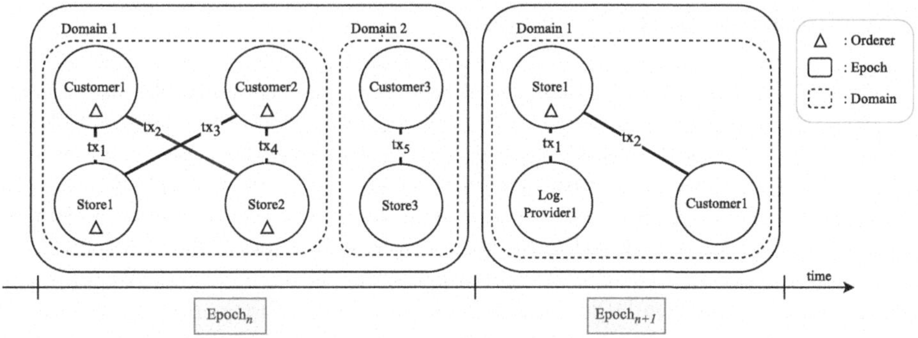

Fig. 4. Interaction graphs within two subsequent epochs.

To demonstrate the possibility of a cycle occurring within the ordering preferences, we take domain 1 (Fig. 4) as an example. Each orderer casts their preference on the ordering sequence. $customer_1$ submits their preference with the sequence $tx_2 > tx_1$, i.e., tx_2 over tx_1, $store_1$ submits the sequence $tx_1 > tx_3$, $customer_2$ submits the sequence $tx_3 > tx_4$, and finally, $store_2$ submits the sequence $tx_4 > tx_2$. The combined sequence of all submissions, based on their submission order, results in the final order $tx_2 > tx_1 > tx_3 > tx_4 > tx_2$. This result serves as a notable instance of the Condorcet paradox [3]. Despite the validity of local orderings, their combination introduces conflict into the collective order.

In case a cycle is detected, the transaction is reverted, and the associated domain is flagged as conflicting. Note that a cycle in the ordering presents a special case: Since all conflicting transactions are reverted, the states of the workflow contracts do not change. Therefore, on the choreography level, the states of the corresponding tasks are not skipped but return to the ready state, as illustrated in Fig. 2. This way, participants have the opportunity to resubmit the conflicting transactions so that they can be buffered and ordered again. Otherwise, if a valid order is found, the task advances to the *ordered* state.

3.3 Execution Phase

In the final stage of the epoch, the ordering contract executes the transactions that do not involve conflicting domains by forwarding them to the workflow contract. According to the lifecycle (Fig. 2), it is still possible for a task to be skipped at this point. This may happen when earlier transactions change the state of the workflow contract so that the transaction of the task no longer satisfies the workflow constraints. Otherwise, the transaction is executed, changing the state of the workflow contract and causing the corresponding task to reach the completed state. A new epoch begins once all phases are completed.

4 Evaluation and Discussion

The evaluation scenario comprises two instances of the choreography introduced in Fig. 1 denoted by C_1 and C_2. The participants are distributed as follows: In instance C_1, the participants are $\{customer_1, log.\ provider_1, store_1\}$, while in instance C_2, the participants are $\{customer_2, log.\ provider_2, store_1\}$. We execute the scenario in 500 epochs, where competition among participants occurs in two ways: across instances and within instances. Across instances, the two customers race for placing orders for the limited supplies at the same store. We limit the scenario to only one available supply item per epoch; thus, only one customer emerges as the winner in each epoch. The competition within an instance emerges at the event-based gateway, where the customer and the logistics provider compete to cast their choice first, i.e., to confirm the shipment or to cancel the order.

4.1 Setup

To conduct the evaluation, we established a network comprising five nodes running Quorum[2] clients. The network operates on the QBFT[3] consensus algorithm, which is an enterprise-grade consensus protocol recommended for permissioned networks. With QBFT as the adopted consensus mechanism, the block proposer is selected in a round-robin fashion. The consensus protocol is configured to generate a new block every five seconds. To ensure a full representation of participants in the network, we allocate each node to a participant, meaning that each participant possesses a dedicated node. Since a participant has complete control over a node, there is potential to inject malicious behavior into the block creation procedure, e.g., to execute front-running attacks.

Given the unpredictability of participants' intentions in a network, we categorize the nodes into three distinct groups:

1. **Honest Nodes**: These nodes operate without malicious intent and propose a block with the order of transactions as locally observed.
2. **Displacement Nodes**: These nodes prioritize their own transactions by displacing the ones submitted by other participants.
3. **Suppression Nodes**: These nodes suppress all transactions competing with their transactions.

Throughout the experiment, $customer_1$ and $log.provider_2$ play the role of malicious nodes attempting to implement both strategies, namely suppression and displacement, individually in two distinct experiments. Across instances, $customer_1$ will target $customer_2$ to obtain the most supplies. Within their own instance, $customer_1$ will target $log.provider_1$ to maximize canceled choices, while $log.provider_2$ will target $customer_2$ to ensure the most confirmed choices. The remaining participants serve as honest nodes.

[2] https://github.com/Consensys/quorum.

[3] https://docs.goquorum.consensys.io/configure-and-manage/configure/consensus-protocols/qbft/.

In our evaluation, we distinguish two setups for invoking choreography tasks: a plain setup, and an ordering contract (OC) setup. In the plain setup, participants send transactions directly to the workflow contract without buffering. In the OC setup, all transactions must pass through the ordering contract, where they are buffered, ordered, and finally executed in the workflow contract. In this experiment, we implemented a straightforward ordering strategy where orderers distribute transactions fairly among participants in a rotating sequence. In each epoch, the priority changes so that each participant's transaction gets a chance to be processed first in turn. Depending on business objectives, the selected orderers could implement more sophisticated strategies. To mitigate the effect of malicious attacks, we implemented a buffer of two blocks for this experiment. This gives honest nodes a window of two blocks to submit their transactions, thus reducing the likelihood of censorship. To have a reproducible experiment, we generate an event log that contains the initial order of interactions and randomly injected delays for 500 epochs. The inclusion of delays serves the purpose of injecting noise into the network, thereby ensuring that the selection of block proposers is not consistently aligned with the order of transactions.

The prototype implementation, including the smart contracts, the event logs, and their analysis results are available on GitHub[4].

4.2 Results

We report the results of the experiment conducted for each setup with the two attack types, i.e., transaction suppression and displacement. In addition, we compare the performance between the setups in terms of latency and cost.

Across Instances. Figure 5 illustrates the results of supply distribution between the two customers. In a plain setup, the malicious node ($customer_1$) could secure the majority of supplies using both types of attacks, primarily through suppression. When employing the ordering contract setup, the supply distribution becomes more evenly distributed between the customers.

Within Instances. Figure 6 and Fig. 7 show the number of times each path following the deferred choice was selected within the instances. The malicious nodes ($customer_1$ in C_1 and $log.\ provider_2$ in C_2) won most of the races in both instances by applying the two types of attacks. On the other hand, the ordering contract setup enforced an even distribution of deferred choices despite the malicious nodes' actions.

Performance. In terms of performance, as illustrated in Fig. 8, the duration of epochs in the ordering contract setups, on average, exceeds that of the plain setups by a factor of four regardless of the attack type. The cost associated with ordering contract setups was at least nine times that of the plain setup. This

[4] https://github.com/Hassan42/Transparent-Transaction-Ordering.

outcome is expected, given that the ordering approach involves the execution of additional on-chain logic.

The results indicate that the ordering contract facilitates a process-aware ordering of transactions by selecting the appropriate orderer throughout the choreography's execution. Even in a malicious node-free environment, the ordering contract leverages the transparency and domain-specific ordering of business transactions. However, the trade-off for achieving process-aware transaction ordering is an increase in latency and cost.

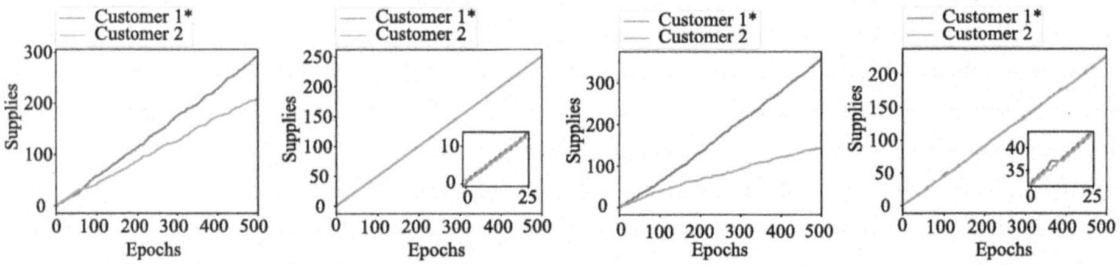

(a) Plain Setup Displacement Attack (b) OC Setup Displacement Attack (c) Plain Setup Suppression Attack (d) OC Setup Suppression Attack
*) $customer_1$ is the malicious participant.

Fig. 5. Distribution of supplies between customers across instances.

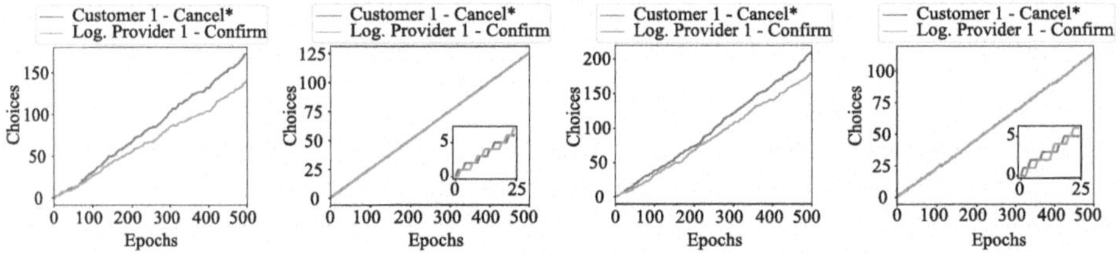

(a) Plain Setup Displacement Attack (b) OC Setup Displacement Attack (c) Plain Setup Suppression Attack (d) OC Setup Suppression Attack
*) $customer_1$ is the malicious participant.

Fig. 6. Distribution of deferred choices within instance C_1.

4.3 Discussion

Based on the experiment's findings, we will discuss the approach regarding liveness, consistency, cost, and latency, with each attribute providing insights into the functionality and constraints of the proposed approach.

Liveness. Complete assurance of liveness is not guaranteed, as certain ordering sequences may result in conflicts with undefined duration. Resolving such conflicts often requires conflicted participants to compromise on their preferences. In

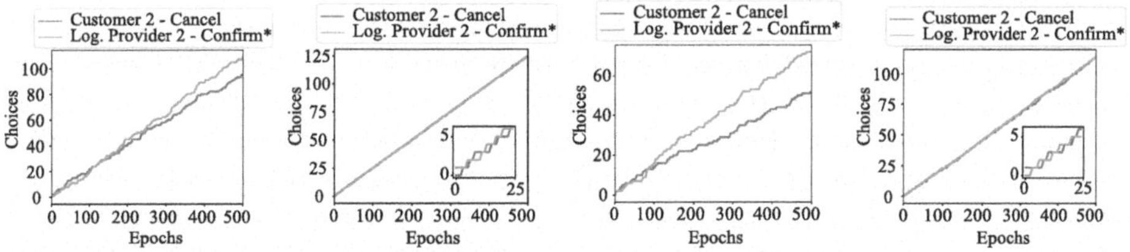

(a) Plain Setup Displacement Attack (b) OC Setup Displacement Attack (c) Plain Setup Suppression Attack (d) OC Setup Suppression Attack

*) $log.provider_2$ is the malicious participant.

Fig. 7. Distribution of deferred choices within instance C_2.

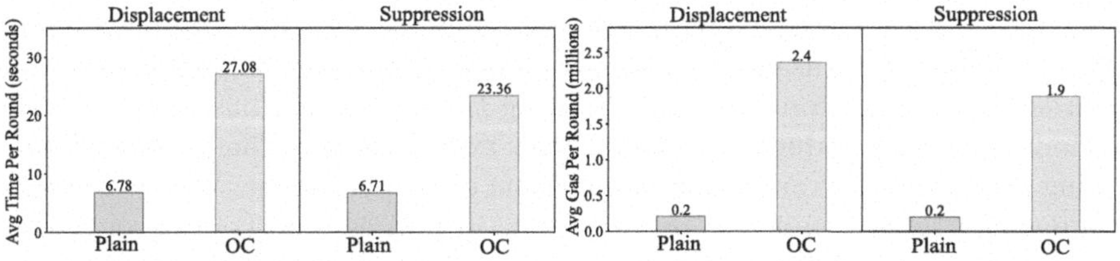

(a) average epoch duration (b) average gas consumption

Fig. 8. Trading-off fairness against performance (a) and gas consumption (b) under suppression and displacement attacks

our approach, the ordering contract emits an event and flags such cases, enabling conflicting participants to resubmit the transactions. To avoid blocking the entire network, the ordering contract minimizes the potential damage by segregating buffered transactions into domains, thereby ensuring partial liveness in the network. Another crucial factor influencing liveness is the continuous generation of blocks by the network hosting the ordering contract, even for empty blocks. Regular block generation is crucial to the buffering phase in order to maintain regular time intervals and the eventual termination of the phase, even if no new transactions are sent.

Consistency. The inherent properties of the blockchain guarantee a consistent ordering of all transactions included in a block. Executing the ordering procedure on-chain ensures a uniform view among all participants. Hence, in our approach, each phase in an epoch remains visible and consistent for all participants.

Cost. Due to the extra on-chain logic introduced by the ordering contract, executing blockchain-enabled choreographies becomes more costly compared to a plain implementation. Therefore, this approach is better suited for private networks rather than public ones, where transaction costs and gas prices are not as significant.

Latency. Incorporating the ordering contract into workflow execution increases latency, primarily caused by transaction buffering. The duration of transaction buffering can be adjusted through a collective agreement among participants as the network progresses. However, reducing the length of the transaction buffer can facilitate displacement or suppression attacks, as fewer block proposers propagate the transactions for an epoch. Further exploration is needed to develop a dynamic approach for tuning the epoch length to balance the trade-off between latency and the level of protection against malicious attacks.

5 Conclusion and Future Work

This paper proposes a novel transparent approach for ordering transactions using contextual information provided by choreography models. The approach mainly relies on buffering transactions and arranging them on-chain before they are forwarded to a smart contract implementing the underlying business logic. The additional measures reduce or in some cases even eliminate the impact of front-running attacks, e.g., transaction suppression or displacement. While the evaluation demonstrates significant improvements in fairness and transparency compared to the baseline, this outcome comes at the expense of increased latency and cost. The suggested approach lays the foundation for additional experiments involving scenarios with a larger number of participants as well as future enhancements and additional investigations into optimizing epoch length selection, exploring new ordering strategies, and embedding process awareness in the underlying consensus protocol.

Acknowledgement. Funded by the Deutsche Forschungsgemeinschaft (DFG, German Research Foundation) - 450612067, and by the Swiss National Science Foundation (SNSF) under project "Flexible Choreographies in Multi-chain Environments" (196958).

References

1. Asayag, A., et al.: A fair consensus protocol for transaction ordering. In: 26th IEEE International Conference on Network Protocols, pp. 55–65 (2018)
2. Dwivedi, V., Norta, A.: Auto-generation of smart contracts from a domain-specific XML-based language. In: Satapathy, S.C., Peer, P., Tang, J., Bhateja, V., Ghosh, A. (eds.) Intelligent Data Engineering and Analytics, vol. 266, pp. 549–564. Springer, Singapore (2022). https://doi.org/10.1007/978-981-16-6624-7_54
3. Gehrlein, W.V.: Condorcet's paradox. Theor. Decis. **15**(2), 161–197 (1983)
4. Kelkar, M., Zhang, F., Goldfeder, S., Juels, A.: Order-fairness for Byzantine consensus. In: Micciancio, D., Ristenpart, T. (eds.) CRYPTO 2020. LNCS, vol. 12172, pp. 451–480. Springer, Cham (2020). https://doi.org/10.1007/978-3-030-56877-1_16
5. Kursawe, K.: Wendy, the good little fairness widget: achieving order fairness for blockchains. In: 2nd ACM Conference on Advances in Financial Technologies, pp. 25–36 (2020)
6. Ladleif, J., Weske, M.: Time in blockchain-based process execution. In: 24th International Enterprise Distributed Object Computing Conference, pp. 217–226 (2020)

7. Ladleif, J., Weske, M.: Which event happened first? Deferred choice on blockchain using oracles. Front. Blockchain **4**, 758169 (2021)
8. Lev-Ari, K., Spiegelman, A., Keidar, I., Malkhi, D.: Fairledger: a fair blockchain protocol for financial institutions. In: 23rd International Conference on Principles of Distributed Systems, vol. 153, pp. 4:1–4:17 (2019)
9. López-Pintado, O., García-Bañuelos, L., Dumas, M., Weber, I., Ponomarev, A.: Caterpillar: a business process execution engine on the ethereum blockchain. Softw.: Pract. Experience **49**(7), 1162–1193 (2019)
10. Mühlberger, R., et al.: Foundational oracle patterns: connecting blockchain to the off-chain world. In: Proceedings of the Blockchain and Robotic Process Automation Forum (at BPM), pp. 35–51 (2020)
11. Nakamoto, S.: Bitcoin: a peer-to-peer electronic cash system. Technical report (2008). https://bitcoin.org/bitcoin.pdf
12. Nassar, M., Rottenstreich, O., Orda, A.: Communication-aware fairness in blockchain transaction ordering. In: 23rd IEEE International Conference on High Performance Switching and Routing, pp. 175–182 (2022)
13. Object Management Group (OMG): Business Process Model and Notation (BPMN), Version 2.0.2 (2014). https://www.omg.org/spec/BPMN/2.0.2/
14. Orda, A., Rottenstreich, O.: Enforcing fairness in blockchain transaction ordering. Peer-to-Peer Netw. Appl. **14**(6), 3660–3673 (2021)
15. Sokolik, Y., Rottenstreich, O.: Age-aware fairness in blockchain transaction ordering. In: 28th IEEE/ACM International Symposium on Quality of Service (2020)
16. Torres, C.F., Camino, R., State, R.: Frontrunner jones and the raiders of the dark forest: an empirical study of frontrunning on the ethereum blockchain. In: 30th USENIX Security Symposium, pp. 1343–1359 (2021)
17. Weber, I., Xu, X., Riveret, R., Governatori, G., Ponomarev, A., Mendling, J.: Untrusted business process monitoring and execution using blockchain. In: La Rosa, M., Loos, P., Pastor, O. (eds.) BPM 2016. LNCS, vol. 9850, pp. 329–347. Springer, Cham (2016). https://doi.org/10.1007/978-3-319-45348-4_19
18. Weske, M.: Business Process Management - Concepts, Languages, Architectures, 3rd edn. Springer, Cham (2019). https://doi.org/10.1007/978-3-662-59432-2
19. Wood, G.: Ethereum: a secure decentralised generalised transaction ledger. Ethereum project yellow paper (2014). https://ethereum.github.io/yellowpaper/paper.pdf
20. Xiao, Y., Zhang, N., Lou, W., Hou, Y.T.: A survey of distributed consensus protocols for blockchain networks. IEEE Commun. Surv. Tutor. **22**(2), 1432–1465 (2020)

Interpreted and Confidential Execution of Process Choreographies on a Blockchain

Oskar Petto[1](✉), Thomas Preindl[2], and Martin Kjäer[2]

[1] TU Wien, Karlsplatz 13, 1040 Wien, Austria
oskar.petto@hotmail.com

[2] Institute of Computer Engineering, TU Wien, Karlsplatz 13, 1040 Wien, Austria
{thomas.preindl,martin.kjaeer}@tuwien.ac.at

Abstract. In our work, we refine the application of blockchain technology for managing inter-organizational business processes without a centralized authority, addressing the issue of data privacy inherent in public blockchains. Building on the proposal of Toldi and Kocsis, which involves recording encrypted instance data and reference hashes in smart contracts and aligning with business process models through zero-knowledge succinct arguments of knowledge (zk-SNARKs), we are able to improve several key metrics. Our approach introduces an efficient blockchain-based system tailored for business process modeling notation (BPMN) process choreographies. Unlike many other approaches in the field of blockchain-based process enactment, our system eliminates the need to deploy process-specific smart contracts. This alteration reduces the cost of instantiating new medium-sized process models by at least 88% compared to Toldi and Kocsis. Additionally, we achieve a greater than 99% reduction in the time required to generate conformance proofs. These enhancements result in a more cost-effective and streamlined approach for the private execution of choreographies on a blockchain.

Keywords: Business process · Choreography · Zero-knowledge proofs · Blockchain · zk-SNARKs

1 Introduction

Collaborative business processes, crucial for enhancing cross-organizational productivity, often encounter challenges like trust deficits among participants. These challenges hinder consensus on process oversight in low-trust environments [26]. Blockchain technology emerges as a potent solution by providing a transparent and decentralized mechanism for executing agreements without needing a trusted third party [7,11]. The use of smart contracts enables the enforcement of business processes, leveraging well-established modeling techniques like BPMN for clarity and effectiveness in execution [18,20].

However, the transparency inherent to public blockchains, such as Ethereum, poses significant privacy risks, revealing sensitive process data and potentially compromising organizational confidentiality [12]. While using private blockchains

C. Di Ciccio et al. (Eds.): BPM 2024, LNBIP 527, pp. 40–54, 2024.
https://doi.org/10.1007/978-3-031-70445-1_3

or encryption on public blockchains offers means to secure data, it simultaneously obscures process execution, limiting public verification capabilities [2,16,17]. This paper presents an innovative approach that enables public verification of private inter-organizational processes using public blockchains. Our method leverages BPMN choreographies, which depict only the direct interactions between participants while abstracting intra-organizational process tasks. We argue that choreographies offer a more suitable model for interactions in collaborative business processes. Instead of emphasizing the internal tasks of individual actors that may be irrelevant to others, choreographies model message exchanges, the primary concern of external parties.

Demonstrating that these interactions have occurred as modeled presents a significant challenge. A promising solution for validating the correctness of private data without revealing it is using zero-knowledge proofs (ZKPs) and their efficient variant, zk-SNARKs. Their expressiveness makes them suitable for various applications, including digital signature schemes, secure electronic voting, and user identification protocols [19]. Our research builds upon the exploration of zk-SNARKs for ensuring business process execution consistency with predefined models while addressing the computational challenges associated with proof generation. We propose a more efficient approach to generating zk-SNARK proofs, which benefits, among other things, from an interpreted approach. This means that, in comparison to current approaches, there is no need to deploy new smart contracts on the blockchain for every new process model [24]. Instead, only a one-time deployment is necessary, which contributes significantly to the domain of confidential blockchain-based business process enactment.

The remainder of this paper discusses the background and related work in Sects. 2 and 3, presents our approach for interpreted and confidential choreography enactment on blockchains in Sect. 4, and covers implementation and evaluation in Sect. 5, before concluding with a summary of our findings and contributions in Sect. 6.

2 Zero-Knowledge Proofs and zk-SNARKs

ZKPs and zk-SNARKs are protocols between a prover and a verifier, where the prover convinces the verifier of the correctness of some computation without revealing its secret inputs. This computation is defined as the decision circuit C of an NP language L, where any $x \in L$ is a true statement, while any $x \notin L$ is a false statement. The process of convincing the verifier involves exchanging a sequence of messages. While ZKPs allow multiple message exchanges between the prover and verifier, zk-SNARKs are non-interactive, which means the prover only sends a single message, referred to as a proof, to the verifier. The high-level security properties of these protocols are:

- Correctness. When a statement is true, the prover can convince the verifier of its truth.
- Soundness. In case of a false statement, no malicious prover can deceive the verifier into accepting it.

- Zero-Knowledge. For true statements, malicious verifiers learn nothing from executing the protocol with the prover except the statement's validity.

For zk-SNARKs, the soundness requirement applies only to provers with polynomial computational power. Furthermore, the length of the proof must be either constant or logarithmic relative to the size of the statement [22].

A zk-SNARK proving system involves three parties. The generator is responsible for the system's setup, which the prover uses to convince the verifier about a statement. According to their role, the parties execute the following algorithms respectively:

- $\mathsf{Gen}(1^\lambda, \mathsf{C}) \rightarrow (\mathsf{pk}, \mathsf{vk})$. Given a security parameter λ and a decision circuit C of an NP language L, the Gen algorithm produces a proving key pk and a verification key vk, which are subsequently made public as protocol parameters.
- $\mathsf{Prove}(\mathsf{pk}, x, w) \rightarrow \pi$. When provided with a proving key pk, a true statement $x \in L$, and a corresponding witness w for x, the Prove algorithm generates a proof π that attests to the truth of $x \in L$.
- $\mathsf{Verify}(\mathsf{vk}, x, \pi) \rightarrow \{0, 1\}$. When given a verification key vk, a statement x and a proof π, the Verify algorithm outputs 1 if $x \in L$.

The statement x is often called public input because it is available to both Prove and Verify. In contrast, the witness w is the private input because w is only available to the Prove algorithm, i.e., only the prover has knowledge or w.

The most efficient zk-SNARKs require that the generator discards any random values used to execute the Gen algorithm. Otherwise, the soundness property does not hold. To increase the trust in this critical step, multiple parties can act as the generator and execute the Gen algorithm as a multi-party computation protocol. This improvement guarantees the soundness of the zk-SNARK as long as at least one participating party discards their random values but increases the setup time and complexity [1].

Two widely recognized zk-SNARKs are Groth16 [10], known for its small proof size but reliant on a trusted setup per decision circuit, and PLONK [8], which generates slightly larger proofs but requires only a one-time trusted setup that different applications can reuse for their circuits.

3 Related Work

Weber et al. [26] were the first to propose a blockchain-based system for implementing process choreographies, translating BPMN choreography diagrams into Solidity smart contracts and enabling automated payments. García-Bañuelos et al. [9] improved on this by implementing a space-efficient approach for storing process states in Solidity smart contracts using Petri nets, significantly reducing gas consumption. Ladleif et al. [12] enhanced BPMN choreography diagrams by integrating features to represent blockchain functionalities, including shared data and decision-making logic encapsulated in smart contracts. Caterpillar, a blockchain-based process execution engine [14], manages processes with subprocesses and multi-instance activities, facilitating direct message exchange among participants through the blockchain. An extension enabled the interpretation of BPMN process models without deploying process-specific smart contracts [13]. Lorikeet [15] uses unique modeling techniques for business processes that interact with fungible and non-fungible assets, showing that its transactions generally require less gas compared to the average transaction within a comprehensive dataset of existing smart contracts.

FlexChain [4] outsources the execution of business process rules to an off-chain processor while maintaining the execution state on a blockchain. Storing these rules off-chain and allowing their modification at runtime provides great flexibility. However, this flexibility prevents smart contracts from enforcing the rules and requires audits of execution traces to ensure correctness.

Mangarone et al. [16,17] utilize attribute-based encryption to regulate user access to process data, thereby enhancing privacy. However, using encryption to conceal process data prevents smart contracts from enforcing rules the process model imposes. The following approaches address this issue by employing zk-SNARKs.

The concept of utilizing zk-SNARKs to verify conformance with a BPMN process model was first explored by Toots [25]. Building on this work, Toldi and Kocsis [24] propose a method that preserves confidentiality while executing business processes on a blockchain. Their method involves storing a hash of the current process state within a smart contract that only permits state updates after verifying a zk-SNARK proof of conformance. These proofs ensure the validity of the hashes, the compliance of state updates with BPMN state transitions, as well as the authority of participants. Upon successful validation, the smart contract updates the current state hash with a new one.

Unlike previous approaches, the method of Toldi and Kocsis achieves complete concealment of the process state on-chain while maintaining auditability by smart contracts. However, each process model requires separate arithmetic circuits and verifier smart contracts, which increases the instantiation costs and the deployment complexity, especially as their use of the Groth16 zk-SNARK requires a distinct trusted setup for each arithmetic circuit. Additionally, generating zk-SNARK proofs for average-sized processes takes approximately 2 min, which we aim to reduce together with the instantiation costs.

4 Confidential Execution of Process Choreographies

We propose a method that allows the confidential execution of BPMN choreography diagrams on a blockchain without requiring process-specific arithmetic circuits or smart contracts. Instead, our approach requires only a single deployment of smart contracts to support arbitrary choreographies up to a maximum size. Before executing a choreography, participants exchange their public keys and the Groth16 zk-SNARK proving keys. Furthermore, they transform the BPMN choreography diagram into an executable representation. The active system components during both a) the transformation process and b) choreography execution are shown in Fig. 1.

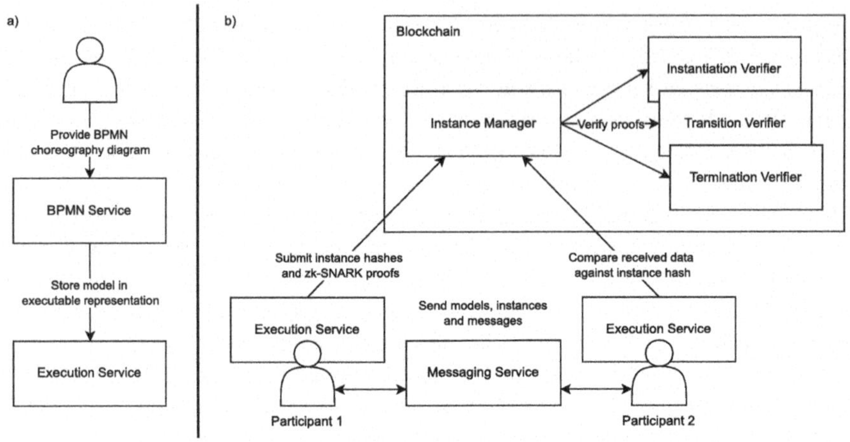

Fig. 1. Active system components during a) the transformation of BPMN choreography diagrams and b) choreography execution

Each participant operates their own Execution Service, responsible for instantiating, transitioning, and terminating choreography instances based on the participant's commands. The Execution Service also generates zk-SNARK proofs and submits them to the Instance Manager smart contract to demonstrate the correctness of these operations. The Instance Manager delegates the verification of these proofs to three separate verifier smart contracts and accepts updates to choreography instances only after successful verification. For each active choreography instance, the Instance Manager stores a single hash of the instance data, which is salted to conceal the data from external observers. The Messaging Service distributes the instance and the choreography model among participants. The hashes in the Instance Manager enable participants to verify the integrity of the received data.

4.1 Choreography and Instance Representation

To simplify process execution, we map the BPMN choreography diagrams to interaction Petri nets [5]. Interaction Petri nets are Petri nets where transitions can represent message exchanges between two partners. This mapping is based on the standard mapping from BPMN process diagrams to Petri nets [6]. The resulting interaction net contains silent transitions, which are transitions that do not correspond to an event or a task in the original process model and needlessly inflate the model's size. While reduction rules exist for Petri nets that can eliminate silent transitions, removing some can cause specific tasks to be executable infinitely many times [9]. In addressing this, we adopt an approach similar to chorpiler[1], a tool recently developed by Stiehle and Weber [21].

[1] https://github.com/fstiehle/chorpiler.

Our approach extends interaction Petri nets by allowing conditional transitions, with the restriction that the messages referenced in the conditions must be integers. This restriction, along with the chosen condition format, ensures that participants can efficiently prove the evaluation of a condition themselves using zk-SNARKs before executing a transition. This method ensures that message contents are not visible on-chain. For a set of integer variables $x_1, \dots, x_k$ and constants $a_1, \dots, a_k, b \in \mathbb{Z}$, each condition has the form

$$a_1 \cdot x_1 + \dots + a_k \cdot x_k + b \odot 0, \tag{1}$$

where $\odot \in \{=, >, <, \geq, \leq\}$.

The proposed system accommodates two types of messages that can be exchanged in choreography tasks: single integers and byte sequences for all other message contents. The tuple

$$m := (h_I, x, r_m, h_m) \tag{2}$$

represents messages of both types, where h_I is a hash of a choreography instance, x is either an integer or a sequence of bytes, and h_m is a hash of the message. In our system, this hash is visible to all participants and since we do not want this hash to leak the message's contents to unintended participants, a salt r is used in hashing. After sampling r and using a hash function H, the message hash h_m is computed according to $m.h_m = H(m) := H(h_I||x||r_m)$, where $||$ denotes string concatenation.

Let n_l be the number of places in the Petri net, n_p be the number of participants, and n_e be the number of messages in the BPMN choreography. Each place is identified by an integer l in the range $[1, n_l]$, each participant by an integer p in the range $[1, n_p]$, and each message by an integer e in the range $[1, n_e]$. A transition t is then defined as

$$t := (L_i, L_o, p_i, p_r, e_i, e_r, c), \tag{3}$$

where L_i is the set of incoming places, L_o is the set of outgoing places, p_i is the optional initiating participant, p_r is the optional responding participant, e_i is the optional initiating message ID, e_r is the optional responding message ID, and c is a condition. If an optional variable is empty, it is assigned a value outside its allowed range. The conditions placed on transitions take the form

$$c := (e_1, \dots, e_k, a_1, \dots, a_k, b, i), \tag{4}$$

where $e_1, \dots, e_k$ are message IDs, $a_1, \dots, a_k$ and b are integer constants as used in Eq. 1, and i is the index of the comparison operator in the tuple $(=, >, <, \geq, \leq)$. If conditions include less than k messages, zero constants ensure that empty messages do not impact the condition evaluation. Hashes of conditions are computed as follows $H(c) := H(a_1|| \cdots ||a_k||e_1|| \cdots ||e_k||b||i)$, whereas the hashes of transitions are computed according to $H(t) := H(L_i||L_o||p_i||p_r||e_i||e_r||c)$.

An executable model M is given by

$$M := (T, L_s, L_e, n_l, n_p, n_e), \tag{5}$$

where T is a set of transitions, L_s is the set of start places, and L_e is the set of end places. Representing the start and end of a choreography, the set of start places contains the incoming places of all transitions resulting from the mapping of start events, and the set of end places corresponds to the outgoing places of all end event transitions. Hashes of models are computed as $h_M = H(M) := H(T||L_s||L_e||n_l||n_p||n_e)$.

An instance I of a model M has the form

$$I := (h_M, f_l, f_p, f_e, r_I, h_I), \tag{6}$$

where h_M is the hash of the model M and $f_p : [1, n_p] \rightarrow S_{\mathsf{pk}}$ is a mapping that associates each participant in the range $[1, n_p]$ with a corresponding public key $\mathsf{pk} \in S_{\mathsf{pk}}$. The mapping $f_l : [1, n_l] \rightarrow \mathbb{Z}$ represents the number of tokens at each place and $f_e : [1, n_e] \rightarrow S_h$ assigns a hash of the message $h_m \in S_h$ to each message ID in the range $[1, n_e]$. The value h_I is a hash of the instance and is published on a blockchain in our architecture, thus representing publicly visible information. A salt r blinds the instance data hash and hides the data from external observers. The salt r is used to compute the instance hash h_I as $h_I = H(I) := H(h_M||f_l||f_p||f_e||r_I)$.

4.2 Instantiation

To start choreography execution, a participant creates an instance I from a model M, providing the public key mapping $I.f_p$ as input. Furthermore, the participant initializes the token count mapping $I.f_l$ according to

$$I.f_l(l) \leftarrow \begin{cases} 1, & \text{if } l \in M.L_s \\ 0, & \text{otherwise,} \end{cases} \tag{7}$$

and creates for all $e \in [1, n_e]$ the message hash mapping $I.f_e$ as

$$I.f_e(e) \leftarrow 0. \tag{8}$$

After sampling the salt $I.r_I$, the participant computes the instance hash $I.h_I$ and, subsequently, creates a signature σ on $I.h_I$ that has to verify with any of the public keys in $I.f_p$. After signing $I.h_I$, the participant generates the instantiation proof and submits it together with $I.h_I$ to the Instance Manager. Upon successful proof verification, the Instance Manager adds $I.h_I$ to the set of instance hashes. Additionally, the participant uses the Messaging Service to share the model M and instance I with the participants responsible for executing the following transitions in the model.

4.3 Transition Execution

During choreography execution, participants execute transitions defined in the model M on an instance I. When a transition $t \in M.T$ is executed on an instance

I, I is updated to I'. While $I.h_M$ and $I.f_a$ remain unchanged, executing a transition samples a fresh salt $I'.r_I$ and thus generates a new hash $I'.h_I$. The token count mapping $I'.f_l$ is computed according to

$$I'.f_l(l) \leftarrow \begin{cases} I.f_l(l) - 1, & \text{if } l \in t.L_i \text{ and } l \notin t.L_o \\ I.f_l(l) + 1, & \text{if } l \in t.L_o \text{ and } l \notin t.L_i \\ I.f_l(l), & \text{otherwise,} \end{cases} \tag{9}$$

where $t.L_i$ and $t.L_o$ represent the sets of incoming and outgoing places of transition t respectively. Given the initiating and responding messages m_i and m_r, the message hash mapping $I'.f_e$ is updated according to

$$I'.f_e(e) \leftarrow \begin{cases} H(m_i), & \text{if } e = t.e_i \\ H(m_r), & \text{if } e = t.e_r \\ I.f_e(e), & \text{otherwise.} \end{cases} \tag{10}$$

Alternatively, participants can pretend to execute a transition by simply sampling a fresh salt $I'.r_I$ and generating a new hash $I'.h_I$ while keeping all other values of I the same. These updates help obscure the timing of genuine transitions, which would otherwise be exposed on the blockchain.

Once the instance I is updated to I', the participant computes a signature σ on $I'.h_I$ that verifies with any public keys in $I.f_p$. If transition t specifies participants $t.p_i$ and $t.p_r$, the execution of t on I requires two signatures σ_i and σ_r on $I.h_I$ that verify with $I.f_p(t.p_i)$ and $I.f_p(t.p_r)$ respectively. These signatures are exchanged using the Messaging Service alongside the initiating and responding messages of the transition:

1. The initiating participant sends the initiating message m_i together with the current instance I, the model M, and the transition t.
2. The responding participant replies with the updated instance I' and their signature σ_r on $I'.h_I$, optionally accompanied by a responding message m_r.
3. Finally, the initiating participant computes their signature σ_i on $I'.h_I$ and the transition proof.

If transition t includes a condition $t.c$, generating the transition proof requires all integer messages $m_1, \ldots, m_k$ referenced in $t.c$ as input. During the generation of the transition proof, $t.c$ is evaluated based on the integer contents of $m_1, \ldots, m_k$, thereby preventing the execution of t if $t.c$ is not satisfied. Following the transition from I to I', the participant submits the hashes $I.h_I$ and $I'.h_I$ to the Instance Manager along with the transition proof. Upon successful verification, it updates the current instance's hash $I.h_I$ with the hash of the updated instance $I'.h_I$. Subsequently, the model M and the updated instance I' are distributed to other participants via the Messaging Service.

4.4 Termination

An instance I of a model M is terminating if $I.f_l(l) = 1$ for some $l \in M.L_e$. As with instantiation and transition execution, a participant's signature on $I.h_I$ is necessary to generate a termination proof. The hash $I.h_I$ is then submitted to the Instance Manager along with the termination proof, which removes $I.h_I$ from the set of instance hashes. While the submission of a termination proof is not mandatory, it signals the completion of the choreography to other participants. It can be used as a trigger for subsequent actions or events.

4.5 zk-SNARK Proofs

We employ zk-SNARK proofs to ensure the conformance of instances to models throughout the instantiation process, the execution of transitions, and the termination of the choreography. Instead of using a single circuit to verify all three operations, we employ three distinct circuits, each with its specific verifier smart contract. While this approach increases deployment costs, it significantly enhances the readability and maintainability of the circuits, as each one is dedicated to a specific purpose. Additionally, the complexity of the circuits for simpler instantiation and termination operations is reduced.

To increase the efficiency of proof generation, the public key mapping in instances, as well as the sets of transitions and end places in models, are each replaced by the root of a Merkle tree containing the same values. While end places can be directly represented as leaf nodes in the Merkle tree, public keys and transitions require hashing before their hashes are included in the Merkle tree. An instance I and a model M are then written as

$$I = (h_M, f_l, \rho^P, f_e, r, h_I) \text{ and } M = (\rho^T, L_s, \rho^L, n_l, n_p, n_e), \tag{11}$$

where ρ^P is the Merkle root of public keys, ρ^T is the root of transitions and ρ^L is the root of end places.

For example, using a Merkle tree representation of the public key mapping $I.f_p$ of an instance I enables a participant to efficiently prove the membership of their public key within $I.f_p$. Participant credentials are then provided as

$$(\sigma, \mathsf{pk}, \pi^P_{\mathsf{pk}}, p), \tag{12}$$

where σ is the participant's signature, pk is their public key and π^P_{pk} is a Merkle proof of membership of pk within the Merkle tree represented by the root $I.\rho^P$. Note that the index p of pk in the Merkle tree always equals the participant ID.

Instantiation Circuit. The instantiation circuit receives an instance hash $I.h_I$ as public input and as private inputs a model M, the instance I, as well as participant credentials $(\sigma, \mathsf{pk}, \pi^P_{\mathsf{pk}}, p)$. It returns true if and only if all the following checks are passed:

1. Check instance and model hashes, i.e. $H(I) = I.h_I$ and $H(M) = I.h_M$.
2. Verify the signature σ on $I.h_I$ using pk.
3. Check the membership of pk in $I.f_p$:
   (a) Verify the Merkle proof π^P_{pk} using the index p.

   (b) Check that $H(\mathsf{pk})$ is the leaf in π^P_{pk}.
   (c) Check that $I.\rho^P$ is the root in π^P_{pk}.
4. Check that the number of tokens in $I.f_l$ satisfy Eq. 7.
5. Check that the message hashes in $I.f_e$ match Eq. 8.

Transition Circuit. The transition circuit receives as public inputs two instance hashes $I.h_I$ and $I'.h_I$ and as private inputs a model M, the instances I and I', a transition t, a Merkle proof π^T_t of t's inclusion in $M.T$, participant credentials $(\sigma_i, \mathsf{pk}_i, \pi^P_{\mathsf{pk}_i}, p_i)$ and $(\sigma_r, \mathsf{pk}_r, \pi^P_{\mathsf{pk}_r}, p_r)$, as well as a list of integer messages $m_1, \ldots, m_k$. It returns true if and only if all the following checks are passed:

1. Check that all required hashes match, which equates to checking $H(I) = I.h_I$, $H(I') = I'.h'_I$ and $I.h_M = I'.h_M$.
2. Check that $I.\rho^P = I'.\rho^P$.
3. Verify the signatures σ_i and σ_r on $I'.h_I$ using pk_i and pk_r, respectively.
4. Check the memberships of pk_i and pk_r in $I.f_p$ using the Merkle proofs $\pi^P_{\mathsf{pk}_i}$ and $\pi^P_{\mathsf{pk}_r}$, as outlined in step 3 of the instantiation circuit
5. Verify that the transition t is in $M.T$:
   (a) Verify the Merkle proof π^T_t.
   (b) Check that $H(t)$ is the leaf in π^T_t.
   (c) Check that $M.\rho^T$ the root in π^T_t.
6. Let b_0 be a boolean variable that is true if and only if the differences between instances I and I' align with the transition t and t's condition holds:
   (a) Evaluate Eq. 9 and Eq. 10.
   (b) To check participant authorization for t, evaluate

$$(t.p_i \in [1, n_p] \Rightarrow t.p_i = p_i) \wedge (t.p_r \in [1, n_p] \Rightarrow t.p_r = p_r). \qquad (13)$$

   (c) Evaluate whether the integer messages $m_1, \ldots, m_k$ match the message IDs referenced in the condition $t.c$. For this, find the message IDs (indices) of $H(m_1), \ldots, H(m_k)$ in $I.f_e$ and compare them to the message IDs in $t.c$.
   (d) Test $t.c$ on the integer contents of $m_1, \ldots, m_n$.
7. Let b_1 be a boolean variable that is true if and only if $I.f_l = I'.f_l$ and $I.f_e = I'.f_e$.
8. Ensure that b_0 or b_1 are true.

Termination Circuit. The termination circuit receives an instance hash $I.h_I$ as public input and as private inputs a model M, the instance I, participant credentials $(\sigma, \mathsf{pk}, \pi^P_{\mathsf{pk}}, p)$, an end place l_e and a Merkle proof $\pi^L_{l_e}$. It returns true if and only if all the following checks are passed:

1. Perform steps 1–3 of the instantiation circuit.
2. Verify the membership of l_e in $M.L_e$:
   (a) Verify the Merkle proof $\pi^L_{l_e}$.
   (b) Check that l_e is the leaf in $\pi^L_{l_e}$.
   (c) Check that $M.\rho^L$ is the root in $\pi^L_{l_e}$.
3. Check that $I.f_l(l_e) = 1$.

5 Implementation and Evaluation

To facilitate the evaluation of our approach, we implement prototypes for several critical components shown in Fig. 1: the transformation of BPMN choreography diagrams, model execution with integrated zk-SNARK proof generation, and the verification of instance hashes within smart contracts. The source code of our prototype implementation is publicly accessible[2].

Our evaluation is twofold. Firstly, we quantitatively measure computational costs for creating and verifying proofs to demonstrate our approach's practical applicability. Secondly, we conduct qualitative and quantitative comparisons against related work, analyze our system's limitations, and identify opportunities for improvement.

5.1 Performance Evaluation

Because arithmetic circuits take a fixed number of inputs, we establish maximum quantities for all model elements. Since smaller values lead to less computational cost, we aim to optimize support for choreographies models of different scales. Following a recent study by Compagnucci et al. [3], we estimate an average occurrence of BPMN elements within choreography diagrams. We create a parameter set called Base, using the nearest powers of two that fit the estimation: 16 places, 16 participants, 16 transitions, one start place, and two end places. To accommodate larger choreographies, we scale these values by powers of two, resulting in additional parameter sets named ×2, ×4, ×8, ×16, and ×32.

Table 1. Proving time for different parameter sets

Circuit	Base	×2	×4	×8	×16	×32
Instantiation	0.09 s	0.15 s	0.27 s	0.49 s	1.03 s	1.85 s
Transition	0.20 s	0.32 s	0.64 s	0.95 s	1.97 s	3.60 s
Termination	0.10 s	0.16 s	0.29 s	0.49 s	0.98 s	1.88 s

Table 1 shows the measured proving times on a system equipped with a 12th Gen Intel® Core™ i7–12800H and 64GB RAM. The proving time of up to 3.60 s is in the same magnitude as Ethereum block times. Hence, the transaction latency experienced by the user remains acceptable. Furthermore, Table 2 depicts the deployment and verification costs measured in Ethereum gas usage. Our results were obtained using Hardhat[3] and show that the one-time deployment can be executed within a single block (15M gas target).

[2] https://github.com/OskarPetto/zk-choreographies.

[3] https://hardhat.org/.

Table 2. Gas usage on Ethereum

	Deployment	Instantiation	Transition	Termination
Gas usage	6,370,066	248,675	256,820	226,728

5.2 Comparison to Related Work

Apart from our approach and the method of López-Pintado et al. [13], all approaches listed in Sect. 3 require deployments of process-specific smart contracts, resulting in significantly higher gas usage for process instantiation. When comparing our gas consumption for process execution to the related work, we find that it falls within the same range.

Due to the similarity with the system of Toldi and Kocsis [24], we emphasize critical differences in Table 3. Their performance metrics are based on six different BPMN process models, with the largest containing 30 tasks and events. We take measurements using the $\times 2$ parameter set, accommodating up to 32 tasks and events, to effectively compare our method to Toldi and Kocsis. While most differences represent improvements over their method, we currently do not support global variables, and our conditions are comparatively less expressive.

Table 3. Key differences of our approach to Toldi and Kocsis [24]

	Our approach	Toldi and Kocsis [24]
System Input	BPMN choreography diagrams	Acyclic BPMN process diagrams
Gateway multiplicity	Any fixed number	2
Deployments & Trusted setups	One for all choreographies	One per process model
Authorization	Mapping from roles to public keys is provided at instantiation	Public keys are attached to BPMN elements
Conditions	Linear combination of integer messages evaluated against zero	Boolean expressions over global variables
Distribution of process state	Shared among participants using a messaging service	Encrypted and stored in the smart contract
Proving time	0.32 s ($\times 2$ parameter set)	53.21–122.47 s
Instantiation gas	248,675 and 6,370,066 one-time	2,078,071–2,408,635
Transition gas	256,820	490,507–548,898

5.3 Limitations and Future Work

To verify the conformance of choreography executions with BPMN choreography diagrams represented as interaction Petri nets, we rely on zk-SNARKs. In our approach, zk-SNARK proofs are crucial for differentiating hashes of instances that align with the choreography from entirely random values. However, we lack a formal proof demonstrating the equivalence of interaction Petri net semantics with our arithmetic circuits. We do not yet support some BPMN concepts, such as sub-choreographies or participant multiplicity.

Another limitation of our approach is the necessity to submit updates to a smart contract after instantiation and each step in the choreography execution. These updates involve storing a 32-byte instance hash and verifying a 256-byte zk-SNARK proof within the smart contract, resulting in considerable transaction costs. To alleviate the issue of high transaction costs, rollups could be employed [23]. Additionally, consolidating multiple transitions into a single zk-SNARK proof or utilizing recursive zk-SNARK proofs could reduce the number of required transactions per choreography.

In the future, exploring termination proofs that prove specific properties of completed choreography instances, like adherence to particular compliance rules, would be valuable. Such proofs could serve various purposes, including supporting hierarchical choreographies by signaling the completion of a sub-choreography to advance a higher-level choreography. Furthermore, they could automatically reward the completion of specific choreographies.

6 Conclusion

We present an architecture for the interpreted and confidential execution of process choreographies on a blockchain. Using an interpreted approach to choreography execution, we aim to reduce instantiation costs compared to methods requiring process-specific smart contracts. The confidentiality of choreography diagrams and instances is ensured by validating the conformance of instance data to models through the verification of zk-SNARK proofs.

We provide a prototype implementation of our methodology, conduct a performance evaluation, and compare related work. The evaluation shows a significant reduction in instantiation costs compared to approaches that require a deployment of process-specific smart contracts. Furthermore, compared to the only other approach that utilizes zk-SNARKs for process execution on a blockchain, our method significantly reduces proving time by 99.39% to 99.73% and decreases the instantiation costs for new processes by 88.03% to 89.67%, depending on the process size.

Future research could focus on reducing the number of blockchain transactions required for executing choreographies or enhancing our termination proofs to include the verification of compliance rules on the executed choreography.

Acknowledgements. This research has been partially supported and funded by the Austrian Research Promotion Agency (FFG) for the research projects "DiCYCLE – Reconsidering digital deconstruction, reuse and recycle processes using BIM and Blockchain" and "USEFLEDS - Unleashing Sector-coupling Flexibility by means of an Energy Data Space" under the contract number 886960 and 905128, respectively.

References

1. Ben-Sasson, E., Chiesa, A., Green, M., Tromer, E., Virza, M.: Secure sampling of public parameters for succinct zero knowledge proofs. In: 2015 IEEE Symposium on Security and Privacy, pp. 287–304 (2015). https://doi.org/10.1109/SP.2015.25
2. Carminati, B., Rondanini, C., Ferrari, E.: Confidential business process execution on blockchain. In: 2018 IEEE International Conference on Web Services (ICWS), pp. 58–65 (2018). https://doi.org/10.1109/ICWS.2018.00015
3. Compagnucci, I., Corradini, F., Fornari, F., Re, B.: A study on the usage of the BPMN notation for designing process collaboration, choreography, and conversation models. Bus. Inf. Syst. Eng. (2023). https://doi.org/10.1007/s12599-023-00818-7
4. Corradini, F., Marcelletti, A., Morichetta, A., Polini, A., Re, B., Tiezzi, F.: A flexible approach to multi-party business process execution on blockchain. Futur. Gener. Comput. Syst. **147**, 219–234 (2023). https://doi.org/10.1016/j.future.2023.05.006
5. Decker, G., Weske, M.: Local enforceability in interaction Petri nets. In: Alonso, G., Dadam, P., Rosemann, M. (eds.) BPM 2007. LNCS, vol. 4714, pp. 305–319. Springer, Heidelberg (2007). https://doi.org/10.1007/978-3-540-75183-0_22
6. Dijkman, R.M., Dumas, M., Ouyang, C.: Semantics and analysis of business process models in BPMN. Inf. Softw. Technol. **50**(12), 1281–1294 (2008). https://doi.org/10.1016/j.infsof.2008.02.006
7. Fridgen, G., Radszuwill, S., Urbach, N., Utz, L.: Cross-organizational workflow management using blockchain technology - towards applicability, auditability, and automation. In: Hawaii International Conference on System Sciences (2018). https://doi.org/10.24251/HICSS.2018.444
8. Gabizon, A., Williamson, Z.J., Ciobotaru, O.: PLONK: permutations over Lagrange-bases for oecumenical noninteractive arguments of knowledge (2019)
9. García-Bañuelos, L., Ponomarev, A., Dumas, M., Weber, I.: Optimized execution of business processes on blockchain. In: Carmona, J., Engels, G., Kumar, A. (eds.) BPM 2017. LNCS, vol. 10445, pp. 130–146. Springer, Cham (2017). https://doi.org/10.1007/978-3-319-65000-5_8
10. Groth, J.: On the size of pairing-based non-interactive arguments. In: Fischlin, M., Coron, J.-S. (eds.) EUROCRYPT 2016. LNCS, vol. 9666, pp. 305–326. Springer, Heidelberg (2016). https://doi.org/10.1007/978-3-662-49896-5_11
11. Khan, S.N., Loukil, F., Ghedira-Guegan, C., Benkhelifa, E., Bani-Hani, A.: Blockchain smart contracts: applications, challenges, and future trends. Peer-to-Peer Netw. Appl. **14**(5), 2901–2925 (2021). https://doi.org/10.1007/s12083-021-01127-0
12. Ladleif, J., Weske, M., Weber, I.: Modeling and enforcing blockchain-based choreographies. In: Hildebrandt, T., van Dongen, B.F., Röglinger, M., Mendling, J. (eds.) BPM 2019. LNCS, vol. 11675, pp. 69–85. Springer, Cham (2019). https://doi.org/10.1007/978-3-030-26619-6_7

13. López-Pintado, O., Dumas, M., García-Bañuelos, L., Weber, I.: Interpreted execution of business process models on blockchain. In: 2019 IEEE 23rd International Enterprise Distributed Object Computing Conference (EDOC), pp. 206–215 (2019). https://doi.org/10.1109/EDOC.2019.00033
14. López-Pintado, O., García-Bañuelos, L., Dumas, M., Weber, I., Ponomarev, A.: Caterpillar: a business process execution engine on the Ethereum blockchain. Softw. Pract. Experience **49**(7), 1162–1193 (2019). https://doi.org/10.1002/spe.2702
15. Lu, Q., et al.: Integrated model-driven engineering of blockchain applications for business processes and asset management. Softw. Pract. Experience **51**(5), 1059–1079 (2021). https://doi.org/10.1002/spe.2931
16. Marangone, E., Di Ciccio, C., Friolo, D., Nemmi, E.N., Venturi, D., Weber, I.: MARTSIA: enabling data confidentiality for blockchain-based process execution. In: Proper, H.A., Pufahl, L., Karastoyanova, D., van Sinderen, M., Moreira, J. (eds.) EDOC 2023. LNCS, vol. 14367, pp. 58–76. Springer, Cham (2024). https://doi.org/10.1007/978-3-031-46587-1_4
17. Marangone, E., Di Ciccio, C., Weber, I.: Fine-grained data access control for collaborative process execution on blockchain. In: Marrella, A., et al. (eds.) BPM 2022. LNBIP, vol. 459, pp. 51–67. Springer, Cham (2022). https://doi.org/10.1007/978-3-031-16168-1_4
18. Mendling, J., et al.: Blockchains for business process management - challenges and opportunities. ACM Trans. Manag. Inf. Syst. **9**(1), 4:1–4:16 (2018). https://doi.org/10.1145/3183367
19. Partala, J., Nguyen, T.H., Pirttikangas, S.: Non-interactive zero-knowledge for blockchain: a survey. IEEE Access **8**, 227945–227961 (2020). https://doi.org/10.1109/ACCESS.2020.3046025
20. Stiehle, F., Weber, I.: Blockchain for business process enactment: a taxonomy and systematic literature review. In: Marrella, A., et al. (eds.) BPM 2022. LNBIP, vol. 459, pp. 5–20. Springer, Cham (2022). https://doi.org/10.1007/978-3-031-16168-1_1
21. Stiehle, F., Weber, I.: Process channels: a new layer for process enactment based on blockchain state channels. In: Di Francescomarino, C., Burattin, A., Janiesch, C., Sadiq, S. (eds.) BPM 2023. LNCS, vol. 14159, pp. 198–215. Springer, Cham (2023). https://doi.org/10.1007/978-3-031-41620-0_12
22. Thaler, J.: Proofs, arguments, and zero-knowledge. Found. Trends® Priv. Secur. **4**(2–4), 117–660 (2022). https://doi.org/10.1561/3300000030
23. Thibault, L.T., Sarry, T., Hafid, A.S.: Blockchain scaling using rollups: a comprehensive survey. IEEE Access **10**, 93039–93054 (2022). https://doi.org/10.1109/ACCESS.2022.3200051
24. Toldi, B.Á., Kocsis, I.: Blockchain-based, confidentiality-preserving orchestration of collaborative workflows. Infocommunications J. **15**(3), 72–81 (2023). https://doi.org/10.36244/ICJ.2023.3.8
25. Toots, A.: Zero-knowledge proofs for business processes. Master's thesis, University of Tartu (2020)
26. Weber, I., Xu, X., Riveret, R., Governatori, G., Ponomarev, A., Mendling, J.: Untrusted business process monitoring and execution using blockchain. In: La Rosa, M., Loos, P., Pastor, O. (eds.) BPM 2016. LNCS, vol. 9850, pp. 329–347. Springer, Cham (2016). https://doi.org/10.1007/978-3-319-45348-4_19

Smart Contracts' Upgradability for Flexible Business Processes

Sidra Malik[1(✉)], H. M. N. Dilum Bandara[1], Nick R. T. P. van Beest[2], and Xiwei Xu[1]

[1] Data61, CSIRO, Eveleigh, Australia
{sidra.malik,dilum.bandara,xiwei.xu}@data61.csiro.au
[2] Data61, CSIRO, Dutton Park, Australia
nick.vanbeest@data61.csiro.au

Abstract. Business process execution and support by blockchain-based systems have received increasing interest, particularly in applications where execution trace immutability and independent verifiability are needed. Recent approaches to automatically convert business process models into smart contracts have significantly streamlined blockchain support for business processes. However, the dynamic nature of business processes requires services to be adaptable to evolving conditions, activities, agreements, or regulations. Subsequent adaptation of the corresponding smart contracts is also a must. Adoption of so-called business process *change patterns* into smart contracts is not straightforward, as smart contracts cannot be modified once deployed. Further, smart contract *upgradability patterns* are not designed with business process applications in mind. In this paper, we present an approach to assess and select the most suitable smart contract upgradability pattern for a given business process change pattern, based on six software quality metrics. Moreover, we provide an assessment of the preferred upgradability patterns for each change pattern using the presented approach.

Keywords: change patterns · flexible business processes · smart contracts · upgradability patterns

1 Introduction

Over the past two decades, Business Process Management (BPM) has evolved from supporting predefined and repetitive units of work in production environments to flexible, loosely-coupled, service-based processes that apply to a wide range of business contexts. Consequently, modern BPM services are required to be flexible and subject to changes in business logic, regulations, or a shift towards different markets. Such changes to the underlying business process models and service compositions are categorized as so-called *Business Process Change Patterns* (BPCPs). BPCPs are characterized as a set of transformation patterns that occur between an original process model P and its modified version P', and are helpful in guiding and managing business process evolution.

Business process execution and support by blockchain-based systems has received increasing interest, particularly in applications where immutability of execution traces and independent verifiability are needed. They are used in various application areas to streamline, automate and optimize business processes. The most common use of smart

C. Di Ciccio et al. (Eds.): BPM 2024, LNBIP 527, pp. 55–70, 2024.
https://doi.org/10.1007/978-3-031-70445-1_4

contracts in BPM is to automate a workflow or a service choreography, triggering the next activity or service when conditions are met. Recent approaches to automatically convert business process models into smart contracts have significantly streamlined blockchain support for business processes [14, 19]. However, the subsequent application of BPCPs into smart contracts is not straightforward, as the blockchain's immutability prevents smart contracts from being modified once deployed. Any update in existing process instances and data requires additional smart contract development, redeployment and data migration. Like any software program, smart contracts may need a change due to a number of reasons, such as bug fixes, business logic changes and performance/cost enhancements. A trivial way to upgrade a smart contract is to deploy a new one. Since the new contract is assigned its own smart contract address, it loses user interactions and access to the previous contract's data. Therefore, the data must either be migrated to the new contract or the previous contract must have pre-built functionality to make it accessible to other contracts. Various *Smart Contract Upgradability Patterns* (SCUPs) have been developed to facilitate future upgrades of smart contracts [1,2,5,11]. SCUPs, however, are not necessarily designed with business process applications in mind.

In this paper, we present an approach to assess and select the most suitable SCUP for a given BPCP based on various software quality metrics such as modifiability, latency, and cost. Moreover, we provide a detailed assessment of the preferred SCUPs for each BPCP using the presented approach. Finally, we discuss how these findings apply to existing tools that automatically convert business process models into smart contracts.

The remainder of the paper is organized as follows: Sect. 2 presents background for BPCPs, along with a brief introduction to smart contracts for Business Processes and SCUPs. Next, related work is presented in Sect. 3. Subsequently, Sect. 4 presents the research method, followed by a thorough analysis of SCUPs against BPCPs in Sect. 5. Finally, challenges and future directions are highlighted in Sect. 6.

2 Preliminaries

2.1 Business Processes Change Patterns

The Business Process Modelling Notation (BPMN) is a business process modelling language specifically developed to be understood by business users [22]. As such, the process flow is represented in a graph-oriented way, where the explicit control-flow is defined by events, activities, and gateways that are connected through sequence flows and message flows [17].

A *design pattern* is a general, reusable solution to a recurring problem within a given context. Workflow patterns [9] enumerate a large collection of business process modeling patterns, providing elementary business process modelling constructs in BPMN notation. *Business Process Change Patterns* (BPCPs) [13,20] are extended from these process modeling patterns to capture business process improvements and changes in dynamic business environments. As such, BPCPs describe a set of transformations between an original process model P and its modified version P'. These patterns assist in managing and guiding the evolution of business processes. Three main categories of these BPCPs are as follows [13]:

Activity Extend/Delete pattern type refers to change patterns that involve either the addition of a new business process activity to temporarily replace an existing set of activities or the permanent removal of certain activities due to changes in the business logic. Figure 1 depicts an example where activity B is deleted permanently or replaced temporarily by activity X.

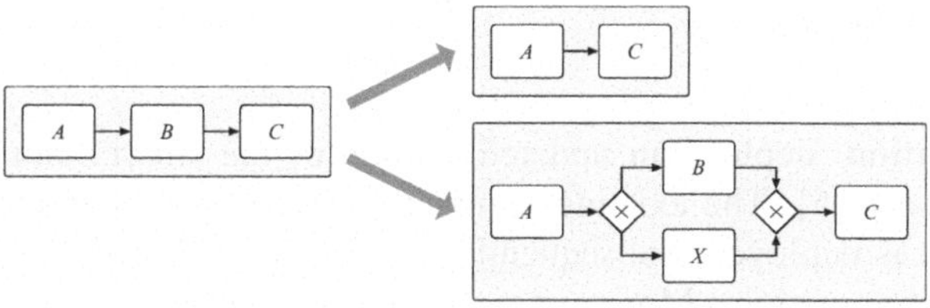

Fig. 1. Activity delete/extend pattern.

Activity Merge/Change/Reposition pattern type does not introduce structural complexities to the business process, but instead merges, replaces, or re-positions existing activities. The activity reposition pattern reorganizes the execution sequence, as shown in Fig. 2. Another example is the activity parallelization pattern, where a sequence of activities is executed in parallel to reduce execution time. Finally, the merging variant combines multiple activities into a single activity.

Fig. 2. Activity reposition pattern.

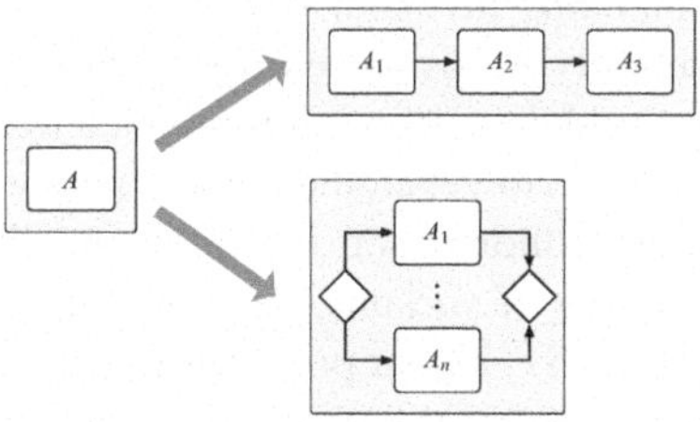

Fig. 3. Activity split change pattern.

Activity Split/Change pattern type decomposes an existing activity into multiple activities. One specific example of this type is the sequential split pattern depicted in Fig. 3. Activity A is decomposed into a series of sequentially linked activities, namely A_1, A_2, and A_3. An activity split could also include gateways where activities A_1, A_2, and A_3 are executed exclusively or in parallel using an OR, XOR, or AND split.

2.2 Smart Contract Upgradability Patterns

Smart contracts are programs stored on a blockchain that execute when specific conditions are fulfilled. Their main purpose is to automate the execution of agreements so that all parties have certainty about the outcome without relying on intermediaries. A number of *Smart Contract Upgradability Patterns* (SCUPs) have been introduced to prepare smart contracts for future upgrades [1,2,5,11]. A brief introduction to each of these patterns is defined below:

Contract Migration deploys an updated version of the smart contract whenever a change is necessary [5]. The existing contract's source code is modified and a new contract instance is deployed. Subsequently, the data is migrated to the new contract using a series of transactions. Moreover, external users and other contracts that call on the previous contract are informed of the new address of the contract.

Data Segregation separates data and business logic into distinct contracts: a logic contract (LC) and a data contract (DC) [11]. In order for the LC to interact with the DC, it is essential for the LC to know DC's address and the specific functions that grant access to the relevant data. Note that the DC is owned by the LC and configured with the LC's address to prevent other addresses from manipulating its state. The DC remains immutable, but the LC can be upgraded by deploying a new one and setting/updating each other's addresses. As such, this pattern only allows updating the business logic, not the data. Data updates would require another pattern-like contract migration. Alternatively, a second DC can be introduced only for the updated data, with the LC managing the complexity of switching between DCs.

Proxy separates a smart contract's interface from its implementation [2]. Transactions call the proxy contract, which forwards them to the implementation contract regardless of the smart contract function being invoked. Whenever the implementation contract is upgraded, the proxy contract only needs to be informed of the new address. While the implementation contract can hold data, it is recommended to adopt the data segregation pattern to store data on the proxy. However, data on the proxy can be manipulated using delegate calls without explicit authorization management like in the data segregation pattern. This pattern has a couple of variants, such as transparent proxy and universal upgradable proxy standard (UUPS) [2] that provide different authorization, ledger storage layout management, and transaction options. For the remainder of this paper, we consider the UUPS pattern as it can deploy a new implementation contract and set its address on the proxy contract within a single transaction.

Diamond is an extension of the proxy pattern that exposes functions from multiple stateless LCs as so-called *facets* [1]. A facet may be exposed through multiple proxies, which hold the data and a mapping of function availability within a facet. Whenever a transaction invokes a function, the proxy checks the mapping to see which facet the function belongs to, before making a delegate call to the respective LC. This allows the proxy's data to be manipulated using the LC's business logic. In contrast to a proxy that has a single LC (which may also have a smart contract size limit), the diamond pattern allows dividing a large LC into multiple smaller LCs, where each LC is within the smart contract size limit. This pattern also enables modular permissions to upgrade the contract, where a function can be upgraded depending on the needs of a specific entity.

3 Related Work

This section presents some related work in the business process and blockchain domain, including some tools for converting business process models into smart contracts, such as Caterpillar [14] and Lorikeet [19].

Caterpillar is a BPMN execution engine built on Ethereum [14]. It accepts a BPMN process model as an input and generates a set of corresponding smart contracts. Caterpillar can instantiate process models, monitor the state of process instances, and execute associated tasks. It can handle process models with sub-processes, advanced BPMN constructs (e.g., boundary events), and multi-instance activities. Caterpillar's architecture includes compiler tools, an execution engine, an event monitor, and a work-item manager. Compilation tools map BPMN to smart contracts containing variables to store the execution state of the process instance. The execution engine on the other hand deploys the smart contracts, creates and manages instances of the process model, and manages external execution calls. The event manager generates notifications whenever a process-related transaction is logged on the blockchain and the work-item manager handles user related tasks.

Lorikeet follows a model-driven approach and generates Solidity smart contracts based on BPMN diagrams and a registry data schema [19]. A key feature of Lorikeet is a graphical user interface (UI) that allows users to design and modify BPMN models, which enables non-technical business users to create smart contracts without extensive blockchain programming expertise. Moreover, the tool allows defining asset types, specifying their attributes, and traceability through tracking asset ownership and transactions. The BPMN and registry modeler UI interacts with the backend via an API gateway. The calls from the BPMN modeler UI are translated into the corresponding Solidity smart contract code. Essentially, a set of smart contract functions are generated for each activity in the BPMN model. The activity flow in the BPMN model is then captured through a set of state variables. On the other hand, the registry modeler UI, is a web-based form collecting customized registry information, i.e., asset attributes stored on the blockchain. The Lorikeet architecture manages the lifecycle actions of a registry record for each process instance separately.

In [7], the authors presented a framework for distributed multiparty choreographies, with a primary focus on auditing in BPM. After the participants are identified, the framework translates a multi-party choreography specification into a Solidity smart contract. In [6], Discrete Event-Hierarchical State Machine (DE-HSM) modeling is used to create methods to identify patterns or DE-HSM sub-models that are suitable to be processed on sidechains instead of the main chain. This enables cost-effective and privacy-preserving sidechains for business process services.

Corradini et al. [8] presented an approach to separate the process state stored on-chain from the execution logic stored off-chain. The process state is represented as data stored within a smart contract, while the execution logic is defined as rules stored in the Inter-Planetary File System (IPFS). This separation enables the modification of execution logic at run-time without modifying the on-chain code.

The above literature [6,7] discusses the use of smart contracts with a primary focus on auditing and privacy-preservation or cost. The work presented in [8] highlights the need of flexibility in blockchains due to dynamically changing business processes by decoupling the logic from smart contract and allowing it to be modified off-chain. Our work extends the same problem area by utilizing the well-studied SCUPs for BPM change/flexibility requirements. We present a granular evaluation of SCUPs against BPCPs on specific metrics. For instance, if frequent activity repositioning is needed as a business requirement, a suitable SCUP can facilitate changes without affecting the smart contract's existing state.

4 Research Method

In this section, we present our method and assessment criteria for choosing an appropriate SCUP for a given process change pattern. Given a business process and a set of anticipated changes to it, our aim is to assist developers and process designers to choose a suitable smart contract structure to encode that business process. Ideally, this should be done in a business process and smart contract implementation agnostic manner. Therefore, we use BPCPs to represent anticipated business process changes and SCUPs to represent smart contract structures. Our research method is depicted in Fig. 4. First, we choose a subset of BPCPs from [13]. We limit our evaluation to one distinct change pattern from each pattern type (see Sect. 2.1) to get an overall understanding. We select the following BPCPs, as they capture the most significant changes in each of the three types of change patterns [13]:

- *Activity AND-split pattern.* New activities are added in parallel to an existing activity and joined by an *AND* gate. This prevents the process from advancing to the next activity without completing all parallel activities.
- *Activity Reposition pattern.* No new activity is added, rather the position of activities is adjusted.
- *Activity Delete pattern.* Existing activities are removed permanently or temporarily.

Moreover, we choose a set of SCUPs from the upgradability patterns discussed by the Ethereum developer community [11] (see Sect. 2.2). We selected four SCUPs, namely contract migration, data segregation, proxy and diamond patterns, which define the most broader categories of the upgradability patterns. We also identify a set of metrics to determine the feasibility, complexity, and cost associated with each change. Next, we assess each change pattern against SCUPs based on the selected metrics. Based on our evaluation, we provide guidance on utilizing a specific SCUP tailored to a BPCP. We analyze the best-fitting SCUP for a given BPCP and provided an upgradability assessment of the SCUPs against generic evaluation metrics.

Given the need to upgrade a smart contract encoding a business process, the most critical question is "Can SCUP u achieve BPCP c?" If so, one needs to understand the complexity of the upgrade process and the associated cost to determine its practicality. To determine such answers, we focus on three high-level software quality attributes

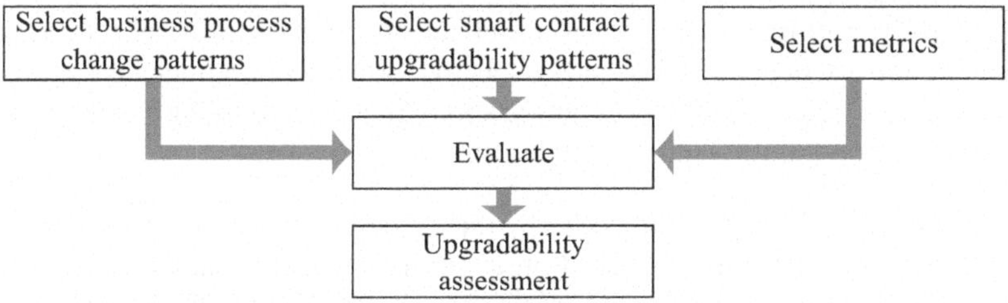

Fig. 4. Research method.

from the ISO/IEC 25010 quality model [12], namely maintainability, performance efficiency, and usability. However, these are composite quality attributes that depend on many other attributes. For example, the authors in [18] list 22 metrics for smart contracts such as lines of code; the number of functions, events, interfaces, contracts, and libraries; and bytecode length that captures maintainability and performance efficiency. Note that most of these metrics evaluate the source code.

In this paper, we abstract away from the source code and select a representative subset of metrics that enable us to evaluate the smart contract structure arising from a given change and upgradability pattern. Figure 5 shows the selected metrics (in yellow) and their relationship to sub-characteristic (in white) and characteristic (in grey) in ISO quality model. For example, the number of contracts impacts modularity, cost and latency to deploy contracts, coupling, and operability. A smart contract's modifiability can then be determined based on its coupling, cyclomatic complexity, and modularity, which in turn determines its maintainability. Next, we explain the chosen metrics and their calculations:

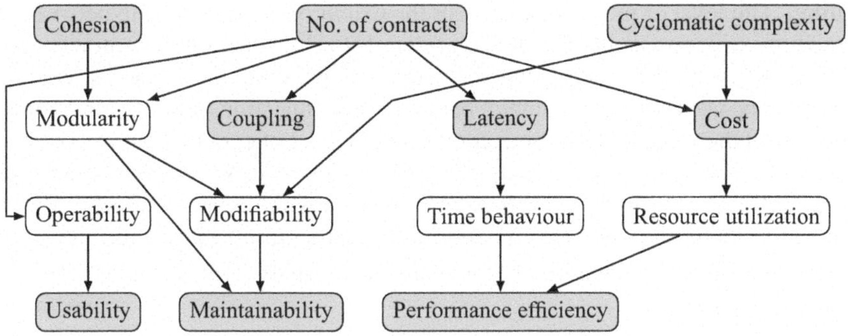

Fig. 5. Graphical overview of the metrics and their relations.

No. of Contracts. SCUPs may use single or multiple contracts. For example, in the data segregation pattern, the contract's data are kept separate from its logic. The diamond pattern has one data and several logic contracts depending on functionality. A module design that splits activities into different contracts enhances upgradability by allowing specific parts of the business process to be modified independently.

Alternatively, multiple contracts can complicate the upgrade process, requiring careful consideration of dependencies and a specific order for contract deployment. As shown in Fig. 5, a higher number of contracts may increase coupling, latency, and cost while decreasing operability and modifiability.

Cost. There is a cost to deploy, upgrade (by redeploying), and execute a smart contract on a public network. The cost is based on a variety of factors such as bytecode size, computation and memory complexity, ledger storage, and bandwidth consumption. The *gas used* in Ethereum captures all such factors; hence, provides an objective measure of the overall overhead of a given upgradability pattern. Such cost information is also useful in private blockchains to optimize performance, e.g., it has been demonstrated that Hyperledger Fabric's throughput depends on smart contract complexity and ledger state [4]. However, the *gas used* metric depends on the specific smart contract implementation and transaction data. Assuming we want to encode and execute the same business process with different SCUPs, we can abstract the transaction cost to be the sum of base transaction cost t and additional costs to deploy a contract of size s, make a call or delegate a call to another contract c, and migrate data of size d.

Cyclomatic Complexity. This indicates code complexity and difficulty in understanding and testing it, by measuring the number of linearly independent paths through the source code [10]. Hence, with respect to change patterns, the amount of parallel paths or decision changes introduced in the business process directly contributes to cyclomatic complexity. When an SCUP segregates data from business logic, the LC needs to call the DC, leading to further branching in the source code. Therefore, we count both the decision points related to the business-process change and the number of calls to other contracts to estimate the cyclomatic complexity.

Coupling. The level of interdependence or the extent of the relationship between multiple modules or components is called *coupling* [16]. Because smart contracts may be coupled in terms of data or logic dependencies (DC-LC coupling or LC-LC coupling), coupling can be used to measure interdependence or connections among smart contracts that are part of the same SCUP. For example, a high level of coupling among smart contracts reflects a design that is difficult to modify or upgrade. We estimate the coupling using the Instability formula [15]. Instability measures the relative susceptibility of a component to changes elsewhere. An instability value close to zero indicates a stable smart contract while a value close to one indicates a smart contract highly susceptible to breaking changes.

Cohesion. The strength of the relationship between functions inside a module or component is indicated by its cohesion [16]. A smart contract with high cohesion is more amenable to upgrades, as modifications can be confined to specific areas without affecting other parts of the logic. While several quantitative cohesion metrics exist, they result in *NaN* values when data and logic are in separate contracts [3]. Therefore, we use the ordinal scale from [16] to measure cohesion. The scale ranges from zero to six (higher is better). Zero indicates the worst cohesion (aka *coincidental* cohesion) when parts of a module are grouped arbitrarily. Six indicates the best cohesion (aka *functional* cohesion) where a module performs a single well-defined function.

Latency. Some SCUPs must follow a specific order of deployment and configuration. For example, in the data segregation pattern DC needs to be deployed before LC because LC needs to know DC's address. This is further complicated on Nakamoto-consensus-based blockchains where a transaction is not finalized until it is included in a block. Hence, for the data segregation pattern, one needs to wait for a long time (known as confirmation blocks) to ensure that DC is persistently stored on the ledger before deploying LC. However, there is considerable variability in the latency between issuing a transaction and achieving finality. This depends on many factors such as the offered transaction fee, network congestion, inter-block time, network throughput and the number of confirmation blocks. It has been demonstrated that latency is a heavy-tailed distribution even for the same transaction fee [21]. As a result, finalizing sequential transactions may take a lot longer than finalizing concurrent transactions. Assuming transaction fees and network conditions remain the same, we can use the time to finality e to get a lower bound on the latency to deploy all smart contracts of a given SCUP.

5 Results

In this section, we assess the four SCUPs using the assessment metrics established in the previous section. Table 1 presents the results. We analyze SCUPs using four different cases. The first is the base case where no upgrade is required for a smart contract. This represents a typical scenario of a newly deployed contract for a business process.

Table 1. Overview of smart contract upgradability complexity.

	No. of contracts	Deployment / upgrade cost	Execution cost	Cyclomatic complexity	Coupling	Cohesion	Latency
Base smart contract							
Contract migration	1	$t+s$	t	1	0	3	e
Data segregation	2	$3t+s$	$t+c$	2	1	3	$3e$
Proxy	2	$2t+s$	$t+c$	2	0.5	3	$2e$
Diamond	$f+1$	$t(f+1)+s$	$t+c$	$f+1$	0.5	6	$2e$
Activity delete pattern							
Contract migration	1	$2t+s+d$	t	1	0	3	$2e$
Data segregation	2	$2t+s$	$t+c$	2	1	3	$2e$
Proxy	2	$t+s$	$t+c$	2	0.5	3	e
Diamond	$f+1$	$2t+s/f$	$t+c$	$f+1$	0.5	6	$2e$
Activity reposition pattern							
Contract migration	1	$2t+s+d$	t	1	0	3	$2e$
Data segregation	2	$2t+s$	$t+c$	2	1	3	$2e$
Proxy	2	$t+s$	$t+c$	2	0.5	3	e
Diamond	$f+1$	t	$t+c$	$f+1$	0.5	6	e
Activity AND split pattern							
Contract migration	1	$2t+s+d$	t	1	0	3	$2e$
Data segregation	2	$3t+s+d$	$t+c$	2	1	3	$3e$
Proxy	2	$t+s$	$t+c$	2	0.5	3	e
Diamond	$f+a+1$	$t(a+1)+as/f$	$t+c$	$f+1$	0.5	6	$2e$

where f = no. of facets, a = new activities, d = no. of data elements, s = transaction cost based on average contract size, t = base transaction cost, c = call/delegate call cost, and e = time to finality.

Subsequently, we examine the impact of the three chosen BPCPs: activity delete, activity reposition, and activity AND split pattern in the following sub-sections. Each metric presented in Table 1 is systematically explored in individual paragraphs, following the order from left to right across the table.

5.1 Base Case

The contract migration pattern always uses a single contract, whereas data segregation and proxy patterns have two smart contracts, LC and DC. The diamond pattern has several contracts that depend on the number of facets (f), the mapping of smart contract functions to the facets that implement them, and a diamond contract holding the data (i.e., DC), hence the total *no. of contracts* become $f+1$.

The single contract used in the contract migration pattern will incur a *deployment cost* of $t+s$, i.e., the base transaction cost t plus the cost s to deploy the bytecode. Splitting a contract into multiple contracts typically increases the bytecode size. However, we can derive a lower bound for deployment cost by ignoring such a minor increase in total bytecode size. The data segregation pattern requires three transactions to deploy DC, LC, and then set LC's address on DC, increasing the overall cost to $3t+s$. Proxy and implementation contract can be deployed with two transactions, i.e., the cost is $2t+s$. The diamond pattern requires f transactions to deploy the facets and another transaction to deploy the proxy while setting function mappings, resulting in a cost of $t(f+1)+s$. Consequently, the diamond pattern has the highest deployment cost.

When business logic and data are in separate contracts, a data write transaction will incur an *execution cost* of $t+c$. Where t is the cost of invoking the first contract and c is the cost to call/delegate call the other contract. However, the execution cost is also a lower bound because it abstracts out the cost of executing smart contract functions and embedding function parameters in the transaction.

The *cyclomatic complexity* is based on the number of calls between the contracts and sub-contracts (e.g., LC-DC and diamond and facets). It is highest for the diamond pattern, as the business logic needs to branch out to f facets.

We assess the *coupling* based on the instability of DC-LC, DC-DC, or LC-LC dependencies. The coupling is highest for the data segregation pattern as the LC relies on the DC for its operation, and any change to the DC may require changes to the LC to maintain compatibility. However, the DC is not susceptible to LC modifications. Alternatively, in the proxy pattern, the LC (i.e., implementation contract) does not have a direct dependency on the DC as it controls the data storage at the proxy using delegate calls. While a diamond has f facets, from each facet's point of view, it is similar to the proxy pattern and has the same coupling value.

We mark the diamond pattern as having a *cohesion* score of 6 (i.e., *functional* cohesion) as each facet can include a single well-defined function representing an activity in the business process. We mark the other three patterns as *procedural* cohesion (awarded a score of 3) since the smart contract functions are grouped in a contract only because they follow a specific arrangement of execution according to business process activities.

Latency increases when contracts need to be deployed sequentially. The data segregation pattern requires deploying DC first, then deploying LC while specifying DC's address, and finally setting LC's address on DC. Hence, it involves three transactions.

As each transaction needs to be finalized before issuing the next one, the lower bound on latency is $3e$, where e is the time to finality since issuing a transaction. With the proxy pattern, two sequential transactions are required to deploy LC and then deploy the proxy with LC's address. All facets in the diamond pattern can be deployed simultaneously. Then only the diamond contract can be deployed while setting the facet mapping. As such, the pattern requires only two sequential transactions, i.e., the latency is $2e$.

5.2 Activity Delete

In the activity delete change pattern, one or more functionalities must be removed from the base smart contract. It may not be trivial to delete a function and update its dependencies. In this case, the contract migration pattern will incur a cost of $t+s$ to redeploy the updated contract (ignoring minor reductions in bytecode size). Another $t+d$ is needed to migrate data of size d, resulting in a total upgrade cost of $2t+s+d$. The upgrade cost for the data segregation pattern is $2t+s$ as it requires redeploying LC and updating DC with the new LC's address. Both the deployment of the modified LC and updating its address on the proxy can be done within a single transaction leading to a cost of $t+s$. For the diamond pattern, as facets are designed to be cohesive, only the facet with the deleted function needs to be redeployed. It will cost $t+s/f$ to deploy the updated facet and another transaction to update the diamond, resulting in a total cost of $2t+s/f$.

The upgrade latency of the contract migration pattern increases as contract deployment and data migration usually happen as two sequential transactions. Due to sequential redeployment of the facet and then updating the diamond with the new address, the diamond pattern's latency remains the same. Upgrade latency in data segregation and proxy patterns is reduced when compared to the base case as only one contract is redeployed. As the activity-delete pattern does not change the number of contracts or their structure, other metrics remain the same as in the base case.

5.3 Activity Reposition

For the activity reposition change pattern, recall that no new activity is added. Rather, the execution sequence of existing activities is changed. Even though a minor change in business logic is required to include an updated activity sequence, all SCUPs except the diamond pattern will need to be redeployed. Therefore, metrics remain the same as in the activity delete pattern. In the diamond pattern, a transaction that invokes the diamond cut function can be used to remove the mapping between a function and its facets. Therefore, it has the lowest upgrade cost among the four patterns.

5.4 Activity Split

The activity split pattern requires dividing a single activity into multiples or adding *a* new parallel or mutually exclusive activities joined by *XOR*, *OR*, or *AND* gateways. These changes may involve the addition of new functions or state variables and may directly impact the upgrade cost and coupling of the smart contract. The DC in the data

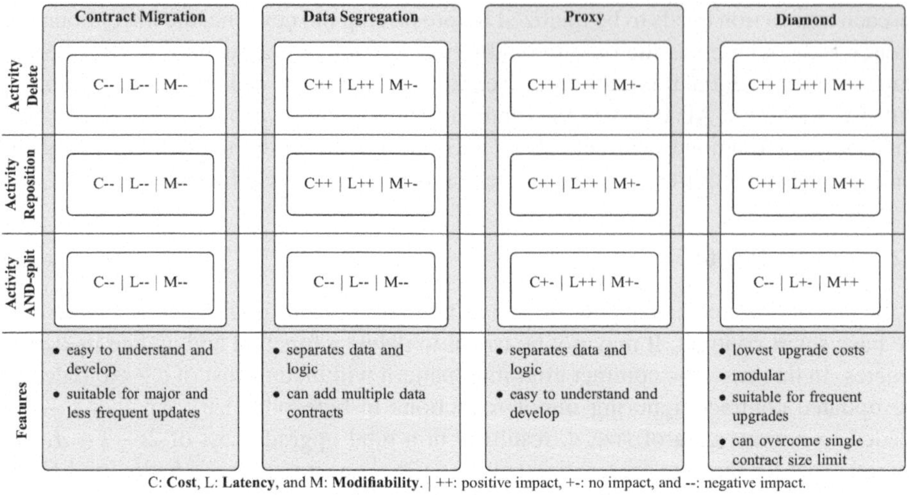

Fig. 6. SCUP assessment for BPCPs.

segregation pattern must also be updated when new data are added. Additionally, data needs to be migrated from the old DC to the new one. Therefore, its cost increases by an additional $t+d$ compared to activity deletion and reposition change patterns. As these three transactions need to be executed in sequence, latency also increases from $2e$ to $3e$. Diamond patterns enable newly added a activities to be placed in new set of facets. Therefore, its upgrade cost includes $a(t+s/f)$ to deploy the facts and another t to update the mapping on diamond. The upgrade cost for proxy pattern remains the same as the LC defines the storage layout through delegate calls though data are stored on the proxy. The cost of contract migration will be similar to the cost of activity deletion and repositioning. Other metrics remain the same for the four SCUPs.

5.5 Smart Contract Upgradability Assessment for BPCPs

Figure 6 presents the SCUP assessment that depicts the relationship between SCUPs and BPCPs based on cost, latency, and modifiability software qualities. It serves as a fundamental, yet intuitive template for developers to comprehend the advantages of various SCUPs for different BPCPs. The metrics in this assessment are chosen based on the findings in Table 1.

Note that metrics like *coupling*, *cohesion*, and *cyclomatic complexity* remain constant for BPCPs. In contrast, *cost*, *latency*, and the *number of contracts* are influenced by BPCPs. The overall modifiability of a smart contract relies on several properties, including cohesion, coupling, the number of contracts, and cyclomatic complexity. Hence, we use modifiability to provide an intuitive representation of all these metrics.

Each SCUP selection u against a corresponding BPCP c is annotated with a certain metric level. The cost (C), latency (L), and modifiability (M) metrics are categorized into three levels. Positive impact ($++$) means that the chosen (c,u) pair is better than the base case. No impact ($+-$) means the (c,u) pair has no effect on C, L, or M. Negative impact ($--$) means the (c,u) pair is worse than the base case. Figure 6 also highlights the dominant features associated with each SCUP.

We can draw the following conclusions from the assessment:

- Splitting a smart contract as an LC(s) and DC to enhance its upgradability does not significantly increase the execution cost as typically $t \gg c$.
- In situations where business processes change frequently, the diamond pattern's higher initial deployment cost is offset by its lower upgrade cost.
- In cases where fewer changes are likely to be required, proxy would be a better option as the cost of redeployment would be lower than the combined cost of deployment and upgrade associated with the diamond pattern.
- Contract migration could be useful when there are few process instances and state variables, as the cost of redeploying a smart contract is lower than upgrading it. In addition, it is the only option when updates are an afterthought.
- The data segregation pattern is only useful when the change pattern does not need a DC upgrade.

Among these patterns, the diamond pattern emerges as a modular and more maintainable approach, with relatively low upgrade costs. Another notable advantage of the diamond pattern is its ability to overcome the maximum bytecode size limitation of a contract, e.g., Ethereum currently has a size limit of 24 KB per contract. By decoupling the functionality into multiple facets, the risk of exceeding the contract size limit is mitigated, as the code can be distributed across multiple contracts. However, it should be noted that the modifiability of the diamond pattern may be limited if existing facets need data changes [1].

The data segregation, proxy, and diamond patterns facilitate business logic and data separation. However, proxy and diamond patterns tend to offer superior maintainability advantages over data segregation. This is primarily due to use of delegate calls that allow the LC to manage the data on the proxy or diamond, without explicit data access functions and access control. This reduces upgrade costs and latency, leading to improved overall performance efficiency. Conversely, the data segregation pattern alone is not useful when DC needs to be upgraded to support new process states or when the existing data schema needs to evolve. Alternatively, the contract migration pattern is the most intuitive to understand, as it involves a single contract and is particularly effective for implementing major updates. However, it might not be feasible in terms of testing and validation effort and data migration cost. When the data segregation pattern needs to be upgraded, it will again be costly due to similar reasons.

Overall, the analysis highlights the strengths and considerations associated with different upgrade patterns, providing valuable insights to developers for selecting the most suitable approach based on their specific business process upgradability requirements.

6 Discussion and Conclusion

Following our assessment, we discuss how our findings can help the business process community select a smart contract structure appropriate for their business processes. We also discuss challenges and directions for future work.

Implications. Our analysis indicates that despite the initial cost of making a smart contract upgradable, it can be offset by lower upgrade costs. For example, when the contract migration pattern is the only option (where the built-in upgradability is absent), it may not be feasible in terms of data migration cost and validation effort. Moreover, building upgradability marginally increases the execution cost from t to $t + c$, where $t \gg c$. Therefore, it would be advantageous for tools such as Lorikeet and Caterpillar – which convert a BPMN model into smart contracts (see Sect. 3) – to build upgradability into the contracts they generate. The tool's architecture and anticipated BPCPs must be considered before this can be done.

Lorikeet, for example, manages assets using data registry models. An access control policy governs the lifecycle management actions of the registry record, such as create, update, delete, and ownership transfer. Such a registry is stored similar to the data-segregation pattern. A change in the business process may not only change the flow of activities but also require a change in an asset's registry model. As such, in the event of registry changes, the DC of the respective asset needs to be redeployed, which will result in the loss of running instances. The only option is to migrate data while dealing with inconsistencies, inefficiencies, and high costs. Changes are also needed to modify the order of state variables that track the activity execution flow, e.g., in the case of the activity re-positioning pattern, in which only the business logic is changed. Both data registry and activity flow changes can be handled more effectively by leveraging a proxy or diamond pattern to avoid data migration issues. Alternatively, Caterpillar has four types of smart contracts to handle various functionalities, namely workflow, worklist, factory, and registry smart contracts. However, Caterpillar's off-chain and on-chain components are highly coupled and a slight change in one component may trigger breaking changes in the above four smart contracts. As the Caterpillar architecture is more modular than Lorikeet, the diamond pattern would be an ideal fit as it offers higher cohesion than other patterns. In contrast to Lorikeet, Caterpillar does not contain an access control mechanism in its worklist handler, meaning anyone can alter the state of running processes. Thus, for any SCUP to be integrated, it is essential to first integrate an access control mechanism into Caterpillar architecture to set permissions for deploying upgrades.

Challenges. Although our evaluation discusses a limited set of BPCPs, slight variations to these BPCP categories would have a similar effect on the evaluation metrics. As our assessment was based on the business logic and data changes needed within a smart contract to accommodate a given BPCP, our findings also apply to updates that do not alter the process structure, such as adding or changing process states and activity scripts. However, business process changes often involve modifications in workflow, data flow, or decision-making logic and can vary widely depending on the industry and organizational context. Each industry may have specific processes tailored to its unique requirements and objectives. Hence, identifying the appropriate SCUPs that align with the specific BPCPs while ensuring that the upgrade effectively captures the intended changes is complex and faces a number of challenges.

When applying SCUPs to update smart contracts executing ongoing process instances, it is essential to ensure backward compatibility to retain seamless interaction with different versions. This is especially challenging when the upgrade involves changes to data schema or function signatures. Similarly, any interdependencies between LC-LC or DC-LC contracts depending on the chosen SCUP must be preserved.

In addition, SCUPs need to undergo thorough testing and validation before being deployed on-chain to ensure the correctness and reliability of the upgraded smart contracts. BPCPs may involve complex scenarios and edge cases that require comprehensive testing, especially for scenarios involving multiple contracts. Therefore, the tests should ensure consistency of the control and data flows of the updated business process. Finally, as SCUPs are relatively new, the adoption and standardization of SCUPs for BPCPs may face challenges in terms of community acceptance, tooling support, and best practices. Collaboration among researchers, practitioners, and industry stakeholders is crucial to address these challenges and establish widely accepted patterns.

Future Work. In this paper, we provided an overview of adopting SCUPs for BPCPs and discussed the complexity and challenges. Future work may address these complexities and challenges in a number of key areas. First, there is a need for tools and frameworks that automate SCUP implementation and management. This may include tools for automatically generating upgradable smart contracts, analyzing contract and data dependencies from both business process and software engineering perspectives, and facilitating seamless upgrades. Second, standardization of smart contract templates with built-in upgradability features for different update scenarios can streamline technical changes at the smart contract level. Third, we plan to extend our analysis, which is mostly based on lower bounds, with an empirical evaluation to determine the cost, latency, and complexity for a set of real-world business process upgradability scenarios.

References

1. ERC-2535: Diamonds, multi-facet proxy. https://eips.ethereum.org/EIPS/eip-2535. Accessed 03 June 2024
2. Proxies - OpenZeppelin Docs. https://docs.openzeppelin.com/contracts/5.x/api/proxy. Accessed 03 June 2024
3. Alexandrov, D., Ismoilov, M., Kozlov, A., Savachenko, A., Zykov, S.: Validating new method for measuring cohesion in object-oriented projects. Procedia Comput. Sci. **192**, 4865–4876 (2021)
4. Androulaki, E., et al.: Hyperledger fabric: a distributed operating system for permissioned blockchains. In: 13th EuroSys Conference, pp. 1–15 (2018)
5. Bandara, H.M.N.D., Xu, X., Weber, I.: Patterns for blockchain data migration. In: European Conference on Pattern Languages of Programs, EuroPLoP 2020, pp. 1–19 (2020)
6. Bodorik, P., Liu, C.G., Jutla, D.: TABS: transforming automatically BPMN models into blockchain smart contracts. Blockchain: Res. Appl. **4**(1), 100115 (2023)
7. Corradini, F., Marcelletti, A., Morichetta, A., Polini, A., Re, B., Tiezzi, F.: Chorchain: a model-driven framework for choreography-based systems using blockchain. In: ITBPM@BPM, pp. 26–32 (2021)
8. Corradini, F., Marcelletti, A., Morichetta, A., Polini, A., Re, B., Tiezzi, F.: A flexible approach to multi-party business process execution on blockchain. Futur. Gener. Comput. Syst. **147**, 219–234 (2023)

9. van Der Aalst, W.M., Ter Hofstede, A.H., Kiepuszewski, B., Barros, A.P.: Workflow patterns. Distrib. Parallel Databases **14**, 5–51 (2003)
10. Ebert, C., Cain, J., Antoniol, G., Counsell, S., Laplante, P.: Cyclomatic complexity. IEEE Softw. **33**(6), 27–29 (2016)
11. ethereum.org: Upgrading smart contracts. https://ethereum.org/en/developers/docs/smart-contracts/upgrading/. Accessed 31 May 2024
12. ISO/IEC 25010:2011: Systems and software engineering - Systems and software Quality Requirements and Evaluation (SQuaRE) - System and software quality models. Standard, International Organization for Standardization, Geneva, CH (2011)
13. Kim, D., Kim, M., Kim, H.: Dynamic business process management based on process change patterns. In: 2007 International Conference on Convergence Information Technology (ICCIT 2007), pp. 1154–1161. IEEE (2007)
14. López-Pintado, O., García-Bañuelos, L., Dumas, M., Weber, I., Ponomarev, A.: Caterpillar: a business process execution engine on the Ethereum blockchain. Softw.: Pract. Experience **49**(7), 1162–1193 (2019)
15. Martin, R.: OO design quality metrics. Anal. Dependencies **12**(1), 151–170 (1994)
16. McConnell, S.: Code Complete. Developer Best Practices Series. Microsoft Press (2004)
17. OMG: Business process model and notation (BPMN) version 2.0 (2011)
18. Tonelli, R., Pierro, G.A., Ortu, M., Destefanis, G.: Smart contracts software metrics: a first study. PLoS ONE **18**(4), e0281043 (2023)
19. Tran, A.B., Lu, Q., Weber, I.: Lorikeet: a model-driven engineering tool for blockchain-based business process execution and asset management. In: BPM (Dissertation/Demos/Industry), pp. 56–60 (2018)
20. Weber, B., Rinderle, S., Reichert, M.: Change patterns and change support features in process-aware information systems. In: Krogstie, J., Opdahl, A., Sindre, G. (eds.) CAiSE 2007. LNCS, vol. 4495, pp. 574–588. Springer, Heidelberg (2007). https://doi.org/10.1007/978-3-540-72988-4_40
21. Weber, I., et al.: On availability for blockchain-based systems. In: 2017 IEEE 36th Symposium on Reliable Distributed Systems (SRDS), pp. 64–73 (2017)
22. White, S.A.: Business process modeling notation (BPMN) version 1.0. Business Process Management Initiative, BPMI.org (2004)

Secure Proof Verification Blockchain Patterns

Tiphaine Henry(✉) and Sara Tucci-Piergiovanni

Université Paris-Saclay, CEA, List, 91120 Palaiseau, France
tiphaine.henry@cea.fr

Abstract. In blockchain networks, transaction data is accessible to all participants by design and hence verifiable. This verifiability feature of data stored on the ledger by any participant fosters trust within data, especially in uncertain environments. However, the public nature of onchain data imposes limitations across various scenarios as subsets of data should be kept private. Zero-knowledge proofs (ZKPs) have emerged as a solution within the literature to overcome this issue. The raw data is not published onchain, only a proof of knowledge of this data is. Hence, the blockchain is used as a trustworthy means for proof verification without requiring data disclosure. Despite their effective use in many scenarios, the formalization of zero-knowledge proof techniques within blockchain settings remains under-explored in current literature, and makes their integration difficult for non-expert blockchain practitioners due to the plurality and complexity of zero knowledge proofs. Software engineering patterns are used in the literature to formalize recurring software engineering practices stemming from the literature and experience of practitioners. Several patterns have been proposed to formalize blockchain-based architecture practices. However, no blockchain patterns tailored to confidential proofs using ZKPs have been designed in the literature. Hence, this paper aims to address this gap by formalizing key blockchain patterns relying on ZKP to handle secure proof verification identified in the literature. We formalize a general pattern called Secure Proof Verification pattern and three related sub-patterns, two of them focusing on efficient or trustless proof verification, and one specifically designed for interval membership verification to aid practitioners in selecting the most suitable non-interactive ZKP design for a blockchain-based application.

Keywords: Blockchain · Software Patterns · Zero-knowledge proofs

1 Introduction

Zero knowledge proofs (ZKPs) have emerged as a key technique in the recent years to share a proof of a knowledge without divulging the inner content of this data [8,26]. This has many practical applications regarding confidentiality-aware scenarios. For instance, in identity verification scenarios, individuals can share

C. Di Ciccio et al. (Eds.): BPM 2024, LNBIP 527, pp. 71–88, 2024.
https://doi.org/10.1007/978-3-031-70445-1_5

a proof of their identity without disclosing additional details such as birth date or location. Similarly, in circular economy contexts, where a manufacturer must demonstrate the non-toxicity of a product to a recycler, the manufacturer can share a proof that its products do not contain a given toxic substance without revealing detailed information about the product recipe, thus safeguarding its industrial know-how [30].

However, in certain scenarios, a lack of trust between the prover and verifiers necessitates a system architecture that ensures the integrity and trustworthiness of zero-knowledge proof generation and verification, guaranteeing that the proof remains untampered [23]. Furthermore, oftentimes, direct interaction between the verifier and the proof issuer is not feasible. For example, a manufacturing company that issued a proof of non-toxicity of a product may have ceased operations, or alternatively, the proof issuer—the manufacturing company—may opt against providing direct access to its information system to unknown verifiers due to cybersecurity concerns. However, subcontracting the custody of the proof to a third party may instill distrust on the verifier's side. Hence, a need arises for an independent, asynchronous, and reliable verification mechanism for ZKPs.

Blockchain has been leveraged as a support for trustworthy proof verification of non-interactive ZKPs in the literature. The non-interactive variant of zero-knowledge proof enables for the asynchronous verification of proofs which fits the blockchain design where participants can interact in an asynchronous fashion. The proof, generated offchain, is stored onchain and can be accessed by any participant of the network. The integrity property of transactions stored onchain ensures that the proof remains tamper proof after its generation. Moreover, as it is by design accessible to any participant for verifiability purposes, this ensures a reliable and asynchronous access to the proof. Moreover, a verification system can be built on top of the blockchain network using smart contracts. Hence, reliability in the verification system can be ensured in a decentralized fashion in uncertain scenarios. Moreover, the integration of non-interactive ZKPs in blockchain based applications addresses key limitation concern of blockchains regarding confidential data handling as any data stored in the ledger is accessible and verifiable by all members of the network [5,24]. Numerous studies integrate non-interactive ZKPs into blockchain architectures to blind transaction content while ensuring verifiability and computational efficiency, primarily utilizing succinct non-interactive arguments of knowledge (SNARKs) [30,32]. However, despite their potential, the scattered nature of information across literature sources and the lack of formalized patterns make selecting suitable non-interactive ZKPs mechanisms challenging for non-expert audiences [8].

Software design patterns are proposed in the software engineering literature to aid practitioners, such as software architects, developers, system administrators, or technical leads, in addressing recurring architectural problems [1,2,17]. These patterns formally describe architectural solutions using standardized formats like GOF or the Alexandrian form format. In recent years, software design patterns have found application in the blockchain field [2,10,36], facilitating streamlined integration of blockchain using best practices. Several papers aim to formalize patterns at various levels, including data management or security patterns [7,42,47]. However, while some papers identify blockchain patterns suitable for confidential data handling [20,47], these collections often lack comprehensiveness. For instance, the encryption pattern introduced by Xu et al. [42,47] remains high-level and does not mention zero-knowledge proof techniques [39,50]. Moreover, these papers often overlook identifying network assumptions and trust models derived from these assumptions, which are crucial in the decision-making process for blockchain application architects. Consequently, there is a research gap in defining blockchain patterns for confidential data handling based on ZKPs. To address this issue, we propose the formalization of the NI-ZKP blockchain integration pattern, that we refer to as the Secure Proof Verification blockchain pattern. We also formalize three corresponding sub-patterns specializing the context of use, two of them focusing on issuing data-related statements, with a focus on proof generation and verification efficiency for the first pattern, and a trustless setup for the second pattern, and the Interval Membership Verification pattern that aims at sharing a proof of a value belonging to a range.

The remainder of this paper is structured as follows. Section 2 introduces key concepts on blockchains and ZKPs. Section 3 discusses related work on blockchain patterns and confidential data handling patterns published in the literature. Section 4 presents the methodology used to construct this paper. In Sect. 5, we present the key blockchain patterns for confidential data handling based on NI-ZKP formalized with the Alexandrian form format. Section 6 concludes the paper with a summary and discussion of the results.

2 Preliminaries

2.1 Blockchains and Blockchain Patterns

A blockchain is a shared immutable ledger based on a peer-to-peer network, i.e., a ledger spread across multiple computing nodes [45]. Blockchains have originally emerged to keep track of digital transfers in financial applications, but they can be used to store any kind of data. Data recorded in a blockchain is replicated at each peer of the network. To add new data to the ledger, users issue transactions containing the data. These transactions are collected by network peers and collaboratively validated through a consensus mechanism before being added to each copy of the ledger. The ledger is organised in blocks of transactions. More precisely, each block holds a block header, a list of verified transactions, and the hash of the previous block to attest the integrity and the ordering of the blocks [40]. This structure makes the transactions immutable and tamper-proof [4]. A smart contract is a program hosted on a blockchain network, activated by transactions [44]. A smart contract comprises both variables and methods. Smart contract methods can execute arbitrary code, access the state of the variables and optionally update them. Generally, executing one of the smart contract's methods results in a new state, meaning a new valuation of its variables. Users interact with a smart contract by submitting transactions that request the execution of one of its methods. For each transaction invoking a smart contract method, the issuer must pay fees similar to those for normal payment transactions.

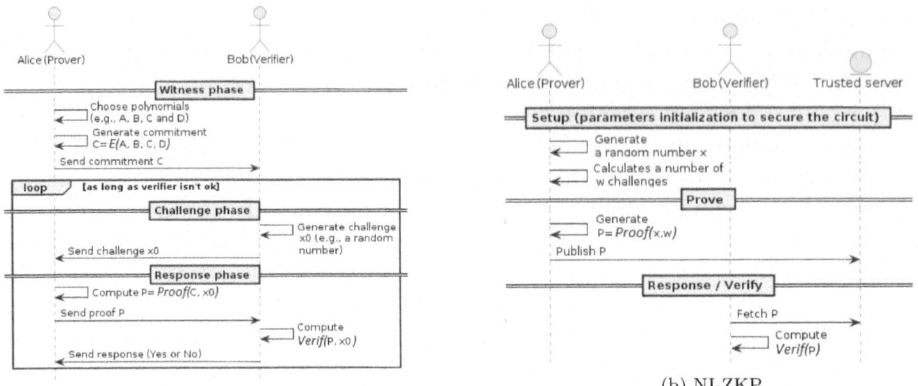

(a) I-ZKP

(b) NI-ZKP

Fig. 1. Illustration of the exchanges between prover and verifier during I-ZKP and NI-ZKP

Software engineering patterns provide well-tested solutions for frequently encountered application development use cases [2]. Standardized formatting such as the Alexandrian Form Format systematically include for each pattern the description, the forces or tradeoffs at stake, the benefits and drawbacks, and their main applications. As mentioned in the introduction, a family of patterns focusing on blockchain patterns has recently emerged in the literature [42,47], formalizing each use such as data management, architectural, or access control strategies.

2.2 NI-ZKPs

ZKPs, formally defined by Goldreich et al. in 1991 [26], serve as cryptographic tools enabling one party to convince another party of the truth of a statement without revealing any additional information beyond the statement's validity. An illustrative analogy often employed to explain ZKPs is the "Alibaba cave" scenario, proposed by Jean-Jacques Quisquater and Louis Guillou in 1989 [38]. In this scenario, consider two individuals, Alice (the prover) and Bob (the verifier), exploring a mysterious cave. Alice possesses knowledge of a secret passage within the cave and aims to convince Bob of her knowledge without disclosing the passage's exact location. Through a series of back-and-forth interactions within the cave, Alice demonstrates her ability to navigate within the cave, convincing Bob of her knowledge without revealing the specific route. The Alibaba cave analogy illustrates the essence of ZKPs: proving knowledge of a fact or secret without revealing the information itself, and without ambiguity.

ZKPs involve providing a *statement* with a validity *proof* to a verifier. Statements can be defined as a formal declaration based on a secret that is subject to verification. An example could be the statement "I know a number x such that $x^2 \equiv 9 \pmod{13}$", which can be verified using modular arithmetic. The proof provides cryptographic evidence of the truth of a statement without revealing underlying information. The proof is computed with the secret data and a commitment to this secret data. For instance, a prover Alice would compute the proof of the statement by using the proof generation algorithm with the secret number x and the commitment to x by computing a commitment $C = E(x)$, where E is a cryptographic hash function. The statement is forwarded with its proof to a verifier Bob. The verifier uses the proof to verify the validity of the statement without accessing the secret data used to build the statement. ZKPs exhibit several key properties essential for ensuring the integrity and confidentiality of verified statements. These properties include *completeness*, which guarantees that an honest statement can convince an honest verifier; *soundness*, which guarantees that the prover cannot convince the verifier of a false statement; and *zero knowledge*, which says that the verifier "learns nothing" other than the fact the statement being proven is true [26]. Some ZKP protocols rely on a trusted setup, which consists of multiple participants generating initial parameters for a cryptographic system, ensuring their integrity. An assumption is made about the honesty of the participants in following each step of the setup. The choice of the ZKP protocol impacts the proof size, the time for proof generation and verification.

Two families of ZKPs exist based on the proof verification design (interactive or non-interactive). Figure 1a illustrates the interactive ZKP scenario. Two parties, the prover and verifier, engage in a series of back-and-forth interactions where (1) the prover sends a commitment using private input data to a verifier, (2) the verifier responds with a challenge, and (3) the prover computes the proof using the commitment and the challenge. These steps loop until the verifier is convinced of the commitment. A non-interactive ZKP (NI-ZKP) scheme extends the concept of ZKPs to scenarios where interaction between the prover and verifier is minimized or eliminated entirely [14]. Figure 1b illustrates the NI-ZKP scheme. Alice generates a proof offline, independently of the verifier Bob. Bob can subsequently verify the proof without further interaction with the prover [28]. The suitability of interactive schemes within blockchain environments has been questioned due to concerns such as communication complexity, network performance degradation, and the necessity for both parties to be online simultaneously [28]. Consequently, NI-ZKPs best apply for blockchain-based scenarios.

2.3 Motivating Example

We now introduce a motivating example anchored in the circular economy to identify key properties needed to handle a proof statement verification scenario that can be addressed by blockchain.

In a circular economy supply chain, participants sometimes need to prove and verify some claims but without the possibility to share or access the content of the statement. For example, a battery manufacturer could need to prove various claims related to the performance, authenticity, and sustainability of lithium-ion batteries without disclosing sensitive information about their formulation, manufacturing process, or environmental impact. These scenarios necessitate a method to verify claims without compromising the confidentiality of the underlying information.

The following key properties are needed in this scenario: *(P1) traceability of exchanges:* tracing proof issuance (e.g., for the verifier to be sure of the identity of the issuer and the integrity of the proof) and tracing proof verification (e.g., in cases where the manufacturer wishes to manage who can verify its claims); *(P2) asynchrony of the verification:* avoid dependency on the proof issuer - separate the verification from the issuing for long term scenarios - e.g., recycling of a product where the manufacturer does no longer exist: how to still access proof that the product is not toxic?; *(P3) trust in the verification step:* being sure that the claim is rightly verified, with correct information.

Blockchain can be a fit to fill these properties. First, transactions recorded onchain offer properties such as transparency and verifiability. Moreover, smart contracts can enforce predefined rules for securing data sharing protocols such as enforcing a predefined access control [16]. Hence, proof issuance and each verification request can be tracked onchain, which addresses the requirements for the traceability of the exchanges *(P1)*. The proof being stored onchain, the verifier does not need to interact directly with the issuer, which answers the requirement for the asynchrony of the verification *(P2)*. Finally, blockchain history is immutable and accessible to all participants of the network, which prevents malicious actors from modifying data or code. As data cannot be tampered, and credentials can be generated regarding issuers for example, this makes the verification step trustworthy *(P3)*. Hence, an interest arise in formalizing integration strategies for NI-ZKP in blockchain-based scenarios.

3 Related Work

The aim of defining blockchain patterns is to summarize practical solutions today identified but scattered in the literature to aid practitioners in selecting appropriate mechanisms while building decentralized applications (dApps). A first set of papers describe formally blockchain patterns. A recent systematic literature review conducted by Six et al. [42] presents a taxonomy of these existing blockchain patterns, with some addressing concerns related to confidential data handling. This work includes the preliminary work by Xu et al. [47] which presents an initial set of data management patterns from which three patterns can be used for confidential data handling, namely "On-chain encryption", "Off-chain data storage", and "State Channels". Six work extends the "Onchain Encryption" pattern group with two patterns: "Commit and Reveal" and "On-chain encryption". However, no mention is made of ZKPs in the identified patterns. Moreover, identified patterns fail to provide trust models or assumptions on network requirements, critical elements in infrastructure design. Another subset of papers describe informally blockchain techniques and decision models for confidential data handling. One paper proposes a decision model for handling sensitive health data in blockchain systems, highlighting encryption strategies but without systematizing encryption techniques through patterns, and missing

ZKP-based techniques [22]. Three works summarize confidential data handling strategies and mention ZKPs, however without in depth description nor following the Alexandrian form format [37,39,50]. Zeiselmair et al. and Qi et al. [37,39,50] summarize verifiable computing techniques with blockchains, however without a proper formalization and with missing details on the integration of ZKPs in blockchain environments. Hence, there is a research gap regarding the formalized description of NI-ZKP blockchain patterns.

4 Methodology

Blockchain data management patterns found in the scientific literature overlook confidential data management with proofs. As such techniques are discussed in both scientific and grey literature sources (e.g., white papers, online news sites, blogs, and videos), we follow a Multivocal Literature Review (MLR) to uncover blockchain patterns for transactional confidentiality. An MLR is an extension of a systematic review utilizing both scientific and grey literature [25]. The MLR process begins with formulating the research question outlined in Sect. 1. From this question, the following search query is constructed: *(((crypto OR zero-knowledge) proof OR ZKP OR SNARK OR STARK OR proof) OR confidentiality OR confidential OR privacy) AND (blockchain OR (blockchain AND pattern) OR dApp OR "smart contract")*. Scientific literature sources are searched through ScienceDirect, ACM Digital Library, IEEE, SpringerLink, and AISEL, while grey literature sources are searched using the Google Search Engine. Snowball sampling is employed to expand the pool of potential sources by tracing citations and links from already identified sources, iterating until theoretical saturation is achieved. We then filter the sources based on inclusion and exclusion criteria, such as whether the technique is rooted in a blockchain-based information system, leverages ZKPs, and is well-defined. Techniques related to identity privacy and access control are excluded. We also exclude protocols focusing solely on blockchain architectures to enhance scalability or interoperability. Indeed our goal is set on identifying techniques allowing the asynchronous verification of uncorrupted proofs as specified in the motivating example. Preference is given to information sources from blockchain platforms and DApp governance bodies in the grey literature over third-party sources. The resulting pool of sources comprises 32 papers from formal literature (11 describing blockchain patterns and corresponding decision models, 12 focusing on literature surveys, and 11 describing techniques for confidential data handling on-chain without software pattern formalization) and 21 web pages from grey literature. Filtering criteria include factors such as articles focusing solely on data protection mechanisms used in the context of a data exchange involving confidential data, and providing detailed explanations. The vetted pool of sources undergoes analysis using open and axial coding, iteratively deriving a general pattern and a set of sub-patterns whose context is specialized. We follow the Alexandrian format [1] as this format aligns with the prevalent structure found in blockchain software pattern papers [42,47]. For each sub-pattern, additional information is provided, including the trust model (honest, honest but curious, or malicious actors), communication hypotheses (synchronous or asynchronous), and confidentiality requirements for the data involved.

5 Secure Proof Verification Blockchain Patterns

5.1 General Secure Proof Verification Blockchain Pattern

Summary: The pattern enables secure and efficient proof validation of sensitive data points referred to as secrets without interaction nor accessing the secret directly. NI-ZKP statements whose proof is generated offchain are verified by a smart contract and stored in the ledger, enabling future verifiers to accept the claim without recomputation (Fig. 2).

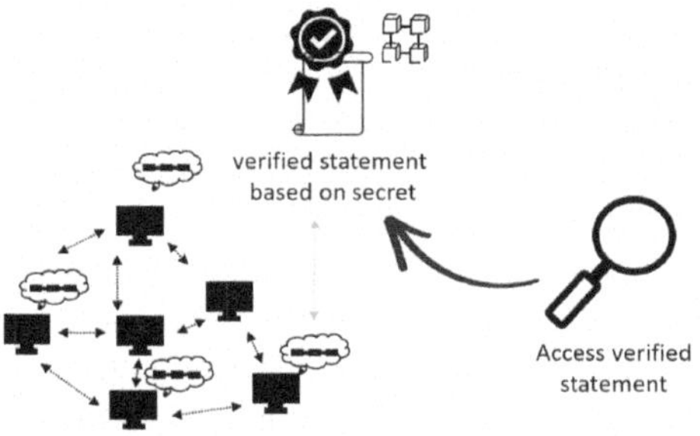

Fig. 2. Illustration of the General Secure Proof Verification blockchain pattern.

Context: A participant needs to share a statement based on a secret onchain without disclosing this secret to blockchain participants.

Problem: The challenge is to publish a statement, derived from a secret, with a proof of its validity on the blockchain. This must be done without revealing the secret or directly interacting with network participants.

Forces: The problem requires balancing the following forces: (i) Statement integrity, availability and verifiability: These three properties can be applied simultaneously to any statement stored onchain which enables the prover to share a statement to any verifier in the long term without a direct interaction; (ii) Confidentiality preservation: Since all data stored on a blockchain is inherently public by design, preserving confidentiality requires careful consideration as it is not possible to store sensitive data onchain directly.

Solution: Depending on the NI-ZKP protocol chosen, the pattern can start with a trusted setup, i.e., an initial step that establishes the security foundation for the ZKP system, relying on trust in the participants to perform the setup honestly and correctly [28]. The purpose of this setup is to generate a set of public parameters whose nature depends on the specific ZKP protocol.

These public parameters may include random elements, mathematical structures needed for proof computation, commitment schemes, or verification keys. After setup, these parameters are made publicly available within the blockchain network for later use during NI-ZKP proof generation or verification. Afterward, a prover generates a proof off-chain using public parameters for the NI-ZKP scheme stored on-chain and its private information. The proof is submitted to the blockchain network. Finally, the verification stage can take place. Network participants (e.g., network nodes, smart contracts, or other participants with access to the blockchain ledger) publicly verify the proof on-chain using cryptographic parameters and the proof itself. The result, typically a Boolean indicating acceptance or rejection, is recorded in the blockchain ledger. If the proof is valid, indicating that the statement made by the prover are correct, the verification process is successful. Future verifiers will then have the possibility to accept the claim without needing to re-execute the verification algorithm. Finally, the pattern entails several requirements. Asynchronous communication between the prover and verifier is allowed, contrasting with interactive ZKPs that rely on synchronicity due to exchanges during proof verification. The statement and the corresponding proof should be public. In case of a trusted setup, the public parameters used for securing the protocol must remain private and discarded at the end of the setup, while public parameters must be public. The algorithms for proof generation and verification should be publicly accessible, allowing any party to generate and verify proofs as needed. At last, the trust model hypothesises that the prover and verifiers do not inherently trust each other. Both prover and verifiers may be malicious: the prover could try to convince the verifier of a wrong proof, and the verifier could try to learn the secret of the prover, hence the use of a smart contract based system [23].

Consequences:

- *Benefits:*
  - Data confidentiality. Data is not stored directly onchain but is embodied into a statement stored onchain.
  - Proof verification asynchrony. The prover does not need to be connected synchronously with the verifier as the proof is stored in the blockchain.
  - Parallel verifications. The proof can be used by several verifiers simultaneously accessing the ledger where the proof is stored.
- *Drawbacks:*
  - Efficiency. Proof generation duration varies depending on the chosen schemes (from ms to hours), which may impact transaction processing efficiency if not chosen appropriately [28,34].
  - Trusted setup. Some schemes require a setup where involved parties are supposed to be honest. This introduces dependencies on these parties and potentially compromises decentralization [20].
  - Tooling. Though toolboxes exist to help implement ZKPs [21,33], the tools have heterogeneous maturities and varying developer community sizes [3].

- Security. Some ZKP protocols rely on a random oracle model hypothesis, which simulates an idealized hash function producing truly random outputs. This hypothesis may introduce vulnerabilities if the implementation of the hash function is not truly random [14].
- Applicability. Not all NI-ZKP solutions handle proofs based on two or more secrets [11].

Related Patterns: This pattern can serve as a foundation for the offchain oracle pattern introduced by Xu et al. [47] to prove the integrity of an API call such as a delegated computation.

Known Uses: This pattern can be used for proving balance thresholds [51], validating computations [11,23], and sharing product properties across supply chains [18].

We describe below the main Secure Proof Verification sub-patterns identified in our literature review, namely (1) Efficient Secure Proof Verification, (2) Trustless Secure Batch Proof Verification, and (3) Interval Membership Verification blockchain sub-patterns.

5.2 Efficient Secure Proof Verification Blockchain Sub-pattern

Summary: A pattern relying on a specific type of NI-ZKP (SNARKs) that offers succinct proofs[1] and an efficient verification. This sub-pattern requires an off-chain trusted setup during which the public parameters required for the proof generation and proof verification are generated. Figure 3b illustrates the pattern.

Context: A participant needs to share a statement based on a secret onchain without disclosing this secret to blockchain participants. The generated proof size should be small and the proof verification should be efficient.

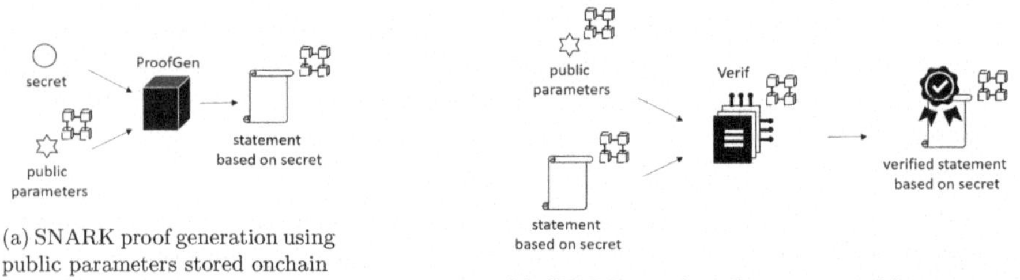

(a) SNARK proof generation using public parameters stored onchain

(b) SNARK proof verification carried by a smart contract

Fig. 3. SNARK blockchain pattern intuition

[1] A succinct proof is a cryptographic construct that enables the verification of complex computations or statements using a compact proof size, typically much smaller than the original computation.

Problem: The challenge is to publish a statement, derived from a secret, with a proof of its validity on the blockchain. This must be done without revealing the secret or directly interacting with network participants. Usability issues may arise as the proof size may be large and the and the proof verification time may require heavy computing power.

Forces: The problem inherits from the forces identified for the general pattern. It additionally requires balancing the following force: (i) Proof verification efficiency: the proof verification should not cause scalability breaks due to the proof size or the verification time.

Solution: The solution involves using a Zcash-style (non-interactive) SNARKS protocol, known as Succinct Non-Interactive Argument of Knowledge which builds on the *Hiding queries to polynomials* strategy and Quadratic R1CS [32][2]. Key actors of this pattern include a trusted authority or a consortium of blockchain participants conducting the trusted setup offchain, the prover, and a smart contract responsible for storing the SNARK public parameters and executing verification. The solution includes the following steps. First, a trusted setup takes place. The public parameters required for SNARKs are generated through a trusted setup process, such as the Power of Tau ceremony, involving the selection of x_0 and computation of homomorphic encryption. x_0 is discarded after setup to maintain security. The setup is conducted off-chain by the SNARK Authority. Then, public parameters generated during setup are published on-chain. Afterward, the prover generates a succinct proof off-chain using private inputs and public parameters. The proof is submitted to the blockchain. Finally, the verifier performs verification of the proof using public parameters stored onchain.

Consequences:

- *Benefits:*
  - Proof Sizing: zk-SNARK proofs are small, requiring less gas for validation compared to other schemes, estimated at only 24% of the gas required for STARKs [3].
  - Efficient Verification: The time required for verification grows very slowly relative to the input size, making it efficient even for large computations (verification time is polynomial in the logarithm of the size of the computation) [30].
  - Non-Interactive: Becomes non-interactive after the trusted setup.
  - Code Maturity: large community and available tools (e.g., Zokrates library and lib-SNARK). Moreover, SNARK integration in Ethereum is considered in EIPs 196 and 197 [3,9,20].
- *Drawbacks:*
  - Lack of Transparency: Users must trust the correct execution of the setup and the destruction of associated secrets [24,27].

[2] Another ZKP building on Hiding queries to polynomials are *Multi-use circuits* (Reducing computational problems into arithmetic circuits). However, no scheme has been identified for a blockchain use.

- High Computational Costs: Generating zk-SNARK proofs is computationally intensive, taking 1 to 2 min per transaction [24,27].
- Non-Succinct Trusted Setup: The setup process is often complex and inefficient, relying on initial participants without guaranteeing security.
- Quantum Assumptions: Vulnerabilities exist due to possible backdoors in elliptic curve random number generators, which are not post-quantum secure [3].
- Side-Channel Attack Vulnerabilities: Mathematical operations and optimizations may inadvertently leak information [3].

Related Patterns: See other Secure Proof Verification patterns.

Known Uses: Zcash employs snarks to ensure transaction privacy, where a transaction sender produces a ZKP of its ability to spend more than or equal to the transaction value, thereby protecting transaction details and ensuring anonymity. Additionally, Zokrates is utilized for processing Know Your Customer (KYC) procedures on-chain [28]. Moreover, ZKPs are employed for proving the correct execution of delegated off-chain computations [20,35], with proofs designed to ensure verification cost independence from the complexity of the off-chained computation.

5.3 Trustless Secure Batch Proof Verification Blockchain Sub-pattern

Summary: This pattern leverages an off-chain proof generator using STARKs and an on-chain verification smart contract to enable the verification of batches of statements within blockchains. By using hash functions, it avoids the need for a trusted setup, enhancing transparency and auditability. The verified proofs are stored on the ledger, making the certified statements accessible to all blockchain participants. Figure 4 depicts the pattern.

Context: A participant needs to share several secrets onchain without disclosing the secrets to blockchain participants. Additionally, participants of the blockchain network may be malicious and have not identified a trusted authority.

Problem: The challenge is to publish a statement, derived from a secret, with a proof of its validity on the blockchain. This must be done without revealing the secret or directly interacting with network participants. Moreover, participants do not or cannot rely on a trusted setup to generate the public parameters, which makes impossible the use of a trusted setup to generate the initial parameters usually required to generate NI-ZKP proofs.

Forces: The problem inherits from the forces identified for the general pattern. It additionally requires balancing the following force: (i) No trusted setup: no trusted intermediaries or centralized authorities should be responsible for the verification.

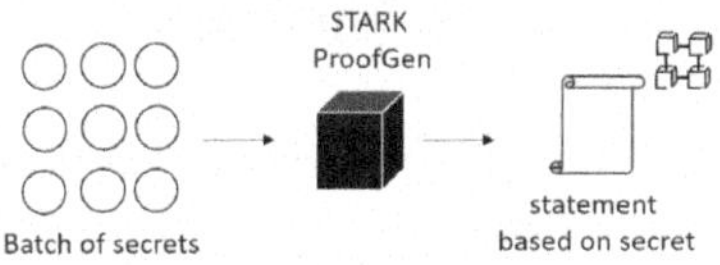

(a) STARK proof generation

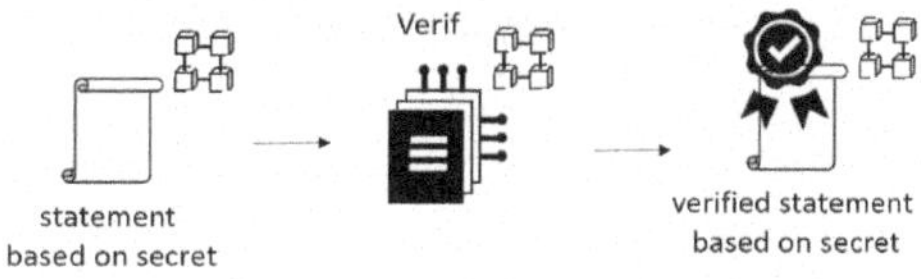

(b) STARK proof verified by a smart contract

Fig. 4. Trustless Secure Batch Proof Verification Blockchain Sub-pattern intuition

Solution: This pattern leverages two bricks, an off-chain proof generator using a STARK which relies on hash functions[3] and an on-chain verification smart contract. The pattern consists in three steps. First, a smart contract integrating the STARK proof verification logic is deployed onchain. It is designed to receive the proof and verify its correctness using the selected STARK library. Then, the prover generates an off-chain proof of a public statement using the secret data and corresponding private commitment. The proof possibly includes a batch of statements. Finally, the smart contract proceeds to the on-chain verification of the proof to validate the correctness of the statement. The outcome of the proof verification is stored on the ledger and accessible to all blockchain participants.

Consequences:

- *Benefits:*
  - Proof generation and verification computational efficiency (linear proving time).
  - Transparency: no trusted setup, minimum trust assumptions.
  - Post-quantum security: hash functions allow for quantum resistance.
- *Drawbacks:*
  - Proof size: Larger than SNARKs (around 100 bytes for SNARKS vs 10–100 KB for STARKs) [41].
  - Verification time: Larger than SNARKs (5–10 ms for SNARKS vs 10–50 ms for STARKs) [41].
  - Code maturity: Lack of developer documentation despite nascent STARK proof generation libraries[4].

Related Patterns: See other Secure Proof Verification patterns.

Known Uses: The Cairo toolchain is used for off-chain proof generation of general computation [33,43]. These proofs and computation results are sent on-chain to a Verifier smart contract and used in applications such as DeversiFi and dYdX on Ethereum, or Immutable on Polygon.

---

[3] Two advantages from using hash functions: (1) quantum resistance, (2) no trusted set-up (more transparent and auditable than zk-SNARKs) [3,8,31].

[4] List of identified STARK libraries at the time of writing: libSTARK, STARKware STARKDEX alpha, STARKExchange, distaff, Cairo.

5.4 Interval Membership Verification Blockchain Sub-pattern

Summary: Publish onchain a statement on a secret integer belonging to a public interval upper and lower bounds without revealing the secret integer to a set of possibly unknown verifiers. Figure 5 depicts the pattern variant with bulletproofs. The proof, generated offchain, consists in defining publicly the upper and lower bounds of the secret. It is verified by a smart contract and the generated certified statement is then accessible to all the participants of the network. In other variants of this pattern, a trusted setup used to generate public parameters is additionally necessary.

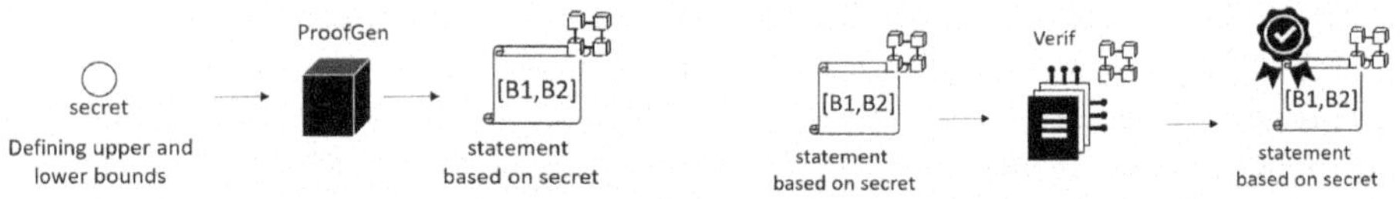

(a) Range proof generation - Bulletproof illustration

(b) Range proof verification - Bulletproof illustration

Fig. 5. Interval Membership Verification Blockchain Pattern intuition

Context: A participant needs to share a statement ensuring that a value falls within a specified range based on a secret without disclosing this secret to blockchain participants.

Problem: The challenge is to publish a statement, derived from a secret, with a proof of its validity on the blockchain. This must be done without revealing the secret or directly interacting with network participants. The proof size should be small.

Forces: The problem inherits from the forces identified for the general pattern. It additionally requires balancing the following forces: (i) trustlessness of the verification mechanisms: no trusted intermediaries or centralized authorities should be responsible for the verification; (ii) proof generation and computation efficiency: the proof should be easily generated and easily verifiable in terms of latency and proof size.

Solution: The pattern aims at proving the validity of a statement claiming a secret data belongs to a specified range. The pattern involves two main approaches, leveraging or not a trusted setup, depending on the chosen range proof protocol and the range size (larger or small) [19,30]. In general, small secrets require less computational effort and resources to generate and verify range proofs. With a trusted setup, the pattern comprises three key steps: deploying a smart contract on-chain to integrate range proof verification logic, generating range proofs off-chain using optimized schemes tailored to different secret sizes [19,30], and verifying these proofs on-chain using the deployed smart contract. The square decomposition protocol is optimized for large secrets while multi-base decomposition is optimized for small secrets. It breaks down secrets into their binary representations, facilitating boolean arithmetic to establish interval membership. We refer to the Efficient Secure Proof Verification Blockchain Sub-pattern for details on the steps. In contrast, without a trusted setup, the pattern adopts a trustless secure batch proof verification approach: first, off-chain proof generation using techniques like Bulletproofs, which do not require a trusted setup [12]; second, on-chain verification of the proofs, ensuring the validity of the data range. This approach enhances transparency and scalability while maintaining security, making it particularly suitable for blockchain environments with minimized trust assumptions. We refer to the Trustless Secure Batch Proof Verification Blockchain Sub-pattern for details on the steps.

Consequences:

- *Benefits:*
  - Proving value within a range without compromise on data confidentiality.
  - Several methods that adapt to several use cases, such as Bulletproofs for succint proofs and multi-base decomposition for small secrets.
  - Reduced communication complexity compared to SNARKS and STARKS. Proof size equivalent to SNARKs [34].
- *Drawbacks:*
  - Computing and storage resources.
  - Quantum attack susceptibility.
  - Reliance on newer, less studied assumptions.
  - Trusted setup depending on the protocol.

Related Patterns: A common use case is to use this pattern for homomorphic commitments: when verifying the legitimacy of a transaction, the range proof protocol is used to verify that a transaction amount is non-negative [19].

Known Uses: Monero uses Bulletproofs to prove that a committed value is within a given range [19]. Mimblewimble employs Pedersen commitments to hide transaction amounts [30]. Provisions prove users' solvency using range proofs combined with Multi-Party Computation (MPC) in Bitcoin [30]. Credential validation is achieved with Hashwires [15]. Additionally, range proofs are used for proving set membership [6,13,48,49].

6 Discussion and Conclusion

In this paper, we propose a new blockchain pattern called Secure Proof Verification Blockchain Pattern that formalizes the integration of NI-ZKPs within blockchain architectures, addressing the need for secure and reliable proof sharing and verification. We moreover identify and formalize three key related subpatterns relying on the SNARKs, STARKs, and range proof ZKP protocols, each serving distinct use cases, respectively providing a proof of knowledge of a value, a batch of values, or a value range verification. We find that the use of the Secure Proof Verification pattern is advocated in situations where the proof must be provided to more than one verifier, and in an untrusted environment (the prover and the verifiers distrust each others, or are not sure they will be able to interact in a peer to peer fashion in the long run).

Secure Proof Verification patterns are to be enriched in the future with complementary sub-patterns for sharing proofs of confidential data onchain. Indeed, zero-knowledge proof protocols are still evolving [34,51], e.g., towards lightweight cryptographic algorithms with stronger privacy under weaker assumptions, or lattice-based NI-ZKP for post quantum era. Moreover, the collection of sub-patterns presented in this paper does not include recent integration architectures such as ZKP-based oracles, zkVMs, zk-rollups, zk-bridges, or private smart contracts [29,46]. Future work should also focus on expanding the categorization of blockchain patterns to include other cryptographic patterns used to manage data onchain (e.g., to share confidential data, to compute over confidential data or to hide computation logic).

References

1. Alexander, C.: A Pattern Language: Towns, Buildings, Construction. Oxford University Press, Oxford (1977)
2. Alexander, C.: The Timeless Way of Building, vol. 1. Oxford University Press, New york (1979)
3. Asher, M.: Zero-Knowledge Proofs: STARKs vs SNARKs (2021). https://consensys.io/blog/zero-knowledge-proofs-starks-vs-snarks. Accessed 10 Oct 2023
4. Ayub, M., Saleem, T., Janjua, M., Ahmad, T.: Storage state analysis and extraction of ethereum blockchain smart contracts. ACM TOSEM **32**(3), 1–32 (2023)
5. Azgad-Tromer, S., Garcia, J., Tromer, E.: The case for on chain privacy and compliance. Stanford J. Blockchain Law Policy **6**(2) (2023)
6. Bai, T., Hu, Y., He, J., Fan, H., An, Z.: Health-zkIDM: a healthcare identity system based on fabric blockchain and zero-knowledge proof. Sensors **22**(20), 7716 (2022)
7. Bandara, H.D., Xu, X., Weber, I.: Patterns for blockchain data migration. In: EuroPlop, pp. 1–19 (2020)
8. Ben-Sasson, E.: A Cambrian explosion of crypto proofs. NAKAMOTO **8** (2020)
9. Ben-Sasson, E., et al.: Zerocash: decentralized anonymous payments from bitcoin. In: IEEE SP, pp. 459–474. IEEE (2014)
10. Benedetti, A., Henry, T., Tucci-Piergiovanni, S.: Gas cost analysis of proxy and diamond patterns: towards trusted smart contract engineering in EVM blockchains. In: FC - WTSC (2024, in press)
11. Benhamouda, F., Halevi, S., Halevi, T.: Supporting private data on hyperledger fabric with secure multiparty computation. IBM J. Res. Dev. **63**(2/3), 3–1 (2019)
12. Bünz, B., Bootle, J., Boneh, D., Poelstra, A., Wuille, P., Maxwell, G.: Bulletproofs: short proofs for confidential transactions and more. In: IEEE SP, pp. 315–334. IEEE (2018)
13. Camenisch, J., Chaabouni, R., Shelat, A.: Efficient protocols for set membership and range proofs. In: Pieprzyk, J. (ed.) ASIACRYPT 2008. LNCS, vol. 5350, pp. 234–252. Springer, Heidelberg (2008). https://doi.org/10.1007/978-3-540-89255-7_15

14. Chaabouni, R., Lipmaa, H., Zhang, B.: A non-interactive range proof with constant communication. In: Keromytis, A.D. (ed.) FC 2012. LNCS, vol. 7397, pp. 179–199. Springer, Heidelberg (2012). https://doi.org/10.1007/978-3-642-32946-3_14
15. Chalkias, K., Cohen, S., Lewi, K., Moezinia, F., Romailler, Y.: Hashwires: hyper-efficient credential-based range proofs. Cryptology ePrint Archive, Paper 2021/297 (2021)
16. Chen, C.L., Deng, Y.Y., Weng, W., Sun, H., Zhou, M.: A blockchain-based secure inter-hospital EMR sharing system. Appl. Sci. **10**(14), 4958 (2020)
17. Chia, S.Y., Xu, X., Paik, H.Y., Zhu, L.: Analysis of privacy patterns from an architectural perspective. In: ICSA-C, pp. 60–67. IEEE (2022)
18. Circularise: Take control of your supply chain with digital product passports (2023). https://www.circularise.com/dpp. Accessed 10 Mar 2023
19. Deng, C., et al.: A survey on range proof and its applications on blockchain. In: CyberC, pp. 1–8. IEEE (2019)
20. Eberhardt, J., Tai, S.: On or off the blockchain? Insights on off-chaining computation and data. In: De Paoli, F., Schulte, S., Broch Johnsen, E. (eds.) ESOCC 2017. LNCS, vol. 10465, pp. 3–15. Springer, Cham (2017). https://doi.org/10.1007/978-3-319-67262-5_1
21. Eberhardt, J., Tai, S.: Zokrates-scalable privacy-preserving off-chain computations. In: iThings. IEEE (2018)
22. Erler, C., Schinle, M., Dietrich, M., Stork, W.: Decision model to design a blockchain-based system for storing sensitive health data. In: ECIS (2022)
23. Ernstberger, J., Chaliasos, S., Zhou, L., Jovanovic, P., Gervais, A.: Do you need a zero knowledge proof? Cryptology ePrint Archive (2024)
24. Feng, Q., He, D., Zeadally, S., Khan, M.K., Kumar, N.: A survey on privacy protection in blockchain system. J. Netw. Comput. Appl. **126**, 45–58 (2019)
25. Garousi, V., Felderer, M., Mäntylä, M.V.: Guidelines for including grey literature and conducting multivocal literature reviews in software engineering. Inf. Soft. Technol. **106** (2019)
26. Goldreich, O., Micali, S., Wigderson, A.: Proofs that yield nothing but their validity or all languages in np have zero-knowledge proof systems. J. ACM (JACM) **38**(3), 690–728 (1991)
27. Huynh, T.T., Nguyen, T.D., Tan, H.: A survey on security and privacy issues of blockchain technology. In: 2019 International Conference on System Science and Engineering (ICSSE), pp. 362–367. IEEE (2019)
28. Konkin, A., Zapechnikov, S.: Privacy methods and zero-knowledge poof for corporate blockchain. Procedia Comput. Sci. **190**, 471–478 (2021)
29. Lavaur, T., Lacan, J., Chanel, C.P.: Enabling blockchain services for IoE with zk-rollups. Sensors (2022)
30. Morais, E., Koens, T., Van Wijk, C., Koren, A.: A survey on zero knowledge range proofs and applications. SN Appl. Sci. **1**, 1–17 (2019)
31. Márquez Solís, S.: Zero trust chain: a design pattern for improved interoperability and security in polkadot. arXiv preprint arXiv:2304.14730 (2023)
32. Nitulescu, A.: zk-snarks: a gentle introduction (2020)
33. Open Zeppelin: Cairo contracts (2024). https://github.com/OpenZeppelin/cairo-contracts
34. Oude Roelink, B., El-Hajj, M., Sarmah, D.: Systematic review: comparing zk-snark, zk-stark, and bulletproof protocols for privacy-preserving authentication. Secur. Priv. (2024)
35. Partisia Blockchain Foundation: Documentation (2023). https://partisiablockchain.gitlab.io/

36. Porru, S., Pinna, A., Marchesi, M., Tonelli, R.: Blockchain-oriented software engineering: challenges and new directions. In: ICSE-C, pp. 169–171. IEEE (2017)
37. Qi, H., Xu, M., Yu, D., Cheng, X.: SoK: privacy-preserving smart contract. High-Confidence Comput. **4**(1), 100183 (2024)
38. Quisquater, J.-J., et al.: How to explain zero-knowledge protocols to your children. In: Brassard, G. (ed.) CRYPTO 1989. LNCS, vol. 435, pp. 628–631. Springer, New York (1990). https://doi.org/10.1007/0-387-34805-0_60
39. Sedlmeir, J., Lautenschlager, J., Fridgen, G., Urbach, N.: The transparency challenge of blockchain in organizations. Electron. Mark. **32**(3), 1779–1794 (2022)
40. Singh, A., et al.: Blockchain smart contracts formalization: approaches and challenges to address vulnerabilities. Comput. Secur. **88**, 101654 (2020)
41. Singh, J.: zk-STARKs vs. zk-SNARKs explained (2022). https://cointelegraph.com/explained/zk-starks-vs-zk-snarks-explained. Accessed 04 July 024
42. Six, N., Herbaut, N., Salinesi, C.: Blockchain software patterns for the design of decentralized applications: a systematic literature review. Blockchain: Res. Appl. **3**(2), 100061 (2022)
43. Starkware: Hello, cairo! (2020). https://medium.com/starkware/hello-cairo-3cb43b13b209
44. Szabo, N.: Formalizing and securing relationships on public networks. First Monday **2**(9) (1997)
45. Wang, H., et al.: Blockchain challenges and opportunities: a survey. Int. J. Web Grid Serv. **14**(4), 352 (2018)
46. Xie, T., et al.: zkbridge: trustless cross-chain bridges made practical. In: ACM SIGSAC (2022)
47. Xu, X., Pautasso, C., Zhu, L., Lu, Q., Weber, I.: A pattern collection for blockchain-based applications. In: Proceedings of the 23rd European Conference on Pattern Languages of Programs, pp. 1–20 (2018)
48. Xu, Z., Chen, L.: DIV: resolving the dynamic issues of zero-knowledge set membership proof in the blockchain. In: ACM SIGMOD, pp. 2036–2048 (2021)
49. Yang, X., Li, W.: A zero-knowledge-proof-based digital identity management scheme in blockchain. Comput. Secur. **99**, 102050 (2020)
50. Zeiselmair, A., Steinkopf, B., Gallersdörfer, U., et al.: Analysis and application of verifiable computation techniques in blockchain systems for the energy sector. Front. Blockchain (2021)
51. Zhang, R., Xue, R., Liu, L.: Security and privacy on blockchain. ACM CSUR (2019)

The Cost of Executing Business Processes on Next-Generation Blockchains: The Case of Algorand

Fabian Stiehle[1(✉)] and Ingo Weber[1,2]

[1] School of CIT, Technical University of Munich, Munich, Germany
{fabian.stiehle,ingo.weber}@tum.de
[2] Fraunhofer Gesellschaft, Munich, Germany

Abstract. Process (or workflow) execution on blockchain suffers from limited scalability; specifically, costs in the form of transactions fees are a major limitation for employing traditional public blockchain platforms in practice. Research, so far, has mainly focused on exploring first (Bitcoin) and second-generation (e.g., Ethereum) blockchains for business process enactment. However, since then, novel blockchain systems have been introduced—aimed at tackling many of the problems of previous-generation blockchains. We study such a system, Algorand, from a process execution perspective. Algorand promises low transaction fees and fast finality. However, Algorand's cost structure differs greatly from previous generation blockchains, rendering earlier cost models for blockchain-based process execution non-applicable. We discuss and contrast Algorand's novel cost structure with Ethereum's well-known cost model. To study the impact for process execution, we present a compiler for BPMN Choreographies, with an intermediary layer, which can support multi-platform output, and provide a translation to TEAL contracts, the smart contract language of Algorand. We compare the cost of executing processes on Algorand to previous work as well as traditional cloud computing. In short: they allow vast cost benefits. However, we note a multitude of future research challenges that remain in investigating and comparing such results.

Keywords: Blockchain · Process Execution · Process Enactment · Workflow · Choreography

1 Introduction

Blockchain can serve as a decentralised platform to ensure the enforcement of process rules and the visibility and integrity of execution traces. The current state of the art executes processes or workflows either through a smart contract that interprets some representation of the process or by transforming a process model into a smart contract implementation [28]. The focus of most earlier works has been on investigating what can be achieved based on so-called first

C. Di Ciccio et al. (Eds.): BPM 2024, LNBIP 527, pp. 89–105, 2024.
https://doi.org/10.1007/978-3-031-70445-1_6

(i.e., Bitcoin) and second-generation blockchain systems (providing turing complete smart contracts) [28,34]. Due to their public and decentralised nature, they have to employ mechanisms to incentivise honest verifiers (e.g., miners) and protect against denial-of-service and Sybil attacks [30]. This necessitates paying fees for transactions. Early work investigating these platforms for process execution demonstrate that these fees are considerable, especially compared to traditional cloud computing [23,26]. With the increasing popularity of these platforms, this problem has since become more pronounced. Stiehle and Weber [28] trace the historic development of transaction fees on public Ethereum and conclude that these have become prohibitively expensive for most process cases [28, p. 16]. To achieve better performance, for a small set of process participants, private blockchain deployments can be employed. However, many of the guarantees a blockchain can offer are a result of decentralisation (see [28] or [33, Chapter 3.2]). Certain high-security use cases may benefit from larger, more decentralised, and resilient deployments [14,28]. Thus, in [28] the authors chart two directions for improving the status-quo: exploring (i) novel next-generation blockchain platforms, and (ii) layer two solutions. Early results on the use of layer-two solutions show that these can considerably reduce transaction fees [29]. However, these solutions build on top of existing blockchain platforms (layer one) and add additional layers of complexity and trade-offs.

So far, blockchain-based cost models are primarily investigated from a protocol perspective, e.g., to identify certain attack vectors [21]. Only few works exist from a system design perspective [22,26]. These models have a heavy focus on Ethereum's cost structure [26] or take an infrastructure perspective, i.e., they remain abstract on the underlying model [22]. We fill a gap in literature, by providing a formalisation and discussion of the cost model of Algorand, and contrast it to the well-studied cost model of Ethereum, which is the basis of existing related work. Our study of Algorand is motivated by its promise of fast finality and low cost [11], as well as—compared to Ethereum—opposing design decisions and qualities. Examples for the latter are Algorand's fixed cost and minimum balance requirement system. To operationalise our formalisations for the use of process execution, we extend an open source compiler for BPMN Choreographies towards multi-platform support and Algorand specifically. To the latter end, we provide a translation from choreographies to TEAL contracts, the native smart contract language of Algorand.

Here, we discuss different design options for TEAL contracts, with different cost implications as indicated by our cost model analysis. We also present an efficient multi-instance execution support. We evaluate our cost model, and compare our design to previous work as well as traditional cloud computing, utilizing previous results of Rimba et al. [26]. Our results highlight the promise of executing processes on next-gen blockchains—in short: they allow vast cost benefits. However, we note the many challenges that remain in investigating and comparing such results. Specifically, our investigation stands as early result showing that investigating novel blockchain protocols is valuable, yet by no means trivial. Many new characteristics must be considered by application developers. For organisations, predictable cost and performance estimates are paramount—we see our study as first contribution to a more holistic understanding. Thus, we believe that our results extend beyond the narrow domain of process execution on blockchain. Following open science principles, we release our implemented artefact as open source software.[1]

2 Background

We assume some familiarity with the Ethereum protocol and focus on introducing Algorand. Nonetheless, we will contrast concepts of Algorand to Ethereum, and retrace certain, more recent developments of Ethereum's evolution; specifically, on its cost model. In the remainder of this manuscript, we discuss cost as consequence of transacting on the public blockchain. Primarily, this is expressed by fees or balance requirements.

2.1 Algorand

Consensus. The Algorand protocol relies on verifiable random functions (VDF), which produce a randomness and proof based on a secret key and seed [11]. The proof enables other participants to verify that the randomness is indeed the result of running VDF with seed and secret key. Each proposer in the network locally runs VDF with a seed determined for each new block and their key. Given that their result is within a given threshold they are a candidate to produce the next block. To achieve scalable byzantine fault tolerance, the VDF is also run to determine membership in a committee. This committee constitutes a random sample of the network, and consensus is executed within this sample. To prevent targeted attacks on committee members (once they are elected), committee members reveal their membership only by participating in the protocol once, after, a new committee is formed. In the common case, the proposed block is finalised and no valid fork can exist in the network. Some edge cases exist where Algorand favours safety over liveness, and additional steps are required. For more details we refer the reader to [11]. To protect against Sybil attacks, the VDF is weighed by stake (i.e., proof-of-stake).

Accounts, Contracts, and Transactions. Algorand[2] accounts can be broadly divided into four types: (i) external accounts, (ii) multisignature accounts, (iii) application accounts, and (iv) smart signature accounts. All accounts can be associated with a balance of *algo* (ALG), assets, and local application state.

---

[1] https://github.com/fstiehle/chorpiler-algorandvm. An archived version is available at: https://doi.org/10.5281/zenodo.12656684.

[2] We describe Algorand based on the Algorand developer documentation (https://developer.algorand.org/docs/get-details, accessed 2024-07-01), but also based on our practical experiments; specifically when documentation remained too ambiguous for our purposes.

Smart contracts fall into two categories: stateful applications and stateless signatures. Signatures can not hold state (but the corresponding account can hold assets, ALG, and local state) and are attached to a transaction. Applications can hold state through different storage types. Throughout Algorand's development, applications became more and more potent, making the use of signatures only necessary for some niche scenarios.[3] We also focus on applications for the remainder of this paper. An application is created through a transaction and requires the specification of an `approval` and `clear` program in the programming language TEAL, as well as the allocation of the required static storage type. Hereby, the approval program implements the main application logic. An application call succeeds when the `approval` program returns with a non-zero value on top of its stack. Otherwise, the transaction is rejected. The `clear` program handles deletion of local state and cannot fail. Beyond the standard application call type, the app can be called by update, delete, and local state transactions. Through local state transactions, accounts can allocate local state within an application (associated with their account). This state can be cleared at any time by the associated account. An application can, in-turn submit transactions of any type, called inner-transactions. Atomic transaction groups can batch together transactions to be executed in an all-or-nothing fashion. Applications can use different storage schemes. Through an associated application account, they can hold ALG and assets. Global and local storage are fixed and must be allocated during application creation. Only box storage can be allocated on-demand by an application. We analyse these storage systems in detail in Sect. 4.

3 Related Work

Cost models in blockchain are often studied from a protocol perspective, where incentives (e.g., [1]) or attack vectors, based on a mismatch of computational effort and cost, are investigated, e.g., [21]. Another line of research is focused on explaining and predicting certain market determinants, e.g., [7]. Cost models from a system design perspective are less investigated, specifically for next-generation blockchains with novel cost structures. Pincheira et al. [22] provide a model from a high abstraction level, to gauge the infrastructure cost of operating an application on a public vs. private blockchain, which they evaluate based on an Ethereum environment.

From a process perspective, Kjäer et al. [14] provide a qualitative discussion of different blockchain technology groups. Our work can be seen as a concrete, primarily quantitative, investigation of the transaction costs concern for a new technology group. Rimba et al. [26] provide a concrete model of the Ethereum cost structure. Our choice of Algorand for this investigation is motivated by the many opposing qualities of its cost model compared to the much studied Ethereum. Furthermore, since Rimba et al., Ethereum's cost model has evolved in many ways, we will review.

[3] https://developer.algorand.org/docs/get-details/dapps/smart-contracts/guidelines/#more-about-smart-signatures-vs-smart-contracts, accessed 2024-07-01.

From the perspective of multi-platform support for process execution, there is work towards enabling the generation and composition of multiple blockchain target implementations [6,8,16]. However, their focus is on second-generation and private blockchain targets. Furthermore, the underlying cost model is not explored. Many aspects of these works are complimentary to our work, such as the blockchain independent access layer described in Falazi et al. [8].

In [35], Xu et al. propose a method to translate DCR graphs to TEAL contracts. In contrast, we study a petri-net based translation technique as proposed in [10], but adapted for BPMN Choreographies. More importantly, our study also differs in scope. We give a formalised treatment of Algorand's underlying cost model, based on which we are able to discuss different design options. We present an architecture for a multi-target compiler, which supports cost efficient execution of multiple instances, and contrast our results to Ethereum's cost model and traditional cloud computing.

4 Preliminary Cost Model Analysis

Algorand's cost model incorporates two aspects: (i) transaction fees, which incur directly for every transaction; and, (ii) minimum balance requirements, which is a complete separate notion, and incur whenever storage slots on the chain are allocated. Balance requirements are not cost *per se*, as it encodes an amount of ALG an account must hold at a minimum.

To study Algorand's cost model in more detail, we introduce Ethereum as a baseline, as the most influential and, arguably, *archetype* of second-gen blockchains. Contrasting new developments against this archetype (and all its problems) can help classifying and contextualising them. Specifically, as Ethereum has come under the most scrutiny—especially for process execution [28].

4.1 Transaction Fees

Transaction cost in Ethereum are a function of the gas units used n_{gas} and the gas price, expressed in gwei (10^{-9} ETH). Since EIP-1559 [2], the gas price can no longer be freely set by a client. The price is now determined by the base fee b_{gas}, which increases or decreases based on the previous block's utilisation. A client can add a priority fee p_{gas}, to incentivise faster inclusion of the transaction. The verifier only receives $n_{\text{gas}} \times p_{\text{gas}}$ as a reward, the rest is made inaccessible (burned). Transactions start with a base gas amount of $n_{\text{gas}} = 21,000$. The additional amount is determined through execution, where each opcode is metered. Opcodes are priced based on their computational complexity and many exhibit dynamic cost, i.e., their cost depend on the current execution context, not just the opcode and its input. Formally, the transaction cost $c^{\text{tx}}_{\text{eth}}$ is

$$c^{\text{tx}}_{\text{eth}} = n_{\text{gas}} \cdot (b_{\text{gas}} + p_{\text{gas}}). \tag{1}$$

In Algorand, transaction cost is at least the fixed value $b_{\text{alg}} = 1,000$, here expressed in malgo (10^{-6} ALG). Should the verifier be congested, the transaction size s_{tx} becomes the determining factor. Formally, transaction cost $c_{\text{alg}}^{\text{tx}}$ is

$$c_{\text{alg}}^{\text{tx}} = \max(b_{\text{alg}}, f_{\text{alg}} \cdot s_{\text{tx}}), \tag{2}$$

where f_{alg} is calculated based on the local level of congestion of a verifier.[4] Thus, contrary to Ethereum, the client can not choose any parameter, congestion is calculated locally rather than globally, and there is no reward payed to the verifier for processing the transaction. Furthermore, there is no incentive the client can provide for their transaction to be prioritised (cf. [20]).

Algorand's transaction fee do not directly account for computational complexity. Rather, a transaction is constrained by an opcode cost limit $l_{\text{op}} = 700$. Opcodes are metered; however most (83%)[5] are priced with a flat fee of 1, and cost is not influenced by the execution context, just the opcode and its input. The opcode limitation can be lifted by batching transactions within an atomic transaction group, e.g., one can issue "dummy" transactions to increase the budget for another opcode heavy transaction in the group. The cost of a transaction group is just the sum of the included transaction costs, i.e., for n_{tx} transactions, the cost is $c_{\text{alg}}^{\text{tx}} \times n_{\text{tx}}$, where $n_{\text{tx}} \leq 16$.[6] Thus, for the perspective of application development, we can write the cost of a singular opcode $c_{\text{alg}}^{\text{op}}$ as

$$c_{\text{alg}}^{\text{op}} = c_{\text{alg}}^{\text{tx}} \cdot (1 + \lfloor \frac{n_{\text{op}} - 1}{l_{\text{op}}} \rfloor), \tag{3}$$

where n_{op} is the metered cost of opcodes used. Applications can issue an additional 256 inner transactions, increasing the pooled opcode limit further.

4.2 Storage Fees

For the purpose of process execution, we focus on persistent storage that can be written and read in the context of a contract call, and is under the control of the (orchestrating) application (layer one storage). Pricing such persistent memory is especially delicate for blockchain platforms, as these grow the replicated storage and require expensive disk operations [21]. In Ethereum, where everything is priced based on metered opcodes, the virtual machine (EVM) defines the `SLOAD` and `SSTORE` opcodes, which load and store byte key value pairs. Following the introduction of EIP-292 [3], these opcodes are metered based on whether storage is accessed for the first time, so called *cold* storage, or not, so called *warm* storage. An access list keeps track of warm slots during the current execution

[4] https://github.com/algorand/go-algorand/blob/v3.24.0-stable/data/pools/transactionPool.go, accessed 2024-06-04.

[5] https://developer.algorand.org/docs/get-details/dapps/avm/teal/opcodes/v10, accessed 2024-07-05.

[6] See https://developer.algorand.org/docs/get-details/parameter, accessed 2024-06-04.

Table 1. Overview of the cost of different storage types in Algorand.

			Storage		
	Account	Contract (2048B)	Box (n_B)	Byte (128B)	Uint (72B)
Cost					
(mb)	100,000	100,000	$2500 + 400\, n_B$	50,000	28,500
(op)	-	-	1	1	1
Refund (of mb)	100%	100%	100%	100%	100%
Limit (B)	-	8,192	-	8,192	4,608

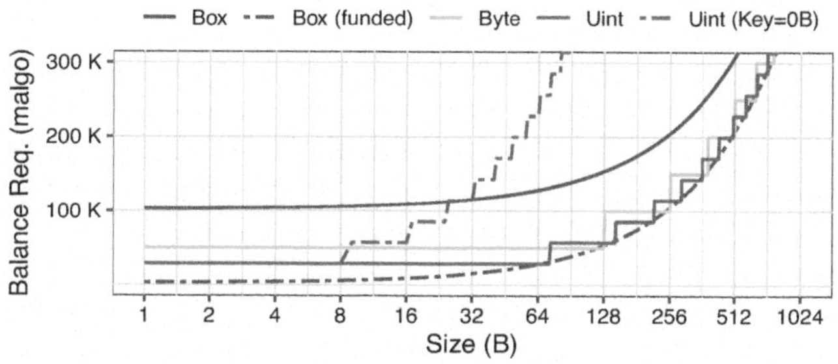

Fig. 1. Balance req. of Algorand storage systems as a function of their size in Byte.

context. A default transaction pre-loads the address of the transaction sender and recipient to the list. An optional transaction parameter can be supplied to specify elements to additionally pre-load. However, due to the current implementation of this list, there is no cost advantage for most contract transactions [13]. In EIP-3529 [4], refunds were drastically reduced. Refunds were introduced to incentivise the economical usage of Ethereum's storage space; specifically for data no longer required. Freeing a contract slot is no longer compensated; freeing a storage slot reimburses 4,800 gas, but only up to a maximum of $\frac{1}{5}n_{\text{gas}}$ of the current transaction.

Algorand has two different storage systems that can be used: global and box storage. Both store key value pairs, where the key is limited to 64B. Global storage can be divided into `Uint`, a 8B unsinged integer value, and `Byte` storage, a max. 128B byte value. In total, global storage is limited to 64 slots (i.e., key value pairs) per application. Box storage allocates a variable size byte value up to a limit of 32KB, the number of boxes is not limited.

Storage in Algorand is priced by the notion of minimum balance, which is defined per account and expresses the balance the account is required to hold at a minimum. This balance is practically frozen, but can be refunded (unfrozen) by freeing the associated storage slot again. The opcodes to store and access storage are also metered, but the cost is a negligible $n_{\text{op}} = 1$ for these operations. See Table 1 for an overview. In contrast to Ethereum, a funded account requires a minimum balance also. Similarly to Ethereum's access list, Algorand has the notion of reference lists. A standard application call transaction pre-loads two references: the caller and the callee. Storage slots outside of the reference lists can not be accessed. The reference list is limited to eight additional entries per transaction. Analogously to the opcode limit, this limit can be increased by pooling transactions. If a contract address is included in the reference list, it also includes all global storage slots associated. Box slots must be included separately and for each 1 KB. E.g., with one transaction, one can access 64 global storage slots, i.e., 8 KB, and 8 KB box storage.

An application architect must choose from the available storage systems. We want to investigate them from a cost perspective. Funding an account $m^{\text{acc}}_{\text{alg}}$ and deploying an application $m^{\text{app}}_{\text{alg}}$ requires a minimum balance of[7]

$$m^{\text{acc}}_{\text{alg}} = m^{\text{app}}_{\text{alg}} = 100,000. \tag{4}$$

To assess the different storage systems, we investigate their cost as a function of bytes we want to store. We illustrate this in Fig. 1, where we show the cost in correspondence to the stored bytes up to 1024B (with a logarithmic x-axis). The minimum balance required by the box storage must be funded by the application, thus, it requires the funding of an additional account. Consequently, box storage is only economical for larger B. However, if an application account is required and already funded (e.g., it needs to be able to send or receive payments anyway), the additional overhead does not exist. Furthermore, if keys are merely used as indices to access storage, i.e., they don't carry any information themselves, packaging different Uint values together will reduce the balance requirements, as the Uint value itself only carries 8B. We illustrate this by an example, and show Uint with only its value storage capacity, so with a key length of 0B. We see that, as soon as an application requires two Uint value slots, it is more economical to package the Uint values in a Byte. We use this insight in Sect. 5. Formally, the required minimum balance $m^{\text{s}}_{\text{alg}}$ per number of bytes n_{B} for the different storage systems is

$$m^{\text{s}}_{\text{alg}} = \begin{cases} 28,500 \cdot (1 + \lfloor \frac{n_{\text{B}}-1}{72} \rfloor) & \text{for Uint,} \\ 50,000 \cdot (1 + \lfloor \frac{n_{\text{B}}-1}{128} \rfloor) & \text{for Byte,} \\ m^{\text{app}}_{\text{alg}} + 2,500 + 400 \cdot n_{\text{B}} & \text{for Box.} \end{cases} \tag{5}$$

In summary, we can express the total balance requirement m_{alg} for n_{acc} amount of accounts and n_{app} amount of apps as

$$m_{\text{alg}} = n_{\text{acc}} \cdot m^{\text{acc}}_{\text{alg}} + n_{\text{app}} \cdot m^{\text{app}}_{\text{alg}} + m^{\text{s}}_{\text{alg}}. \tag{6}$$

5 From BPMN Choreographies to Teal Applications

In this section, we describe multi-target Chorpiler, an extension of Chorpiler's architecture introduced in Stiehle and Weber [29]. Our aim is to facilitate quick implementation of novel blockchain targets, such as Algorand.

---

[7] Some additional cost apply to oversized contracts, see https://developer.algorand.org/docs/get-details/dapps/smart-contracts/apps#minimum-balance-requirement-for-a-smart-contract, last accessed 2024-06-04.

Since García-Bañuelos et al. [10], the state of the art in optimised code generation for smart contracts is to parse a process model into a petri net, which can then be reduced according to equivalence rules. In the target output, the marking of this net is then encoded as a bit array (the net is 1-safe), which allows efficient storage as an unsigned integer $i \in \mathbb{N}_0$. Then, the binary digits of i correspond to all markings with tokens, e.g., 2^1 would translate to a token in marking M_1.[8] To perform transitions, bit shifts can be executed on i; these can be usually performed very efficiently, and carry a negligible opcode cost. Chorpiler uses this technique to translate BPMN Choreographies to Solidity smart contracts, using a template engine. We modified Chorpiler's architecture to support multi platform targets, a goal of our architecture is a minimal core component, that assists in implementation of new blockchain targets. A specific requirement is also to support rigorous software testing, as the goal is to produce well-tested contract code. This is paramount for blockchain-based applications [17]. This requirement is also informed by the need of practitioners of traditional workflow execution systems, which note that this aspect is often neglected in academic artefacts [27]. Our separation of concerns avoids bloating the core package with blockchain platform specific dependencies, environments, and test suites.

We sketch the key components of our architecture in Fig. 2. We show the Chorpiler core package on the right side. Here, a key enabler is the introduction of a `Process Encoder` component, which generates the previously described marking encoding from a petri net. A new target implementation can implement the `Template Engine` interface, and can make use of the provided encoder. For the purpose of this study, we made use of this to implement a target for the Algorand virtual machine (AVM). The AVM's native language is TEAL, an assembly-like stack-oriented language. Other language bindings that compile to TEAL, e.g., PyTeal, or TealScript exist. However, PyTeal does not allow native

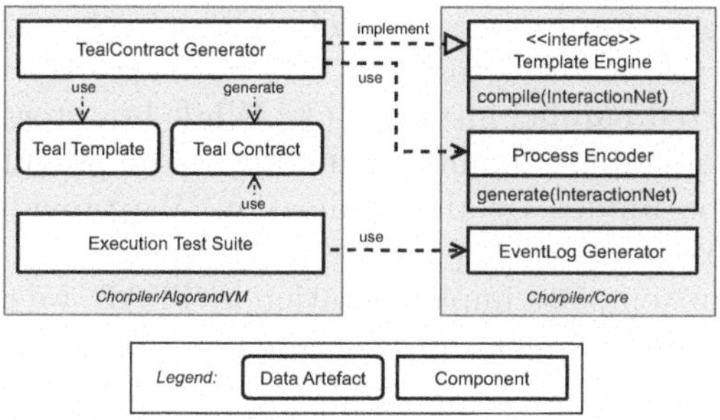

Fig. 2. Architecture excerpt of the refined modular Chorpiler.

[8] *Remark.* The number of supported places is limited by the virtual machine's supported unsinged integer size. In Ethereum, this is 256 bit; in Algorand this is 64 bit. Trivially, another integer can be allocated in Ethereum [10]. As shown in Sect. 4, this is not efficient in Algorand. In Sect. 5.1, we outline a technique for multi instance support using variable packaging. This technique can also be applied to increase the limitation on places.

Python expressions and TealScript is still in development and has not yet been scrutinised in security audits. Using the Process Encoder architecture allowed us to implement the new target with almost no overhead. The codebase is dedicated to the back-end: the TEAL contract template and testing the produced outputs. Importantly, a new target implementation is not forced to use the Process Encoder, and could even use a different template engine, allowing a developer a high level of flexibility to adopt to new targets. To facilitate testing new targets, we also introduce an `EventLog Generator`. In Weber et al. [31], the authors introduced a methodology to fuzzy test smart contract outputs for a given model. To do so, non-conforming events are generated, which are then replayed on the smart contract, which consequently must reject them. The EventLog Generator allows the generation of such replay logs from an *eXtensible Event Stream* (XES) file, containing conforming events. We make use of this in our correctness tests, which we describe in Sect. 6 in more detail.

5.1 Multi-instance Execution in Teal

To support the cost efficient interleaved execution of multiple process instances, we introduce an approach based on variable packaging, which allows a (theoretical) unlimited instantiations of the process model, without deploying a new application contract. We have shown in Sect. 4, that allocating global storage slots is inefficient when the key is predictable from a certain threshold. For Uint storage, this is already for $n > 1$.

Mutli-instance execution requires the correlation of any incoming message to a receiving instance's data object [18]. For each instantiation, our design assigns a new key $i \in \mathbb{N}_0$, in an ascending order, which uniquely identifies an instance C_i. Here, i is used as offset, to determine the storage slot in a packaged variable. Instances can be packaged into byte or box storage. Specifically, for a storage requirement of p places for a process model, we require $k = 1 + \lfloor \frac{p-1}{8} \rfloor$ bytes. C_i is then allocated to the next k bytes from $i \times k$ in the storage up to a reasonable maximum of supported parallel instances C_n. While box storage is theoretically unlimited, a reasonable upper bound should always be provided; specifically, as box storage must be funded by the application. We provide a more detailed discussion in Sect. 6.

For details on our template implementation in TEAL, we refer the interested reader to our open source artefact in Footnote 1.

6 Evaluation

To evaluate our implementation and cost models, we performed correctness and cost benchmarks on well-known process models often used in related work. Namely, the supply chain [31] (adapted from [9]) and incident management [19] case. It also allows us to compare our work to previous work on cost models, i.e., Rimba et al. [26] and previous work on Chorpiler, i.e., Stiehle and Weber [29].

We implemented a new chorpiler package as sketched in Sect. 5. The package is capable of generating TEAL contracts from BPMN Choreographies. The compiler supports our in Sect. 5.1 described multi-instance execution by packaging the AVM's `uint64` types into byte or box storage. We used an algorand node to deploy our contracts and simulate and execute transactions in a local sandbox test network. Our test environment runs `go-algorand` in version `3.24.0`, which is the latest stable release at the time of writing.[9]

We extended the chorpiler core functionalities to be able to parse XES logs and generate non-conforming traces. To benchmark the correctness of our implementation, we followed the methodology outlined in [31]. For each case, we replayed all possible conforming traces (two for supply chain and four for the incident management case) and generated 2000 non-conforming traces. To generate a non-conforming trace, Chorpiler randomly manipulates a conforming trace by a range of operations: adding events (by evaluating other conforming events), removing an event, and swapping the position of two events. It then removes any coincidentally created conforming trace. We replayed these traces locally in our environment using the Algorand JavaScript SDK. We performed these conformance tests on all our implemented contracts. All conforming traces were accepted and lead to the end event, and all non-conforming traces were rejected.

To assess cost, we extracted from each transaction: the opcode cost, the change in minimum balance requirement for caller and contract, and the transaction fee. We report on the results for all replayed conforming traces on both cases in Table 2. We show cost in the form of average opcode cost per transaction, average transaction fee per process case, and minimum balance cost per case, where the minimum balance requirement is shown exclusive of the balance requirement of the involved external accounts. We show the executions for all analysed storage systems. For both cases, a storage size of 8B, the smallest data type natively supported by the AVM, is sufficient. When considering singular case executions, Uint is unsurprisingly the most cost efficient implementation. Byte and Box contracts implement the byte packaging we introduced; however, they do not pay off and lead to higher opcode cost and balance requirements. The Box implementation also requires an additional transaction to fund the account. However, under more realistic conditions, when multiple cases are executed in parallel, the Uint implementation requires a re-deployment of the contract for every new instance, which quickly increases the minimum balance requirement. We illustrate this in Fig. 3, where we show the minimum balance requirement when multiple instances C_n of supply chain or incident management are executed. Our benchmark results closely follows our analysis in Sect. 4, where we discussed that allocating additional Uint beyond one is inefficient. We also see that for applications with already funded application accounts (e.g., as they implement an escrow functionality), box storage is also economical for smaller C_n. We did not encounter any cost which we could not account using Eq. 3 and Eq. 6. Notably, while we report our findings excluding the required balance to fund participant's external accounts—as they are stable costs that all implementations share—these are considerable. For the small process cases supply chain

[9] Our evaluation environment is available at the provided link in Footnote 1.

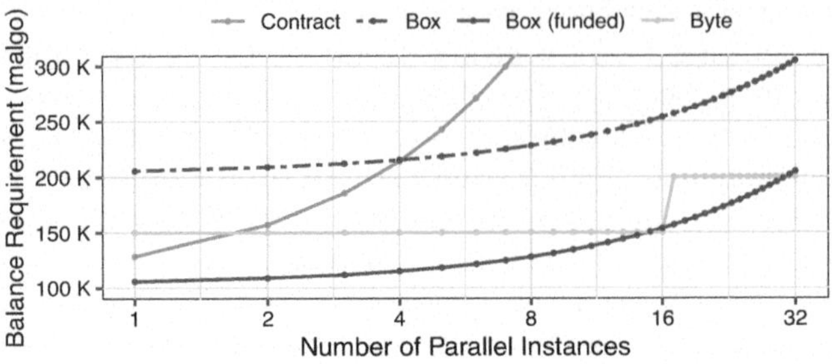

Fig. 3. Minimum balance requirement for multiple parallel instance executions.

Table 2. Benchmark result for the generated process cases supply chain (SC) and incident management (IM), excluding external account balance requirements.

	SC ($p = 11$)			IM ($p = 8$)		
Cost	Uint	Byte	Box	Uint	Byte	Box
(avg/tx) op	67	83	83	95	111	110
(avg/case) tx	11,000	11,000	12,000	7,000	7,000	8,000
(case) mb	128,500	150,000	206,100	128,500	150,000	206,100

Table 3. Execution cost comparison of the incident management case. We show cost in the native units: gas and malgo. We show how this translates to $US with different exchange rates. The avg. prices for ETH are $US 178.94, 1744, and 3156, and for ALG 0.53, 0.16, and 0.20. SWF is reported from [26], with a data retention setting of one day and 99 years.

		Exchange Rate in $US			
	Cost	1	avg. 2019	avg. 2023	avg. 2024
EVM	(gas)				
Chorpiler [29]	599,463	0.020	3.599	35.075	63.473
AVM	(malg)				
(tx)	11,000	0.011	0.006	0.002	0.002
(incl. mb)	928,500	0.939	0.494	0.147	0.189
SWF [26]	($US)				
(24 hours)	0.0009				
(99 years)	0.1816				

(9 participants) and incident management (8 participants), a total balance of 900K and 800K of malg is required, respectively. When including them in the cost calculation, they would amount to 85% of the total execution cost of Uint.

We now want to put these results into context and make use of the fact that the incident management process model has been investigated in the context of process execution on Ethereum and traditional cloud computing. Comparing costs across different blockchain platforms is challenging, as it has long been understood that market capitalisation and exchange rates (e.g., ETH-US) are affected by various different market dynamics [7,15]. We follow the approach by Rimba et al., which conducted a benchmark of the Amazon Simple Workflow Service (SWF) to contrast the cost of blockchain execution [26]. SWF is a process-oriented software suite capable of executing workflows and offered as part of Amazon Web Services (AWS). They report on an implementation of the incident management case and the respective cost factors. Put simply, SWF is priced based on the workflow elements executed, the execution time, and the data transfer and retention rates. We refer the reader to Rimba et al. [26] for more details. As they provide detailed pricing information, the recency of their results can be assessed by examining the most recent documentation provided by SWF. Indeed, we did not encounter any changes to their pricing model or technical changes rendering Rimba et al.'s results incompatible. Therefore, we deemed it unnecessary to reproduce the benchmark outlined in Rimba et al. We think it serves as a valuable illustrative baseline. And, as we show, cost of blockchain execution is still multitudes higher, likely rendering small unobserved technical changes to SWF negligible. We summarise our comparison in Table 3. We show the results from Stiehle and Weber [29], deploying and executing the incident management process in an EVM environment. We were able to reproduce these results. For the AVM, we show the cost of the Uint implementation, as most efficient implementation for singular executions of incident management. We show how these costs translate to $US for various exchange rates of $US to ETH and ALG, respectively, for the average exchange rate in 2019 (the earlierst available data point for Algorand), 2023, and 2024.[10] For SWF, we show different data retention settings. SWF fees increase when process data needs to be retained longer. In theory, blockchain data is stored indefinitely. Notably, for Algorand, we show transaction fees separate from balance requirements. Balance requirements include the funding of all participant's accounts. Unsurprisingly, the main driver for cost in Algorand is also storage. A driver for smaller applications, which have minimised their storage requirements through engineering efforts, is the minimum balance requirement for accounts. Specifically, for our investigated process cases, with a high ratio of participants, external account's balance requirements can make up up to 85% (for Uint) of the total cost.

7 Discussion and Future Directions

Comparing the execution costs of Algorand to Ethereum and to SWF, specifically for higher levels of data retention, shows the promise of process enactment on next-gen blockchains. However, the evolution of exchange rates for public

[10] Based on the average gas price in 2023 of 33.55 gwei, determined based on data from Etherscan (https://etherscan.io/chart/gasprice, accessed 2024-06-03). The ETH and ALG prices are the averaged daily closing prices for a given year, as reported by Yahoo Finance (https://finance.yahoo.com/quote/ETH-USD/, accessed 2024-06-03).

blockchains is by no means stable or predictable. We illustrated the steep climb of ETH from 2019 on—calculations which are in line with previous work [28]. Algorand has charted a somewhat reverse trend. Both platforms exhibit opposing qualities in many other aspects. While Ethereum utilises a gas fee market, Algorand defines fixed fee costs. While this increases predictability and usability in the immediate sense, should Algorand rise in popularity as a speculative asset, the fixed fee nature would force protocol changes to adapt to a changed exchange rate. Currently, Algorand facilitates important protocol changes by a governance model, where votes are open for a period of three months.[11] In many ways, Algorand has not yet come under such scrutiny as Ethereum, whose operators were forced to make many adjustments to its cost model based on discovered or exploited attack vectors in the past (see e.g., [21]). With EIP-1559, Ethereum reduced the level of user agency in the fee market in favour of better usability. Furthermore, certain market dynamics have introduced unintended consequences, which forced drastic reductions in gas refunds [4]. Conversely, the minimum balance system of Algorand operates independently of the transaction fee system. While we treated minimum balance requirements as cost, they essentially resemble invested or frozen assets, where notions like opportunity cost apply. The balance requirement is also exposed to devaluation and appreciation effects as the exchange rate of ALG to other currencies changes. As we have shown, minimum balance requirements can constitute for up to 85% of the total cost, so these considerations are paramount when considering Algorand. Finally, while Ethereum incentivizes honest verifiers by rewards, Algorand does not currently distribute rewards based on blocks or transactions. Rather, rewards are distributed by so-called reward programs.[12]

Our results show the promise of blockchain-based process execution on next-gen blockchains. However, our study also stands as early evidence that adapting existing approaches to novel protocols is by no means trivial. Protocol designs can differ greatly, and involve novel trade-offs, implications, and attack vectors. Superficial understanding of new protocols can in principle lead to poor design decisions and in consequence to security vulnerabilities, wasted effort, and cost.

Highlighting the above design differences illustrates the need for a systematic investigation of the design space. Conducting a kind of design archaeology [5], and unearthing different design principles as well as constructing taxonomies, could bring immense value to both users and designers of decentralised protocols—not just in the design of such systems, but also in communicating design decisions [32].

While we have illustrated how concrete cost models can influence the design of an approach *a priori*, and help weigh different design options, the AVM results shown are all based on static local experiment runs. We see opportunity in automated simulation approaches to assess existing implementations in more realistic environments that are capable to simulate e.g., different levels of congestion.

---

[11] https://governance.algorand.foundation, accessed 2024-06-04.

[12] https://www.algorand.foundation/200-million-algo-staking-rewards-program, accessed 2024-06-04.

Simulation can facilitate fast and secure redesign of different process implementations [24]. In the case of large decentralised systems, this constitutes a major challenge. Specifically, studying systems under load in an economical manner remains a challenge. Current benchmark systems [12,25] focus on the system as a whole. They do not natively support reactive applications like choreographies, where participants observe the blockchain, and execute contracts as a consequence of observed events.

8 Conclusion

Blockchain-based process systems must deal with platform specific limitations, cost models, and security considerations. This is especially pronounced in blockchain, where design knowledge is not mature. We have introduced a compiler architecture, which supports implementing new blockchain targets with minimal overhead, and demonstrated its use for the Algorand blockchain. We presented a comprehensive analysis and formalisation of the Algorand cost model. Our presented approach, a method to compile BPMN choreographies to TEAL contracts, makes immediate use of these insights. Our manuscript illustrates that assessing cost for novel blockchain protocols is by no means trivial. However, for organisations, predictable cost and performance estimates are paramount—without an understanding of operational cost, conducting a cost-benefit analysis is difficult if not impossible. We see our study as a contribution to a more holistic understanding of novel blockchains.

Acknowledgements. The authors wish to disclose the use of generative AI to assist in the editing of the manuscript and the implementation of the artefacts. Generated output was never taken "as-is", it was reviewed and verified by the authors.

References

1. Abbasi, M., Manshaei, M.H., Rahman, M.A., Akkaya, K., Jadliwala, M.: On algorand transaction fees: challenges and mechanism design. In: ICC, pp. 5403–5408. IEEE (2022)
2. Buterin, V., Conner, E., Dudley, R., Slipper, M., Norden, I., Bakhta, A.: EIP-1559 (2019). https://eips.ethereum.org/EIPS/eip-1559
3. Buterin, V., Swende, M.: EIP-2929 (2020). https://eips.ethereum.org/EIPS/eip-2929
4. Buterin, V., Swende, M.: EIP-3529 (2021). https://eips.ethereum.org/EIPS/eip-3529
5. Chandra Kruse, L., Seidel, S., vom Brocke, J.: Design archaeology: generating design knowledge from real-world artifact design. In: Tulu, B., Djamasbi, S., Leroy, G. (eds.) DESRIST 2019. LNCS, vol. 11491, pp. 32–45. Springer, Cham (2019). https://doi.org/10.1007/978-3-030-19504-5_3
6. Corradini, F., et al.: Model-driven engineering for multi-party business processes on multiple blockchains. Blockchain: Res. Appl. **2**(3), 100018 (2021)
7. Donmez, A., Karaivanov, A.: Transaction fee economics in the ethereum blockchain. Econ. Inq. **60**(1), 265–292 (2022)

8. Falazi, G., Hahn, M., Breitenbücher, U., Leymann, F., Yussupov, V.: Process-based composition of permissioned and permissionless blockchain smart contracts. In: EDOC, pp. 77–87. IEEE (2019)
9. Fdhila, W., Rinderle-Ma, S., Knuplesch, D., Reichert, M.: Change and compliance in collaborative processes. In: IEEE SCC, pp. 162–169 (2015)
10. García-Bañuelos, L., Ponomarev, A., Dumas, M., Weber, I.: Optimized execution of business processes on blockchain. In: Carmona, J., Engels, G., Kumar, A. (eds.) BPM 2017. LNCS, vol. 10445, pp. 130–146. Springer, Cham (2017). https://doi.org/10.1007/978-3-319-65000-5_8
11. Gilad, Y., Hemo, R., Micali, S., Vlachos, G., Zeldovich, N.: Algorand: scaling byzantine agreements for cryptocurrencies. In: SOSP, pp. 51–68 (2017)
12. Gramoli, V., Guerraoui, R., Lebedev, A., Natoli, C., Voron, G.: Diablo: a benchmark suite for blockchains. In: EuroSys, pp. 540-556. ACM (2023)
13. Heimbach, L., Kniep, Q., Vonlanthen, Y., Wattenhofer, R., Züst, P.: Dissecting the eip-2930 optional access lists. arXiv preprint arXiv:2312.06574 (2023)
14. Kjäer, M., Preindl, T., Kastner, W.: Towards an understanding of trade-offs between blockchain and alternative technologies for inter-organizational business process enactment. In: Kopke, J., et al. (eds.) BPM: Forum, LNBIP, pp. 36–50. Springer, Heidelberg (2023). https://doi.org/10.1007/978-3-031-43433-4_3
15. Kukacka, J., Kristoufek, L.: Fundamental and speculative components of the cryptocurrency pricing dynamics. Finan. Innov. **9**(1), 61 (2023)
16. Ladleif, J., Friedow, C., Weske, M.: An architecture for multi-chain business process choreographies. In: BIS, vol. 389, pp. 184–196 (2020)
17. Lu, Q., Weber, I., Staples, M.: Why model-driven engineering fits the needs for blockchain application development. IEEE Blockchain Newsl. **2** (2018). https://blockchain.ieee.org/newsletter/september-2018/why-model-driven-engineering-fits-the-needs-for-blockchain-application-development. Accessed 03 June 2024
18. Meyer, A., Pufahl, L., Batoulis, K., Fahland, D., Weske, M.: Automating data exchange in process choreographies. Inf. Syst. **53**, 296–329 (2015)
19. OMG: BPMN 2.0 by Example, Version 1.0 (2010). https://www.omg.org/cgi-bin/doc?dtc/10-06-02. Accessed 29 Mar 2023
20. Öz, B., Rezabek, F., Gebele, J., Hoops, F., Matthes, F.: A study of mev extraction techniques on a first-come-first-served blockchain. In: Proceedings of the 39th ACM/SIGAPP Symposium on Applied Computing, pp. 288–297 (2024)
21. Perez, D., Livshits, B.: Broken metre: attacking resource metering in evm. arXiv preprint arXiv:1909.07220 (2019)
22. Pincheira, M., Donini, E., Vecchio, M., Giaffreda, R.: An infrastructure cost and benefits evaluation framework for blockchain-based applications. Systems **11**(4), 184 (2023)
23. Prybila, C., Schulte, S., Hochreiner, C., Weber, I.: Runtime verification for business processes utilizing the Bitcoin blockchain. FGCS **107**, 816–831 (2020)
24. Pufahl, L., Wong, T.Y., Weske, M.: Design of an extensible BPMN process simulator. In: Teniente, E., Weidlich, M. (eds.) BPM 2017. LNBIP, vol. 308, pp. 782–795. Springer, Cham (2018). https://doi.org/10.1007/978-3-319-74030-0_62
25. Rezabek, F., Glas, K., Von Seck, R., Aroua, A., Leonhardt, T., Carle, G.: Multilayer environment and toolchain for holistic network design and analysis. arXiv preprint arXiv:2310.16190 (2023)
26. Rimba, P., Tran, A.B., Weber, I., Staples, M., Ponomarev, A., Xu, X.: Quantifying the cost of distrust: comparing blockchain and cloud services for business process execution. Inf. Syst. Front. **22**(2), 489–507 (2020)

27. Ruecker, B.: Practical Process Automation. O'Reilly Media Inc., Newton (2021)
28. Stiehle, F., Weber, I.: Blockchain for business process enactment: a taxonomy and systematic literature review. In: BPM: Forum. LNBIP, vol. 459, pp. 5–20 (2022)
29. Stiehle, F., Weber, I.: Process channels: a new layer for process enactment based on blockchain state channels. In: Di Francescomarino, C., Burattin, A., Janiesch, C., Sadiq, S. (eds.) BPM. LNCS, pp. 198–215. Springer, Heidelberg (2023). https://doi.org/10.1007/978-3-031-41620-0_12
30. Tschorsch, F., Scheuermann, B.: Bitcoin and beyond: a technical survey on decentralized digital currencies. IEEE Commun. Surv. Tutor. **18**(3), 2084–2123 (2016)
31. Weber, I., Xu, X., Riveret, R., Governatori, G., Ponomarev, A., Mendling, J.: Untrusted business process monitoring and execution using blockchain. In: La Rosa, M., Loos, P., Pastor, O. (eds.) BPM 2016. LNCS, vol. 9850, pp. 329–347. Springer, Cham (2016). https://doi.org/10.1007/978-3-319-45348-4_19
32. Williams, K., Chatterjee, S., Rossi, M.: Design of emerging digital services: a taxonomy. EJIS **17**(5), 505–517 (2008)
33. Xu, X., Weber, I., Staples, M.: Architecture for Blockchain Applications. Springer, Cham (2019). https://doi.org/10.1007/978-3-030-03035-3
34. Xu, X., et al.: A taxonomy of blockchain-based systems for architecture design. In: ICSA, pp. 243–252. IEEE (2017)
35. Xu, Y., Slaats, T., Düdder, B., Debois, S., Wu, H.: Distributed and adversarial resistant workflow execution on the algorand blockchain. In: Matsuo, S., et al. (eds.) FC 2022. LNCS, pp. 583–597. Springer, Heidelberg (2022). https://doi.org/10.1007/978-3-031-32415-4_35

BPMS Blockchain Technology Soft Integration For Non-tamperable Logging

Johannes Loebbecke[1(✉)], Thomas van Loo[1], Juergen Mangler[1], Zhendong Ma[2], Tomáš Pitner[3], and Stefanie Rinderle-Ma[1]

[1] TUM School of Computation, Information and Technology, Technical University of Munich, Garching, Germany
{johannes.loebbecke,juergen.mangler,stefanie.rinderle-Ma}@tum.de
[2] Luxoft/DXC Technology, Munich, Germany
zhendong.ma@plus.ai
[3] Faculty of Informatics, Masaryk University, Brno, Czech Republic
pitner@muni.cz

Abstract. In contemporary business process management (BPM), the need for secure and transparent interactions between organizations is paramount. Integrating business process management systems (BPMS) with blockchain technology can offer protection against malicious actors and increase trust, e.g., in process mining results, by allowing organizations to share execution traces and associated information. However, hard integration approaches, which we define as execution of processes on the blockchain as smart contracts, have not found widespread application. A hard integration approach, e.g., by utilizing a BPMN to solidity compiler as Caterpillar, has a high barrier of entry since it requires a company to restructure its infrastructure and processes, especially pertaining to potential immutability of smart contracts, and the integration of outside information through oracles. In this work, we instead want to explore a soft integration approach, which allows companies to continue using their existing BPMS infrastructure and integrate blockchain technologies without having to change the overall architecture of the system. The goal is to produce distributed non-tamperable logs, created through the use of private blockchains, serving as basis for compliance checking and, thus, secure and trustworthy execution. In this work, we present two soft integration approaches, which are discussed, implemented, and evaluated regarding their integration complexity and performance. The results suggest that a soft integration approach of blockchain technology can enhance the reliability and traceability of existing BPM systems with low integration effort, thus pointing towards a path for high acceptance within organizations.

Keywords: BPMS · Non-Tamperable Logging · Hyperledger Fabric · Processes and Mining · Blockchain Technology Integration

1 Introduction

Since the conception and deployment of Bitcoin [14], adoption of blockchain technology has been on the rise [2]. Blockchain technology provides an immutable ledger

C. Di Ciccio et al. (Eds.): BPM 2024, LNBIP 527, pp. 106–120, 2024.
https://doi.org/10.1007/978-3-031-70445-1_7

without having to rely on a central authority to maintain data integrity. In contemporary business process management (BPM), the need for enhanced security and transparency between organizations is paramount. Accordingly, there has been widespread interest from both academia and industry to integrate blockchain technology into Business Process Management Systems (BPMS) [4, 12]. The immutable nature of a blockchain's ledger allows observing the process execution and, subsequently, the creation of trustworthy, non-tamperable logs, which store the execution history of processes, including API calls, responses, and data exchanges.

In a traditional system, the execution log is created by a central authority, which becomes a single point of failure and makes it susceptible to tampering by malicious actors. However, for a process that includes multiple business partners, trust among these partners and the accuracy of recorded process transactions is paramount [12]. Furthermore, a tamper-proof log also ensures that malicious actors which interfere with process execution tracking, can not cover their tracks by manipulating execution logs. The popularity of process mining technology [5], including the inclusion of sensor data for process analysis and optimization, further increases the need for trust in execution logs [10, 11].

However, while these advantages are well known, blockchain technology is not widely integrated with, or used for process management. Currently, Business processes are often supported and enacted through traditional Business Process Management Systems (e.g., Camunda[1], Tim Solutions[2], Aristaflow[3], ...). BPMS allow organizations to design, model, execute, and monitor processes via modeling tools, as well as to integrate with other enterprise systems, simulation engines, execution engines, visualization, and optimization tools. We identified two different approaches in which blockchain technology can be integrated into Process Management Systems: (1) the process itself is stored and executed on the blockchain (like, Caterpillar[4]) or (2) a traditional BPMS (like Camunda) is utilized for modeling and execution, while transaction information is stored on the chain. We call (1) a hard integration approach and (2) a soft integration approach. While (1) requires a paradigm shift in how processes are executed and relies on a separate and different engine, (2) allows to partially and gradually support blockchain technology and continue to utilize existing engines.

For this paper, we discuss the potential of the soft integration approach. For this purpose we analyzed different contributions that applied blockchain technology to BPM problems and implemented two different soft integration prototypes using the Hyperledger Fabric[5] and the Cloud Process Execution Engine CPEE[6]. The different prototypes are evaluated regarding their integration complexity and performance. The results suggest that a soft integration approach allows for the creation of tamperproof logs without significant integration complexity and sufficient performance. The complete Implementation details can be found in the Github Repository[7].

---

1 https://camunda.com, last access 2024-05-2024.
2 https://tim-solutions.de/, last access: 2024-04-10.
3 https://artistaflow.com, last access: 2024-04-10.
4 https://github.com/orlenyslp/Caterpillar, last access: 2024-04-10.
5 https://hyperledger.org/projects/fabric, last access: 2024-04-10.
6 https://cpee.org/, last access: 2024-05-28.
7 https://github.com/thomasvloo/fabric-testnet, last accessed: 2024-04-10.

The remainder of this paper is structured as follows: in Sect. 2 related work is discussed. In Sect. 3 the properties of hard and soft integration approaches are elaborated in detail. In Sect. 4 considerations regarding the implementation of a soft integration approach are discussed, including the selection of private blockchains and how they allow for different execution scenarios. Finally, in Sect. 5, we discuss the performance of the soft integration approach, as well as its performance properties regarding certain design decisions. The paper concludes with a summary of findings and proposed future work (see Sect. 6).

2 Related Work

While the implementation of cryptocurrencies has led to widespread interest in blockchain technology due to its influence on the financial industry, both academia and industry have been exploring applications in other fields. BPM entails the design, execution, monitoring, and optimization of business processes. In contemporary business process management, the need for enhanced security and transparency amongst inter-organizational process executions is paramount. Accordingly, there has been significant interest in integrating BPMS with blockchain technology.

In 2020, Garcia et al. [4] conducted a systematic literature review of blockchain technology in collaborative business processes. The authors used the methodology proposed by Kitchenham [6] to identify 34 papers addressing BPM using blockchain technology. Out of these studies, 71% are conceptual and theoretical proposals describing how blockchains could be integrated with process management and what advantages and challenges such an integration has. The authors concluded that, while interest of the scientific community is growing, current integration efforts are still in early stages.

Another systematic literature by Stiehle et al. [15] focuses on technical contributions for blockchain integration. They identified 30 studies, out of which they found that 86 percent enforced control flow on the chain which indicates a hard integration approach. Furthermore, they specifically highlight that current integration efforts struggle with flexibility and scalability, due to the immutable nature of the blockchain and performance limitations imposed by the consensus mechanism. In contrast, our approach tries to tackle exactly these two challenges, and thus improve industry acceptance.

The challenges and opportunities for the intersection of BPM and blockchain technology are extensively discussed by Mendling et al. [12], especially regarding inter-organizational processes. They present various technical challenges, including throughput, latency, and bandwidth. As major opportunities, they state three main advantages that blockchain integration can bring: (1) A blockchain acts as an immutable ledger that serves as a trustworthy transaction history for all involved parties. (2) Smart contracts can ensure compliance with the process and offer independent global monitoring. (3) Blockchain technology allows for easy encryption to limit data readability for contexts where such a restriction is necessary.

A soft integration approach is presented by Li et al. in [7]. They present the architecture of their proposal without implementation details, which blockchain network can potentially be used, or which required or delivered performance characteristics their solution has. For their approach, blockchain technology is loosely coupled with the process engine. Each operation conducted by the engine is logged by a separate component to create data reliability and secure transactions between process tasks and connected services. Compared to the approach presented in this paper, Li et al. propose only a single transaction for interaction between the engine and a connected service. Thus data tampering can only be confirmed when a global view is assumed, not independently from the engine's perspective and a potential partner (represented by the connected service).

A major research stream is focused around hard integration with the Caterpillar platform described and developed by López-Pintado et al. [9]. Caterpillar[8] is an open-source business process management system running on the Ethereum Blockchain[9]. Caterpillar provides functionality similar to traditional BPMSs, allowing users to model and execute instances of business process, but additionally uses the Ethereum network to preserve the execution instances on the blockchain and handle workflow routing. The Solidity smart contracts are generated using a BPMN-to-Solidity compiler to represent the workflow routing logic of given process models. Subsequently, an on-chain execution engine is used to deploy these models and create immutable execution logs. Several contributions have extended the Caterpillar project, such as [8] and [13]. López-Pintado et al. [8] proposed dynamically relating actors with collaborative business processes and generating smart contracts to verify role restrictions. Meanwhile, Ercenne et al. [13] outline Blockchain Studios as an extension of Caterpillar, with a focus on process models based on human tasks to incorporate role restrictions. Their hard integration approach allows creation of immutable logs as well as secure and transparent process execution, however, it requires organizations to adapt their entire process management infrastructure and additionally involves significant operating costs due to the use of transaction fees on the Ethereum Platform.

A similar tool, Lorikeet, is presented by Tran et al. [16]. Like Caterpillar, Lorikeet includes the translation of BPMN models to Solidity smart contracts while also incorporating a registry data schema. They also describe a backend component to compile, deploy and interact with the generated smart contract, however low level implementation details are not addressed. Compared to Caterpillar, Lorikeet supports fewer BPMN elements but supports asset control and participation selection [3]. Conclusively, Lorikeet focuses on generating and deploying smart contracts in Solidity from BPMN models, however, it also requires organizations to adapt their existing BPMS leading to a high integration effort, and is not backend agnostic, since Solidity is not used by every blockchain.

Alves et al. [1] present a softer integration approach using the process orchestration platform Camunda and the permissioned blockchain network Hyperledger Fabric[10]. The processes are modeled and executed using the Camunda BPMS, while smart contracts on the Hyperledger network are used to register and manage generated information. Javascript clients and Hyperledger SDKs are used to integrate the smart contracts with the Camunda outputs. This contribution addresses the soft integration approach and creates immutable data logs of process executions. However, technical details of the implementation are not presented and evaluated. Furthermore, their approach cre-

[8] https://github.com/orlenyslp/Caterpillar, last access: 2024-04-09.

[9] https://ethereum.org/en/, last access: 2024-05-27.

[10] https://www.hyperledger.org/projects/fabric.

ates additional risks in introducing special proxy services between the process engine and the called services, which are responsible for communicating with the blockchain. Thus, again, tampering might occur before data arrives at the proxy service, and it is hard to confirm by the connected components.

Conclusively, the majority of contributions are theoretical or focused on hard integration, while there are only a few soft integration implementations. However, we argue that a soft integration approach has several advantages, which are presented in the following section.

3 Hard vs. Soft Integration

While integrating blockchain technology with BPMS has clear advantages (e.g., decentralized non-tamperable logging) and disadvantages (e.g., integration effort) regardless of how it is implemented, different integration approaches lead to additional advantages and disadvantages. This section presents advantages and disadvantages of using a soft integration approach, where modeling and execution is handled by a traditional BPMS, compared to a hard integration approach, where processes are stored and executed on the chain. The analysis of both approaches is summarized in Table 1.

Table 1. Advantages and Disadvantages of Soft and Hard Integration. *Improved* is in comparison with no blockchain integration, while *Strong* can be regarded as stronger than *Improved*

Soft Integration	Hard Integration
+ Immutable Execution Log	+ Immutable Execution Log
+ Improved Security	+ Strong Security
+ Improved Compliance Assurances	+ Strong Compliance Assurances
+ Advantages of BPMS (editable, etc.)	– Immutable processes
+ Low integration complexity	– High integration complexity
+ Low additional infrastructure costs	– New infrastructure costs
– Additional complexity in process engine	– Trusted triggers needed
– Engine/Communication needs to be secured	

3.1 Hard Integration

For the hard integration approach, the processes are stored and executed on the chain. A major advantage of blockchain technology, the immutable nature of the chain also leads to a major disadvantage for a hard integration approach: The immutable nature of the chain also means that deployed smart contracts, which in the hard integration approach represent the processes, can not be paused, edited nor continued without violating the non-tamperable control flow by allowing migration of the process into a new smart contract. For many domains, an executable business process requires many implementation iterations and a major advantage lies in how it can be easily extended to deal with changes and unforeseen circumstances. Furthermore, additional complexity is added by deploying processes on the chain, and organizations have to adjust their entire infrastructure to work with the chain. Finally, while a purpose of blockchain integration is to increase security and trust among process partners, a hard integration approach leads to additional security vulnerabilities since the triggers (or oracles) that use information from outside the chain have to be trusted.

The advantages of the hard integration approach are the creation of non-tamperable logs and better compliance assurances due to unchangeable processes as long as the triggers are trusted.

3.2 Soft Integration

For the soft integration approach, the processes are modeled, executed, and monitored with a traditional Business Process Management System. This means the advantages of a modern BPMS are kept: processes can be stopped, modified, and continued from their previous execution state. However, it still retains the main advantage of the hard integration approach since a non-tamperable log can be created on the chain. Furthermore, a soft integration approach also comes with additional advantages, most notably, the simpler and less costly integration. Furthermore, for a collaborative process, not all partners have to participate in the blockchain and smart contracts can still be used for independent compliance checking.

If private blockchains are used, the soft integration approach's main disadvantage is added complexity in the process or the process engine (depending on the exact implementation) for initializing the chain and passing info to a chain communication component or direct invocation of chain smart contracts. Furthermore, the process engine and any intermediate components between the process engine and blockchain can be regarded as new single points of failure that need to be secured. A final limitation of the soft integration approach is, that it does not directly enable tamper-proof execution, but only tamper-proof logging, which ensures non-repudiation and traceability.

For soft integration, we propose the two approaches depicted in Fig. 1. If a process engine supports creating transactions in a ledger (Fig. 1(a)), then every execution of an activity for a particular instance leads to a transaction ② that contains all the data that is sent to the implementation of the activity (e.g., a service). The service itself creates its own transaction ③, that contains both the received data and the data returned to the process engine. After receiving the data, the process engine again creates a transaction ④ with the data it received. ① represents an optional interaction between the network and the process engine, that potentially initializes the ledger.

The smart contract can check if the data has not been tampered with by having separate transactions ②–④. This approach, of course, assumes a secure connection between both the process engine and services with the blockchain network. For an attacker, it is arguably difficult to consistently tamper with multiple connections, i.e., the ones between a process engine and a service, as well as the ones to the blockchain network.

Figure 1(a) shows a blockchain-enabled process engine with both a blockchain-enabled service as described above as well as a non-blockchain-enabled service. Such legacy services continue to work fine, ② and ④ are stored in the ledger (i.e., the non-tamperable log), but consistency checks comparing sent and received data are only possible for a blockchain-enable service.

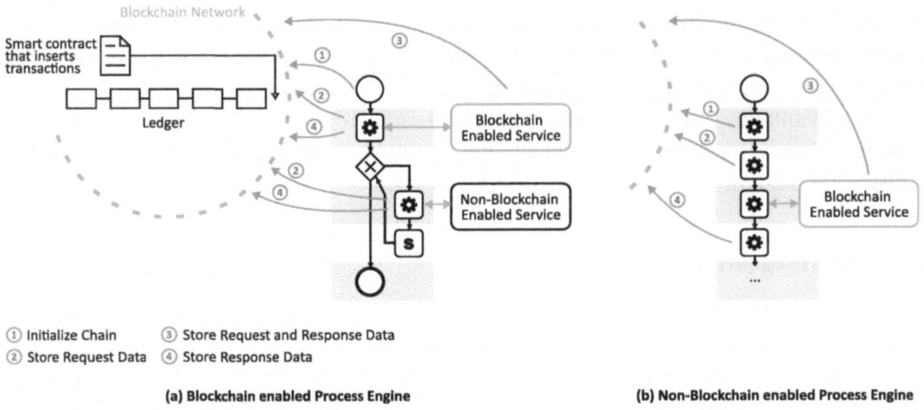

Fig. 1. Soft Integration With and Without Engine Support

Figure 1(b) shows the potential implementation for a non-blockchain-enabled process engine. Here additional tasks (potentially automatically generated for given process models), ensure the interactions ②-④ as described above. I.e., the activity from Fig. 1(a), is now represented by 4 activities. This inflates the process model but ensures the broadest possible compatibility.

4 Implementing a Soft Integration Approach

4.1 Blockchain Platform Selection

The first challenge for a soft integration approach is selecting a fitting blockchain platform. For an initial selection, 20 potential candidates are considered based on their academic and industry relevance as well as smart contract capabilities. The complete list of initial candidates can be found in Table 2. It's important to note that very high claimed transaction speeds are theoretical and often rely on unrealistic expectations, such as all participating nodes trusting a single validating node or an extremely high number of validating nodes.

To better structure the selection process, a series of key criteria were identified:

- **Public vs Private:** A public blockchain allows anyone to join and participate in the network, while in a private blockchain, access is restricted to a specific group of participants. For integrating blockchain technology with BPMS, a private blockchain is generally preferred since organizations have a higher level of trust among each other compared to the general public, and smaller chains with higher trust allow for the use of better-performing consensus mechanisms. A downside of private blockchains is that they have to be deployed, configured, and maintained in contrast to the most well-known and most studied blockchains (Bitcoin, Ethereum). In addition, a private blockchain is in general more vulnerable to attacks that require control of a certain number of nodes (e.g., over 51%), due to their smaller size. However, for soft integration, the participating actors are well-known to each other, which would make such an attack very difficult.

Table 2. Initial List of Potential Blockchain Platforms; Etherum and Hyperledger have been referenced before. For all platforms referenced below, the last access date is 2024-04-24

Platform	Type	Smart Contracts	Active Maint.	Strong Community	Enterprise Focus	Stated Max Transactions per Second
algorand.com	Public	Yes	Yes	Yes	No	7500
avax.network	Public	Yes	Yes	Yes	No	4500
bnbchain.org	Public	Yes	Yes	Yes	No	100
cardano.org	Public	Yes	Yes	Yes	No	1000
corechain.tech	Private	No	No	No	No	Not Available
eosnetwork.com	Public	Yes	Yes	Yes	No	10,000
Ethereum	Public	Yes	Yes	Yes	No	15
Hyperledger Besu	Private	Yes	Yes	Yes	Yes	400
Hyperledger Fabric	Private	Yes	Yes	Yes	Yes	2000
near.org	Public	Yes	Yes	Yes	No	2000
nemproject.github.io	Public	Yes	Yes	Yes	No	4000
neo.org	Public	Yes	Yes	Yes	No	10000
polkadot.network	Public	Yes	Yes	Yes	No	1000
consensys.net/quorum/	Private	Yes	Yes	Yes	Yes	750
r3.com/products/corda	Private	Yes	Yes	Yes	Yes	1700
ripple.com	Public	No	Yes	Yes	No	1500
solana.com	Public	Yes	Yes	Yes	No	2000
stellar.org	Public	No	Yes	Yes	No	4000
tezos.com	Public	Yes	Yes	Yes	No	170
zilliqa.com	Public	Yes	Yes	Yes	No	2800

- **Permissionless vs Permissioned:** In a permissionless blockchain, every participant can participate in all blockchain activities, including transaction validation and block creation. Meanwhile, in a permissioned blockchain, permissions can be assigned to different roles within the network, restricting access to certain activities. For a soft integration approach, we deem a permissioned blockchain to be more suitable since it allows for the creation of access roles, enabling more trust between organizations.
- **Consensus Mechanism:** The choice of consensus mechanism heavily depends on the previous two criteria. In public, permissionless networks, anyone can create and validate blocks, meaning the consensus mechanisms need to be stricter to prevent malicious behavior. Accordingly, in these open networks, the proof of work or proof

of stake mechanisms are widely used, while in permissioned networks, other mechanism, like the Practical Byzantine Fault Tolerance (PBFT), are regarded as sufficient. The choice of consensus mechanism influences several key factors, which are summarized in Table 3. It's important to note that several permissioned blockchains permit the use of different consensus mechanism and Byzantine Fault Tolerance is just one example of a deterministic consensus mechanism, however, the differences when comparing proof of work/stake to other deterministic consensus mechanism are largely consistent.

– **Academic and Industry Relevance:** Finally, the chosen blockchain technology needs to be well-known and widely used to maximize acceptance by organizations participating in process orchestration. This criterion was evaluated by analyzing community size, media presence, enterprise focus, and the popularity of smart contracts for each platform.

Table 3. Comparison of PoW, PoS, and PBFT Consensus Mechanisms

Feature	Proof of Work (PoW)	Proof of Stake (PoS)	PBFT
Mining/Validation	Computational mining	Staking of assets	Rotating primary node
Finality	Probabilistic	Probabilistic	Deterministic
Throughput	Lower	Moderate	Higher
Security	High	High	High in smaller networks
Suitability	Permissionless networks	Permissionless networks	Permissioned networks

Having decided on these key criteria, five suitable prominent distributed ledger platforms can be identified (out of the initial 20 candidates): Ethereum, Hyperledger Besu[11], Hyperledger Fabric, Quorum[12], and R3 Corda[13]. The analysis results of the five final candidates and additional scoring based on the criteria to decide on a final choice can be found in Table 4. The scores range from 1-10 and are derived based on the specific requirements for a soft integration with BPMS and the available consensus mechanism. The scoring would be different for different use cases and depending on the choice of consensus mechanism. Based on our scoring, we found the Hyperledger Fabric platform to be the most suitable for a soft integration approach.

[11] https://www.hyperledger.org/projects/besu, last access: 2024-05-28.

[12] https://consensys.io/, last access: 2024-04-10.

[13] https://r3.com/products/corda/, last access: 2024-04-10.

Table 4. Scoring of final 5 candidates for soft integration of blockchain platform with BPMS

Platform	Ethereum	Besu	Fabric	Quorum	R3 Corda
Type	Public	Both	Private	Private	Private
Permissioned	No	Both	Yes	Yes	Yes
Security	9/10	8/10	8/10	8/10	8/10
Privacy	5/10	8/10	9/10	8/10	9/10
Performance	4/10	7/10	8/10	6/10	6/10
Scalability	4/10	7/10	8/10	6/10	6/10
Cost	5/10	8/10	9/10	8/10	9/10

4.2 Solution Architecture

Our proposed soft integration implementation aims to devise a reliable method that ensures the entire execution history of processes, including API calls, responses, and data exchanges, is securely logged on the blockchain without significant integration complexity. For this purpose, two different architectural approaches were developed after testing Hyperledger Fabric's functionalities: (1) The REST API-Based Approach and (2) the direct-invocation Approach. For complete implementation details, refer to the GitHub page[14].

REST API-Based Approach. The REST API-Based approach consists of several key components: The Hyperledger Fabric (test) network, a smart contract, a REST API server, and the client BPM system. The idea is to use the API server as middleware to communicate between the BPMS and the blockchain, creating a flexible, language-agnostic, and highly decoupled architecture. Since existing BPM systems regularly call API requests for process execution, integrating the API-Based Approach with the processes or BPMS (i.e., Fig. 1) is simple, as knowledge of how to invoke REST services securely is widespread. An overview of the approach can be seen in Fig. 2.

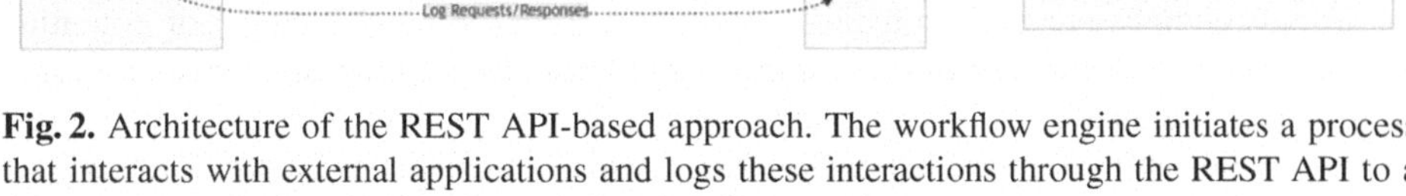

Fig. 2. Architecture of the REST API-based approach. The workflow engine initiates a process that interacts with external applications and logs these interactions through the REST API to a smart contract deployed on the Hyperledger Fabric network.

However, while this approach is extremely easy to integrate with various process management systems, it also adds a new single point of failure in the REST API server

[14] https://github.com/thomasvloo/fabric-testnet, last accessed: 2024-04-10.

similar to [1]. Furthermore, the server adds additional latency to the execution, which is already a major downside of integrating blockchains with BPMS. The direct-invocation approach aims to solve these problems.

Direct-Invocation Approach. The direct-invocation approach facilitates a direct connection between BPMS and Blockchain, bypassing the REST API. Since access to the blockchain is not abstracted through the API interface, integration with the process engine is more challenging. The process engine needs to be extended such that every activity also invokes a blockchain call before and after execution with sufficient authentication. This strong coupling also means that any changes to the BPMS or the process might require adjustment to the integration code. However, removing the middleware between the engine and the blockchain network can enhance the performance, and security risks related to adding an additional component (i.e., the API server) can be minimized. Figure 3 shows an overview of the approach.

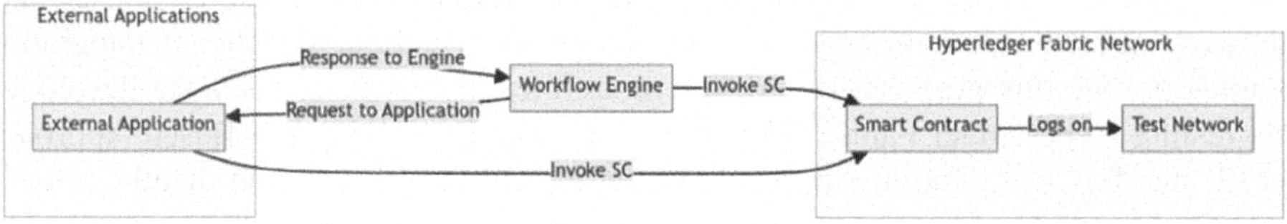

Fig. 3. Architecture of the direct-invocation approach. The workflow engine initiates a process that interacts with external applications and logs these interactions by directly invoking a smart contract deployed on the Hyperledger Fabric network.

While this approach offers the desired enhanced performance, security, and reduced latency, it requires additional integration and maintenance effort since the different systems are now strongly coupled and language-specific. An extensive evaluation was conducted to evaluate the performance differences, which is presented in the following section.

5 Performance Evaluation

As process engines are used in many domains that rely on a high number of potential transactions, it is important to explore potential limitations to the number of transactions that can be made. For example, in the manufacturing domain, processes interact at high velocities and produce high data volume (which translates to transaction payload volume).

5.1 Methodology

The performance of both the direct-invocation approach and the API-based approach was evaluated by simulating API requests to the Hyperledger Fabric test-network. The objective was to measure the throughput in terms of transactions per second (TPS) when faced with increasingly larger batches of transaction submissions. In addition, the payload sizes were varied between 1000 and 10000 characters to simulate the variance in data size typically observed within production environments. Other aspects, such as integration complexity, flexibility, and security, were considered qualitatively through discussions with domain experts.

5.2 Performance Evaluation

The performance evaluation results can be seen in Fig. 4. While the REST-API-based approach maintains a relatively consistent TPS rate, the direct-invocation approach shows consistently higher throughput, albeit with higher variability. These results reflect the prior expectations since the latency added by the middleware is removed in the directly connected direct-invocation approach. The results also point towards a maximal throughput of around 170 TPS when using the hyperledger test network, however this is likely a limit of the hyperledger test network, considering that hyperledger themselves state a maximal throughput of 2946.7 TPS[15]. That being said, 170 TPS can be considered more than sufficient for process orchestration in many domains.

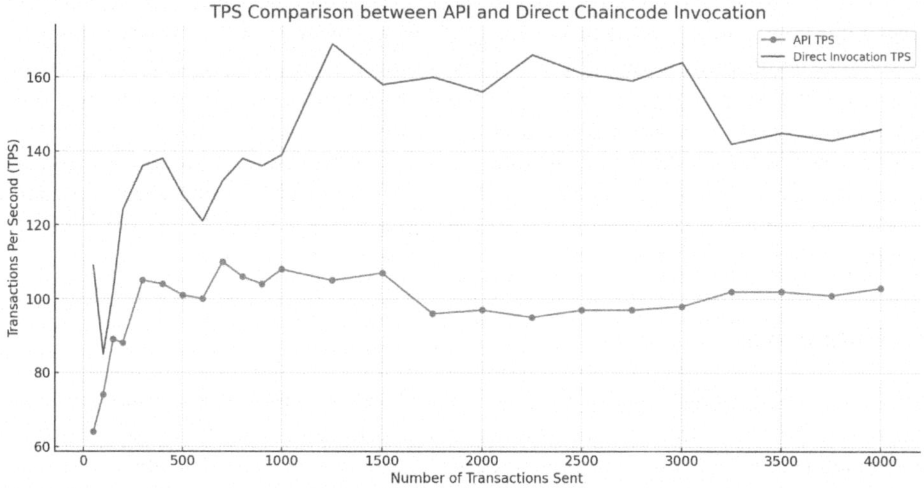

Fig. 4. Transactions Per Second (TPS) comparison between API and Direct Chaincode Invocation approaches.

5.3 Complexity, Flexibility, and Security

The main benefit of integrating blockchain technology with business process engines is to enable decentralization with enhanced security, transparency, and trust amongst participating organizations. Furthermore, a blockchain-capable BPMS opens many possibilities for potential decentralized use cases and application scenarios. The main reason

[15] https://www.hyperledger.org/blog/2023/02/16/benchmarking-hyperledger-fabric-2-5-performance, last access: 2024-05-28.

for a soft-integration approach is to reduce the barrier of entry by having very little integration complexity. Accordingly, these factors are analyzed in this section. Table 5 summarizes the results of this analysis.

Table 5. Comparison

	REST-API	Direct-Invocation
Ease of Integration	+	–
Flexibility	+	–
Performance	–	+
Security	=	+

The direct-invocation approach has better performance and as discussed in Sects. 2 and 4 is also more secure since it has fewer failure points. Here it is important to note, that the security of the REST-API approach can be extended with well-known methods, however, this could in turn lead to a further decrease in performance. Meanwhile, when it comes to integration complexity and flexibility, the REST-API approach is preferred since it has low coupling, is BPMS technology stack agnostic, and is already supported by some existing process engines, while the direct-invocation approach needs to be adjusted to the used engine and requires significant knowledge of blockchain technology and smart contract coding. Overall, these results stress that the correct soft integration approach depends on the application domain. Furthermore, both soft integration approaches are still significantly easier and less costly than a hard integration approach.

We found that the transaction payload size (up to sizes of 100's of MiB) has no unexpected impact on TPS.

6 Conclusion

While blockchain technology has become widely known both in industry and academia due to its influence on the financial domain, there has been no widespread integration of blockchain technology with business process management systems. We analyze related work and identify the high barrier of entry in transitioning the existing BPMS an organization is already using onto the blockchain as well as the disadvantages of immutable processes as the main reasons for this low adaption rate. Therefore, we propose a soft integration approach, which allows for the creation of immutable execution logs with low integration complexity.

Our work aims to use blockchain technology to create distributed, non-tamperable process execution logs without significant integration efforts. We discuss the advantages and disadvantages of using a soft integration approach, where the process is modeled, executed, and monitored using a traditional BPMS, compared to a hard integration approach, where the process is executed on the chain in the form of a smart contract. By logging the entire execution history, including API calls, responses, and data exchanges on a permissioned blockchain, we create an immutable execution log that can be used as a reliable basis for compliance management, process mining, and optimization and also improves trust amongst organizations participating in the process. Furthermore, security is improved since malicious actors can not hide tampering with data that is sent over the wire during process execution (which leads to manipulated process logs). In addition, existing functionalities and tools, such as modeling, execution, error handling, and monitoring/dashboarding, can still be used. We analyzed 20 different blockchain platforms and found 5 suitable platforms on which we used an extensive scoring system to determine the Hyperledger Fabric platform as most suitable for our research. The Cloud Process Execution Engine (CPEE) was selected for process execution and connected to Hyperledger test networks through both, a REST server as well as direct chaincode invocation. Both the REST API and the direct invocation approach are easier to integrate and the interaction costs less compared to a hard integration approach. Through a quantitative evaluation, we show that the performance of both approaches is sufficient for process execution. We also discuss strengths and weaknesses of both implementation strategies. While the direct invocation approach has noticeably higher performance, it requires significantly more integration effort than the REST API-based approach. All created software artifacts are provided for future research and testing.

A limitation of the soft integration approach is, that it does not directly enable tamper-proof execution. However, immutable processes, as a result of hard integration, have considerable disadvantages due to their static nature as discussed in Sect. 3. In addition, tamperproof execution is difficult to achieve even with a hard integration approach for many processes that rely on information from outside of the chain provided via oracles. If tamperproof execution is still required for well-established processes with little changes and reliance on outside information, related work [9, 16] can be used to integrate a BPM to Solidity Translator with the process engine. A further limitation is that the tests were done using the Hyperledger test network. Accordingly, throughput values measured during the performance evaluation should be seen as lower bounds rather than upper bounds of the throughput possible with the Hyperledger Fabric platform. However, the observed performance was already sufficient for process orchestration for typical application domains (e.g., we never saw more than 300 TPS for real-world processes in the manufacturing domain).

In the future, we plan to test the prototypical implementations in a production environment and demonstrate mechanisms that automatically detect deviations between immutable log, client-side log, and expected execution behavior to detect compliance violations or tampering. Furthermore, moving away from the Hyperledger test network to a custom, production-grade network would allow for a more accurate evaluation.

References

1. Alves, P.H.C., et al.: Exploring blockchain technology to improve multi-party relationship in business process management systems. In: ICEIS, no. 2, pp. 817–825 (2020)
2. Clohessy, T., Acton, T., Rogers, N.: Blockchain adoption: technological, organisational and environmental considerations. In: Treiblmaier, H., Beck, R. (eds.) Business Transformation through Blockchain, pp. 47–76. Springer, Cham (2019). https://doi.org/10.1007/978-3-319-98911-2_2

3. Di Ciccio, C., et al.: Blockchain support for collaborative business processes. Informatik Spektrum **42**, 182–190 (2019)
4. Garcia-Garcia, J.A., Sánchez-Gómez, N., Lizcano, D., Escalona, M.J., Wojdyński, T.: Using blockchain to improve collaborative business process management: systematic literature review. IEEE Access **8**, 142312–142336 (2020). https://doi.org/10.1109/ACCESS.2020.3013911
5. Grisold, T., Mendling, J., Otto, M., vom Brocke, J.: Adoption, use and management of process mining in practice. Bus. Process. Manag. J. **27**(2), 369–387 (2021)
6. Kitchenham, B., Charters, S.: Guidelines for performing systematic literature reviews in software engineering, vol. 2 (2007)
7. Li, M., Huang, G.: Blockchain-enabled workflow management system for fine-grained resource sharing in e-commerce logistics. In: 2019 IEEE 15th International Conference on Automation Science and Engineering (CASE), pp. 751–755 (2019). https://doi.org/10.1109/COASE.2019.8843250
8. López-Pintado, O., Dumas, M., García-Bañuelos, L., Weber, I.: Dynamic role binding in blockchain-based collaborative business processes. In: Giorgini, P., Weber, B. (eds.) CAiSE 2019. LNCS, vol. 11483, pp. 399–414. Springer, Cham (2019). https://doi.org/10.1007/978-3-030-21290-2_25
9. Lopez-Pintado, O., Garcia-Banuelos, L., Dumas, M., Weber, I.: Caterpillar: a blockchain-based business process management system. In: International Conference on Business Process Management (2017). https://api.semanticscholar.org/CorpusID:3629238
10. Mangler, J.: Datastream xes extension: embedding IoT sensor data into extensible event stream logs. Future Internet **15**(3), 109 (2023)
11. Mangler, J., et al.: From internet of things data to business processes: challenges and a framework (2024)
12. Mendling, J., et al.: Blockchains for business process management-challenges and opportunities. ACM Trans. Manag. Inf. Syst. (TMIS) **9**(1), 1–16 (2018)
13. Mercenne, L., Brousmiche, K.L., Hamida, E.B.: Blockchain studio: a role-based business workflows management system. In: 2018 IEEE 9th Annual Information Technology, Electronics and Mobile Communication Conference (IEMCON), pp. 1215–1220. IEEE (2018)
14. Nakamoto, S.: Bitcoin: a peer-to-peer electronic cash system. Decentralized Bus. Rev. 21260 (2008)
15. Stiehle, F., Weber, I.: Blockchain for business process enactment: a taxonomy and systematic literature review. In: Marrella, A., et al. (eds.) BPM 2022. LNCS, pp. 5–20. Springer, Heidelberg (2022). https://doi.org/10.1007/978-3-031-16168-1_1
16. Tran, A.B., Lu, Q., Weber, I.: Lorikeet: a model-driven engineering tool for blockchain-based business process execution and asset management. In: BPM (Dissertation/Demos/Industry), pp. 56–60 (2018)

Robotic Process Automation (RPA) Forum

Preface of the Robotic Process Automation (RPA) Forum

Robotic Process Automation (RPA) is a maturing technology in the field of Business Process Management (BPM). RPA allows organizations to automate high-volume and repetitive tasks – also referred to as routines – performed by human users. The enactment of these routines is emulated by means of a software robot that works on the applications' user interfaces (UIs) in the same way as the original human operators did. Examples of those routine tasks include data transfer between applications, automated email query processing, and collation of payroll data from different sources.

At this point in time, RPA has reached a certain level of technological maturity and organizational adoption. This means researchers now have the opportunity to examine RPA from a broader perspective, encompassing three distinct viewpoints:

1. Low-code automation: RPA is part of a development towards low-code automation aimed at building and automating processes with off-the-shelf software solutions that do not require extensive programming skills.
2. RPA for smart automation: RPA can be combined with other technologies, such as process mining, OCR, or chatbots, with the goal of a more flexible, holistic process automation.
3. AI-empowered RPA: a new level of automation targeted to inject AI, ML or LLM capabilities to identify routine tasks from multiple session recordings that are suitable for automation. This goes beyond conventional rule-based approaches, allowing for a more intelligent and adaptable automation strategy.

Topics like these are studied by a broad range of researchers across communities. While computer scientists might focus on the technological aspects, management researchers study the impact of RPA on organizations. Information systems scholars might focus on the socio-technological nature.

The RPA Forum of the 22nd International Conference on BPM aims to provide a forum for this broad range of researchers to exchange ideas, fuel new research, and start corresponding collaborations. In total, the forum attracted 9 international submissions on a diverse range of topics around RPA. All submissions were reviewed by three Program Committee members which led to an acceptance of five papers. These five papers show the full range of topics mentioned above.

Martínez-Rojas et al. propose a novel technique that, first, integrates embeddings from both screenshots and screen text obtained through OCR and, second, clusters the UI log events using these combined features to identify the activity. Their results show that this method enhances activity identification, outperforming current state-of-the-art techniques in RPA, and demonstrates promising improvements in accuracy and reliability.

François et al. explore artifact reuse in no-code and low-code environments, where citizen developers are not typically engaged in traditional IT roles. To achieve this, authors review the literature on automation reuse to identify current challenges. Subsequently, they employ Transaction Cost Theory (TCT) as a basis to outline design principles that organizations can adopt to enhance the reusability of their artifacts.

Völker et al. draw inspiration from the related fields of software programming and business process modeling to introduce complexity metrics for RPA bot models. These allow to objectively measure and compare the complexity of automation workflows, and can thus, for example, provide an indication of where to start refactoring models in the bot repository.

Hohenadl et al. conduct a semi-structured interview study with RPA professionals to explore the factors influencing developers' decisions when selecting and translating manual actions into automated processes. The findings from the interview have been the basis to show their implications on RPM tools and to give guidance for the improvement of RPM tools.

Finally, Hohenadl et al. survey and structure the possible types of user interactions that can be recorded by analyzing the literature on Robotic Process Mining (RPM) and the related field of user behavior mining (UBM). They foster a unified understanding of the key terms and concepts in RPM by proposing a conceptualization that draws upon ontologies of software, user interfaces, and RPA to embed this comparably new research area. To explore the practical applicability of the conceptualization, authors conclude with a number of concrete usage scenarios that strengthen the design and development of RPA bots.

We would like to thank all authors who submitted their work to the RPA forum. Moreover, we also express our gratitude to the Program Committee who carefully reviewed the submission. Last, we thank the BPM 2024 chairs and organizers for their work on the overall conference and generous chance to host the RPA forum.

Organization

Program Chairs

Simone Agostinelli	Sapienza University of Rome, Italy
Daniel Amyot	University of Ottawa, Canada
Henrik Leopold	Kühne Logistics University, Germany

Program Committee

Aleksandre Asatiani	University of Gothenburg, Sweden
Bernhard Axmann	Technische Hochschule Ingolstadt, Germany
Adela del Río Ortega	University of Seville, Spain
José González Enríquez	University of Seville, Spain
Peter Fettke	German Research Center for Artificial Inteilligence (DFKI) and Saarland University, Germany
Hannu Jaakkola	University of Tampere, Finland
Christian Janiesch	TU Dortmund University, Germany
Andrés Jiménez-Ramírez	University of Seville, Spain
Abderrahmane Leshob	UQAM, Canada
Fabrizio Maria Maggi	Free University of Bozen-Bolzano, Italy
Andrea Marrella	Sapienza University of Rome, Italy
Massimo Mecella	Sapienza University of Rome, Italy
Dan O'leary	University of Southern California, USA
Teijo Peltoniemi	University of Turku, Finland
Petr Průcha	Technical University of Liberec, Czech Republic
Jana-Rebecca Rehse	University of Mannheim, Germany
Yara Rizk	IBM Research Cambridge, USA
Rehan Syed	Queensland University of Technology, Australia
Lucinéia Heloisa Thom	Federal University of Rio Grande do Sul, Brasil
Inge van de Weerd	Utrecht University, The Netherlands
Maximilian Völker	University of Potsdam, Germany
Moe Thandar Wynn	Queensland University of Technology, Australia

From Screenshots to Process Models: Improving Activity Identification Through Screen Text

A. Martínez-Rojas(✉), J. L. Alonso-Rocha, A. Jiménez-Ramírez, and J. G. Enríquez

Department of Computer Languages and Systems, University of Seville, Avenida Reina Mercedes, S/n., 41012 Seville, Spain
{amrojas,jalonso2,ajramirez,jgenriquez}@us.es

Abstract. The analysis for executing Robotic Process Automation (RPA) projects increasingly relies on monitoring user activities through Robotic Process Mining (RPM) techniques. Traditional approaches capture direct information using loggers that capture UI logs, i.e., sequences of events that include data from (1) the keyboard, (2) the mouse, and (3) the application elements, such as its name, the Excel cell, the clicked button, etc. Although the latter is highly relevant for identifying the activity that is being performed, this information is not accessible in virtualized environments; only screenshot data is available. This limitation necessitates activity identification based on screenshots alone. A significant challenge with this method is its sensitivity to minor interface changes, such as different zoom levels or notifications, which can cause detection failures. To address this, we propose a novel approach that, first, integrates embeddings from both screenshots and screen text obtained through OCR and, second, clusters the UI log events using these combined features to identify the activity. Our results show that this method enhances activity identification, outperforming current state-of-the-art techniques, and demonstrates promising improvements in accuracy and reliability.

Keywords: Robotic Process Automation · UI Logs · Screen-based Task Mining · Activity Identification

1 Introduction

Robotic Process Automation (RPA) uses graphical user interfaces (GUIs) to automate and integrate systems, allowing software robots to mimic human

This research was supported by the EQUAVEL project PID2022-137646OB-C31, funded by MICIU/AEI/10.13039/501100011033 and by FEDER, UE; the DISCOVERY project (2021/C005/00148631), funded by Unión Europea NextGeneration EU and "Plan de Recuperación, Transformación y Resiliencia" of the Ministry of Economic and Digital Transformation; and the grant FPU20/05984 funded by MICIU/AEI/10.13039/501100011033 and by FSE+.

C. Di Ciccio et al. (Eds.): BPM 2024, LNBIP 527, pp. 125–137, 2024.
https://doi.org/10.1007/978-3-031-70445-1_8

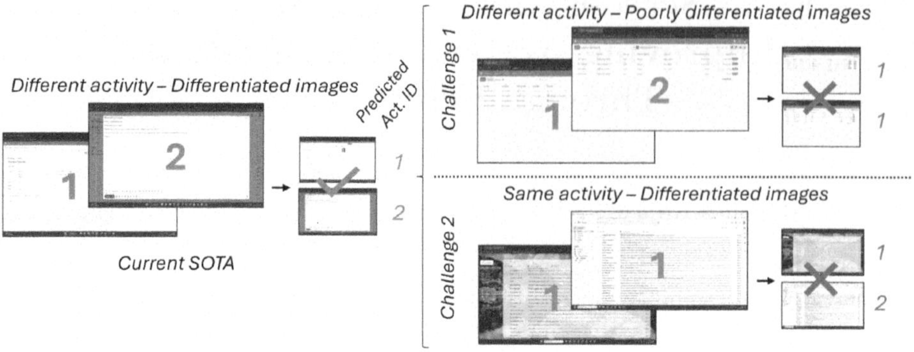

Fig. 1. Motivation problem.

interactions with computer applications [1]. It has significantly expanded across various industries, particularly in Business Process Outsourcing [4]. This environments often involve multiple enterprises interacting within shared information systems, requiring high levels of security through secured or virtualized connections [8]. These connections pose distinct challenges for process analysis techniques, especially Robotic Process Mining (RPM) or Task Mining (TM), which focus on recording user interaction activities based on observable events and actions at the user interface (UI) level [7].

Traditional RPM methods identify UI activities directly from user interactions with applications through action loggers [3,9]. In virtualized environments, however, access is limited to mouse and keyboard inputs and screenshots, which lack structured information. This complicates the collection of activity data, requiring alternative approaches for accurate activity identification.

Jiménez-Ramírez et al. [8] propose the identification of activities (i.e. the Activity ID) based on visual information, by grouping events using screenshot fingerprints (i.e., a hash). However, this approach has several challenges (cf. Fig. 1). First, activities with similar screen appearances but different actions can be confused, relying solely on-screen fingerprints. Second, the same activities can be misidentified due to small GUI changes, such as color adjustments, zoom levels, or minor GUI customizations. This issue is especially pronounced in multi-user contexts, where recordings come from different computers and users that perform the same processes.

To address these challenges in virtualized environments, this paper poses the following research question (**RQ**): *Does the combination of textual with visual information on the screen improve activity identification in multi-user contexts?* This RQ aims to explore whether including the extracted text from the screen may enhance activity identification when combined with the raw visual information.

Therefore, the rest of the paper is organized as follows. Section 2 outlines the theoretical background. Section 3 details our approach that combines visual and textual information on the screen to identify the activity. Section 4 presents our evaluation methodology and results. Section 5 discusses the findings and implications. Section 6 reviews similar proposals in the literature. Finally, Sect. 7 shows the conclusions and future directions for this research.

2 Background

2.1 Robotic Process Mining and UI Logs

UI logs are files used to store information gathered from recording processes during human-machine interaction. This UI log introduces attributes that describe actions during interactions with the information system. These attributes can be classified into three groups: process attributes (e.g., timestamp, event ID), application attributes (e.g., app name, screenshot), and user input attributes (e.g., clicks, keystrokes) [2,12].

One of the disciplines that leverages this UI log as input is RPM. RPM is a family of techniques to discover repetitive routines that can be automated using RPA technology. RPM analyzes interactions between one or more workers and one or more software applications during the performance of one or more tasks in a business process [6]. In our previous work [10], we define a framework to apply RPM in virtualized environments. This approach is available under the name ScreenRPA[1] and illustrated in Fig. 2. It starts with monitoring user behavior to obtain a UI log of the process (Fig. 2, step 1). Each event recorded in the UI log has an assigned screenshot. From these screenshots, a Screen Object Model (SOM) is extracted (Fig. 2, step 2.1), which is a data structure that represents the visual and the text information on a screenshot, and hierarchical relationship between these UI components [11]. This SOM detection implies Computer Vision and Optical Character Recognition (OCR) [15] techniques to extract all UI components and plain text, respectively.

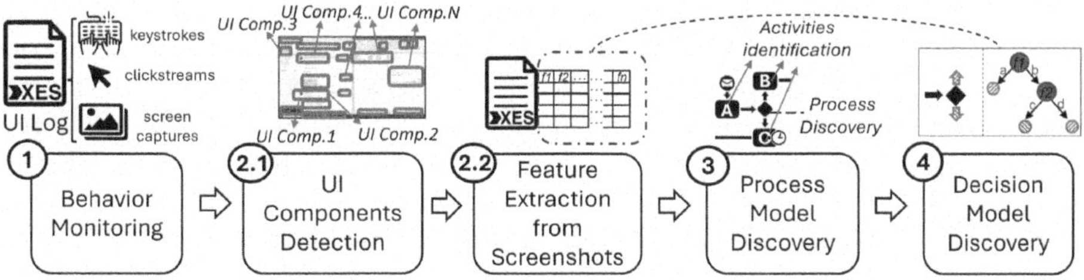

Fig. 2. Application of activity identification in RPA (extracted and edited from [10]).

Then, UI Feature Extractors (UIFE) are applied (Fig. 2, step 2.2) to extract information from the SOM, i.e. structured screenshots [10]. This additional information is used to perform process model discovery (Fig. 2, step 3). This is the step where our proposal is contextualized since *activity identification* techniques are applied to associate each UI log event with an activity. Thus, they can be used by *process discovery* algorithms to obtain the process model. Finally,

[1] ScreenRPA Framework: https://github.com/RPA-US/screenrpa.

decision model discovery techniques are applied (Fig. 2, step 4) to extract the rules affecting decision points in this process model.

2.2 Image Hashing and the CLIP Model

There are currently two main approaches for extracting information from screenshots. The first is image hashing, a traditional technique for comparing screenshots through their hashes. Jiménez-Ramírez et al. [8] employ it to calculate screenshot similarity using Hamming distance, extracting the fingerprint of screenshots in two real business processes.

An alternative method is the use of the CLIP (Contrastive Language-Image Pre-training) model by OpenAI, introduced by Radford et al. [13]. CLIP processes images and text together, generating compatible vector representations. This solution simplifies experimentation by avoiding the errors associated with using separate models for image and text feature extraction. CLIP has proven to be a robust and efficient approach for both image processing and text modeling.

3 Approach

This section outlines our approach for identifying activities based on the content of screenshots.

Our method (cf. Algorithm 1) assumes the context detailed in Sect. 2.1 to integrate visual information, such as visual-level data with textual information from the screenshot, to create a one-dimensional feature vector, i.e. embeddings, for each event, which is subsequently clustered.[2] Initially, a copy of the UI log to be analyzed is created, and an empty list for the embeddings associated with each event is initialized (cf. Algorithm 1 lines 1–2). For each event in the UI log (cf. Algorithm 1 line 3), the following steps are performed.

First, the screenshot content is structured by extracting the SOM (cf. Algorithm 1 line 4), capturing both the UI component hierarchy and associated text. Next, a text feature extractor is applied to the structured SOM, to obtain the specific text to combine in the embedding (cf. Algorithm 1 line 5). This text could comprise anything from all on-screen text to text from specific UI components, such as application headers or the browser's tabs. Subsequently, the function `textImageToEmbedding` combines the extracted text with the screenshot image using the CLIP method [13] to obtain an embedding. To merge them, weights are assigned to each type of information to include more features from the textual or visual part (cf. Algorithm 1 line 6). The resulting embedding is appended to our list (cf. Algorithm 1 line 7).

Once all events have been processed, a clustering algorithm is applied to the embeddings to group events by their features (cf. Algorithm 1 line 9). Various clustering methods can be implemented; in this proposal, we use hierarchical agglomerative clustering from the Scikit-learn library.[3] The optimal number of

[2] Approach source code in: https://github.com/RPA-US/processdiscovery.

[3] Agglomerative Clustering Scikit-learn: https://scikit-learn.org/stable/modules/generated/sklearn.cluster.AgglomerativeClustering.html.

Algorithm 1. Activity Identification based on screenshots

```
Require: UI log uiLog, UIFE textExtractor, Float imageWeight, Float textWeight
Ensure: UI log processedUiLog
processedUiLog ← uiLog.copy()
eventsAsEmbeddings ← [ ]
for all event ∈ uiLog do
    SOM som ← detectUIElements(event.screenshotPath)
    String text ← applyTextFeatureExtractor(som)
    List[Float] eventEmb ← textImageToEmbedding(text, imageWeight, textWeight)
    eventsAsEmbeddings ← eventsAsEmbeddings + [eventEmb]
end for
processedUiLog["activity_id"] ← clusterEvents(eventsAsEmbeddings)
return processedUiLog
```

clusters is determined by testing between 2 and 10 clusters and selecting the one with the best Silhouette score based on [14]. Each event is assigned a cluster ID, which represents the activity ID of the event. Finally, the UI log is returned with the activity ID associated with each event, which is indicated as an additional attribute.

4 Evaluation

This section presents the results of activity identification experiments in various processes. We evaluate different settings of our approach and compare them with the image hashing method to benchmark against the state-of-the-art [8].[4] Sect. 4.1 outlines the evaluation setup, while Sect. 4.2 analyzes the results.

4.1 Set-up

For the evaluation, two processes inspired by real process models were used. From each real process model, two synthetic UI logs were generated, containing 58 events representing 8 different activities. This content resulted from executing multiple instances of the same process.

The first process is called **invoice resolution**. It is based on a process model from a real payment management context within Cenit[5], an IT Spanish consultancy company. The process involves 8 activities: (1) initial view of the invoices module, (2) review of the invoice list, (3) list of customer payments, (4) record customer payments, (5) review of customer payments, (6) resolve pending customer invoices, (7) navigate between customers, and (8) review customer details. It uses only screenshots from Odoo[6] views.

The second process is called **payment notification**. It is the same process as the real case study analyzed previously in [10]. The process consists of the

[4] Evaluation input and output data: https://doi.org/10.5281/zenodo.11368319.

[5] Cenit S.L: https://www.cenitcon.com/.

[6] Odoo is an open-source ERP system accessible at https://odoo.com.

following activities: (1) review the email inbox, (2) open an email, (3) navigate between customers, (4) review customer details, (5) notify customers of overdue payments, (6) resolve pending customer invoices, (7) open notification ticket for customers, and (8) delete a customer.

These processes have been instantiated as UI logs with different properties to evaluate the proposal in different contexts that we call *problems*:

- **Problem 1: Invoice resolution - Customer details with ID.** The screenshots of this UI log represent different activities that are *poorly differentiated* visually, e.g., lists of invoices and payments which are quite similar in Odoo. This similarity suggests that the screenshots belong to the same clusters, yet it is difficult to distinguish between them. This UI log is characterized by the fact that in the eighth activity (review customer details), the text contained in the UI component of the browser's tab is the identifier of each customer, e.g. *Azure Interior - Brandon.*
- **Problem 2: Invoice resolution - Customer details generalized.** This UI log differentiates itself from the previous one because all customer identifiers of the eighth activity have been generalized to *Customer View*, which represents the text in the browser's tab. This was done looking to analyze how event clustering is performed when this browser's tab is not identifying the case object, i.e. not identifying the customer.
- **Problem 3: Payment notification - Single user recording.** This UI log includes well-differentiated screenshots originating from the recording of a single user, i.e., the same computer. Therefore, the screenshots representing the same activities retain a GUI customization, maintaining a similar color scheme and appearance for the different cases in the UI log.
- **Problem 4: Payment notification - Multi-user recording.** In this UI log the key difference comes from the recording of two different users, i.e. two different computers, which means differences in the GUI. Specifically, one user displays the browser's content at normal size, while the other has a custom zoom of 150%.

Based on these problems, we address the study of the **RQ**: *Does the combination of textual with visual information on the screen improve activity identification in multi-user contexts?* For this analysis, we vary the following parameters to execute specific configurations of each problem:

- **Technique**: indicates the technique to be used to extract the feature vector or embedding that will subsequently be clustered. In the case of *Hash*, boolean vectors, and in the case of *Our approach*, it is obtained as indicated in Algorithm 1.
- **Text**: denotes the feature extractor passed as the *textExtractor* parameter to Algorithm 1. For *Hash*, no value is assigned. In the case of *All On-screen*, all extracted text from the screen is used, extracting text from all UI components via the SOM, i.e., all on-screen text. For *Browser's Tab*, only the visible text string of the browser's tab is used, with the SOM extracting text from UI components of type Browser's tab.

- **Weights:** refers to the values assigned to the image weight and text weight input parameters, respectively, within Algorithm 1. In the case of *Hash*, only the image is considered by definition, hence the weights are set (100, 0), following the format $(ImageWeight, TextWeight)$.

Setting the aforementioned parameters, executions are designed for the *Hash* and our technique. For ours, weights are ranged from (100,0) to (0,100), decreasing by 0.2 increments, resulting in a total of 15 configurations for each Problem: 1 for *Hash* technique, 7 for ours using *All On-screen* as text, and 7 for ours using *Browser's Tab* as text. Thus, each execution will be evaluated with the following classification metrics:

- **Precision**: is the ratio of correctly identified events for a specific activity to the total number of events identified as belonging to that activity.
- **Recall**: is the ratio of correctly identified events for a specific activity to the total number of events that should have been identified as belonging to that activity, i.e. *ground truth.*
- **F1-Score**: is the harmonic mean of precision and recall, balancing the relative contribution of precision and recall.

4.2 Results

This section presents the results obtained for each problem (cf. Table 1). This evaluation has been executed on an Intel Core i7-13700KF 3.40 GHz CPU, 32 GB of RAM, and Windows 10 as the operative system. An analysis of the results for each process evaluated is provided below.

Table 1. Evaluation results. Weights follow the format (ImageWeight,TextWeight).

Problem	Technique	Text	Weights	F1-Score	Precision	Recall
Problem 1	Hash	–	(100,0)	0.66	0.75	0.62
	Our approach	All on-screen	(100,0)	0.67	0.72	0.67
			(80,20)	0.67	0.72	0.67
			(60,40)	0.85	0.89	0.83
			(50,50)	0.85	0.89	0.83
			(40,60)	0.85	0.89	0.83
			(20,80)	0.83	0.89	0.80
			(0,100)	0.83	0.89	0.80
	Our approach	Browser's Tab	(100,0)	0.67	0.72	0.67
			(80,20)	0.81	0.89	0.78
			(60,40)	0.85	0.89	0.83
			(50,50)	0.85	0.89	0.83
			(40,60)	0.85	0.89	0.83
			(20,80)	0.85	0.89	0.83
			(0,100)	0.85	0.89	0.83

(*continued*)

Table 1. (*continued*)

Problem	Technique	Text	Weights	F1-Score	Precision	Recall
Problem 2	Hash	–	(100,0)	0.66	0.75	0.62
	Our approach	All on-screen	(100,0)	0.67	0.72	0.67
			(80,20)	0.67	0.72	0.67
			(60,40)	1.00	1.00	1.00
			(50,50)	0.85	0.89	0.83
			(40,60)	0.84	0.89	0.81
			(20,80)	0.85	0.89	0.83
			(0,100)	0.85	0.89	0.83
	Our approach	Browser's Tab	(100,0)	0.67	0.72	0.67
			(80,20)	0.81	0.89	0.78
			(60,40)	1.00	1.00	1.00
			(50,50)	1.00	1.00	1.00
			(40,60)	1.00	1.00	1.00
			(20,80)	1.00	1.00	1.00
			(0,100)	1.00	1.00	1.00
Problem 3	Hash	–	(100,0)	0.72	0.75	0.71
	Our approach	All on-screen	(100,0)	0.91	0.92	0.95
			(80,20)	0.77	0.76	0.83
			(60,40)	0.70	0.72	0.73
			(50,50)	0.76	0.78	0.74
			(40,60)	0.76	0.78	0.74
			(20,80)	0.76	0.78	0.74
			(0,100)	0.76	0.78	0.74
	Our approach	Browser's Tab	(100,0)	0.91	0.92	0.95
			(80,20)	0.83	0.89	0.80
			(60,40)	0.76	0.78	0.74
			(50,50)	0.76	0.78	0.74
			(40,60)	0.64	0.66	0.63
			(20,80)	0.64	0.66	0.63
			(0,100)	0.64	0.66	0.63
Problem 4	Hash	–	(100,0)	0.35	0.45	0.36
	Our approach	All on-screen	(100,0)	0.51	0.58	0.53
			(80,20)	0.51	0.58	0.53
			(60,40)	0.64	0.73	0.59
			(50,50)	0.64	0.68	0.61
			(40,60)	0.76	0.79	0.74
			(20,80)	0.76	0.78	0.74
			(0,100)	0.76	0.78	0.74
	Our approach	Browser's Tab	(100,0)	0.51	0.58	0.53
			(80,20)	0.51	0.58	0.53
			(60,40)	0.79	0.87	0.74
			(50,50)	0.61	0.66	0.59
			(40,60)	0.64	0.66	0.63
			(20,80)	0.64	0.66	0.63
			(0,100)	0.64	0.66	0.63

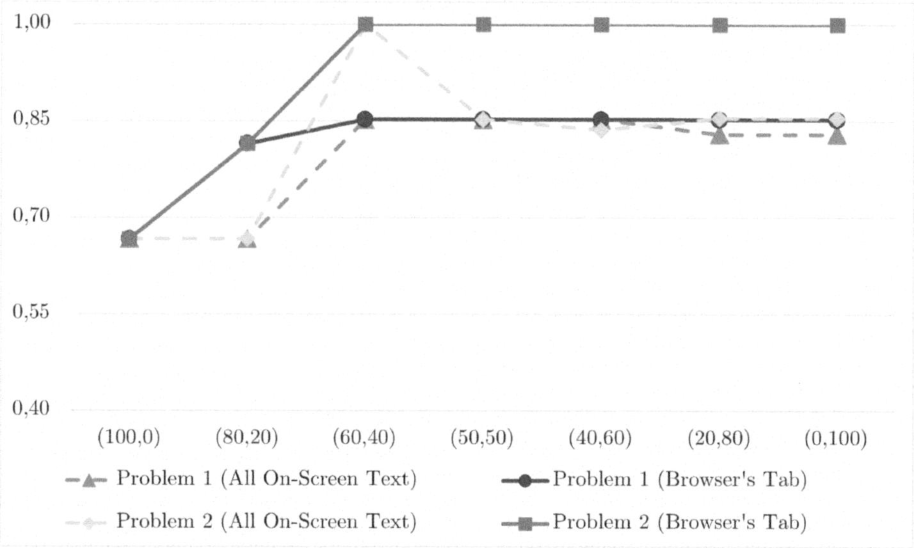

Fig. 3. Evolution of F1-Score (y-axis) for invoice resolution problems while increasing text weight. The x-axis denotes (*ImageWeight*, *TextWeight*).

Invoice Resolution. The results shown in Fig. 3 demonstrate that proposals incorporating text outperform those that rely only on visual information when the UI log contains poorly differentiated images (cf. Fig. 1, challenge 1). This improvement becomes prominent with a text weight of 40% or higher. The results indicate a superior F1-score for Problem 2 as compared to Problem 1. This disparity is attributable to the *customer details* activity screenshots in Problem 1, which include a unique identifier for each customer inside the Browser's Tab, e.g. *Gemini Furniture - Oscar*. This leads to the error of grouping by case, i.e. customers, instead of by activity. In contrast, Problem 2 comprises *customer details* events under a unified browser's tab *Customer View*, which facilitates their grouping as the same activity.

The results obtained for this process indicate that both options improve upon proposals based solely on visual information. Furthermore, we observe that the variations between choosing the browser's tab or all on-screen text do not exceed 10%, and its variation is strongly influenced by the nature of the text selected, i.e. whether the tab text identifies the activity or the specific case.

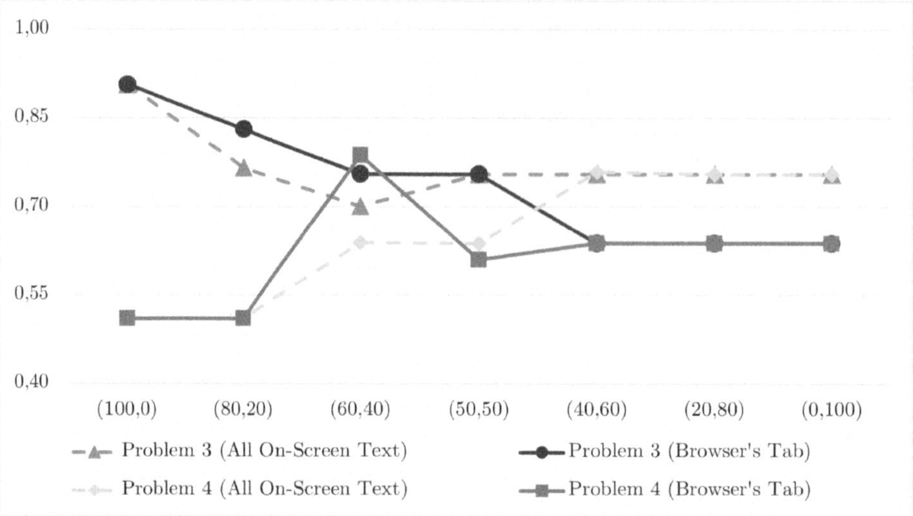

Fig. 4. Evolution of F1-Score (y-axis) for payment notification problems while increasing text weight. The x-axis denotes (*ImageWeight*, *TextWeight*).

Payment Notification. Subsequently, Fig. 4 presents the results from the experiments with Problems 3 and 4. In this case, each one has distinct recording conditions. The F1-Score demonstrates a significant difference when comparing the use of visual information in single-user UI logs versus in multiple-user UI logs, corresponding to Problems 3 and 4 respectively.

In Problem 3 (single-user), which involves a single-user recording, optimal results are derived from utilizing only visual information, i.e. (100,0), compared to all executions including text. This is attributed to the UI log containing activities with well-differentiated screenshots (cf. Fig. 1, left side). So, including text that can contain data specific to each case, detracts from accurate activity identification. Nevertheless, exclusive reliance on visual information becomes less effective in environments containing minimal UI variations between screenshots of the same activity. This observation is particularly evident in Problem 4, characterized by multi-user recordings with varying screen zoom levels (cf. Fig. 1, challenge 2). Here, the incorporation of textual information significantly improves the results, while approaches based purely on visual information see up to a 40% reduction in F1-Score compared to Problem 3. These findings underscore that relying solely on visual information is insufficient, as minor UI screenshot differences make activity identification more difficult. However, as this Problem UI log is also affected by displaying case-specific information in the browser's tab text, can be said that integrating all on-screen text typically yields superior outcomes.

Hence, an analysis of the two processes reveals that, in this case, the most effective results are achieved with a 60% weighting for images and 40% for text. This fact suggests that a hybrid approach combining both visual and textual information is more advantageous than relying solely on one modality. It enhances overall robustness, especially in multi-user UI logs contexts, achieving approximately a 30% improvement in the F1-Score. Therefore, in response to our **RQ**, it is concluded that the combination of image with text in a 60,40 ratio for event clustering significantly improves activity identification in multi-user contexts.

5 Discussion

In this section, we discuss the limitations and shortcomings identified in our proposal.

Firstly, the selection of processes for evaluation lacks a systematic methodology, which potentially undermines the generalizability of our findings. A more structured selection method, coupled with an increased diversity and volume of test cases, would enhance the reliability of the conclusions. Regarding the text sources used, our research only relied on text extractors for *all on-screen* text and the *browser's tab*. Enriching our array of text extractors, as well as examining their interplay across a broader spectrum of processes, could provide deeper insights.

Another limitation concerns the image-text combination method for vector representations, used in our study. We exclusively utilized the CLIP method, which is currently the most promising technique in the state-of-the-art for this, as discussed by Radford et al. in [13]. Other proposals for image-text combination could be explored.

Lastly, our study was restricted to using only agglomerative clustering for the identification of activities. This limitation opens up avenues for future research to explore different types of hierarchical clustering, such as divisive clustering. In the same manner, their setting parameters should be tested in a broader evaluation.

6 Related Work

Several approaches have been proposed in the area of activity identification. Bala et al. [5] highlighted the early attempts in traditional process mining for activity identification, emphasizing the need for structured data and known identifiers for effective monitoring and analysis of business processes. Their work focuses on a semi-automatic technique for discovering event relations based on an event log perspective rather than a UI log.

Agostinelli et al. [3] and Leno et al. [9] explored UI log-based methods where the focus was on extracting activity identifiers directly through loggers, i.e., concept: name. These methods did not consider images, relying instead on structured logs to identify activities. This approach, while effective in certain scenarios, is limited by its dependence on well-structured and comprehensive logs, which are not always available, particularly in virtualized or remote environments [10].

To address these limitations, Jiménez-Ramírez et al. [8] extended the domain of process discovery to include image data. This work is closely related to our proposed method. They introduced an approach that utilizes image hashing to extract fingerprints from screenshots, which are then clustered to identify different activities. By associating clusters of these image fingerprints with specific activities, they could perform activity identification based solely on visual data. This method showed promise in environments where capturing structured logs is challenging, but it remained sensitive to minor interface changes, such as having the browser in dark mode or the Windows taskbar in a different color.

Building upon the limitations and advancements in these previous works, our approach introduces a novel method that integrates information from both visual and textual data. This hybrid approach not only enables activity identification in virtualized environments, as the text is recognized from the image but also introduces text as an additional resource of information. This makes the clustering less dependent on the image and improves the robustness of activity identification against UI variations.

7 Conclusions and Future Work

This paper proposes a method to identify the activity of events in a process event UI log. For this, it combines visual features (i.e., raw screenshots) with text extracted from the screen. This approach is meant to be reliable in multi-user environments where the UI log may contain events of the same process but with different graphical settings.

In this multi-user scenario, image-based proposals (i.e., only considering visual features) tend to fail. However, our algorithm has been tested on two different processes using up to four UI logs, with varying weights assigned to visual features versus text. The results show that, although there are extreme configurations that may work for concrete scenarios, a balanced configuration of the weight results in a general improvement compared to the current state-of-the-art approach.

Although the presented proposal provides promising results, we envisage several directions for future research.

On the one hand, considering the different perspectives of the processes, two lines are distinguished. First, the problem can be approached at different levels of abstraction for a multi-granular view, where different business expert profiles are considered. Second, the specific action associated with each screenshot can be studied, as user actions, e.g., clicking "delete" instead of "accept" differ significantly.

On the other hand, about the technical aspects of the approach, two additional research lines are distinguished. First, regarding the clustering algorithm outcome, instead of grouping actions by numerical labels for identification, they could be defined through a semantic context. This could be addressed by using summaries generated by large language models. Second, to expand the options in response to the observed text issues, we could explore Machine Learning algorithms for text selection dynamically. This would consider the process context and utilize advanced Natural Language Processing techniques to maximize information extraction from different text sources.

References

1. van der Aalst, W.M.P., Bichler, M., Heinzl, A.: Robotic process automation. Bus. Inf. Syst. Eng. **60**(4), 269–272 (2018)
2. Agostinelli, S., Lupia, M., Marrella, A., Mecella, M.: Automated generation of executable RPA scripts from user interface logs. In: Asatiani, A., et al (eds.) Business Process Management: Blockchain and Robotic Process Automation Forum, pp. 116–131 (2020)
3. Agostinelli, S., Lupia, M., Marrella, A., Mecella, M.: Reactive synthesis of software robots in RPA from user interface logs. Comput. Ind. **142**, 103721 (2022)
4. Aguirre, S., Rodriguez, A.: Automation of a business process using robotic process automation (RPA): a case study. In: Applied Computer Sciences in Engineering: 4th Workshop on Engineering Applications, WEA 2017, Cartagena, Colombia, 27–29 September 2017, Proceedings 4, pp. 65–71 (2017)
5. Bala, S., Mendling, J., Schimak, M., Queteschiner, P.: Case and activity identification for mining process models from middleware. In: Buchmann, R.A., Karagiannis, D., Kirikova, M. (eds.) PoEM 2018. LNBIP, vol. 335, pp. 86–102. Springer, Cham (2018). https://doi.org/10.1007/978-3-030-02302-7_6
6. Dumas, M., Rosa, M.L., Leno, V., Polyvyanyy, A., Maggi, F.M.: Robotic process mining. In: van der Aalst, W.M.P., Carmona, J. (eds.) Process Mining Handbook. LNCS, vol. 448, pp. 468–491. Springer, Heidelberg (2022). https://doi.org/10.1007/978-3-031-08848-3_16
7. El-Gharib, N.M., Amyot, D.: Robotic process automation using process mining - a systematic literature review. Data Knowl. Eng. **148**, 102229 (2023)
8. Jimenez-Ramirez, A., Reijers, H.A., Barba, I., Del Valle, C.: A method to improve the early stages of the robotic process automation lifecycle. In: Giorgini, P., Weber, B. (eds.) CAiSE 2019. LNCS, vol. 11483, pp. 446–461. Springer, Cham (2019). https://doi.org/10.1007/978-3-030-21290-2_28
9. Leno, V., Augusto, A., Dumas, M., La Rosa, M., Maggi, F.M., Polyvyanyy, A.: Discovering data transfer routines from user interaction logs. Inf. Syst. **107**, 101916 (2022)
10. Martínez-Rojas, A., Jiménez-Ramírez, A., Enríquez, J., Reijers, H.: A screenshot-based task mining framework for disclosing the drivers behind variable human actions. Inf. Syst. **121**, 102340 (2024)
11. Martínez-Rojas, A., Reijers, H.A., Jiménez-Ramírez, A., Enríquez, J.G.: What are you gazing at? an approach to use eye-tracking for robotic process automation. In: Köpke, J., et al. (eds.) Business Process Management: Blockchain, Robotic Process Automation and Educators Forum, pp. 120–134 (2023)
12. Martínez-Rojas, A., Jiménez-Ramírez, A., Enríquez, J.G., Reijers, H.A.: A tool-supported method to generate user interface logs, vol. 2023-January, pp. 5472–5481 (2023)
13. Radford, A., et al.: Learning transferable visual models from natural language supervision, vol. 139, pp. 8748–8763 (2021)
14. Shahapure, K.R., Nicholas, C.: Cluster quality analysis using silhouette score, pp. 747–748 (2020)
15. Troller, M.: Practical OCR system based on state of art neural networks. Ph.D. thesis, Czech Technical University in Prague Dejvice, Czech Republic (2017)

Designing the Organizational Reuse Environment - Enabling Citizen Developers to Reuse Process Automation Artifacts

Peter A. François[1(✉)] and Ralf Plattfaut[2]

[1] South Westphalia University of Applied Sciences, Lübecker Ring 2, 59494 Soest, Germany
francois.peter@fh-swf.de

[2] Chair of Information Systems and Transformation Management, University of Duisburg-Essen, Universitätsstraße 2, 45141 Essen, Germany
ralf.plattfaut@icb.uni-due.de

Abstract. With the increasing proliferation of Robotic Process Automation and other low-code approaches in automating business processes come new challenges for reusing automation artifacts. Business logic is implemented in multiple technologies and developed decentrally across the organization. Especially citizen developers, who do not have extensive programming knowledge, face issues regarding reuse. Based on a thorough literature analysis and informed by Transaction Cost Theory, this research in progress identifies five challenges for reuse. Following design science research, we develop five principles that help organizations set up reuse environments that encompass both low-code and traditional development regarding automation artifacts. In addition, we describe how we plan to validate the principles in the next stage of our research.

Keywords: Reuse · Low-Code · Process Automation · Citizen Developer · Technological Rule · Design Science

1 Introduction

Organizations use various technologies to automate their business processes [1], including Robotic Process Automation (RPA) and other low- or no-code approaches. Some of these automation efforts are carried out by Citizen Developers. Citizen developers are regular employees aiding in process automation. They can apply their in-depth process knowledge during optimization and automation, however, they have little or no formal knowledge regarding professional techniques – such as reuse [2, 3]. This "lack" of knowledge impairs awareness regarding the benefits and pitfalls of reuse.

The benefits of reuse range from lower implementation cost over faster implementation time to consistent (high) artifact quality and overall reduced maintenance needs [4–6]. Simultaneously, reuse brings a set of issues or even dangers. A recent example is the reuse of Log4j, a logging utility used for logging in Java programs. The vulnerability allowed malicious entities to inject and execute code in multiple thousand software systems that (re)used the framework (see, e.g., [7]). With the advent of citizen developers,

C. Di Ciccio et al. (Eds.): BPM 2024, LNBIP 527, pp. 138–153, 2024.
https://doi.org/10.1007/978-3-031-70445-1_9

reuse – and therefore both its benefits and issues – are made available to various business users [8]. They then face similar issues as traditional, trained developers in achieving the benefits and mitigating potential issues and dangers of reuse.

Since reuse holds such benefits and can be so hard to conduct correctly, organizations would benefit from design principles for creating reuse environments that, on the one hand, unlock the potential of reuse for all employees conducting Business Process Optimization (regular and citizen developers) and, on the other hand, shield the organization (and individuals) from potential damage from improper reuse or improper artifacts. These principles must be easily understandable, not only by regular IT developers but also by citizen developers. At the same time, the design principles must be applicable to all process automation artifacts at all stages of development, both in low-code and traditional development to harness the full potential of all reusable artifacts.

The participation of citizen developers and the use of varying technologies (traditional programming and RPA or low- and no-code technologies) bring significant challenges to business process automation. The decentral nature of citizen development and the citizen developers' lack of development knowledge can hinder reuse. Simultaneously, important and valuable business logic now exists within traditional and also low-code artifacts. This leads to the following research questions:

- What are the challenges associated with the reuse of process automation artifacts, especially considering a citizen developer involvement?
- How should organizations design organizational environments from a managerial standpoint to facilitate the reuse of distributed process automation artifacts in different technologies?

This paper is structured in 7 sections. We first describe the design science method used to answer these questions. We chose to place the background section thereafter since the concepts described there directly inform our design principles. Next, we set out to develop design principles for the organizational reuse environment. To this end, we rely on the Transaction Cost Theory (TCT). As a first rigor-cycle, we employ challenges regarding reuse based on a structured literature review. Finally, we derive design principles for the organizational reuse environment. We end with a concluding discussion outlining our future research approach to evaluate the design principles.

2 Method

This paper follows a design science research approach [9, 10]. Figure 1 gives an overview of the research process. We first (1) conducted a thorough systematic literature review to identify existing research on automation reuse using "automation reuse" without quotation marks as the search term in the AIS eLibrary, Senior Scholar's Basket of Eight, and the proceedings of the BPM conference. The literature review consists of 2075 original hits. We chose these outlets to both get a broad overview of the field of reuse in process automation [11] and also a good idea of how the phenomena of low-code development and citizen developership impact it. We then filtered the articles, first by title, then by abstract. The filtered search resulted in 136 articles. After conducting a selective backward search, we included 170 articles in total. We then applied grounded

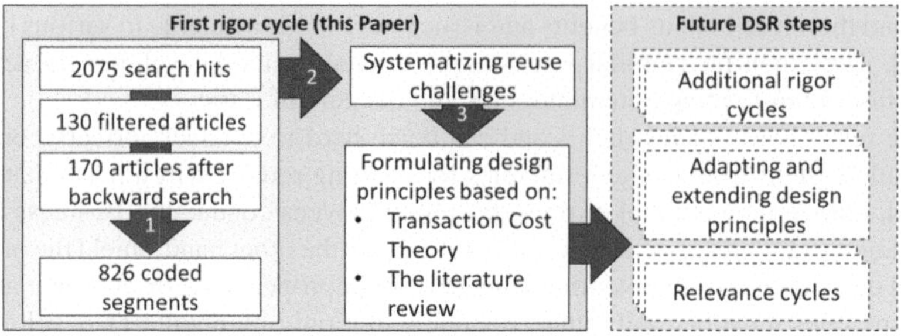

Fig. 1. Research Process

theory [12] to the 170 selected papers, resulting in 826 coded segments. Based on these results, we applied axial coding, resulting in 89 aggregated codes. The literature review process is described in further detail in [13].

As a second step, we systematized the challenges regarding reuse based on the literature. The results of this step are described in Sect. 5.

Based on the findings of the entire literature review, as a third step, we applied a view of TCT and developed a set of organizational design principles following van Aken [14]. According to van Aken [14], such principles are to be viewed as an "exemplar" that aids in achieving a specific state in a specific situation, meaning that adjustment is required when aims or situations diverge. For this paper, we assume that every design principle aims to maximize the potential of reuse by decreasing transaction cost while mitigating potential pitfalls. Following the work of Gregor et al. [10], who take the work of van Aken [14] into consideration, we detail the design principles with the relevant justificatory knowledge (e.g., TCT as our kernel theory).

The effectiveness of the proposed principles will be evaluated in a later stage in our research. To this extent, we plan to validate our findings in multiple organizations (so far being in contact with 7) and refine them utilizing the practitioner's feedback.

3 Background

3.1 Low-Code Development and Citizen Developers

Low-code platforms allow developers to create software with pre-defined components that include specific functionality while using little to or even no code in the traditional sense [3]. The use of the pre-defined automation components allows the (re)use of basic functionality provided by the platform vendor, which can then be configured to fit the process and specific requirements [15]. The use of components and the quick reconfigurability can lead to high development efficiency [15]. However, low-code systems can reach a complexity similar to regular systems [16]. The quick artifact development and current knowledge of process execution improvement potentials allow quick process innovation, which is one driver for businesses to utilize such systems [17].

RPA is one technology in this category that allows the automation of digital tasks that were previously manually conducted on the graphical user interface [32].

Low-code development is usually performed in a visual editor and is said to require little to no formal training, which also allows "regular" (non-IT) employees to create automation solutions utilizing them [15]. These so-called Citizen Developers have little to no formal programming training [3], but being close to the process execution allows them to bring a deep understanding of business processes, allowing for process and IT alignment [3, 18].

3.2 Reuse for Citizen-Developers

The idea of partial or full reuse of artifacts such as libraries or standard software has been a part of information systems (IS) development for a long time (see, e.g., [19, 20]). Reuse is the act of utilizing an existing artifact in new ways or utilizing (parts of) existing artifacts in new development [19]. Reuse may range from copying and pasting parts of artifacts (see, e.g., [19]) over modular or model-driven systems (see, e.g., [21, 22]) up to the reuse of partial or entire architectures [23]. Reuse is applied not only to purely technical artifacts but may also encompass specifications such as business process models (see, e.g., [24]) or knowledge (see, e.g., [25]). Due to its benefits, it is routinely applied in software engineering practice (see, e.g., [26, 27]) with some modern developing methodologies even having reuse as one of the core concepts of development (e.g., service-oriented architectures [19, 26]).

Allowing non-IT personnel to develop low-code or no-code applications has enabled companies to develop IT artifacts more quickly and to bring IS development closer to the business knowledge manifested in organizational routines [2, 8]. Traditional programming knowledge – however, is not necessary to become a citizen developer [2]. The citizen developers creating these low-code applications now face similar questions regarding reuse that professional IT workers face without having undergone the same comprehensive training. Therefore, methods and technologies to help citizen developers with reuse without having to undergo vast training are needed.

The literature on the reuse of low-code and citizen developer software is sparse. Table 1 offers a selection of the seminal literature.

Since artifacts developed by citizen developers are naturally created decentrally (by employees across the organization), finding suitable artifacts for reuse across a vast number of private artifact repositories may increase the effort for each instance of reuse. At an extreme, this may lead to constant redevelopment due to this high effort or even biases like the "not-invented-here syndrome" (e.g., [19, 30]). A high resulting number of artifacts further complicates finding "the right" artifact [31]. Thus, artifact development in a citizen developer setting can lead to further inefficiencies in software reuse.

The described additional inefficiencies in the reuse process can make reuse unviable, especially when compared to the reduced initial or re-developing cost citizen developer systems bring (compare, e.g., [32] and [19]). In contrast, however, Lacity et al. [28] identify reusing low-code solutions as a major driver for scaling low-code solutions. Noppen et al. [33] additionally describe the need for reuse in low-code applications due to high maintenance efforts caused by a large number of similar components.

Table 1. Excerpts from the seminal literature on low-code and citizen developer reuse

Source	Stance on reuse
[28]	The authors propose reuse as one measure to allow quick scaling and cost-saving in low-code projects. They describe that the existence of a large number of components reduces assembly and delivery cost in new development. However, they do not describe how to set up a reuse environment
[16]	Reuse of low-code applications is severely limited by a lack of professional development features that aid it, such as modules (depending on the technology). This can lead to opportunistic reuse and, thus, inefficient reuse and propagating errors
[29]	After scaling RPA operations, the case company used reusable components to enable the automation of more processes. This is supported by a central component repository allowing citizens to search for components and track their automation ideas/wishes. The authors describe that the repository eases maintenance

3.3 Transaction Cost Theory

Coarse [34, 35] argued that there are costs associated with performing transactions in a market. According to Coarse [34], a resource is traded until either A) the value of a good is maximized within the market or B) the transaction cost of performing further trades outweighs the trade's added value. The formation of organizational structures can be used to minimize transaction cost within the organization. A large availability of goods on the market, therefore, allows to satisfy more customers and leads to lower purchasing cost. However, Coarse [34] also describes a negotiation cost associated with organizations agreeing on the terms of trading a specific good.

Regarding software artifacts, the value of an artifact is not necessarily decreased when multiple (non-malicious) actors have access to the artifact. Especially in the same organization, the value of an artifact can be harnessed several times, gaining additional value from each (re)use [19]. In theory, an artifact can bring infinite value if there is a reuse opportunity that brings higher benefits than transaction costs. In practice, there is a higher cost associated with creating a reusable artifact than a traditional artifact in the first place or adapting a reusable artifact to be reusable in a new context [19].

4 Taking a Transaction Cost Perspective on Reuse

We argue that organizational reuse environments can include several 'markets' for reusable components and thus have several boundaries that can add transaction cost. Especially in citizen development (usually conducted decentrally), "silos" can form.

Figure 2 shows exemplary artifact markets existing in an exemplary organizational reuse environment and the relative associated transaction cost. Each time reuse (i.e. a transaction) is considered, the reusing actor (RA) has the option of accessing markets 'offerings', choosing and reusing one of the artifacts. These artifacts may be developed by citizen developers or "professionals" in low-code or traditional programming.

First, the RA may have access to a private repository that holds their working files. The RA should be roughly aware of the artifacts within, leading to a low transaction

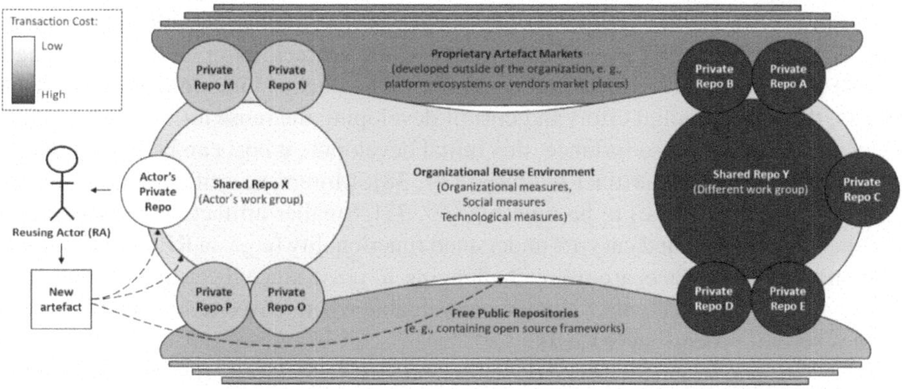

Fig. 2. Reuse artifact markets inside an organizational reuse environment

cost regarding reuse. In the case of a shared workgroup repository the RA has access to ("Repo X" in Fig. 2), an additional element of evaluation cost applies since the repository holds artifacts created by other actors which must be examined and understood before they can be reused adding to the transaction cost.

Further personal or shared repositories (e.g., group folders of other groups) may exist within the company. The RA may not have access to these ('infinite' transaction cost). Only if the RA has some prior knowledge about which sort of artifacts may be within the repositories can the RA request access to artifacts from these repositories. Alternatively, an actor with access to these repositories could voluntarily offer the RA artifacts stored within them. For this to happen, the participants must create some form of connection. The transaction cost would still be fairly high for both artifacts, requiring direct coordination, possibly manual component transfer, and trust between both parties. Especially when the RA is a citizen developer, additional hurdles lay in understanding and evaluating the artifacts and making them compatible with, for example, RPA.

Additionally, there may be an organizational environment designed to aid reuse, including a repository of components for reuse across the organization, where the organizational measures and social constructs within the organization influence reuse.

Another possible source for reusable artifacts outside the organization is proprietary artifact markets or free public repositories. Proprietary Artifact markets allow actors to buy artifacts from vendors [18]. In the case of a trusted vendor, this transaction can be seen similarly to a transaction within the organization, plus the buying price for the artifact. Free public repositories allow access to artifacts without purchase [36]. In the case of an unknown artifact provider, an additional evaluation cost may occur. Both artifact markets and free public repositories may, however, include a large number of competing vendors offering similar components. This high number of components can then further increase the cost of discovery and evaluation.

5 The Challenges of Reuse

From the literature, we identify five major challenges that must be overcome to conduct reuse successfully. We will describe them in the following.

C1 Designing Reusable Artifacts: Challenges in developing reusable artifacts arise in the development of the basic functionality of the artifact, its granularity, making it reusable in several contexts, and finding all relevant contexts. First, in creating reusable artifacts, there is the initial effort and cost of developing the reuse artifact. To calculate the cost of a specific reuse instance, this initial development cost can be divided by the number of instances the artifact is reused [19, 38]. Moreover, artifacts must have the "correct granularity" (size) to be reusable [19, 37]. Smaller artifacts are easy to reuse since they have simple and easy-to-understand functionality (e.g., an RPA bot to log into a system). Larger artifacts are more challenging to reuse since they include context or process-specific requirements, yet are more valuable when reused (e.g., an RPA bot that executes several process steps) [19].

There can also be an additional development cost to make an artifact reusable (e.g., additional requirements like interfaces or effort for generalizing; see [38, 39]). The field of Software Engineering has developed various procedures to reduce reusability cost by designing artifacts for reuse, such as standardized interfaces like REST or design principles for reusable artifacts (see, e.g., [22, 40, 41]). Citizen developers, however, may not be aware of these concepts, especially as RPA enables automation via the user interface. These concepts may also not be easily applicable to low-code technologies.

Artifact creators must consider all possible contexts in which an artifact may be used during its lifetime to be able to cover requirements introduced by the use in that context [42]. Imagining all potential contexts is almost impossible at the time of artifact creation. Creating artifacts that are reusable in different contexts can be costly in itself. Adaptation (see C3 below) can be necessary when reuse in a specific context has not been considered during initial development.

C2 Accessing and Evaluating Reusable Artifacts: When deciding whether to reuse, one must weigh reuse against redevelopment [19, 38]. Thus, an actor has to find the relevant reusable artifacts (e.g., in one of many repositories), gain access to them (following formal or informal access processes), and evaluate the available artifacts regarding their fit in the desired context (e.g., by examining documentation).

Consequently, similar to the concept of coordination cost in TCT, there is the cost of finding and gaining access to reusable artifacts. Especially with small artifacts, the cost of finding them may offset the benefit of reuse [31]. Artifacts must be searched for amongst additional artifacts, making the discovery process tedious. This is especially true if artifacts are not kept in a central repository but must be retrieved from various undocumented repositories throughout the organization. Access may be complicated or unachievable when artifacts are stored in multiple repositories and exist in several versions. The decentral nature of citizen development can further this issue.

In addition, there is the problem of evaluating the artifact. Each of the artifacts found must be understood (to some extent) and evaluated regarding its value potential for this specific reuse instance. The actor has to evaluate if the artifact fits the new context. The artifact's fit for the new context (and thus the amount of adaptation required) plays an essential part in the cost of reuse and, thus, the decision between reuse and remake [38].

For citizen developers, a lack of formal training in software engineering and understanding of code may further complicate assessing components.

C3 Adapting and integrating Reusable Artifacts: When an artifact is found to be applicable for use, adaptation and integration can be necessary to enable the artifact to fit into the new environment and achieve its maximum potential. This adaptation can, for example, include making artifacts compatible or adapting an artifact to fit a different context [19, 41]. Increasing the effort spent in Evaluating Reusable Artifacts (C2) can influence the effort in Adapting it (C3); it can aid in identifying more suitable artifacts that require less adaptation and show which adaptation will be necessary. If high adaptation rates are necessary, redevelopment can be more cost or time-efficient [19, 43]. Regardless of whether the artifacts have been adapted, it must be integrated into the new environment. The integration of artifacts can be laborious and error-inducing [4]. Therefore, the adaptation and integration must be tested, leading to further effort [19].

C4 Ensuring Security and Maintainability: The reuse of artifacts can necessitate additional maintenance. If adaptations and maintenance operations are not propagated upwards through the reuse pedigree, several versions of the same artifact will further increase the need for (now separately conducted) maintenance and allow for additional security and reliability issues (see, e.g., [7, 44]). Security and maintenance for reuse must therefore be considered in two directions: It is hard to track where a specific artifact has been reused (e.g., 'Who uses our purchase bot? We want to change it.') and vice versa, which artifacts have been reused in a specific artifact (e.g., 'Which artifacts (low-code and traditional) does our purchase bot rely on?'). The additional effort in security and maintenance leads to cost that may be small for generalized and often reused parts but larger when specialized artifacts with extensive adaptation are reused.

As citizen developers rarely have in-depth IT knowledge, performing such security assessments and maintenance can be especially difficult for them. Reusing components without sufficiently understanding them can introduce additional issues [5].

C5 Managing Reuse: Suitable governance mechanisms are critical for successful reuse [45]. Too little guidance will hinder security and maintainability (see C4) and will not allow the organization to harness the full potential of reuse [46]. It can also make reuse inefficient: too little knowledge, too few reuse mechanisms, and too few best practices will be available to the reuse actors. Too strict guidance will increase the transaction cost as reuse processes become more complex (tracking reuse, asking for permission, asking for requirements of other departments, coordinating update schedules). Introducing citizen developers can further complicate governance efforts [47].

6 Designing the Reuse Environment to Reduce Transaction Cost

The challenges mentioned contribute to the cost of each reuse transaction, making reuse less viable. We argue that transaction cost is a roadblock for reuse for low-code automation. Even in traditional programming, transaction costs should be reduced as much as possible to reap the full benefits of reuse. However, since artifacts created using low-code technologies are cheaper and quicker to produce, there appears to be a reduced benefit from reusing such an artifact compared to reusing more expensive traditional artifacts. Reuse has been identified as critical to getting low-code projects up to scale

[28]. In the following, we develop five design principles (DPs) for organizational reuse environments based on the abovementioned challenges, using transaction cost as the kernel theory. We deem these DP viable both for low-code and traditional development.

DP1 Enable Actors to build Reusable Artifacts. For reuse to be possible, suitable reusable artifacts must be available in the reuse market. As citizen developers are often not formally trained in developing artifacts with reuse in mind, this is a considerable challenge due to the decentral nature of artifact development. Therefore, to enable reuse actors to build a meaningful number of reusable artifacts…:

Give All Artifact Creators (including Citizen Developers) Clear Guidance on How to Perform Reuse. Knowledge on reuse is a critical fac*Give all artifact creators (including citizen developers) clear guidance on how to perform reuse.* tor in being able to perform reuse effectively and efficiently [19, 43]. Since artifacts must be reusable in other departments, developers must know the concept of reuse, how it is done in the organization (standards and conventions), and how to create reusable artifacts [43, 45, 46].

Give Artifact Creators the Knowledge Required to Decide Which Artifacts Should be Designed with Reuse in Mind and made Available. Since making artifacts reusable in different contexts is associated with an additional cost [19, 42] and large repositories can lead to high search, retrieval, and evaluation cost [19, 31], RAs must decide which artifacts they should prepare for reuse and make available. To be able to make this decision, artifact creators need to be enabled to estimate the reuse potential of components. As citizen developers usually are close to the point of process knowledge creation (see, e.g. [17, 48]), this value potential will include knowledge on how processes are (or should be) conducted (see, e.g., [3]).

Offer Reusable Artifacts in Different Granularity (where Possible). Finding the right granularity is a balancing act between large components that bring the most benefits and small components that can fit a variety of contexts without adaptation [37]. The benefit of reusing artifacts that are too small may be countered by the transaction cost [49]. Therefore, artifacts that are available in different granularities or are logically dividable (e.g., assembled from several different artifacts) should be offered as such. If low-code applications are used as a temporary fix (see e.g. [17]) this approach can also help transition parts of the automation to "regular" backend systems [50].

DP2 Empower Artifact Access and Evaluation. To ensure that all (potential) reuse actors are able to quickly evaluate potential reuse artifacts and deal with the number of available artifacts…:

Ensure that Every (potential) Reuse Actor can Easily Find Potential Reuse Artifacts Across Technologies and Departments. According to Coarse [34], every market transaction requires finding potential market partners, building contact with them, negotiating the terms of the trade, conducting the trade, and evaluating the goods. Coarse states:"These operations are often extremely costly, sufficiently costly at any rate to prevent many transactions that would be carried out in a world in which the pricing system worked without cost" ([34]). In the context of reusing artifacts Rothenberger and Kulkarni [51] phase this as: "The ideal retrieval method would neither require any prior knowledge of the repository nor any informal communication among developers to find the components needed." [51] We, therefore, argue that any potential developer should have A) easy access to reusable components, with B) a standardized way of accessing them, with C) a mostly automated "contract negotiation" (i.e., access rights), preferably in D) one central artifact repository that allows searching and filtering as artifact repositories have been identified as essential for reuse (see, e.g., [19, 51, 52]).

Give Relevant Metadata on the Available Artifacts. Evaluating artifacts for their reusability and need for adaptation requires a major effort. This effort increases with the number of artifacts available [31, 41]. Relevant metadata aids evaluation [53].

Name Trusted and Untrusted External Sources of Reuse Components. As described above, the reuse of artifacts created outside of the organization is another viable option. In this case, trust and security become more important in artifact evaluation [41]. When artifacts are trusted, and the characteristics of the artifacts are sufficiently evident, developers can utilize the artifacts in a black box approach (not examining the inside working of the artifacts), leading to additional time savings [20]. Trusted artifacts should be marked as such. The "Not invented here syndrome" may additionally hinder reuse from outside repositories and vendors [19, 30]. We therefore suggest naming trusted artifact suppliers.

DP3 Enforce Traceability and Lifecycle Control. To allow all reuse actors to trace bugs and find security issues in their artifacts...:

Ensure that Reuse Artifacts can be Traced throughout the Entire Reuse Pedigree. Changes to software artifacts can impact other artifacts [53]. Therefore, changes must be tracked and (if necessary) applied to other contexts in which the artifact is used. Otherwise, security flaws and bugs remain in these contexts. Otherwise, there will be multiple versions of the same artifact, leading to increased maintenance and reuse (evaluation) cost. If security issues arise and are inherited to the new context a faulty system is reused in, these issues must be trackable throughout the entire reuse pedigree.

Observe the Lifecycle of Reuse Artifacts. Make its status and changes known to the RAs. Each software should conform to a software lifecycle to achieve conformance, quality, and security [54]. As low code governance can be centrally or decentrally organized [47], mechanisms should be in place that allow either central or decentral lifecycle tracking in order to be able to find obsolete artifacts in the reuse pedigree and update all (possibly adapted) artifact instances. Research must still be done on how the lifecycle can be effectively tracked in such environments.

DP4 Manage Knowledge and Reuse Through Communities. To allow all reuse actors to resolve issues, communicate about reuse, build, and share knowledge...:

Connect RAS through Communities of Practice and Allow Them to Motivate Each Other. The attitude of individual reuse actors and trust between RAs can impact the success of reuse activities [55]. If there is no bond and collegiality between RAs, reluctance to share code with or use code by others can arise [43, 56]. Communication and connection is necessary for the success of reuse [25]. Therefore, organizations should create reuse communities, forming bonds and trust between reusing developers.

Allow for the Building of Cross-sectional Knowledge by Implementing Suitable Knowledge Management. Knowledge (on reuse or in general) can also be reused. However: "Many organisational and professional cultures reward – sometimes unconsciously – knowledge creation over knowledge reuse" [57]. Organizations should, therefore, enhance the reuse of knowledge through appropriate measures of knowledge management, like central knowledge repositories. The resulting community can also help citizen developers who are new to reuse since new reuse actors face a steep learning curve [43]. Hallikainen et al. [29] found knowledge sharing as crucial to scale low-code efforts.

Advance the Reuse on Knowledge by Utilizing Appropriate Training and Providing Teaching Material. The amount of knowledge required necessitates adequate training in reuse [43]. Therefore, organizations should provide general training regarding reuse and their specific reuse processes, considering the needs of different developer roles.

DP5 Build Strategy and Governance. To enable all reuse actors to align their reuse procedures with the company's goals and to further reuse…:

Clarify the Organization's Strategy and Goals Regarding Reuse. This includes whether reuse should be seen as a goal in itself (e.g., to reduce the number of artifacts to maintain) or if reuse should only be conducted when financially advantageous. Middle managers, in particular, can oppose efforts to set up reuse programs if a strategic view is missing. Reuse efforts are often lengthy and do not show immediate benefits in their personal performance indicators. On the contrary, efficiency may go down at first [5, 19]. A clearly defined strategy and goals will also aid developers in making reuse decisions (e.g., reuse vs. remake, which artifact to choose).

Reward Actors Whose Artifacts Are Reused. Kim and Stohr [19] describe that lower-level management, when judged on their individual (and often short-term performance), may not see the benefits of reuse. Similarly, in the spirit of TCT, Coarse [34] states that when the value-adding processes of one actor (e.g., the reusing developer) diminishes value or introduces cost (e.g., maintenance cost) at a second actor, the second actor should be compensated by the first actor for the additional cost. Since a large portion of the cost for reusable artifacts arises at the department of the creating actor (development cost, continued maintenance, issue resolution), some form of compensatory mechanism is recommendable that lessens the burden of providing reusable artifacts.

Build Suitable Governance Mechanisms to Guide Reuse Processes yet do not Over-Restrict Reuse. Governance mechanisms must not be too restrictive since "For the software reusability technique to be effective, it should be easier to reuse the devices than to create them from scratch." [58]. Yet, mechanisms must ensure security, compliance, and compatibility between artifacts. While there are first insights on low-code governance (e.g. [47, 59]), future research should determine suitable mechanisms for reuse.

7 Concluding Discussion

In this paper, we developed five design principles for the organizational reuse environment to enable citizen developers to reuse process automation artifacts. We based these principles on common challenges in reuse and applied a transaction cost perspective.

Research has called for mechanisms that aid organization-wide reuse for both low-code and BPM [6]. The revised BPM capability framework calls for "Multi-purpose Process Design" [60]. Research on RPA has found "design[ing] bots for scalability and later migration" as a critical success factor for RPA [61]. Similarly, Syed et al. describe scaling RPA as a challenge and call for methods to aid it [62]. The proposed design principles can help operationalize these calls. With our paper, we aid in furthering reuse through a managerial approach. Researchers can build upon the principles and translate the design-oriented insights into contributions to more general theories. Furthermore, we extend the literature on process automation reuse [19]. Potentially, our insights could inform existing theories in the domains of BPM, socio-technical change or new development approaches for low-code development. For example, François et al. [63] propose a method for anchoring reuse in the BPM lifecycle. Van Looy and Rotthier [64] describe creating a set of reusable "building blocks" to aid process automation. Průcha and Madzík [65] describe a technical approach to finding similar low-code components. Such endeavors could benefit from our proposed approach. In addition, we contribute to the Low-Code and Citizen Developer literature. For example, Bock et al. call for "methods for guiding lay developers in the use of LCPs and classical software development facilities" [15]. Our design principles can inform such methods and enable organizations to implement managerial methods to allow process automation reuse.

Practitioners can use the challenges and design principles to implement or improve process automation reuse initiatives, especially when relying on low-code or no-code development. By doing so, they can circumvent the challenges of reuse systematized above. By reducing the effort required for each act of reuse, we aid in maximizing the overall value of reuse and making reuse accessible for citizen developers.

We plan a thorough in-vivo evaluation within a recently started four-year research project with multiple private sector organizations. Here, we plan to implement the design principles in these organizations, conduct interviews with the responsible personnel, and implement an overarching reuse environment between the six participating organizations to aid cross-organizational reuse. We have conducted the first rigor cycle [9] in this paper. The challenges of traditional software reuse have been stringently researched from a theoretical and practical perspective [19]. Together with the challenges faced in scaling low-code development like RPA [62], this underlines the practical relevance of our proposed organizational design principles. Our planned evaluation will also lead to iteratively refining them based on the empirical insights [66].

In addition, more research should be carried out on citizen developer reuse. So far, it is unclear if the motivation to provide reuse components or to reuse components by others is different for Citizen Developers who may not have extensive programming knowledge and thus may be less able to assess artifacts regarding their reuse benefit.

We are aware that our research also comes with limitations. We acknowledge that the "products" (components) in the market can, in theory, be used indefinitely without losing value, which is not in the original spirit of TCT. However, the theory seems to hold for other digital goods. In addition, while we conducted a broad literature review and have implemented our design principles stringently based on this, there may be additional relevant concepts in the literature or practice.

In summary, reuse has a rich history in Software Engineering. However, citizen developers face additional challenges in reuse, such as low formal training and decentral development. Organizations should create an environment that allows for both the reuse of traditional and low-code artifacts in order to maximize value. In this paper, we developed design principles for creating organizational reuse environments. To this aim, we systematized common issues in reuse and applied a transaction cost perspective, minding to the needs and limitations of citizen developers.

Acknowledgements. This work has been developed for the project KEBAP at South Westphalia University of Applied Sciences. The project (reference number: 13FH034KX0) is partly funded by the German Federal Ministry of Education and Research (BMBF).

References

1. Dumas, M., La Rosa, M., Mendling, J., et al.: (2018) Fundamentals of Business Process Management, 2nd edn. Springer, Berlin Heidelberg (2018)
2. Woo, M.: The rise of no/low code software development—no experience needed? Engineering **6** (2020)
3. Novales, A., Mancha, R.: Fueling digital transformation with citizen developers and low-code development. MISQ Exec. **22** (2023)
4. Apte, U., Sankar, C.S., Thakur, M., et al.: Reusability-based strategy for development of information systems: implementation experience of a bank. MISQ **14** (1990)
5. Allen, G., Parsons, J.: Is query reuse potentially harmful? anchoring and adjustment in adapting existing database queries. Inf. Syst. Res. **21**, 56–77 (2010)
6. Beerepoot, I., Di Ciccio, C., Reijers, H.A., et al.: The biggest business process management problems to solve before we die. Comput. Ind. **146**, 103837 (2023)
7. Xie, W., Iyer, L,, Simpson, S.J.: Agile software development vulnerabilities and challenges: an empirical study. AMCIS 2022 Proc. **15** (2022)
8. Li, Y., Huang, R.: Participating in citizen development: theory of planned behavior. AMCIS 2022 Proc. **2** (2022)
9. Hevner, A.R., March, S.T., Park, J., Ram, S.: Design science in information systems research. MISQ **28** (2004)
10. Gregor, S., Chandra Kruse, L., Seidel, S.: Research perspectives: the anatomy of a design principle. JAIS **21** (2020)
11. Gogan, J.L., McLaughlin, M.-D., Thomas, D.: Critical incident technique in the basket. ICIS 2014 Proc. (2014)
12. Wolfswinkel, J.F., Furtmueller, E., Wilderom, C.P.M.: Using grounded theory as a method for rigorously reviewing literature. EJIS **22**, 45–55 (2013)
13. François, P.A., Plattfaut, R.: The Reuse of Business Process Automation Artefacts. LNI 337 (2023)
14. van Aken, J.E.: Management research based on the paradigm of the design sciences: the quest for field-tested and grounded technological rules. J. Manag. Stud. **41** (2004)
15. Bock, A.C., Frank, U.: Low-Code Platform. BISE **63**, 733–740 (2021)
16. Lethbridge, T.C.: Low-code is often high-code, so we must design low-code platforms to enable proper software engineering. In: Margaria, T., Steffen, B. (eds.) Leveraging Applications of Formal Methods, Verification and Validation, vol. 13036. Springer, Cham (2021)

17. François, P.A., Borghoff, V., Plattfaut, R., et al.: Why companies use RPA: a critical reflection of goals. Bus. Process Manag. BPM 2022, LNCS **13420** (2022)
18. Naqvi, S.A.A., Zimmer, M.P., Syed, R., et al.: Understanding the socio-technical aspects of low-code adoption for software development. ECIS 2023 Res. Papers **357** (2023)
19. Kim, Y., Stohr, E.A.: Software reuse: survey and research directions. J. Manag. Inf. Syst. 14 (1998)
20. McIlroy, M.D.: Mass produced software components. In: NATO Software Engineering Conference (1968)
21. Delgado, A., Ruiz, F., García-Rodríguez de Guzmán, I.: A reference model driven Architecture linking Business Processes and Services. HICSS 2018 Proc. 4651–4660 (2018)
22. Huang JC, Henfridsson O, Liu MJ (2022) Extending digital ventures through templating. Inf Syst Res 33
23. Li, S., Zhang, H., Jia, Z., et al.: Understanding and addressing quality attributes of microservices architecture: a systematic literature review. Inf. Softw. Technol. **131** (2021)
24. Becker, J., Delfmann, P., Knackstedt, R: Adaptive reference modeling: integrating configurative and generic adaptation techniques for information models. In: Reference Modeling: Efficient Information Systems Design Through Reuse of Information Models. Physica (2007)
25. Asatiani, A., Penttinen, E., Rinta-Kahila, T., et al.: Organizational implementation of intelligent automation as distributed cognition: six recommendations for managers. ICIS 2019 Proc. (2019)
26. Baskerville, R., Cavallari, M., Hjort-Madsen, K., et al.: Extensible architectures: the strategic value of service oriented architecture in banking. ECIS 2005 Proc. (2005)
27. Bjørnstad, S.: A research programme for object-orientation. EJIS **3** (1994)
28. Lacity M, Willcocks LP, Craig A (2015) Robotic Process Automation: Mature Capabilities in the Energy Sector. The Outsourcing Unit Working Research Paper Series
29. Hallikainen, P., Bekkhus, R., Pan, S.: How OpusCapita used internal rpa capabilities to offer services how opuscapita used internal RPA capabilities to offer services to clients to clients. MISQ Exec. **17** (2018)
30. Katz, R., Allen, T.J.: Investigating the Not-Invented-Here (NIH) syndrome: a look at the performance, tenure, and communication patterns of 50 R&D project groups. R&D Manag. **12** (1982)
31. Nazareth, D.L., Rothenberger, M.: Does the 'Golidlocks Conjecture' apply to software reuse? JITTA **8** (2006)
32. Lacity, M., Willcocks, L.P.: Innovating in service: the role and management of automation. In: Willcocks, L.P., Oshri, I., Kotlarsky, J. (eds.) Dynamic innovation in outsourcing: Theories, cases and practices. Palgrave Macmillan (2018)
33. Noppen, P., Beerepoot I., van de Weerd, I, et al.: How to keep RPA maintainable? business process management. BPM 2020, Lect. Notes Comput. Sci. **12168** (2020)
34. Coarse, R.: The problem of social cost. JLE **3** (1960)
35. Coarse, R.: The nature of the firm. Economica **4** (1937)
36. Stallman, R.M.: Free software, free society: Selected essays of Richard M. GNU Press, Boston, Stallman (2002)
37. Subramanyam, R., Ramasubbu, N., Krishnan, M.S.: In search of efficient flexibility: effects of software component granularity on development effort, defects, and customization effort. ISR **23** (2012)
38. Nazareth, D.L., Rothenberger, M.A.: Assessing the cost-effectiveness of software reuse: a model for planned reuse. J. Syst. Softw. **73** (2004)
39. Heinrich, B., Huber, A., Zimmermann, S.: Make-and-sell or buy of web services a real option approach. ECIS 2011 Proc. (2011)

40. Anguswamy R, Frakes WB (2013) Reuse Design Principles. DReMeR '13 - International Workshop on Designing Reusable Components and Measuring Reusability Picture held in conjunction with the 13th International Conference on Software Reuse
41. Ren, M., Lyytinen, K.J.: Building enterprise architecture agility and sustenance with SOA. CAIS **22** (2008)
42. Bērziša, S., Bravos, G., Gonzalez, T.C., et al.: capability driven development: an approach to designing digital enterprises. BISE **57** (2015)
43. Sherif, K., Menon, N.M.: Managing technology and administration innovations: four case studies on software reuse. JAIS **5** (2004)
44. Poulin, J.S., Caruso, J.M., Hancock, D.R.: The business case for software reuse. IBM Syst. J. **32** (1993)
45. Joachim, N., Beimborn, D., Weitzel, T.: The influence of SOA governance mechanisms on IT flexibility and service reuse. JSIS **22** (2013)
46. Becker, A., Widjaja, T., Buxmann, P.: Value potentials and challenges of service-oriented architectures. BISE **3** (2011)
47. Borghoff, V., Plattfaut, R.: Steering the robots: an investigation of IT governance models for lightweight IT and robotic process automation. In: Business Process Management: LNBIP, vol. 459 (2022)
48. Mirispelakotuwa, I., Syed, R., Wynn, M.T.: Is RPA causing process knowledge loss? Insights from RPA experts. In: Köpke, J., López-Pintado, O., Plattfaut, R. et al. (eds) Business Process Management: Blockchain, Robotic Process Automation and Educators Forum, vol. 491. Springer Nature (2023)
49. Rothenberger, M.A., Jain, H., Sugumaran, V.: A platform-based design approach for flexible software components. JITTA **18** (2017)
50. Strothmann, A., Schulte, M.: Migrating from RPA to backend automation: an exploratory study. In: Business Process Management: Blockchain, Robotic Process Automation and Educators Forum. BPM 2023 Proceedings (2023)
51. Rothenberger, M.A., Kulkarni, U.R.: Software reuse in information systems development. AMCIS 2000 Proc. (2000)
52. Witman, P.D.: Tracing patterns of large-scale software reuse. AMCIS 2006 Proc. (2006)
53. Juhrisch, M., Thies, G.: Service management beyond UDDI - a design science approach. PACIS 2008 Proc. (2008)
54. (2017) ISO/IEC/IEEE International Standard - Systems and software engineering - Software life cycle processes (ISO/IEC/ IEEE 12207)
55. Ryan, T.J., Walter, C., Alarcon, G., et al.: The influence of personality on code reuse. HICSS 2019 Proc. (2019)
56. Schreieck, M., Wiesche, M., Krcmar, H.: Governing innovation platforms in multibusiness organisations. EJIS (2022)
57. Davenport, T.H.: Process management for knowledge work. In: vom Brocke, J., Rosemann, M. (eds.) Handbook on Business Process Management 1: Introduction, Methods, and Information Systems, 2nd ed. Springer (2015)
58. Montecinos, F.: An experience in the establishment of a software reuse culture in a real environment. AMCIS 1996 Proc. (1996)
59. Kedziora, D., Penttinen, E.: Governance models for robotic process automation: the case of Nordea Bank. JITTC **11** (2021)
60. Kerpedzhiev, G.D., König, M.U., Röglinger, M., et al.: An exploration into future business process management capabilities in view of digitalization. BISE **63** (2021)
61. Plattfaut, R., Borghoff, V., Godefroid, M.E., et al.: The critical success factors for robotic process automation. Comput. Ind. **138**, 103646 (2022)
62. Syed, R., Suriadi, S., Adams, M., et al.: Robotic process automation: contemporary themes and challenges. Comput. Ind. **115**, 103162 (2020)

63. François, P.A., Kampmann, M., Plattfaut, R., et al.: Systematically embedding automation reuse in business process management projects. LNI **340** (2023)
64. van Looy, A., Rotthier, S.: Kiss the documents! How the city of Ghent digitizes its service processes. In: vom Brocke, J., Mendling, J. (eds.) Business Process Management Cases. Springer International Publishing (2018)
65. Průcha, P., Madzík, P.: SiDiTeR: Similarity discovering techniques for robotic process automation. In: LNBIP, vol. 491. Springer Nature Switzerland, Cham (2023)
66. Hevner, A.R.: A three cycle view of design science research. SJIS **19** (2007)

Measuring Complexity of Bot Models in Robotic Process Automation

Maximilian Völker(✉) and Mathias Weske

Hasso Plattner Institute, University of Potsdam, Potsdam, Germany
{maximilian.voelker,mathias.weske}@hpi.de

Abstract. Robotic process automation (RPA), a technology to automate structured tasks on computers in a light-weight fashion, thrives on graphical models, enabling business users and citizen developers to create automation workflows without programming knowledge. However, due to the created flows often not being optimized and the fine-grained nature of the RPA instructions, these models quickly become complex and extensive, complicating their maintenance and comprehensibility. In this paper, we draw inspiration from the related fields of software programming and business process modeling to introduce complexity metrics for RPA bot models. These allow to objectively measure and compare the complexity of such workflows, and can thus, for example, provide an indication of where to start refactoring models in the bot repository.

Keywords: Robotic Process Automation · Modeling · Complexity · Metrics

1 Introduction

Robotic process automation (RPA) enables not only technical users but also business users to create automations for day-to-day tasks performed on computer systems [29], such as for data retrieval and transfer. This technology utilizes small software scripts, also called bots, to operate computer applications in a manner similar to a human user [1] or, if available, by calling the appropriate programming interfaces [14], thus automating structured and data-based tasks on computers in a lightweight fashion [18]. As the scripts are typically defined through a visual, user-friendly interface [29], business users are also encouraged to create such automations, thereby becoming "citizen developers" [4]. In addition to manual modeling, many RPA tools allow users to record a sequence of interactions that can then be automated, i. e., replayed, by an RPA script [8]. Building on this concept, robotic process mining (RPM) aims to automatically generate such scripts based on user interactions recorded over an extended period of time, from which it is attempted to identify potentially automatable sequences [16].

Irrespective of the method by which an RPA bot model is obtained, without further optimization, it is likely not optimal [9,22]. For instance, users often switch back and forth between applications to transfer data, which may be

C. Di Ciccio et al. (Eds.): BPM 2024, LNBIP 527, pp. 154–168, 2024.
https://doi.org/10.1007/978-3-031-70445-1_10

reflected in the model but could be optimized in RPA. Additionally, the "building blocks" of RPA bots are highly granular, which quickly results in extensive models [28]. Considering that organizations have numerous RPA bots in place, particularly when employees are allowed to create automations on their own, it can be challenging to identify which bots may benefit from refactoring or optimization.

In this paper, we introduce complexity metrics for RPA bot models to be able to better assess the complexity of implemented automations. To this end, we review corresponding metrics from the related areas of software development and business process modeling and derive metrics for the RPA domain. These metrics allow to objectively measure and thus compare various aspects of complexity for RPA bot models. Among other aspects, this may assist in the identification of potential candidates for improvement, or to classify and cluster models according to specific characteristics.

In the following, we first examine various complexity metrics for software and process models in Sect. 2. In Sect. 3, we present our adaptations and new metrics for robotic process automation models that cover different aspects and are tailored to the specifics of RPA. Additionally, we have implemented the metrics in a vendor-independent RPA modeling platform as described in Sect. 4. Finally, the paper is concluded in Sect. 5.

2 Related Work: Complexity in Software and Processes

Robotic process automation with its visual, process-like approach for creating automation scripts borrows concepts from both, software engineering and business process modeling. On the one hand, its "building blocks" represent actual work instructions to be carried out on the system on specific software and data, like clicking a button or entering text in a text box, or represent other software tasks, such as testing for a regular expression or assigning a value to a variable [25]. On the other hand, these blocks are usually visually arranged and connected to define their execution order [29], resembling business process models with their activities connected by control flow [3]. In the following, we refer to these models as *RPA (bot) models*, whose nodes, i. e., building blocks, are *RPA operations* (cf. [25]).

At the same time, there are similarities between business processes and software [5] and various metrics[1] for software have been adapted and applied to business processes and their models (e. g., [5,10,21]).

Cardoso defines "process complexity as the degree to which a process is difficult to analyze, understand, or explain" [7] and distinguishes four different perspectives on process complexity: (i) activity complexity, related to the number of activities; (ii) control-flow complexity, considering the different execution paths; (iii) data-flow complexity, associated with data structures and parameters; and (iv) resource complexity, based on the entities required for activity execution.

[1] As in related work, we use the term *metric* to also refer to measures (e.g., [6,19]).

According to a structured literature review by Polančič and Cegnar [19], the most frequently discussed metrics in the literature include the *Number of Activities*, *Control-Flow Complexity*, and the *Nesting Depth*, which will be briefly introduced in the following. Additionally, many complexity metrics for business process models are concerned with the quantity of certain model elements, like the *Total Number of Gateways* [19].

Based on the *Lines of Code* metric commonly used in software, Cardoso et al. [5] introduce a metric for the total *Number of Activities* in a process, which can be extended to additionally include the number of control-flow elements in well-structured processes or the number of joins and splits.

Another metric inspired by software metrics is the *Control-Flow Complexity* [6], which is defined as the sum of the complexity scores of all control-flow splits (XOR, OR, AND) in the process, where the score corresponds to the number of different states a split introduces, e.g., the complexity score of an XOR-split corresponds to the number of its outgoing arcs.

However, the Control-Flow Complexity metric may not always be suitable for representing how difficult it is to understand a model [11]. Therefore, Gruhn and Laue [11] suggest another metric, again based on preexisting software metrics, that does not consider the complexity of all splits in isolation, but how many decisions must be made to reach a certain activity, i.e., its *Nesting Depth*. Based on the nesting depths of all activities in the model, the maximum and mean values can then be calculated for the model.

Besides other proposed metrics for the total numbers of various modeling elements and their ratios [21], a more sophisticated metric uses cognitive weights assigned to each modeling element [10]. Additionally, Cardoso et al. [5] adapted the complexity metrics for software by Halstead [13] to fit the business process context, resulting in the *Halstead-based Process Complexity* metrics for computing estimates for process length, volume, and difficulty. Like their original counterpart, they are calculated based on four basic measures that, for the business process domain, comprise the "number of unique activities, splits and joins, and control-flow elements" [5] and their total number, as well as the number of different data objects involved in the process and their total number of uses.

3 Complexity Metrics for RPA Bot Models

In this chapter, we introduce various metrics tailored to RPA bot models that assess their complexity regarding different aspects, partly by revising the metrics presented for software and business process models and adapting them to the RPA context.

In terms of the process-complexity perspectives presented by Cardoso [7], there are several characteristics that contribute to complexity in RPA bot models, as described in this chapter and also illustrated by the two toy examples given in Fig. 1. Bot models in RPA consist of (i) operations, corresponding to activities in business process models; which are put in order via (ii) control flow

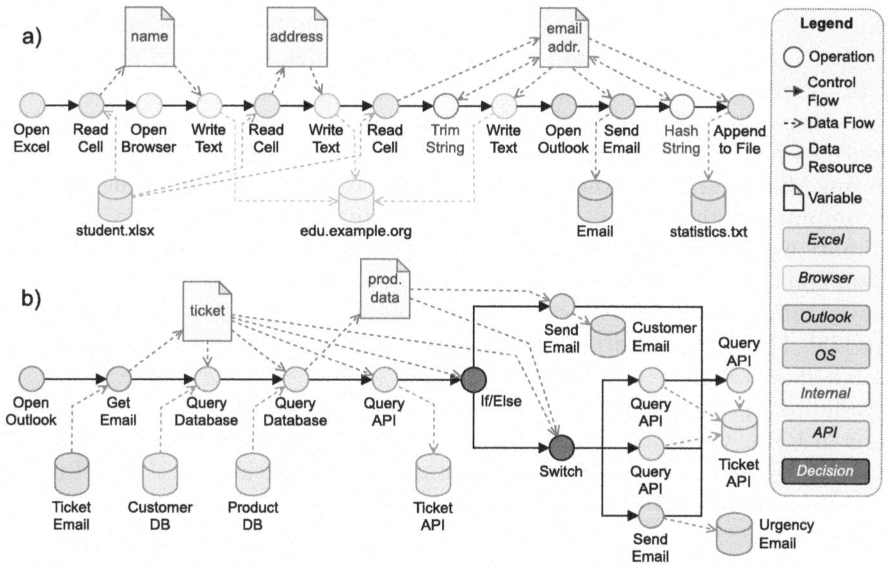

Fig. 1. Two simplified examples of RPA bot models in a generic notation featuring the operations and their ordering, as well as the internal variables and accessed data resources. The color coding is for illustration purposes only. (Color figure online)

by linking operations; (iii) data in the form of variables, defining how information is cached in the bot; and (iv) resources, from which information is retrieved or output to [26,27].

Model a) in Fig. 1 describes a commonly used example of a data transfer of personal information from an Excel workbook to a web form (e. g., [2,15,16]), extended by sending an email to confirm the transfer and a file used for keeping record. The other example, model b), describes a bot that triages support tickets received via email by enriching information and either directly replying to the customer or routing the ticket to an appropriate agent.

3.1 Basic Complexity Metrics

In the following, we introduce basic metrics for RPA bot models, inspired by the complexity metrics for software and business process models as introduced in Sect. 2 and structured according to the process-complexity perspectives of Cardoso [7].

Activity complexity plays an important role in RPA, as with the fine-grained, code-like operations that constitute an RPA workflow, the model dimensions quickly increase, causing even smaller automations to result in large models [28].

Foremost, based on the widely known metric measuring the lines of code in software or the number of activities in processes respectively, we introduce the

Number of Operations (NOO). NOO counts the number of building blocks an RPA bot model comprises, and thus provides a rough estimation of the overall size of the model.

As there are different types of operations, i. e., building blocks, for RPA models, we furthermore introduce the *Number of Automation Operations (NOA)*, counting all operations in the model that directly operate on and thus automate the computer system and its applications, like `Read Cell` or `Get Email`. Besides automation operations, RPA scripts also include operations that do not affect the system and only indirectly contribute to the automation, such as `Trim String` to remove trailing whitespace from a string, making decisions, or other preparatory steps [25]. These are nevertheless necessary and also contribute to the complexity of a bot model.

Building upon these two metrics and considering the different types of operations in RPA, we introduce the *Ratio of Automation Operations (RAO)*. RAO calculates the ratio of the number of automation operations in a model (NOA) to the total number of operations (NOO) in it:

$$RAO = \frac{NOA}{NOO}$$

This metric indicates whether the bot model features complex internal logic, as few automation operations compared to the total number of operations will result in a low RAO, or whether it is a more straightforward automation with only few auxiliary operations, indicated by a RAO close to 1.

Control-flow complexity in RPA, as in business processes, depends on the specific use case. While rather simple, linear automations do not exhibit a high control-flow complexity, RPA tools certainly also allow modeling complex behavior, especially regarding loops and decisions [24] that can be employed to create more complex and context-sensitive automations.

To be able to judge how many decisions are part of an RPA bot model, the *Number of Decisions (NOD)* can be determined. This can indicate whether it is a more linear automation with little variation, like in model a), or a more complex automation where the execution path may change at several points, as in model b). However, this does not consider the complexity of the decisions involved, i. e., regarding the decision logic or the number of decision branches. For the latter, an adaptation of the more complex *Control-Flow Complexity* metric, briefly outlined in Sect. 2, can be employed. It considers the number of possible states introduced by a decision, and will be described in more detail in Sect. 3.2.

Furthermore, the *Nesting Depth (ND)* metric can be applied to RPA bot models as well. It determines the maximum and mean number of decisions that must be made before an operation in the automation can be executed, reflecting the complexity and effort required to track and comprehend the control flow of the bot model (cf. [11]). For example, in model b), sending the urgent email requires two preceding nested decisions, resulting in a nesting depth of 2 for this operation.

Data-flow complexity may be more significant in RPA models than in traditional business process models, since one of the main purposes of RPA is to work with data, e. g., its retrieval, validation, and transfer [14,22]. While data is generally important in business processes as well, it is not necessarily specified in the model, which often consists of more abstract tasks. RPA, in contrast, requires a detailed specification of the data involved in each automation step, more specifically, the definition and usage of variables that cache data internally and that act as input or output for operations [27], as also illustrated in Fig. 1.

A basic measure related to data-induced complexity is the *Number of Variables (NOV)*, which counts how many variables, that cache data internally, are used in a bot model. Although it does not consider the complexity of the variables themselves, such as simple strings compared to complex tabular data or even objects, it can indicate the overall extent of the internal data structure.

A common aspect of RPA is data transformation [15], where information is read from a source and internally transformed by the bot to fit the format required by the target source. In model a), for instance, the email address retrieved from the Excel workbook is first trimmed of whitespace and later hashed before being added to the file, which considerably alters the original content. These transformations contribute to the complexity, as the information obtained and cached in a variable undergoes changes throughout the execution and consequently does not equal the information written back at a later stage. Therefore, for each variable, the number of operations that transform it, i. e., that read from it and write to it, can be determined. Based on these numbers, the maximum and mean *Number of Variable Transformations (NOVT)* can be computed for a model. If new variables are created at each step of transformation, they would not be accounted for in the NOVT but would increase the NOV instead.

Resource complexity is closely related to the data flow in RPA. There are various resources RPA can work with to extract or manipulate data, such as files, websites, or databases [27]. Additionally, the software and information systems RPA can interact with and operate on, such as Microsoft Excel, can be considered as resources, since they are a prerequisite for the automation's execution.

The *Number of Automated Software (NOS)* states how many different software applications are utilized and automated by an RPA bot.

Similarly, the *Number of Data Resources (NODR)* is the total number of different data resources accessed by the bot. This can be further subdivided into the *Number of Data Resources Read (NODRr)* and the *Number of Data Resources Written (NODRw)*, depending on the type of access performed on a data resource. A data resource may contribute to both numbers if the bot reads from it and writes to it.

Based on these measures, the *Ratio of Data Resource Access (RODRA)* can be calculated. It relates the number of data resources read (NODRr) to the number of data resources written (NODRw) in the following way:

$$RODRA = \frac{NODRr - NODRw}{NODRr + NODRw}$$

This metric always yields a value between −1 and +1 and allows determining whether a certain access type prevails. The more data sources are read from compared to written to in a bot model, i. e., NODRr > NODRw, the closer RODRA approaches 1, indicating that the bot likely performs a data compilation task. The closer RODRA is to −1, the more data sources are written to compared to read from, i. e., NODRr < NODRw, indicating that the bot is distributing data. If RODRA is approximately 0, the number of data sources written to and read from is comparatively balanced.

The data resource that is being worked on and the corresponding application that is utilized together define the context in which an automation operation of an RPA bot is performed [28]. For example, the Excel application and the `student.xlsx` workbook jointly define the context for the `Read Cell` operations in model a).

The *Number of Contexts (NOC)* totals the number of different contexts, i. e., unique pairs of application and data, that are present in the bot model. The *Number of Context Switches (NOCS)* quantifies how often the bot switches between different contexts, i. e., how often the active application or the accessed data changes. For example, switching from Excel to the browser constitutes a context change. However, operations that are not automation operations and are consequently not performed in a context, do not affect this count. For instance, the internal string operations in model a) or the decisions in model b) are ignored during the context analysis and thus do not influence these metrics.

By dividing the total number of contexts (NOC) by the number of context switches (NOCS)[2], the *Ratio of Context Switches (ROCS)* can be computed:

$$ROCS = \frac{NOC}{NOCS}$$

If ROCS equals 1, the bot visits each context exactly once. However, if ROCS is below 1, it means that contexts have to be revisited by the bot. A low ROCS indicates that the bot may exhibit human-like "back and forth" behavior, as may be observed in recordings, and thus may yield potential for optimization.

It is important to note that some of the presented metrics, the ratios in particular, are initially not control-flow sensitive. In model b), for example, either the customer email or the urgent email can be written, or even neither. However, in its basic form, the RODRA metric would simply count all accesses to data resources, without considering the logical constraints imposed by the control flow. One potential solution is to compute the mean value of the metric over all possible execution paths.

3.2 Compound Complexity Metrics

The metrics presented so far are primarily straightforward to determine, encompass different complexity perspectives, and can thus quickly provide an overview. Nevertheless, more complex metrics, as described in the following, may lead to

[2] Entering the first context is also considered a context switch, i. e., NOCS $\geq$ NOC.

more precise measurements of specific aspects or cover multiple perspectives at once.

Halstead-Based Process Complexity for RPA. The Halstead complexity metrics for software [13] have been adapted for process models by Cardoso et al. as Halstead-based Process Complexity (HPC) [5]. In the following, we similarly aim to adapt these metrics to the domain of RPA bot models.

The original metrics for software are based on four different measures that must first be determined for the given algorithm or script. These are: n_1, which represents the number of different operators present in the given piece of software, such as function names, brackets, or arithmetic operators; n_2, which is the number of different operands, such as the variables and constants involved; N_1, which denotes the total number of operators; and N_2, which signifies the total number of operands [12].

We define the terms operator and operand in the case of RPA as follows. *Operators* are, similar to software, the operations, i. e., the building blocks, used in an RPA bot. These blocks cover, as discussed before, automation operations, internal operations, and control-flow-altering operations like decisions (cf. [25]). This definition is analogous to the one used for the Halstead-based Process Complexity that refers to operators as activities, gateways, and other control-flow constructs [5] (cf. Sect. 2). *Operands* in the RPA domain are the variables and data resources used by the bot, again similar to software with its variables and constants. This is also consistent with the definition of operands for the HPC as "data variables that are manipulated by the process" [5]. With these definitions, the basic measures of the HPC in the context of RPA can be determined as follows:

n_1 = number of different operations in the model
n_2 = number of data resources and variables used in the model $(NOV+NODR)$
N_1 = number of all operations in the model (NOO)
N_2 = number of occurrences of (or access to) every data resource and variable

By applying the formulas of the Halstead-based Process Complexity [5], the following metrics can be obtained:

Bot Vocabulary: $n = n_1 + n_2$, the total number of unique elements in the model.
Bot Length[3]**:** $N = N_1 + N_2$, the total number of occurrences of each element in the model.
Bot Volume: $V = N * \log_2(n)$, intended to denote the implementation size, assuming that each distinct element is represented by a binary number [12].
Bot Difficulty: $D = \frac{n_1}{2} * \frac{N_2}{n_2}$, related to error sensitivity, which increases with the number of different operations used and the frequency of data reuse, i. e., if $N_2 > n_2$.

[3] In contrast to Cardoso et al., we use the original length formula, rather than the formula for the estimated program length [12], since we want to assess exactly the present model.

Table 1. Control-flow constructs in RPA and their Control-flow Complexity

Construct	CFC
if *(optional execution)*	1
if-else	2
if-elseif *(n times)*-else	$1 + n + 1$
switch *(with n cases)*	n
parallel	1

Control-Flow Complexity for RPA. As mentioned in Sect. 3.1, the basic measure for the number of decisions does not account for the complexity of the decisions involved. In this context, the *Control-flow Complexity (CFC)* for processes has been introduced in [6], which was later validated in a study that demonstrated the CFC to be highly correlated with the complexity of the control flow perceived by users [7]. This metric, inspired by McCabe's Cyclomatic complexity measure for software code, is based on an analysis of the model's control-flow structure and reflects the number of different states a process can reach, induced by the different types of control-flow splits, such as AND, XOR, and OR. It aims to quantify the amount of mental states a process designer needs to keep in mind during modeling or viewing a process model [6].

Given that RPA bot models are similar to process models as previously discussed, the Control-flow Complexity metric can also be applied to RPA, with some specific considerations, as, regarding decisions, RPA appears to be more comparable to software code than to business processes. For one, to the best of our knowledge, no RPA tool supports inclusive decisions (OR) that would allow executing multiple branches simultaneously. In a similar manner to software, once a condition is evaluated to true, the respective control-flow branch is executed, and the others are discarded. Exclusive decisions (XOR) in turn are widely supported, often in the form of "if-else", "if-elseif-else", and "switch" operators [24], which are analogous to the common software concepts. The parallel execution of operations (AND) is not commonly supported [26], but it is nevertheless considered in the following.

Table 1 lists the different types of control-flow operations in RPA and assigns them a Control-flow Complexity in accordance with the definitions in [6], i. e., each construct is assigned the number of possible execution paths initiated by it. For example, the switch in model b) has three cases that, correspondingly, induce three different potential execution paths (or states) that the bot could continue on. With regard to parallel execution, there is only one possible state that can be reached, as all outgoing branches are executed in any case, resulting in its CFC always being 1 (cf. [6]).

Based on these observations, the Control-flow Complexity of an RPA bot model can be determined by summing up the CFCs of all control-flow constructs in the model (cf. [6]).

Table 2. Selected metrics calculated for the two bot models given in Fig. 1. Values marked with * are means across all possible execution paths.

Metric	NOO	RAO	NOD	ND_{max}	CFC	NOV	$NOVT_{max}$	NOS	NODR	RODRA
Model a)	13	0.85	0	0	0	3	2	4	4	–0.5
Model b)	12	0.83	2	2	5	2	0	4	6	0.35*
Metric	NOC	NOCS	ROCS	n_1	N_1	n_2	N_2	Length	Vol.	Diff.
Model a)	4	8	0.5	9	13	7	20	33	132	12.86
Model b)	4	5*	0.8	7	12	8	18	30	117.21	7.88

3.3 Application to Example

In Table 2, the two RPA bot models shown in Fig. 1 are compared regarding selected metrics introduced in this paper. Given that the two examples are of a similar size, the observed differences in the metrics are not particularly significant, yet they point out noteworthy details and properties of the models.

Both models feature a similar RAO, i. e., they do not have an overly complex internal logic. However, in contrast to model a), model b) contains nested decision logic ($NOD = ND_{max} = 2$) and consequently might require more attention to its control flow when reviewing the model. In return, model a) includes at least one variable that is updated twice ($NOVT_{max}$), thus potentially requiring special attention to its data flow.

A considerable difference can be observed in their data resource access. Model a) has a RODRA of –0.5, indicating that data is written to more resources than read from and that it consequently likely distributes data. This can be confirmed in the model, as the student's data in the Excel file is transferred to a web application, used to send an email, and appended to a file. In contrast, the RODRA of model b) slightly hints at a data compilation task with a value of 0.35 (averaged across all execution paths).

Another interesting aspect is the ratio of context switches (ROCS). While for model b) it is relatively close to 1, the ROCS for model a) is 0.5, indicating that the bot needs to revisit contexts multiple times and consequently might yield potential for improvement. Looking at the model, the "back and forth" behavior while transferring data from Excel to the Browser could be optimized.

Finally, regarding the Halstead-based metrics, the bot length and volume are comparable, but model b) performs better with respect to the bot difficulty (Diff.). This is likely due to its slightly smaller size and the lower number of data accesses performed, despite having one more operand.

4 Ontology-Based Complexity Analyses

We have implemented the aforementioned metrics in the OntoRPA platform, which enables users to model RPA bots in a vendor-independent way based on a conceptualization [26], the ontology of RPA operations (ORPAO). Building on this foundation, each element in a bot model is associated with the corresponding concept in the ontology, providing semantic information about it.

Table 3. Complexity metrics and their definition based on the ontology, as well as the covered complexity perspectives (**a**ctivity, **c**ontrol flow, **d**ata flow, **r**esource). Concepts from the ORPAO are set in `typewriter`. #(`class`) corresponds to the number of elements of the (super-)type `class` in the model.

Metric	Calculation	A	C	D	R
Number of Operations (NOO)	#(`RPAOperation`)	X			
Number of Automation Operations (NAO)	#(`AutomationOperations`)	X			
Number of Decisions (NOD)	#(`DecisionSplit`)		X		
Nesting Depth (ND)	$\triangleq$		X		
For each `RPAOperation` its nesting level in the CF structure formed by `ControlFlowOperations`					
Number of Variables (NOV)	#(`TransientData`)			X	
Number of Variable Transformations (NOVT)	$\triangleq$			X	
For each `TransientData`, #(`RPAOperation`) that `requires` and `yields` this `TransientData`					
Number of Automated Software (NOS)	$\triangleq$				X
#(distinct `Software`) that are `automated` by at least one `RPAOperation` in the model					
Number of Data Resources (NODR)	#(`DataResource`)				X
Number of Data Resources Read (NODRr)	$\triangleq$	X			X
#(`DataResources`) that are at least once related to an `RPAOperation` via a `reads` association in the model					
Number of Data Resources Written (NODRw)	$\triangleq$	X			X
#(`DataResources`) that are at least once related to an `RPAOperation` via a `writes` association in the model					
Number of Contexts (NOC)	$\triangleq$	X		X	X
# of distinct tuples of the `Software automated` and `DataResources accessed` by an `RPAOperation` in the model					
Halstead n_1 (Operators)	#(distinct `RPAOperations`)				
Halstead n_2 (Operands)	NOV + NODR				

The ontology catalogs the available "building blocks" for RPA and relates them to different other aspects, such as the software automated and the kind of data accessed [25]. Additionally, it includes a knowledge base that allows to map existing RPA bots, e. g., implemented in UiPath or Robot Framework, to a conceptual bot model in OntoRPA, and vice versa.

The complexity metrics introduced in this paper can be universally determined for all existing RPA bot implementations, at least manually. To be able to compute the metrics automatically in OntoRPA, we associated them with the corresponding concepts in the ORPAO, as illustrated in Table 3. `RPAOperation` denotes the superclass of all available building blocks, and they may be associated with certain `DataResources`, such as files, or `TransientData`, i.e., bot-internal variables, they access [27]. In accordance with Table 3, the *Number of*

Data Resources Read (NODRr), for example, can be computed by counting all `DataResources` that are part of the RPA bot model at hand and that are connected to at least one `RPAOperation` in the model via a `reads` relation. In model b), this would apply to the ticket email, the customer database, and the product database, for example.

The metrics component[4], which has been incorporated into the open-source OntoRPA platform, enables users to compute various metrics presented in this paper for bot models stored in the repository. The component leverages the process tree-like structure that OntoRPA uses internally, as well as the semantic information about each node in a model provided by the ORPAO, as described previously and in Table 3.

5 Discussion and Conclusion

In this paper, we introduced metrics that measure different aspects of complexity in RPA bot models and that were partly inspired by existing complexity metrics for software and business process modeling. These metrics are described and defined in a general manner and are thus not bound to a specific RPA tool but can be determined for any RPA implementation. Additionally, we provided an implementation that can automatically calculate these metrics for models within the OntoRPA platform, drawing on their semantic annotations.

Being able to assess the complexity of RPA bots in an automated and objective way allows not only comparing different models on various aspects, but also to categorize them based on certain characteristics, such as whether the automation rather distributes or compiles data. Additionally, it enables the identification of bots with potential for improvement, such as those that feature unnecessary context switches. These metrics may improve the management of large bot repositories by providing insights, categorizing bot models, and highlighting complex implementations in an automated manner.

So far, the metrics provided in this paper cover primarily rather simple aspects of RPA bot models. More research is necessary to better understand and assess the perceived complexity and difficulty of such models by users. Most importantly, the proposed metrics should be validated in future research, particularly the compound metrics, to test if further RPA-specific adjustments are required. Additionally, user studies could be conducted to derive "cognitive weights" for the different types of elements in RPA bot models, similar to the work by Gruhn and Laue for process models [10].

Previous work demonstrated that certain model metrics correlate with the occurrence of errors, and thus can be used to predict the probability of errors in the model [17]. It is conceivable that such correlations also exist with RPA, which would further increase the importance of metrics for RPA models, and consequently should be investigated in the future.

[4] https://github.com/bptlab/onto-rpa-platform/tree/main/components/metrics.

Besides complexity, there are also various other aspects that could be measured, such as quality, changeability, and fault tolerance as for business process models [23]. In addition, runtime measurements should be considered, for instance, cycle times and average durations of the individual automation steps. Such information may provide relevant information for improvements and for finding more efficient ways of automation, similar to the experiments conducted by Průcha and Skrbek [20].

References

1. van der Aalst, W.M.P., Bichler, M., Heinzl, A.: Robotic process automation. Bus. Inf. Syst. Eng. **60**(4), 269–272 (2018)
2. Agostinelli, S., Lupia, M., Marrella, A., Mecella, M.: Reactive synthesis of software robots in RPA from user interface logs. Comput. Ind. **142**, 103721 (2022). https://www.sciencedirect.com/science/article/pii/S016636152200118X
3. Aguirre, S., Rodriguez, A.: Automation of a business process using robotic process automation (RPA): a case study. In: Figueroa-García, J.C., López-Santana, E.R., Villa-Ramírez, J.L., Ferro-Escobar, R. (eds.) WEA 2017. CCIS, vol. 742, pp. 65–71. Springer, Cham (2017). https://doi.org/10.1007/978-3-319-66963-2_7
4. Bock, A.C., Frank, U.: Low-code platform. Bus. Inf. Syst. Eng. **63**(6), 733–740 (2021)
5. Cardoso, J., Mendling, J., Neumann, G., Reijers, H.A.: A discourse on complexity of process models. In: Eder, J., Dustdar, S. (eds.) BPM 2006. LNCS, vol. 4103, pp. 117–128. Springer, Heidelberg (2006). https://doi.org/10.1007/11837862_13
6. Cardoso, J.: Control-flow complexity measurement of processes and weyuker's properties. In: 6th International Enformatika Conference. Transactions on Enformatika, Systems Sciences and Engineering, pp. 213–218. Enformatika, World Enformatika Society (2005)
7. Cardoso, J.: Business process control-flow complexity. Int. J. Web Serv. Res. **5**(2), 49–76 (2008)
8. Correia, C., Rodrigues da Silva, A.: Platform-independent specifications for robotic process automation applications. In: Pires, L.F., Hammoudi, S., Seidewitz, E. (eds.) MODELSWARD 2022, pp. 379–386. SciTePress (2022)
9. Flechsig, C., Lohmer, J., Lasch, R.: Realizing the full potential of robotic process automation through a combination with BPM. In: Bierwirth, C., Kirschstein, T., Sackmann, D. (eds.) Logistics Management. LNL, pp. 104–119. Springer, Cham (2019). https://doi.org/10.1007/978-3-030-29821-0_8
10. Gruhn, V., Laue, R.: Adopting the cognitive complexity measure for business process models. In: Yao, Y. (ed.) 5th IEEE International Conference on Cognitive Informatics, pp. 236–241. IEEE Computer Society (2006)
11. Gruhn, V., Laue, R.: Complexity metrics for business process models. In: Abramowicz, W. (ed.) Business Information Systems - 9th International Conference on Business Information Systems (BIS 2006), pp. 1–12. LNI, Gesellschaft für Informatik e.V. (2006), https://dl.gi.de/items/e0000ba6-71b2-4889-a664-4c07cc124ed3
12. Halstead, M.H.: Natural laws controlling algorithm structure? ACM SIGPLAN Not. **7**(2), 19–26 (1972)

13. Halstead, M.H.: Elements of Software Science, Operating and Programming Systems Series, vol. 2. Elsevier, New York (1977)
14. Hofmann, P., Samp, C., Urbach, N.: Robotic process automation. Electron. Mark. **30**(1), 99–106 (2020)
15. Leno, V., Dumas, M., La Rosa, M., Maggi, F.M., Polyvyanyy, A.: Automated discovery of data transformations for robotic process automation. In: Zhang, D., Freitas, A., Tao, D., Song, D. (eds.) Proceedings of the AAAI-20 Workshop on Intelligent Process Automation (IPA-2020). AAAI Press (2020). https://minerva-access.unimelb.edu.au/handle/11343/241635
16. Leno, V., Polyvyanyy, A., Dumas, M., La Rosa, M., Maggi, F.M.: Robotic process mining: vision and challenges. Bus. Inf. Syst. Eng. **63**, 301–314 (2021)
17. Mendling, J., Neumann, G., van der Aalst, W.: Understanding the occurrence of errors in process models based on metrics. In: Meersman, R., Tari, Z. (eds.) OTM 2007. LNCS, vol. 4803, pp. 113–130. Springer, Berlin (2007). https://doi.org/10.1007/978-3-540-76848-7_9
18. Penttinen, E., Kasslin, H., Asatiani, A.: How to choose between robotic process automation and back-end system automation? In: Bednar, P.M., Frank, U., Kautz, K. (eds.) ECIS 2018 Proceedings. AIS (2018)
19. Polančič, G., Cegnar, B.: Complexity metrics for process models - a systematic literature review. Comput. Stand. Interfaces **51**, 104–117 (2017). https://doi.org/10.1016/j.csi.2016.12.003
20. Průcha, P., Skrbek, J.: API as method for improving robotic process automation. In: Marrella, A., et al. (eds.) BPM 2022. LNCS, vol. 459, pp. 260–273. Springer, Heidelberg (2022). https://doi.org/10.1007/978-3-031-16168-1_17
21. Rolón, E., Ruiz, F., García, F., Piattini, M.: Applying software metrics to evaluate business process models. CLEI Electron. J. **9**(1) (2006)
22. Syed, R., et al.: Robotic process automation: contemporary themes and challenges. Comput. Ind. **115** (2020)
23. Sánchez González, L., García Rubio, F., Ruiz González, F., Piattini Velthuis, M.: Measurement in business processes: a systematic review. Bus. Process Manag. J. **16**(1), 114–134 (2010). https://doi.org/10.1108/14637151011017976
24. Völker, M., Siegert, S., Weske, M.: Adding decision management to robotic process automation. In: González Enríquez, J., Debois, S., Fettke, P., Plebani, P., van de Weerd, I., Weber, I. (eds.) BPM 2021. LNBIP, vol. 428, pp. 23–37. Springer, Cham (2021). https://doi.org/10.1007/978-3-030-85867-4_3
25. Völker, M., Weske, M.: Conceptualizing bots in robotic process automation. In: Ghose, A., Horkoff, J., Silva Souza, V.E., Parsons, J., Evermann, J. (eds.) ER 2021. LNCS, vol. 13011, pp. 3–13. Springer, Cham (2021). https://doi.org/10.1007/978-3-030-89022-3_1
26. Völker, M., Weske, M.: Ontology-supported modeling of bots in robotic process automation. In: Ralyté, J., Chakravarthy, S., Mohania, M., Jeusfeld, M.A., Karlapalem, K. (eds.) Conceptual Modeling, pp. 239–254. Springer, Heidelberg (2022). https://doi.org/10.1007/978-3-031-17995-2_17
27. Völker, M., Weske, M.: On the nature of data in RPA bots. In: Monti, F., Rinderle-Ma, S., Ruiz Cortés, A., Zheng, Z., Mecella, M. (eds.) Service-Oriented Computing. LNCS, vol. 14420, pp. 29–37. Springer, Cham (2023). https://doi.org/10.1007/978-3-031-48424-7_3

28. Völker, M., Weske, M.: Ontology-based abstraction of bot models in robotic process automation. In: Almeida, J.P.A., Borbinha, J., Guizzardi, G., Link, S., Zdravkovic, J. (eds.) Conceptual Modeling, pp. 239–256. Springer, Cham (2023). https://doi.org/10.1007/978-3-031-47262-6_13
29. Willcocks, L.P., Lacity, M., Craig, A.: The IT function and robotic process automation: The Outsourcing Unit Working Research Paper Series (15/05), London, UK (2015). http://eprints.lse.ac.uk/64519/

Decision-Making in Robotic Process Automation Programming and its Influence on Robotic Process Mining

Tom Hohenadl[1,2](✉), Bernhard Axmann[2], and Christian Stummeyer[2]

[1] KU Eichstätt-Ingolstadt, Ostenstr. 28a, 85072 Eichstätt, Germany
tom.hohenadl@stud.ku.de

[2] Technische Hochschule Ingolstadt, Esplanade 10, 85049 Ingolstadt, Germany
{bernhard.axmann,christian.stummeyer}@thi.de

Abstract. Robotic Process Automation (RPA) technology leverages software bots to automate repetitive tasks within digital environments. While RPA offers significant efficiency gains, current bot programming processes remain heavily manual, requiring extensive effort from developers and domain experts. A current research gap exists in the factors influencing developer decisions during the low-code programming of RPA bots, specifically the choices they make when implementing a selected process. This limited knowledge hinders efforts to automate the bot programming process itself utilizing Robotic Process Mining (RPM) methods. This research study aims to address this knowledge gap by investigating the decision-making of RPA developers. A semi-structured interview study with RPA professionals was conducted to explore the factors influencing developers' decisions when selecting and translating manual actions into automated processes. The findings from the interview have been the basis to show their implications on RPM tools and to give guidance for the improvement of RPM tools.

Keywords: Robotic Process Automation · Robotic Process Mining · Task Mining · Process Automation

1 Introduction

Robotic Process Automation (RPA) has become a ubiquitous capability within the modern business landscape. As organizations strive to streamline operations and improve efficiency, RPA technology offers a compelling solution for automating repetitive, rule-based tasks [11]. Building on this foundation, Robotic Process Mining (RPM) emerges as a complementary technology that further optimizes RPA capabilities [8].

Despite the promise of automatic process selection and bot generation through recorded user behavior, RPM faces a critical hurdle: replicating human decision-making. Current RPM approaches heavily rely on pre-defined rules [2,15], limiting their ability to capture the dynamic choices developers

C. Di Ciccio et al. (Eds.): BPM 2024, LNBIP 527, pp. 169–184, 2024.
https://doi.org/10.1007/978-3-031-70445-1_11

make when translating user actions into RPA operations [26]. While RPM tools hold immense potential for streamlining RPA development, their ability to mimic human developers' decision-making capabilities remains limited. To achieve truly autonomous automation, where machines can not only identify tasks but also make informed decisions during implementation, RPM requires a deeper understanding of the factors influencing developer choices. This understanding can guide the development of more intelligent RPM tools that can not only automate process selection but also RPA bot design, ultimately achieving a higher level of automation that surpasses current limitations. This technological gap presents a significant research opportunity. To address this critical need and pave the way for the next generation of intelligent RPM tools, this study answers the following research questions:

Research Question 1: What factors influence how manual actions are being automated? - Identifying the considerations guiding developer implementation of RPA bots can yield valuable insights into the development decision-making process.

Research Question 2: What is the impact of these findings on the RPM framework? - Mapping identified factors influencing developer decisions to the current state of RPM can assess limitations of existing tools and pinpoint potential areas for improvement. This analysis will provide a vital theoretical foundation for future RPM development efforts, empowering the creation of more intelligent and adaptable RPM tools that can effectively support businesses in the automation journey.

The paper is structured into Sect. 2 outlining the RPA project lifecycle next to the RPM framework, Sect. 3 containing our interview study approach and the interview participant demographics, followed by Sect. 4 answering the first research question. Afterwards, a discussion of the theoretical challenges identified in section Sect. 5 is presented before concluding the paper in Sect. 6.

2 Related Work

RPA projects focus on the development of bots and follow a phase based project lifecycle. Multiple RPA project frameworks have emerged, such as a meta-synthesis in Syed et al. [24], a phase based framework from Herm et al. [11], or a lifecycle approach from Jimenez-Ramirez et al. [13].

The approach focus ourselfs on the lifecycle approach for RPA projects, as it is on the same level as the RPM model described by Dumas et al. [8]. To compare both approaches we aligned the respective phases and create a flowchart, which is displayed in Fig. 1. On the right side of the figure the RPM pipeline is visualized. The left side contains a visualization of the lifecycle stages of a bot as outlined in [13]. Both approaches however start with the selection of the process scope and a context analysis. This includes the screening of tools and implementing a proof of concept [11]. Once RPA is selected as a business project to implement, suitable process candidates have to be selected. At this point the two approaches of manual RPA development in the RPA project lifecycle of Jimenez-Ramirez et al.

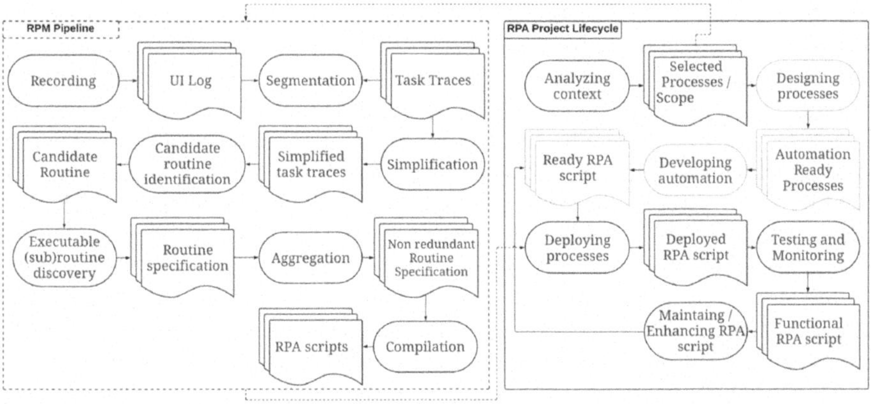

Fig. 1. RPA-to-RPM interrelation overview based on [8,13]

[13] and the automated RPM pipeline of Dumas et al. [8] diverge. While in the manually conducted process business have to select suitable candidates based on human experience and screening of processes. Detailed research on process selection criteria and the benefitfactors can be found in [4,9,12]. Once a process has been selected in the manual framework the design phase, including rework and optimization [24] are conducted. The optimized process is developed using the available tools in a RPA team, often referred to as center of excellence [11], or through business users, so called citizen developers [20]. Once the manual development the script is deployed as RPA environments are characterized by the lack of testing environments [13]. At this point the RPM pipeline also does follow the RPA project lifecycle again. Before this conjunction however, RPM follows a different, automated approach.

A set of users are recorded using a logging tool, such as Action Logger [16], smartRPA [3], or screenRPA [17]. The UI log is passed into a segmentation process, which identifies task traces. A task is a sequence of user actions with information systems [1]. Once the tasks are identified the data is filtered for noise, such as redundant copy activities without pasting or simply browsing the web [16]. The UI log then further is processed for candidates to be automated. Not all processes can be automated as RPA heavily relies on if-then-else rules [9]. Dumas et al. refer to this as determinism, which refers to a bots ability to always determine the next RPA operation to be operated [8]. If this cannot be mined from the task traces, the bot currently cannot be compiled into a working RPA script. Furthermore, the RPM tool has to evaluate the frequency as a focus criterion as well, because frequently executed tasks make a more suitable candidate routine [12].

Once the routine is discovered its actions have to be mapped to executable RPA operations through a transformation methodology. This currently involves hard-coded rules on how to transfer actions or action sequences into executable RPA operations as inspectable in the source code of smartRPA [3]. Actions that cannot be automated by design will therefore not be transferred into the routine specification at the current point in time. The last step before bot compilation is the aggregation of equal events into an optimal routine for automation [8]. Redundant variants may be the filling of a web form in a different order in two recorded task executions. Finally, the mapping is conducted and a executable process specification in the format of an RPA bot is created [2]. At this point, the same state as after the manual development of a bot is achieved through the RPM pipeline. The output of the pipeline can be used by the RPA experts and the final manual phases of the RPA project lifecycle can be fulfilled. These include the go-live, testing in the runtime environment, as well as the bots monitoring. Lastly, the RPA bots have to be maintained in case a process, some user interface or application programming interface, or the RPA software change [18].

As highlighted, a critical challenge lies in the rigid translation of user actions into RPA operations. This difficulty stems not only from the limitations of identifying all possible actions upfront, but also from the continuous evolution of RPA software functionalities, business processes, and the businesses' technology landscape. Having established the key steps within the RPM pipeline and RPA project lifecycle, particularly the identification of variants and their automated compilation, we now shift our focus to the interview study to identify aspects in manual process development, that can reduce the before mentioned difficulties.

3 Methodology

This section contains a description of the design, execution, and coding cycles of the qualitative study conducted based on a semi-structured interview process. Figure Fig. 2 outlines the phases of the research approach including the interview study as well as the mapping of the findings on the RPM pipeline, which is shown in Fig. 1.

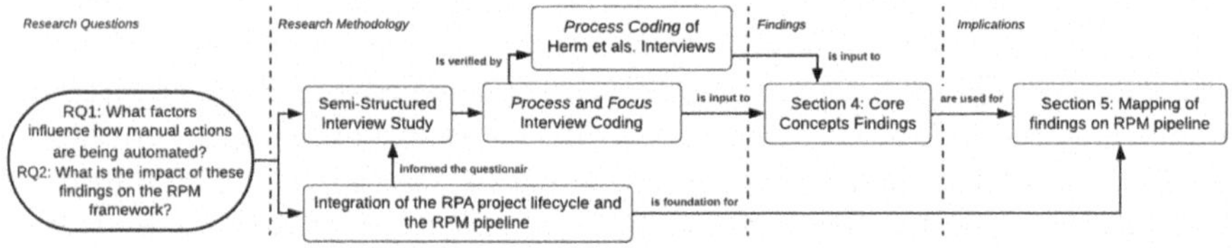

Fig. 2. Research Methodology and Section Outline

3.1 Interview Study Design

Drawing upon Padgett's guidelines [19], the interview protocol was constructed with three key sections. The first section gathered data on participants' general RPA experience for statistical comparison and validity purposes. The second

Table 1. Demographics of original interview study participants

	Sector	Role	RPA Exp.*	TM Exp.**
P1	Process Mining	Research Business Consultant	6/0	6
P2	RPA Vendor	Senior Solution Engineer	5/10	1
P3	RPA Vendor	Senior Solution Architect	10/50-100	NA
P4	RPA Vendor	Lead Automation Solution Engineer	7/20	1
P5	RPA Consultant	Project Manager	7.5/50+	7.5
P6	RPA Developer	Senior RPA Developer	5/50	1
P7	RPA Developer	Junior RPA Developer	1/1	NA
P8	RPA Developer	Competence Lead	6/25-30	NA
P9	RPA Developer	Senior RPA Consultant	4/50-100	16
P10	RPA Developer	RPA Developer	4/40-45	NA
P11	RPA Developer	KI Model Developer	4/Few	NA
P12	Academia	PhD Candidate	3/0	2
P13	Academia	PhD Candidate	3/0	3
P14	Academia	PhD Candidate	1/0	4
P15	Academia	Assistant Professor	6/100+	6

*RPA = Experience Years/Developed Bots, ** Task Mining Experience in Years

section explored the documentation and implementation of user actions within their experienced RPA projects. Finally, the third section investigated participants' understanding of the relationship between RPA and task mining, user interaction log processing, and their experience with these areas. To achieve the research objective, a strategic selection process identified three specialist groups: RPA developers, RPA businesses, and RPA researchers. Participants had to either (i) be an active RPA developer, (ii) have researched Task Mining, RPM, or RPA in the last 5 years, or (iii) be responsible for development or solution architecture at an RPA software vendor. The interviewees were selected from the authors' professional network and from network contacts, ensuring diversity in their affiliation, job role, and position based on the participant inclusion criteria. Yet, a focus on developers was leading the recruitment process as they implement the user actions into the RPA bots. The participant pool comprises six developers, four RPA software vendor participants, one RPA consultant, and four research participants. Table 1 summarizes the demographic characteristics of the 15 interview participants.

3.2 Interview Study Data Collection

The semi-structured interview phase involved conducting interviews with all 15 participants to comprehensively explore the research topic. The interviews were recorded using a web-based conference system. The recordings were transcribed, anonymized, and prepared for first-cycle coding. While the interviews provided sufficient evidence, the analysis of the coding revealed an opportunity to further solidify our understanding of RPA development decisions. To achieve this, a follow-up survey was conducted with all participants after the interviews. This survey aimed to gather more specific data on a particular aspect of RPA operations, namely the capability of RPA tools to replay tasks utilizing the RPA operations offered by RPA technology as described by Völker and Weske [26]. Six interviews from Herm et al. [10] exploring RPA project stages were incorporated to complement the analysis. This diversified the spectrum of views further to generate a broad foundation for the qualitative coding process. The participant demographics from these interviews are available in Table 2.

Table 2. Demographics of Herm et al. interview participants [10]

	Sector	Role	RPA Experience (Years)
A1	Banking	Internal Consultant	1–2
A2	Consulting	External Consultant	3–5
A3	Marketing	Process Owner	6–8
A4	Sales	Internal Consultant	1–2
A5	Process Management	Process Lead	2
A6	Not Available	Team Lead CoE RPA	2

3.3 Qualitative Coding Process

After all, interviews were conducted, the first coding round for the newly conducted interviews was carried out by *process coding* [23]. Process coding is conotating action in the data to capture observable activity. The selection of this initial coding cycle method was driven by the objective of uncovering decision-making activities and their influencing factors within real-world RPA development practices. Overall interviews 689 coded segments emerged during the first coding cycle. These 689 segments were coded utilizing 39 low-level process codes that were developed through the first cycle of coding. The codes were transferred into the qualitative data analysis software MAXQDA. During the transfer, *focus coding* was used to generate meaningful concepts of the low-level codes and create our coding scheme [7]. The focus coding round resulted in five themes displayed in Fig. 3, which are made up of 39 low-level codes. The number in Fig. 3 outlines the overall quantity of coded segments per theme. Incorporating the data from Herm et al. [10] provided a valuable additional resource. Coding these interviews confirmed the validity of the codes identified in the original 15 interviews. This is particularly significant because Herm et al.'s research had a different focus.

3.4 Thematic Saturation and Analysis Concepts

The analysis utilized a rigorous approach to achieve thematic saturation for the research question. This involved iterative coding of interview data. New low-level codes emerged steadily until the analysis of 13 interviews. Subsequent interviews yielded no new codes, suggesting saturation had been reached and the core themes relevant to the research question had been captured [5]. Equally, no new codes emerged from including additional interviews from Herm et al. [10] in the analysis pool. This signifies that no new themes or significant insights emerged from the later interviews, indicating a comprehensive understanding of the participants' experiences.

4 Findings

The iterative coding process yielded six distinct conceptual themes. These themes represent central ideas or patterns identified within the interview data. While not mutually exclusive, some themes may exhibit overlap. In such cases, a nuanced discussion will delineate the boundaries of each concept and illuminate their interplay. All interview quotes presented here are clean verbatim for improved readability and comprehension.

The core concepts were further clustered into three broader categories, which have been mentioned repeatedly in the related literature [2,4,9,13]: personal experiences with RPA, technological considerations for implementation, and the organizational context surrounding RPA adoption.

Figure 3 visually represents how the following described concepts map to these broad categories and how they overlap. This visualization complements the textual description and provides a clearer picture of the multifaceted nature of RPA development decision dependencies and what was prioritized throughout the interviews and is relevant for programming decisions as the coding quantity displays (number in brackets).

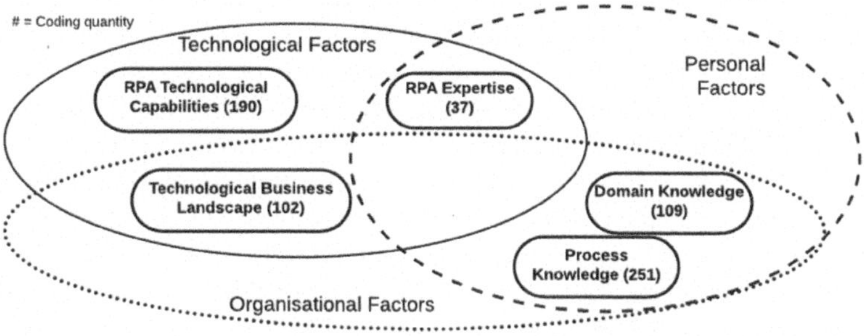

Fig. 3. Venn Diagram of concept interrelation including code quantity

4.1 RPA Expertise

This section explores the concept of RPA experience, which refers to the accumulated knowledge base of individuals responsible for automation initiatives.

Seasoned developers possess a nuanced understanding of the specific capabilities offered by the RPA tools they use. This allows them to leverage the full potential of the technology for various automation scenarios. By leveraging this expertise, developers are empowered to make informed decisions that align with the overall RPA strategy and the desired automation goals. P5 specifically states to "have an understanding of individual tools [to] understand the client side as well as the discussions, the thought process, the mindset of a client with respect to automation." P7 answered that expertise allows developers to understand what RPA building blocks "are more used for and [..] why they are better developed" leading to a higher resilience of RPA bots. P6 even states "I probably have something in my mind that I look for, but I haven't thought about it that much. I think it comes like automatically." Therefore, the expertise does intuitively influence the developers when implementing bots. Seasoned developers possess the ability to make a broader range of automation choices, drawing upon their experience with previously developed RPA solutions, "because they know what they built" (P10) and possessing a cognitive implementation repository. They have likely encountered a wider variety of RPA building blocks within the chosen tool-set compared to citizen developers or those with limited development experience. P8 particularly outlines that their developers "have a background in business information systems. So, they understand the business as well as the IT", whereas the citizen developers are primarily business experts.

4.2 RPA Technology Capabilities

A critical point of distinction for RPA tools lies in their approach to external system integration. "Leveraging building blocks for low-level actions in RPA expedites the automation development process. These pre-built components enable the user to quickly assemble automation workflows without the need to code every action from scratch" (P1). These variations necessitate user development decisions regarding the most appropriate approach for a given automation scenario. Developers must consider the technical possibilities offered by the chosen and available RPA tool when mapping actions within the automation process. They further have to select the best fit for automating, if there are multiple solutions available. All participants mentioned some form of building blocks in their RPA development because it is "more convenient to use the dedicated RPA operation for certain types of activity" (P4) than building a workaround.

However, "in some cases, generic building blocks alone are insufficient to implement application-specific logic" (P12). Thus, the developers face the decision of how to work around this limitation posed by the available functions of the RPA tool as they "tend to understand what is possible through the bot and where exactly the bot will get stuck" (P5). RPA is an enabler for UI automation but is limited in its functionality in specific scenarios. P7 compares RPA to a "normal programming language" and "felt restricted because you could only use the building blocks the engine provides". The capabilities of RPA tools, particularly their approach to external system integration and the availability of pre-built building blocks, significantly influence development decisions.

4.3 Technological Business Landscape

Beyond the capabilities of the chosen RPA tool itself, developers must navigate the complexities of the surrounding business process environment. This environment, referred to as the Business Process Ecosystem (BPE), presents several factors that influence its suitability for RPA automation. Developers must assess the UI design and automation capabilities of the target system. "For example [...] Salesforce. The UI is not so automation-friendly, [because] sometimes there are overlapping UI elements and it's hard to build a stable automation on it" (P10). In this case, an application limits the developer from implementing resilient RPA. The developers have to find workarounds, such as using the API (P10), to fulfill the process requirements. However, the BPE can provide additional tools, such as Generative AI, natural language processing, or already-developed automated solutions. Thus, in the case RPA falls short, "it has to be clubbed with other ancillary technologies" (P5) to achieve the automation goal.

The runtime environment, where the RPA bots are deployed and executed, introduces another layer of decision-making for developers. A critical factor is whether the bot will operate in an unattended manner or require some level of human intervention. This unattended vs. attended automation approach has a significant impact on how error handling is designed and the overall stability of the automated process is ensured. "[RPA] will not be able to take care of certain exceptions which require human intelligence" (P5). Additionally, the concept of Human-in-the-Loop comes into play as the developer needs "to understand whether this particular solution should have a human in the loop" (P5).

Finally, technical aspects of the runtime environment, such as resolution settings and click coordinates, must also be considered, as they can influence how UI interactions have to be programmed within the bot itself. Similarly, if the runtime environment should allow certain currently not automate-able behavior, in P4s case multi-factor authentication (MFA), it sometimes needs "that discussion with the IT team and the security team [...] to accommodate this MFA" (P4) to allow automating the process using RPA.

4.4 Process Knowledge

Once a suitable process has been selected by the business or the business analysts, the automation developers often improve and adapt the process. Because in RPA "it makes sense to have a hybrid role and that hybrid role is a mixture between analyst and developer" (P8). Therefore, the developers transfer the process documentation into a solution documentation while focusing on improvement. This primarily includes the factors of speed, process stability, and reduction of development effort. Generally speaking, an automated process "could be a little bit more lean than what it used to be" (P3). Especially, the identification of irrelevant behavior or redundant behavior is an influencing factor during development. For example, actions that users perform manually could be executed faster at runtime and within the RPA bot, because "it's much faster than doing the same thing from within the front end" (P4).

While process documentation often reflects the ideal, streamlined happy path from a business perspective, it may not capture the full picture for RPA development. In some cases, variants and decision points are documented through "executing a couple of scenarios which are exceptions" (P4), allowing developers to implement if-then-else rules for the observed behavior. However, interviewee P9 highlights a crucial challenge: "Real-world process execution often involves further deviations and exceptions." Documentation often falls short in capturing these complexities, particularly regarding decision points, which are often described as "not so clear" and the "trickiest part in development" (P9).

Furthermore, the understanding of the data manipulated during a process is paramount. Developers leverage this knowledge to utilize variables and process data effectively within the bot (P9). This can involve, for example, gathering relevant values and performing calculations within the RPA bot itself, rather than relying on calculations within the target application. As interviewee P10 emphasizes, "it's important to have the exact logic behind the process," and data is a key driver of that logic. Process knowledge empowers developers to identify recurring patterns within this logic. By leveraging this understanding, they can reduce coding overhead by utilizing RPA operations with loops or nested cases instead of coding all scenarios sequentially.

4.5 Domain Knowledge

While process knowledge is crucial for efficiently and stably translating individual tasks into RPA bots, domain knowledge plays a distinct, yet equally essential role. Similar to the BPE, domain knowledge focuses on the surrounding factors that influence development decisions. However, unlike technical considerations, domain knowledge examines how work is performed within the specific RPA use case and aligns with the broader RPA strategy goals. By leveraging the domain knowledge, developers can evaluate the documented process and prioritize steps based on their relevance to the automation objectives. This knowledge can even empower them to critically assess the necessity or relevance of actions in the business use case, potentially leading to the removal of irrelevant tasks from the automation workflow. P15 outlines that "you need to know a priori what are the user actions that are not useful for your goal" which further can be explained because "the developer has zero knowledge of the process and you want to put every point in there [...] that may arise from the development point of view" (P3). Furthermore, domain knowledge enables developers to consider the bigger picture, encompassing the upstream conditions leading into the process and the downstream steps that follow. "But it's quite difficult to actually get the sense of the end-to-end process" (P9). Through domain knowledge, the developer can add the domain semantics to the input data and provide the output of the process in the way that the next link in an end-to-end process chain needs this output.

5 Implications on Robotic Process Mining

Following a case study approach informed by Runeson et al. [22], a further comparative analysis of the revealed findings and available RPM and Task Mining tools was conducted. The findings of Sect. 4 presented in Fig. 3 were aligned to the phases of the RPM pipeline, which is outlined in Fig. 1. The following subsection are therefore named after the RPM pipeline phases described in Sect. 2. The analysis examined the decision-making processes involved in manual RPA development, as revealed through expert interviews, alongside the capabilities of currently available RPM tools. Existing, publicly available RPM artifacts such as ActionLogger [16], SmartRPA [3], screenRPA [17], and task mining solutions from the industry leaders UiPath [25] and Celonis [6] served as the basis for this comparison. By mapping interview-derived insights onto the functionalities of these RPM tools, a head-to-head comparison along the model in Fig. 1 is established. This comparison aims to assess the potential for RPM to replicate and potentially surpass human decision-making capabilities within the context of RPA development.

5.1 Segmentation, Routine-Identification, and -Discovery Phases

A key limitation of traditional manual process documentation lies in its inherent inability to capture all process variants comprehensively. As highlighted in Sect. 4, this can lead to incomplete process understanding and hinder effective RPA development. RPM tools offer a more robust alternative through their integrated process discovery methods based on the recorded UI logs, which can be segmented into individual task traces [2]. The RPM approach leverage process discovery algorithms to analyze the recorded process executions, enabling the identification of process variants across multiple execution instances from different persons or points in time [14]. This capability eliminates the need for developers to possess in-depth domain knowledge to uncover all variations. Consequently, the variant detection methodology inherent to RPM tools surpasses the limitations of traditional manual process documentation. A critical challenge remains in automatically generating RPA bots that handle process variants. While current RPM tools can identify multiple variations within a process [2], they struggle to translate that identification into variant-specific bot functionality. This limitation stems from the difficulty of automatically extracting the reasoning behind variant selection from UI logs. Unlike humans, who can leverage contextual cues and domain knowledge to understand why a process deviates, current RPM tools automate one process path into a single bot. This hinders their ability to replicate the dynamic decision-making that occurs during execution path selection at runtime. Consequently, automated bot creation for processes with inherent variants remains elusive with existing open-source RPM tools.

5.2 Simplification and Aggregation Phases

As with any automation initiative, an inefficient process automatized will be an inefficient automated process. Thus, another challenge for RPM tools lies in their current inability to replicate the human ability to identify and address process inefficiencies during development. While humans can analyze process documentation or user behavior and rely on domain knowledge to identify areas for improvement, RPM tools currently focus solely on automating existing actions. This limitation prevents RPM from achieving process optimization, which is a key step in RPA initiatives [9]. For example, a human developer might recognize the inefficiency of manually calculating data in Excel when a bot could perform the calculation using RPA operations (P5). Therefore, future RPM tools should incorporate functionalities that go beyond simple replay-automation. By analyzing user behavior and process data, RPM tools could potentially identify alternative automation approaches, e.g., using RPA functions instead of user actions, or implement API RPA operations for improved efficiency. Furthermore, similar to the previous dynamic bot compilation process, RPM tools could leverage data from previously automated processes to identify best practices and suggest potential optimizations based on successful past implementations.

5.3 Bot Compilation - RPA Operation Selection

While the RPM tools demonstrate promise in process discovery and segmentation from UI logs, their current automation capabilities are limited in scope and functionality. The key bottleneck lies in their reliance on a pre-defined, hard-coded translation process [3,16]. These rules often utilize generic RPA tool actions to avoid application-specific dependencies. This presents a significant limitation. While developers focus on resilience and technological capabilities, the RPM solutions are not evaluating any of these factors. Only the single hard-coded rule for translation is considered once the UI log was preprocessed and provided for bot compilation. RPA tools already offer a wide range of building blocks beyond those currently incorporated into the hard-coded translation rules of existing RPM solutions. Developers have the flexibility to leverage these diverse options within their chosen RPA development environment as highlighted in Sect. 4. This manual development approach empowers developers to select the better suited RPA operations based on their experience and the automation goals, resulting in a potentially more resilient and efficient bot.

To overcome the limitations of hard-coded translation, future RPM development could explore a more dynamic approach to mapping recorded user actions to RPA operations. This approach could involve enriching the conceptualization of RPA operations proposed by Völker and Weske [26] by incorporating explicit knowledge about developer decision-making processes. By understanding the rationale behind developer choices, RPM tools could become more adept at selecting the most appropriate RPA operation for a given user action. Alternatively, RPM tools could leverage existing successful implementations. Similar to the comparison mechanisms explored by Průcha and Madzík [21], the system could identify pre-built bots that closely resemble the user actions being processed. Instead of relying solely on pre-defined rules, the RPM tool could analyze user actions and compare them against a library of already implemented and operational bots.

5.4 Bot Compilation - Run-Time Environment

Another noticeable difference between human and RPM development lies in their approach to the development environment. RPM tools typically capture the user's execution environment during recording. This data includes details like file paths, installed software, and screen resolution. The resulting automated bot relies on these configurations, as the data is directly translated into RPA operations. However, as highlighted in the previous findings, the runtime environment where the bot actually executes can differ significantly from the user's environment. This discrepancy can lead to issues if the bot encounters unexpected variations in file paths, data availability, screen resolution, or available applications.

In contrast, human developers leverage their understanding of the target run-time environment [13]. They often develop bots on a system that closely resembles the intended deployment environment. This allows them to anticipate potential issues and design the bot's logic based on the run-time environment, not just the initial recording environment. Two solutions seem possible to overcome this limitation of RPM tools. On the one hand, RPM tools could allow users to specify the intended run-time environment alongside the user's actual environment. On the other hand, RPM tools could integrate functionalities to emulate the target environment, which removes the difference between the recording environment and the run-time environment.

6 Conclusion

Through in-depth interviews with RPA developers, researchers, and business experts, this study gained insights into the factors influencing developer decision-making during RPA bot development. However, it is important to acknowledge the inherent limitations of interview-based qualitative research. Interviewer and participant bias can influence the data collected. Additionally, coding interview data to identify themes can be subjective, potentially overlooking nuances or introducing interpretations not reflected in the raw interviews. Nevertheless, the findings illuminate the complex interplay of technological capabilities, organizational contexts, and individual developer experiences that influence RPA bot development. The goal of this research is to pave the way for truly autonomous RPA development, where machines can not only identify tasks for automation but also make informed decisions during implementation. The head-to-head comparison between human developer decisions and the capabilities of current RPM tools reveals a critical gap - the inability of RPM to replicate the dynamic decision-making inherent in human expertise to generate robust automation.

While humans create automation with a focus on resilience, RPM tools rely on static translation rules for low-level routines. Further research could identify common use cases, highlight areas where RPA operations are frequently applied, and pinpoint situations where developers resort to workarounds to overcome technical limitations. By analyzing these patterns, criteria can be established to evaluate the strengths and limitations of RPA software. This additionally provides insights for developing a framework to assist RPM tools in selecting the most suitable RPA operations for a specific automation task. The research could identify techniques to analyze developer choices and successful automation workflows to integrate these solutions into newly compiled bots in RPM. By enabling RPM to learn from human expertise and adapt its building block selection, RPM moves closer to achieving true automation autonomy in RPA development.

References

1. Abb, L., Rehse, J.R.: Process-related user interaction logs: state of the art, reference model, and object-centric implementation. Inf. Syst. 102386 (2024). https://doi.org/10.1016/j.is.2024.102386
2. Agostinelli, S., Marrella, A., Mecella, M.: 11 automated segmentation of user interface logs. In: Czarnecki, C., Fettke, P. (eds.) Robotic Process Automation, pp. 201–222. De Gruyter STEM, Berlin/München/Boston (2021). https://doi.org/10.1515/9783110676693-011
3. Agostinelli, S., Lupia, M., Marrella, A., Mecella, M.: SmartRPA: a tool to reactively synthesize software robots from user interface logs. In: Nurcan, S., Korthaus, A. (eds.) CAiSE 2021. LNBIP, vol. 424, pp. 137–145. Springer, Cham (2021). https://doi.org/10.1007/978-3-030-79108-7_16
4. Axmann, B., Harmoko, H., Herm, L.-V., Janiesch, C.: A framework of cost drivers for robotic process automation projects. In: González Enríquez, J., Debois, S., Fettke, P., Plebani, P., van de Weerd, I., Weber, I. (eds.) BPM 2021. LNBIP, vol. 428, pp. 7–22. Springer, Cham (2021). https://doi.org/10.1007/978-3-030-85867-4_2
5. Baker, S.E., Edwards, R.: How many qualitative interviews is enough? https://eprints.ncrm.ac.uk/id/eprint/2273/4/how_many_interviews.pdf
6. Celonis: Task mining (2024). https://docs.celonis.com/en/task-mining.html
7. Charmaz, K.: Constructing Grounded Theory. Introducing Qualitative Methods, 2nd edn. SAGE, Los Angeles (2014)
8. Dumas, M., La Rosa, M., Volodymyr, L., Polyvyanyy, A., Maggi, F.M.: Robotic process mining. In: van der Aalst, W.M.P., Carmona, J. (eds.) Process Mining Handbook, pp. 468–491. Springer, Cham (2022). https://doi.org/10.1007/978-3-031-08848-3_16
9. Eggert, M., Moulen, T.: Auswahl von geschäftsprozessen zur anwendung von robotic process automation: Vergleich relevanter kriterien aus theorie und praxis. In: D'Onofrio, S., Meinhardt, S. (eds.) Robotik in der Wirtschaftsinformatik, pp. 107–129. Edition HMD, Springer, Wiesbaden (2023). https://doi.org/10.1007/978-3-658-39621-3_6
10. Herm, L.-V., et al.: A consolidated framework for implementing robotic process automation projects. In: Fahland, D., Ghidini, C., Becker, J., Dumas, M. (eds.)

BPM 2020. LNCS, vol. 12168, pp. 471–488. Springer, Cham (2020). https://doi.org/10.1007/978-3-030-58666-9_27
11. Herm, L.V., Janiesch, C., Helm, A., Imgrund, F., Hofmann, A., Winkelmann, A.: A framework for implementing robotic process automation projects. Inf. Syst. e-Bus. Manag. (2023). https://doi.org/10.1007/s10257-022-00553-8
12. Hofmann, A., Prätori, T., Seubert, F., Wanner, J., Fischer, M., Winkelmann, A.: 4 process selection for RPA projects. In: Czarnecki, C., Fettke, P. (eds.) Robotic Process Automation, pp. 77–90. De Gruyter STEM, Berlin/München/Boston (2021). https://doi.org/10.1515/9783110676693-004
13. Jimenez-Ramirez, A., Reijers, H.A., Barba, I., Del Valle, C.: A method to improve the early stages of the robotic process automation lifecycle. In: Giorgini, P., Weber, B. (eds.) CAiSE 2019. LNCS, vol. 11483, pp. 446–461. Springer, Cham (2019). https://doi.org/10.1007/978-3-030-21290-2_28
14. Leno, V., Augusto, A., Dumas, M., La Rosa, M., Maggi, F.M., Polyvyanyy, A.: Identifying candidate routines for robotic process automation from unsegmented UI logs. In: ICPM 2020 Proceedings, pp. 153–160 (2020). https://doi.org/10.1109/ICPM49681.2020.00031
15. Leno, V., Deviatykh, S., Polyvyanyy, A., La Rosa, M., Dumas, M., Maggi, F.M.: Robidium: automated synthesis of robotic process automation scripts from UI logs (2020)
16. Leno, V., Dumas, M., La Rosa, M., Maggi, F.M., Polyvyanyy, A.: Action logger: enabling process mining for robotic process automation. In: Proceedings of the Dissertation Award, Doctoral Consortium, and Demonstration Track at BPM 2019. CEUR Workshop Proceedings, vol. 2420, p. 5. CEUR-WS (2019). https://hdl.handle.net/10863/19700
17. Martínez-Rojas, A., Jiménez-Ramírez, A., Enríquez, J.G., Reijers, H.A.: Analyzing variable human actions for robotic process automation. In: Di Ciccio, C., Dijkman, R., Del Río Ortega, A., Rinderle-Ma, S. (eds.) Business Process Management, vol. 13420, pp. 75–90. Springer, Cham (2022). https://doi.org/10.1007/978-3-031-16103-2_8
18. Noppen, P., Beerepoot, I., van de Weerd, I., Jonker, M., Reijers, H.A.: How to keep RPA maintainable? In: Fahland, D., Ghidini, C., Becker, J., Dumas, M. (eds.) BPM 2020. LNCS, vol. 12168, pp. 453–470. Springer, Cham (2020). https://doi.org/10.1007/978-3-030-58666-9_26
19. Padgett, D.: Qualitative Methods in Social Work Research, 3rd edn. SAGE, Los Angeles (2017)
20. Prinz, N., Huber, M., Leonhardt, J., Riedinger, C.: Unleash the power of citizen development: leveraging organizational capabilities for successful low-code development platform adoption. In: HICSS 2024. https://hdl.handle.net/10125/106444
21. Průcha, P., Madzík, P.: Siditer: similarity discovering techniques for robotic process automation. In: Plattfaut, R., Rehse, J.R., et al (eds.) BPM: Blockchain, Robotic Process Automation and Educators Forum, vol. 491, pp. 106–119. Springer, Cham (2023). https://doi.org/10.1007/978-3-031-43433-4_7
22. Runeson, P., Host, M., Reiner, A., Regnell, B.: Case Study Research in Software Engineering: Guidelines and Examples, 1st edn. Wiley, Hoboken (2012). http://site.ebrary.com/lib/alltitles/Doc?id=10538737
23. Saldaña, J.: The Coding Manual for Qualitative Researchers, 4th edn. SAGE, Los Angeles (2021)
24. Syed, R., Wynn, M.T.: How to trust a bot: an RPA user perspective. In: Asatiani, A., et al. (eds.) BPM 2020. LNBIP, vol. 393, pp. 147–160. Springer, Cham (2020). https://doi.org/10.1007/978-3-030-58779-6_10

25. UiPath: Uipath: Task mining (2022). https://www.uipath.com/product/task-mining
26. Völker, M., Weske, M.: Conceptualizing bots in robotic process automation. In: Ghose, A., Horkoff, J., Silva Souza, V.E., Parsons, J., Evermann, J. (eds.) ER 2021. LNCS, vol. 13011, pp. 3–13. Springer, Cham (2021). https://doi.org/10.1007/978-3-030-89022-3_1

Democratizing Robotic Process Mining: A Conceptual Framework for User Actions, Tasks, and RPA Bots

Tom Hohenadl[1,2](✉), Maximilian Völker[3], Christian Stummeyer[2], and Mathias Weske[3]

[1] Katholische Universität Eichstätt-Ingolstadt, Eichstätt, Germany
tom.hohenadl@stud.ku.de
[2] Technische Hochschule Ingolstadt, Ingolstadt, Germany
christian.stummeyer@thi.de
[3] Hasso Plattner Institute, University of Potsdam, Potsdam, Germany
{maximilian.voelker,mathias.weske}@hpi.de

Abstract. While Robotic Process Automation (RPA) enables companies to automate tasks in a comparable lightweight manner, the creation of such automations still requires a considerable manual effort. Robotic Process Mining (RPM) tries to mitigate this effort by analyzing recorded user interactions with a computer system to identify routines automatable by RPA and to derive corresponding RPA scripts. The rapid evolution of both RPA and RPM has led to ambiguities in the terminology used across research, prototypes, and industry. In this paper, we survey and structure the possible types of user interactions that can be recorded by analyzing the literature on RPM and the related field of user behavior mining. Additionally, we aim to foster a unified understanding of the key terms and concepts in RPM by proposing a conceptualization that draws upon ontologies of software, user interfaces, and RPA to embed this comparably new research area. To explore the practical applicability of the conceptualization, we conclude with a number of concrete usage scenarios that strengthen the design and development of RPA bots.

Keywords: User Behavior Mining · Robotic Process Automation · Robotic Process Mining · Ontology

1 Introduction

User Behavior Mining (UBM) explores the world of user interactions with information systems, aiming to analyze patterns and extract valuable insights [25]. Robotic Process Mining (RPM), a subfield of user behavior mining, focuses on automating repetitive tasks discovered within user interaction logs [6]. This automation potential arises from the core functionalities of the RPM framework: recording user actions, abstracting low-level actions into meaningful tasks, and automatically generating Robotic Process Automation (RPA) scripts that automate these tasks [3].

C. Di Ciccio et al. (Eds.): BPM 2024, LNBIP 527, pp. 185–199, 2024.
https://doi.org/10.1007/978-3-031-70445-1_12

With the advancing research (e.g., [1,6,13]), various terms and concepts related to RPM emerged, such as *(inter-)action*, *task*, and *routine*. Likewise, the prototypes created for RPM or for logging user interactions (e.g., [3,12,14]) use their own notions of these terms and are based on individual decisions about which events to record and at what level of abstraction, resulting in custom approaches for aggregating and transforming observed user interactions into RPA scripts.

In this paper, we aim to provide a structured overview of possible user interactions relevant to RPM in order to establish a foundation for RPM-related logging tools. For this purpose, a systematic literature review focused on literature on robotic process mining and the related field of user behavior mining is conducted. Based on the results, a taxonomy of user interactions is derived that structures the types of recordable user interactions, which may guide and support the development of respective tools and provides a more profound understanding of low-level user actions that contribute to the automation of manual business processes through RPM.

Furthermore, we seek to consolidate the terms used in RPM by embedding them in a broader context. While most works focus on relating the terms directly, we draw upon existing conceptualizations, such as the ontology of RPA operations [29], an ontology of user interaction and interfaces [21], and an ontology of software [18], to demonstrate the relation of RPM and its concepts to these fields. With the resulting conceptualization, we want to bridge the gap between task mining, RPM, and RPA to foster a common and broader understanding of RPM and its impact on RPA. Additionally, we point out new possibilities enabled by the conceptualization, such as an ontology-based mapping of recorded interactions and building blocks of RPA scripts, lifting the need for hard-coded transformations as currently employed by tools.

More specifically, the paper addresses two key research objectives:

Research Objective 1 Structured consolidation of user interactions: We aim to expand the current understanding of low-level user actions by creating a taxonomy based on a systematic literature review on RPM solutions and user interaction logs.

Research Objective 2 Holistic RPM conceptualization: We aim to foster a common understanding of RPM and its terms by proposing a holistic conceptualization that relates this field of research to Robotic Process Automation and software in general.

The remainder of this paper is structured as follows. In Sect. 2, the necessary background information on RPM and user interaction logs is provided. In Sect. 3, we describe the literature review process and the resulting taxonomy of low-level user actions. Based on the review of actions and the available related RPM literature, we then conceptualize RPM in Sect. 4. In Sect. 5, the implications of the conceptualization are described before concluding the paper in Sect. 6.

2 Related Work and Example

Robotic Process Automation has been adapted across industries to automate repetitive low-level processes executed by white-collar workers [7,28]. This technology utilizes bots that replay the human manual path taken to fulfill tasks to achieve process goals [27]. Traditionally, RPA has been seen as ideal for automating workflows that are repetitive, well-structured, and rule-based [28]. However, recent RPA improvements utilizing artificial intelligence have eased the limitation of strictly rule-based RPA and allowed businesses to use machine learning to automate further routines [27]. To automate tasks, RPA bots execute sequences of RPA operations, the "building-blocks" offered for automation by the respective RPA software vendors [29].

Current research in the broad scope of RPA has identified the potential to automate and streamline the bot creation process. This research stream is called Robotic Process Mining (RPM) and was initially defined by Leno et al. [13]. The main objective of the RPM framework is to automatically identify candidate routines from low-level user interactions and automatically emulate these candidates into RPA bots [6]. Specifically, RPM tools focus on identifying repetitive routines and variations of those routines that are executed multiple times. RPM utilizes user interaction (UI) logs, which provide a high-resolution recording of low-level manual activities executed during task fulfillment [3]. These logs enable a detailed understanding and analysis of how users interact with the system [1].

Abb and Rehse proposed a reference model for UI logs, which defines the core concepts and properties specific to this type of event data [1]. A key distinction between UI logs and traditional process mining logs lies in the absence of a case notion. Process mining logs typically link events to specific process instances, i.e., the case notion, whereas UI logs represent a recording of user actions across one or multiple tasks. Each event within a UI log corresponds to a low-level user activity. The core elements of such an activity are the user, the low-level action, and the target object, as displayed in Fig. 3 on the left side [1].

Table 1. UI log based on the running example in [2]

case	time:timestamp	Category	Application	concept:name
1	10:29:32	Office	Excel	readCell
1	10:29:33	Clipboard	Clipboard	copy
1	10:29:35	Browser	Chrome	mouseClick
1	10:29:36	Browser	Chrome	selectField
1	10:29:37	Clipboard	Chrome	paste
1	10:29:40	Browser	Chrome	submit

Once captured, user interaction logs serve as input to RPM tools which segment these logs into traces, identify suitable candidates for RPA, and emulate these routines into RPA bots [6]. Current tools, such as Robidium [12], Desktop Activity Mining [15], and SmartRPA [4], provide capabilities for the automatic emulation of routines and the translation of recorded UI actions into executable RPA operations. This translation process relies on pre-defined, hard-coded rules within the RPM tools [12] that establish a direct mapping between recorded user activities and corresponding RPA operations.

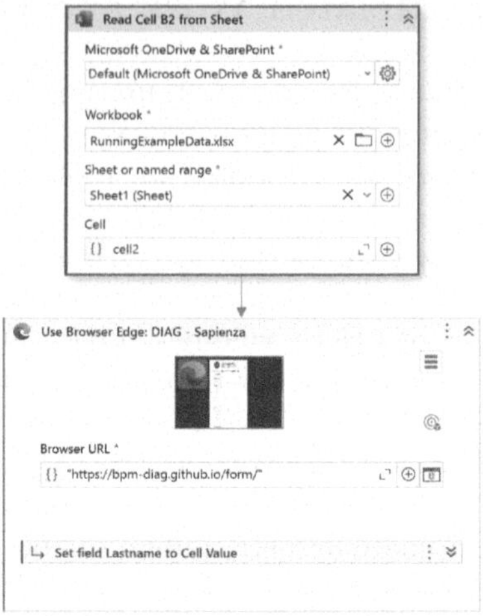

Fig. 1. RPA bot in UiPath studio

To illustrate the concept of transforming a user interaction log into an RPA bot, we consider the running example by Agostinelli et al. [2], in which users transfer travel request data from an Excel file into a web submission form. In Leno et al. [12], a similar example is provided, including data transfer from a spreadsheet to a web form. The relevant part of the UI log is displayed in Table 1. The example contains copy-and-paste steps that are necessary to fill out the form and successfully submit it for further processing. The process involves two user interfaces: the Excel spreadsheet and the web form. An RPM tool would now analyze the log, detect routines, and try to find the appropriate RPA operations to emulate the bot. In the case of the log depicted in Table 1, the RPM tool transforms the sequence of focussing the cell and copying its value into a `Read Cell` RPA operation, exemplary displayed as UiPath RPA operations in Fig. 1. Equally, the browser events of selecting the form field and pasting the content were transformed into a `Use Browser` RPA operation. However, translating sequences of actions in an application to a corresponding RPA operation currently relies on hard-coded mappings. Therefore, RPM tools are limited to the types of events and actions captured in the logs and the pre-programmed translation rules. Introducing new functionalities within the target application, e.g., new click types not captured by the recording, necessitates manual intervention to modify the mapping logic, hindering the adaptability of the generated RPA bots.

3 A Taxonomy of User Interactions

As task mining [16], user behavior mining [25], or robotic process mining [6] have surfaced alongside the ongoing RPA improvement [7], so have new research and industry terms and definitions, particularly regarding the way user actions are described and utilized. As a consequence, different UI recorders and RPM tools log events at different levels of granularity and use different terms for the observed actions.

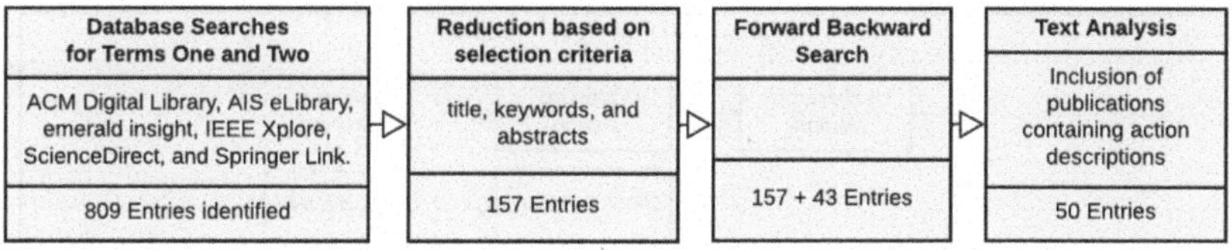

Fig. 2. Literature review process for low-level action descriptions

In the following, we want to improve the current understanding of low-level user actions and their impact on the user interface. To address this first research objective, we have developed a taxonomy based on a literature review, which was conducted following the methodology outlined by Kitchenham et al. [10]. Figure 2 outlines the review process, which is described in the following subsection.

3.1 Literature Review on Low-Level User Actions

Due to the focus on UI logs and their relevance in UBM, RPM, or task mining, the following search phrase for the literature review was used: "("log" OR "event" OR "trace") AND ("user behavior mining" OR "task mining" OR "robotic process mining" OR "desktop activity mining")". These keywords relate to different user behavior or UI log analysis methods, including "user behavior mining" [25], "task mining" [1], "robotic process mining" [6], and "desktop activity mining" [15]. The search was conducted using the following databases: ACM Digital Library, AIS eLibrary, emerald insight, IEEE Xplore, ScienceDirect, and Springer Link. This initial search revealed the term "user behavior log" and thus, a second search was carried out across all mentioned libraries with this term. All relevant publications from both searches were then combined to form the final set of 809 candidate publications for the literature review.

The initial screening process, based on titles and abstracts, identified a set of 157 publications potentially relevant to the research topic. Publications excluded during this stage fell outside the scope of user actions and RPM, often belonging to entirely different research areas or containing server-side event recordings, particularly from web usage mining research. To further refine the selection and ensure a more focused analysis, a full-text review of the remaining 157 publications was conducted. This involved reading the full text of each paper for content related to UI logs, user actions, or methodologies for recording user actions. Only publications defining or describing low-level user actions were included in the literature review. In this process, a set of 43 further candidates for inclusion in the literature review was identified through a Backward-Forward search. After completing the review process, a final selection of 50 publications released in the period from 2006 to 2024 was made[1]. All publications provided insights for conceptualizing the user interactions model described in the subsequent section.

Inspired by the taxonomy development method in Nickerson et al. [17], we intuitively constructed a concept-based taxonomy for user actions in RPM. This

[1] Complete list and mapping: https://zenodo.org/doi/10.5281/zenodo.12635361.

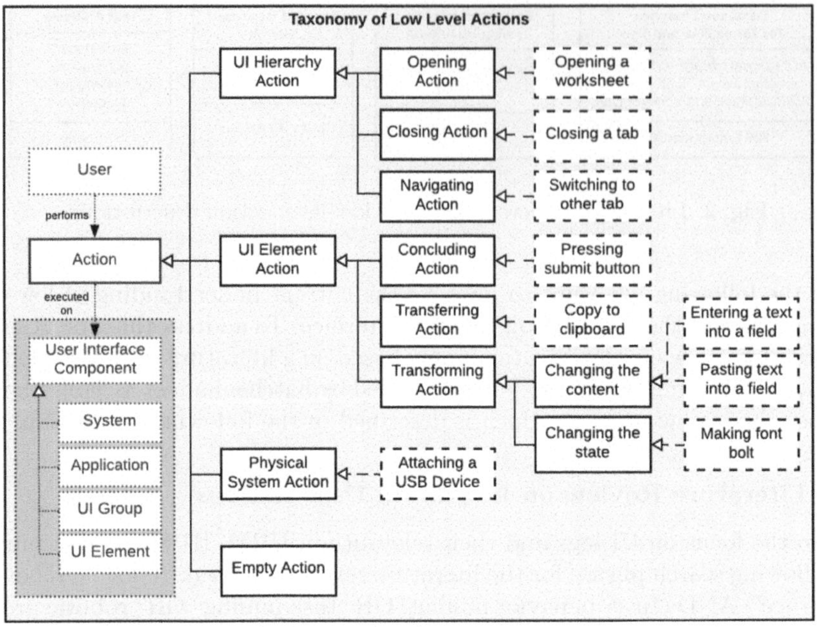

Fig. 3. Reference Model from [1] enriched with own taxonomy of actions

approach summarizes the description of the various action types outlined in the existing literature and thus prepares an abstract classification of user actions as well as a summary of the impact of various action types. This taxonomy groups actions based on their effect on the user interface, resulting in four core action types detailed in the center of Fig. 3.

3.2 Conceptualization of Low-Level Actions

Based on the reviewed literature, we identified the following four overarching concepts, of which two are further detailed in six descriptive action types.

Hierarchy Actions. The first concept, `hierarchy action`, relates to all actions changing the displayed graphical elements on the screen, i.e., the UI hierarchy shown in Fig. 3.

Specifically, there are `opening actions` to open an `application` on `operating system` level or a `UI group` within an `application`. Opening actions are the foundational desktop interactions to open applications concerning the task fulfillment [26]. All actions creating a new process on the system level displaying a new object are also considered opening actions [3].

Within the proposed taxonomy, the concept of `closing actions` refers to actions that terminate currently active instances of either `applications` or `UI groups`. In a broader computer context, closing actions typically signify the termination of a running process. The equivalent on a mobile device is closing the app through system functions, e.g. swiping. Only in [5] closing is treated as a particular action type. However, for RPA, the closing action as a `software control action` [29] is necessary for unsupervised settings. Unsupervised RPA requires terminating user sessions, i.e., closing or logging out, to prepare the virtual machine for the next process, likely involving different users.

`Navigating actions` serve to navigate the `UI hierarchy` [1]. On the one hand, a `navigating action` can occur between applications. These refer to a change of the currently active `application` or the `system` [19], leading to another `application` or `system` in focus. On the other hand, navigating actions can change the presented `UI group` and display another `UI group` in the same `application` [6], e.g., a change of sheets in Excel, or switches selecting different cells or fields in a `UI group` [11]. However, it is important to consider the nuanced role of scrolling within navigation actions. Navigating actions alter the displayed UI hierarchy, such as moving between applications. Scrolling primarily affects the visibility of UI elements within a specific group or view, without changing the underlying structure.

UI Element Actions. While the previously mentioned actions apply to the user interface generally, the second overarching type of `UI element actions` does change the state of a `UI element`. Our analysis identified three distinct action types: `transforming`, `transferring`, and `concluding` actions. While these actions directly relate to `UI elements`, they generally do not affect the overall structure of the displayed `UI group`, `application`, or `system`.

A `transforming action` is executed on a `UI element` and changes its current state. Most publications refer to transforming actions as edit actions. However, there are differences regarding the impact of the action. While some transforming actions edit the content of a `UI element`, others just change generic properties of a `UI element`, e.g., making the font bold. In [11], two cases of transformation through user actions are distinguished: implicit and explicit actions. Implicit actions are not deterministic because human interaction can vary in every task execution, e.g., entering a VPN authentication code. Explicit actions are deterministic and thus can be gathered from analyzing the UI log, e.g., pasting [3]. Explicit actions are key for automation, as RPA capitalizes on deterministic if-then-else rules [27].

All publications referring to `transferring actions` name copy and paste in some form. A transferring action is an action that takes the content of a `UI element` and stores it in the clipboard. While the transforming action outlines the change of visible `UI elements`, the transferring action is concerned with the non-tangible data perspective. Each UI log entry has a subset of parameters describing the data perspective [12], i.e., the current data values processed. One of these parameters is the clipboard. If a transfer action occurs, the clipboard value and the corresponding data parameter are appended in the UI log.

As the last action concept on UI elements, we define the `concluding action` as an umbrella term for actions completing a task, further possibly indicating the beginning of a new task. While navigating actions, such as clicking menus or buttons to switch between screens, prepare the user interface for subsequent interactions, concluding actions represent the culmination of a sequence of user manipulations within a specific user interface hierarchy. These manipulations typically involve modifying or interacting with UI elements to achieve a specific task objective. The concept of concluding actions holds particular significance in segmenting user interaction logs into distinct tasks, as demonstrated in Rebmann et al. [24]. The literature frequently mentions specific examples of such actions, like `pressing submit buttons` [12], which signals the user's intent to finalize the current task and commit the undertaken actions within the application.

Physical System Actions. The analysis of user interaction logs revealed a category of events extending beyond direct interactions with UI elements at the application level. While the focus of this literature review is on automatable actions in the context of RPM, such physical system actions indicate a routine that cannot be automated without a human in the loop. These events, while not purely digital in nature, offer valuable insights into user workflows. They are identified as physical system events, encompassing user interactions with external devices that connect to the digital system. Examples include attaching USB sticks or headsets via cable. Notably, these events can be captured not only by specialized RPM tools like SmartRPA [4] but also by broader system diagnostic tools such as Microsoft Diagnostics [20].

Empty Actions. Finally, another special concept was identified during the literature review: `empty actions` that do not change anything within the `UI hierarchy` and do not relate to a specific `UI element`. Particularly, [11] states that mouse clicks on the background without interaction with any `UI element` are empty actions and should not be recorded.

In conclusion, our work establishes a foundational taxonomy for user actions within the context of RPM and UI logs. The taxonomy captures the different types of user interactions, particularly focusing on their impact on the `UI hierarchy`. It differentiates between meaningful actions that modify the UI hierarchy in general or UI elements specifically, physical human actions triggering events, and empty actions that represent extraneous clicks or keystrokes without an effect recorded by logging solutions.

4 Conceptualization of UI Logs and RPA Bots

In the following, we address our second research objective by embedding the notion of actions and other RPM-related concepts frequently used in the literature into software concepts and connect them with concepts of user interfaces and RPA, to provide a comprehensive view of RPM terms and their relation to adjacent fields and concepts.

The provided conceptualization is based on four existing ontologies and their extensions: (i) the Ontology of Descriptions and Situations (DnS) [8], (ii) the Core Software Ontology (CSO, COSC) [18], (iii) the ontology of user interfaces UI_2Ont (UI) [21], and (iv) the ontology of RPA operations (RPA) [29]. Both, the ontology of RPA operations and the ontology of user interfaces, are built on

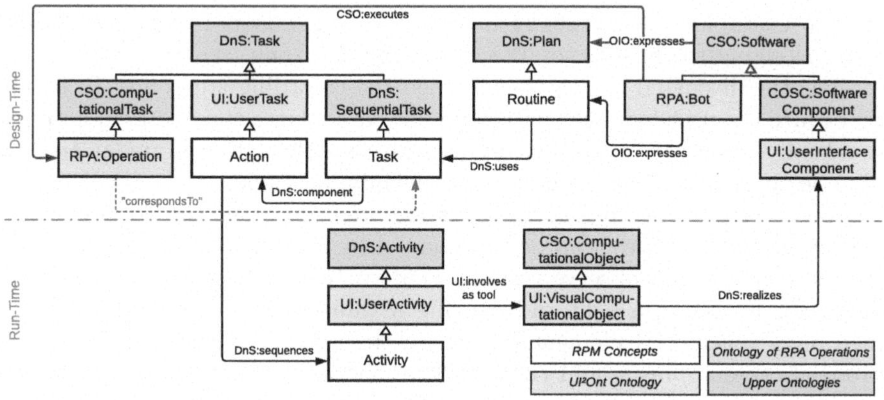

Fig. 4. Conceptual connections of RPM-related terms (no prefix), RPA ("RPA:"), and user interfaces ("UI:"), and their embedding in software concepts ("CSO:", "COSC:")

top of the CSO, which in turn utilizes DnS, an extension of the foundational ontology DOLCE.

As indicated in Fig. 4, the conceptualization of RPM distinguishes the design time, i.e., concepts that are general and descriptive, and the run time with concepts that represent specific instances that can be observed.

At run time, during the recordings of a user interaction log, specific `Activities` can be observed. These are a kind of `UI:UserActivities` that *involve* specific `UI:VisualComputationalObjects`, i.e., instances of `UI:UserInterface-Components`. These components can be simple UI elements, such as a text box, but also a composition or group of elements [21]. Furthermore, they are part of `CSO:Software` and thus, jointly, resemble the concept of a target object and its subclasses UI element, UI group, Application, and System as introduced by Abb and Rehse [1].

Taking up the example presented in Table 1, each log entry describes one `Activity`, characterized by a timestamp, the UI element it was performed on, and more recorded context information. For example, at 10:29:33, an `Activity` in Excel was observed where the user copied the value of the cell that was focused during the previous activity.

An `Activity` is the result of a user[2] executing a certain type of `Action`, which is a type of `UI:UserTask` that describes a possible interaction with the user interface [21]. The `Actions`, like pasting text into a field, can be classified according to the taxonomy presented in Sect. 3.2. In other words, while the `Activity` is actually carried out, the `Action` rather describes what can be done and thus belongs to the design time.

[2] The concept `User` is not part of this conceptualization. However, UI$_2$Ont, CSO, and DnS already include the notion of a user so that this aspect can be incorporated in future work.

In the example, the Activity observed at 10:29:33 is a *copy to clipboard* action, belonging to the class of Transferring Actions (see Sect. 3.2).

So far, the discussed concepts concerned the recording step only. To be able to reproduce the user behavior with RPA, RPM analyzes the observed activities to eventually generate an RPA script, and for this, the rather granular Actions need to be grouped and combined to more meaningful Tasks (cf. [1,6]). With that mapping, a certain sequence of Actions executed on a specific UI/application (resulting in Activities), can be represented by a more abstract Task.

In the example, we observe two consecutive Activities that represent the Actions for focussing a field and copying a value to the clipboard, both performed on a specific UI element (a cell) in the same application (Excel). The Actions-Task-mapping would yield that this specific action-sequence performed in the application Excel can be represented a more abstract *Read Cell In Excel* task.[3] Furthermore, it is important to note that there can be different ways, i.e., sequences of Actions, for performing a specific Task. In this example, the value could also be copied using the context menu or the dedicated copy-button in the application. Both would result in a different sequence of observed Activities but they have the same effect and thus semantically belong to the same Task (cf. [6]).

Furthermore, a UI log does not only contain the sequence of activities for a single task but for many. One of the challenges of RPM is to identify a reoccurring sequence of tasks in a UI log that forms a Routine [6]. On the conceptual level, a Routine is a DnS:Plan conceived by a user and consisting of one or more Tasks, that pursues a certain goal (cf. [8]), like transferring data from a worksheet to a web form. While the Routine is rather conceptual, it can, provided that its Tasks are automatable (cf. [6]), be expressed (implemented) by an RPA:Bot. For this, the Tasks present in the Routine need to be mapped to RPA:Operations, i.e., to the building blocks that are offered and predefined by RPA tools [29] and that can be used to automate the observed Task.

In the example, we can observe two Tasks. First, *Read Cell in Excel*, followed by *Write Text to Field in Browser*. As both tasks can be automated using RPA, the mapping would result in the RPA operations RPA:ExcelReadCell and RPA:ChromeWriteField that can be used to automate the observed activities.

In other words, based on recorded Activities involving "instances" of UI:-UserInterfaceComponents that are the execution of possible Actions by a user, RPM aims to abstract the Actions to Tasks in order to detect automatable Routines. For a Routine, an RPA:Bot expressing it is to be created, i.e., it executes the RPA:Operations corresponding to the Routine's Tasks and is thus capable of automating it.

[3] The granularity of log entries created by different tools can vary as they already may perform some post-processing, like combining focussing the UI element and entering text.

The presented conceptualization highlights the relations among RPA terms but also to other related concepts. Furthermore, it demonstrates the different levels of abstraction that are inherent to RPM, from `Activities` observed at run time to abstract `Tasks` and `Routines` that represent the conceptual design time.

Besides this rather documentary purpose, the mappings of `Actions` to more general `Tasks` and of `Tasks` to `RPA:Operations` that can be used to automate them, enable a practical use of the conceptualization as the rules for abstraction do not need to be hard-coded anymore to every RPM tool but can be incorporated in this framework, which is flexibly extendable. The following section briefly discusses further points of application.

5 Usage Scenarios of the Conceptualization

This section explores the practical applications and use cases enabled by the introduced conceptualization within the context of RPA and RPM. While acknowledging its limitations as a work in progress, the work presented in this paper is tool-agnostic and may be used to empower new types of RPM software, build on a common ontological understanding and foundation.

5.1 Identify Equal Processes in Different Applications

Current methods for selecting RPA processes based on UI logs often struggle when the same task is performed across different applications. This is because they primarily rely on application attributes within the logs, meaning even slight UI variations are treated as distinct processes. With the enriched, ontology-based understanding, the mapping from tasks to individual building blocks offered by RPA vendors yields more semantic information. Based on the taxonomy of operations included in the ontology of RPA operations the RPM tasks are mapped to (cf. Section 4), executions of the same task across different applications can be identified. With this level of abstraction, RPM tools could detect semantically similar or even equal task traces, even though its specific actions and tasks differ.

Imagine a user performing the same task in a web browser application and a desktop client application, as shown in Fig. 5. Traditionally, these would be seen as separate processes due to the distinct UI hierarchies. However, by recognizing the underlying task of reading a cell, the system can now link these seemingly different actions together. This opens doors for broader application of interchangeable building blocks, where a web-based task can be executed in the desktop client, or vice versa, offering greater flexibility and efficiency in RPA development. Thus, the conceptualization not only offers flexibility regarding the RPA tool used (cf. [29]), but also provides further flexibility through task interchangeability.

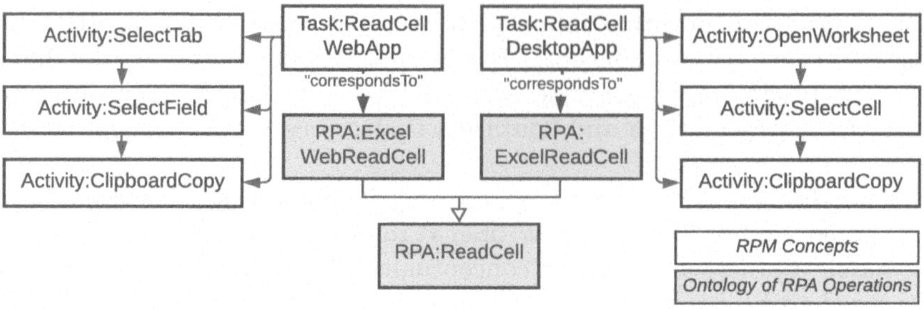

Fig. 5. Mapping of similar tasks to building blocks from [29] with the same abstract semantics

5.2 Reusability of Implemented RPA Bots

Currently, companies struggle to build repositories of implemented RPA solutions for reusability and scalability [22]. Especially, companies with citizen development can increase the return on investment if implementations are reused. Citizen development is the usage of the regular workforce in a business not specialized in automation development, which can create individual automation solutions for their use cases without depending on a central entity, e.g., a center of excellence [22]. Průcha and Madzík already proposed SiDiTer for the comparison of RPA bots [23], however, in this case, the bots and the effort for development already manifested.

With the presented conceptualization that connects tasks in RPM with RPA operations, RPM tools could automatically check an RPA bot repository for bots that are similar to the discovered routine. Reusing the already implemented bots or parts of it would reduce the development overhead and avoid the detection and generation of bots that were already implemented.

5.3 Generalization of RPM Logs

One of the significant challenges in utilizing UI recordings for process automation is the initial requirement for extensive domain knowledge, as highlighted by Gao et al. [9]. This knowledge is crucial for identifying high-level tasks within the low-level UI interactions captured in the recordings. Currently, switching between any application in the broader RPA scope can impact the level of detail, format, and naming conventions used. This inconsistency often necessitates significant adjustments when transitioning between tools. However, the proposed conceptualization offers a solution to this limitation.

As more actions and tasks are incorporated into the conceptualization framework, it establishes a common ground for user interactions and tasks. It may enable the translation of UI logs between different recorders and thus facilitate a smoother transition between different RPM solutions, reducing the limitations associated with vendor lock-in for RPM implementations, similar to the knowledge base for RPA operations in the ontology of RPA operations [29].

Besides translating between different tools, it enables the development of new, ontology-based RPM solutions that can draw directly from the concepts and relations in the conceptualization, e.g., to be able to detect the same routines in different but similar applications (such as Google Sheets and Microsoft Excel), and to easily create RPA scripts for different vendors (cf. [29]).

6 Discussion and Conclusion

This paper presents a novel conceptualization of low-level user actions and an enhanced ontology of RPA and RPM based on a systematic literature review. This action categorization enables vendor- and research-artifact-independent mappings using the reference model for user interaction logs. We bridge the gap between Task Mining research and Robotic Process Automation by introducing a conceptualization that connects these domains. By leveraging existing literature on RPA bots, RPA operations, user interaction logs, and RPM, we provide an overview of how RPA operations can be used to automate user tasks. The extension through user interaction logs and task-to-plan mapping establishes a clear connection between user activities and their corresponding RPA operations.

The presented conceptualization is based on the work from three key publications [1,6,29] and focuses on the integration of the central RPM concepts into a comprehensive conceptualization. Its focus is to structure low-level user actions in RPM and the relations between UI logs and RPA bots. Therefore, many other aspects of user behavior mining, task mining, or RPA are not considered, such as process selection, data privacy regulations when recording users, or the variant discovery from user interaction logs.

While a systematic literature review provides a robust foundation for building knowledge, it inherently possesses certain limitations. In the context of this study, a key concern is the potential existence of additional UI logs or design science artifacts containing valuable user interaction data that might not have been identified during the literature search. Consequently, the proposed taxonomy, which is based on the user actions identified within the reviewed publications, may not encompass the full spectrum of user interactions that exist in practice. Although some literature on human-computer interaction, mobile app testing, and web usage mining was covered by the search phrases in the review, other research domains may not have been considered due to the focus set on RPM.

To further enhance the utility of RPM tools for robotic and generic process automation, future research should explore the decision-making process behind selecting which tasks are automated by specific operations. While prior research has extensively investigated the theoretical success factors for RPA implementation, the technical considerations influencing low-level action automation through RPA operation selection remain unexplored. Our work addresses this gap by outlining the process of translating tasks into specific RPA operations. However, further investigation is needed to explore the "how" and "why" behind selecting a particular automation approach for a given task.

References

1. Abb, L., Rehse, J.R.: Process-related user interaction logs: state of the art, reference model, and object-centric implementation. Inf. Syst. **124**, 102386 (2024). https://doi.org/10.1016/j.is.2024.102386
2. Agostinelli, S., Lupia, M., Marrella, A., Mecella, M.: Automated generation of executable RPA scripts from user interface logs. In: Asatiani, A., et al. (eds.) BPM 2020. LNBIP, vol. 393, pp. 116–131. Springer, Cham (2020). https://doi.org/10.1007/978-3-030-58779-6_8
3. Agostinelli, S., Lupia, M., Marrella, A., Mecella, M.: SmartRPA: a tool to reactively synthesize software robots from user interface logs. In: Nurcan, S., Korthaus, A. (eds.) CAiSE 2021. LNBIP, vol. 424, pp. 137–145. Springer, Cham (2021). https://doi.org/10.1007/978-3-030-79108-7_16
4. Agostinelli, S., Marrella, A., Mecella, M.: 11 automated segmentation of user interface logs. In: Czarnecki, C., Fettke, P. (eds.) Robotic Process Automation, pp. 201–222. De Gruyter (2021). https://doi.org/10.1515/9783110676693-011
5. Devaurs, D., Rath, A.S., Lindstaedt, S.N.: Exploiting the user interaction context for automatic task detection. Appl. Artif. Intell. **26**(1-2), 58–80 (2012). https://doi.org/10.1080/08839514.2012.629522
6. Dumas, M., La Rosa, M., Leno, V., Polyvyanyy, A., Maggi, F.M.: Robotic process mining. In: van der Aalst, W.M.P., Carmona, J. (eds.) Process Mining Handbook. LNBIP, vol. 448, pp. 468–491. Springer, Cham (2022). https://doi.org/10.1007/978-3-031-08848-3_16
7. Fersht, P., HFS Research: Robotic process automation (RPA) market revenues worldwide from 2017 to 2023. https://www.statista.com/statistics/740440/worldwide-robotic-process-automation-market-size/
8. Gangemi, A., Borgo, S., Catenacci, C., Lehmann, J.: Task taxonomies for knowledge content: Metokis deliverable D07 (2004)
9. Gao, J., van Zelst, S.J., Lu, X., van der Aalst, W.M.P.: Automated robotic process automation: a self-learning approach. In: Panetto, H., Debruyne, C., Hepp, M., Lewis, D., Ardagna, C.A., Meersman, R. (eds.) OTM 2019. LNCS, vol. 11877, pp. 95–112. Springer, Cham (2019). https://doi.org/10.1007/978-3-030-33246-4_6
10. Kitchenham, B.: Procedures for performing systematic reviews (2004)
11. Leno, V., Augusto, A., Dumas, M., La Rosa, M., Maggi, F.M., Polyvyanyy, A.: Discovering executable routine specifications from user interaction logs. http://arxiv.org/pdf/2106.13446v1
12. Leno, V., Deviatykh, S., Polyvyanyy, A., La Rosa, M., Dumas, M., Maggi, F.M.: Robidium: automated synthesis of robotic process automation scripts from UI logs. In: van der Aalst, W.M.P., et al. (eds.) BPM 2020 Best Dissertation Award, Doctoral Consortium, and Demonstration & Resources Track. CEUR Workshop Proceedings, vol. 2673, pp. 102–106. CEUR-WS.org (2020)
13. Leno, V., Polyvyanyy, A., Dumas, M., La Rosa, M., Maggi, F.M.: Robotic process mining: vision and challenges. Bus. Inf. Syst. Eng. **63**, 301–314 (2021). https://doi.org/10.1007/s12599-020-00641-4
14. Leno, V., Polyvyanyy, A., La Rosa, M., Dumas, M., Maggi, F.M.: Action logger: enabling process mining for robotic process automation. In: Depaire, B., et al. (eds.) BPM 2019 Dissertation Award, Doctoral Consortium, and Demonstration Track. CEUR Workshop Proceedings, vol. 2420, pp. 124–128. CEUR-WS.org (2019)
15. Linn, C., Zimmermann, P., Werth, D.: Desktop activity mining - a new level of detail in mining business processes. In: Czarnecki, C., et al. (eds.) Workshops der

INFORMATIK 2018. LNI, Köllen Druck+Verlag GmbH (2018). https://dl.gi.de/handle/20.500.12116/17225
16. Mayr, A., Herm, L.V., Wanner, J., Janiesch, C.: Applications and challenges of task mining: a literature review. In: Beck, R., et al. (eds.) 30th European Conference on Information Systems (ECIS 2022). AIS (2022). https://aisel.aisnet.org/ecis2022_rip/55
17. Nickerson, R.C., Varshney, U., Muntermann, J.: A method for taxonomy development and its application in information systems. Eur. J. Inf. Syst. **22**(3), 336–359 (2013). https://doi.org/10.1057/ejis.2012.26
18. Oberle, D., Grimm, S., Staab, S.: An ontology for software. In: Staab, S., Studer, R. (eds.) Handbook on Ontologies. IHIS, pp. 383–402. Springer, Heidelberg (2009). https://doi.org/10.1007/978-3-540-92673-3_17
19. Oliver, N., Smith, G., Thakkar, C., Surendran, A.C.: Swish: semantic analysis of window titles and switching history. In: Edmonds, E., et al. (eds.) Proceedings of the 11th International Conference on Intelligent User Interfaces, pp. 194–201. ACM, New York (2006). https://doi.org/10.1145/1111449.1111492
20. Park, S., Lee, S.: Diaganalyzer: user behavior analysis and visualization using windows diagnostics logs. Forensic Sci. Int. Digit. Investig. **43**, 301450 (2022). https://doi.org/10.1016/j.fsidi.2022.301450
21. Paulheim, H.: Ontology-Based Application Integration. Springer, New York (2011). https://doi.org/10.1007/978-1-4614-1430-8
22. Plattfaut, R., Borghoff, V., Godefroid, M., Koch, J., Trampler, M., Coners, A.: The critical success factors for robotic process automation. Comput. Ind. **138**, 103646 (2022). https://doi.org/10.1016/j.compind.2022.103646
23. Průcha, P., Madzík, P.: SiDiTeR: Similarity discovering techniques for robotic process automation. In: Köpke, J., et al. (eds.) Business Process Management: Blockchain, Robotic Process Automation and Educators Forum. LNBIP, vol. 491, pp. 106–119. Springer, Cham (2023). https://doi.org/10.1007/978-3-031-43433-4_7
24. Rebmann, A., van der Aa, H.: Unsupervised task recognition from user interaction streams. In: Indulska, M., et al. (eds.) Advanced Information Systems Engineering. LNCS, vol. 13901, pp. 141–157. Springer, Cham (2023). https://doi.org/10.1007/978-3-031-34560-9_9
25. Rehse, J.R., Abb, L., Berg, G., Bormann, C., Kampik, T., Warmuth, C.: User behavior mining. Bus. Inf. Syst. Eng. (2024). https://doi.org/10.1007/s12599-023-00848-1
26. Schmidt, B., Kastl, J., Stoitsev, T., Mühlhäuser, M.: Hierarchical task instance mining in interaction histories. In: Protopsaltis, A., et al. (eds.) Proceedings of the 29th ACM International Conference on Design of Communication, pp. 99–106. ACM, New York (2011). https://doi.org/10.1145/2038476.2038495
27. Syed, R., et al.: Robotic process automation: contemporary themes and challenges. Comput. Ind. **115**, 103162 (2020). https://doi.org/10.1016/j.compind.2019.103162
28. van der Aalst, W.M.P., Bichler, M., Heinzl, A.: Robotic process automation. Bus. Inf. Syst. Eng. **60**(4), 269–272 (2018). https://doi.org/10.1007/s12599-018-0542-4
29. Völker, M., Weske, M.: Conceptualizing bots in robotic process automation. In: Ghose, A., Horkoff, J., Silva Souza, V.E., Parsons, J., Evermann, J. (eds.) ER 2021. LNCS, vol. 13011, pp. 3–13. Springer, Cham (2021). https://doi.org/10.1007/978-3-030-89022-3_1

Central and Eastern European (CEE) Forum

Preface of the Central and Eastern European (CEE) Forum

The main objective of the CEE Forum is to provide a discussion and dissemination platform for researchers and practitioners conducting BPM research relevant to CEE countries. This forum is important for exploring and discussing innovative ideas on BPM methods and approaches applied in Central and Eastern Europe (CEE) countries presenting numerous specific challenges due to cultural, political, regional, or organizational differences and the various nature of the business processes. Associated with the BPM conference, this year's CEE Forum emphasizes differences stemming from the challenges posed by the business environment, IT-related specifications, IT legacy, country-specific legislature and regulations, and the differing levels of digital maturity in the context of digital transformation. Although organizations from CEE countries apply the same BPM methods, technologies, and approaches as non-CEE countries, while responding to the specific challenges, they create alternative experiences that might be valuable for better understanding BPM in various contexts and regional conditions. Specifically, this year, the forum received twelve submissions, of which six were carefully selected based on rigorous reviews for presentation at the CEE Forum and inclusion in the proceedings.

The paper "Adapting to the Dynamic Nature of Business Processes in the Digital Age" explores the intricate nature and diversity of business processes, recognizing their adaptation to the evolving economic landscape. Authors discuss the nature of unstructured processes and unstructured processes, arguing their increasing relevance and but also investigating significant correlations between the implementation of IT tools and the prevalence of unstructured processes.

In the paper "Extending CMMN for effective management of data in knowledge-intensive processes" authors build their contribution on the understanding that contemporary processes are characterised by a high degree of flexibility, creativity, and complexity, and that they usually rely on different techniques and technologies compared to traditional "well-structured" processes. Based on a preliminary analysis of the shortcomings of Case Management Model and Notation (CMMN) propose an extension of the element Case File Item and its associated elements, in order to overcome some of the identified shortcomings.

Advocating for a paradigm shift from traditional healthcare models, emphasising the importance of collaboration among healthcare professionals, patients, and caregivers contributes to the changing roles of stakeholders in digital environments. The study "Telemedicine and BPM in Cardiac Rehabilitation: A Comprehensive Post-Myocardial Infarction Care Program Case Study" examines this transformative potential of integrating telemedicine and Business Process Management (BPM) in the healthcare industry.

Continuing the importance of digital competencies, the research addressed in the paper "Adapting to Change: Employees Ambidexterity as a Driver for Operational

Adaptability and Organizational Development" contributes to the topic by arguing that employee ambidexterity plays a critical role in achieving operational adaptability and the development of an organization in a dynamic environment, challenging the role of technology in achieving organizational adaptability and market success and confirming the necessity of research related to the social implications of agile BPM.

The authors of the paper "Enhancing Business Process Management Through Nature Assessment: Development and Deployment of the Business Process Nature Assessment Tool" address the challenges small and medium-sized enterprises (SMEs) face when evaluating their processes prior to implementing Business Process Management (BPM). The research is oriented on underscoring the critical role of BP assessment in organisational success and aiming to provide support in that sense authors propose a solution by developing and demonstrating an online tool which allows for quick and effective evaluations of process unpredictability and knowledge intensity, aiding in the selection of suitable BPM methodologies, Knowledge Management (KM) and Human Resource Management (HRM) strategies, and supporting IT systems.

The survey "The Role and Application of OMG Notations in IT: Analysis and Survey on the Effectiveness of Business Notations among IT companies in Poland" identified the most commonly used platforms for OMG notations, but also showed that no direct link between tool choice and diagram creation could be found in the selected sample, confirming that there is no single universal modeling tool applicable at all stages of the process of creating modern software.

We extend our gratitude to everyone involved in making the CEE Forum 2024 successful: the authors who diligently submitted their papers, the Program Committee members for meticulously reviewing each paper, and the speakers who shared their valuable work. We also thank the BPM 2024 chairs and organizers for their support in preparing the CEE Forum.

Organization

Program Chairs

Michal Krčál	Masaryk University, Czech Republic
Monika Malinova Mandelburger	TU Wien, Austria
Gregor Polančič	University of Maribor, Slovenia
Katarina Tomičić-Pupek	University of Zagreb, Croatia

Program Committee

Ahad Zareravasan	Masaryk University, Czech Republic
Andrzej Niesler	Wroclaw University of Economics and Business, Poland
Mateja Bule	University of Maribor, Slovenia
Dalia Suša Vugec	University of Zagreb, Croatia
Damian Dziembek	Czestochowa University of Technology, Poland
Daniela Garbin Praničević	University of Split, Croatia
Dragana Stojanović	University of Belgrade, Serbia
Igor Pihir	University of Zagreb Faculty of Organization and Informatics, Croatia
Jan Mendling	Humboldt-Universität zu Berlin, Germany
Katarzyna Gdowska	AGH University of Krakow, Poland
Katja Kous	University of Maribor, Slovenia
Krzysztof Kluza	AGH University of Krakow, Poland
Ljubica Milanović Glavan	University of Zagreb, Croatia
Marite Kirikova	Riga Technical University, Latvia
Martin Macák	Masaryk University, Czech Republic
Martina Tomicic Furjan	University of Zagreb Faculty of Organization and Informatics, Croatia
Natalia Potoczek	Institute of Economics, Polish Academy of Sciences, Poland
Oktay Turetken	Eindhoven University of Technology, The Netherlands
Saša Brdnik	University of Maribor, Slovenia
Tomasz Gzik	Military University of Technology, Poland

Telemedicine and BPM in Cardiac Rehabilitation: A Comprehensive Post-Myocardial Infarction Care Program Case Study

Marek Szelągowski[1(✉)], Justyna Berniak-Woźny[1], Audronė Lupeikienė[2], and Jerzy Szewczyk[1,2]

[1] Systems Research Institute, Polish Academy of Sciences, Warsaw, Poland
marek.szelagowski@dbpm.pl

[2] Institute of Data Science and Digital Technologies, Vilnius University, Vilnius, Lithuania

Abstract. Cardiovascular diseases (CVD) remain the leading cause of mortality globally, accounting for approximately 30% of all deaths. Despite Poland's exemplary system for managing myocardial infarctions, the country faces one of the highest mortality rates in Europe during the first year post-heart attack. This paper examines the transformative potential of integrating telemedicine and Business Process Management (BPM) in the treatment of chronic diseases, with a particular focus on cardiac telerehabilitation. Through an in-depth case study of the Comprehensive Post-Myocardial Infarction Care Program (KOS-Zawał), the research demonstrates how a BPM approach supported by telemedicine can enhance treatment quality and reduce mortality rates for post-heart attack patients. The study advocates for a paradigm shift from traditional healthcare models, emphasising the importance of collaboration among healthcare professionals, patients, and caregivers. It highlights the ability of telemedicine to deliver personalised care, reduce healthcare costs, and improve patient outcomes. Cardiac telerehabilitation not only significantly lowers costs and provides greater accessibility compared to traditional rehabilitation but also provides preventive benefits, reducing the incidence of acute conditions such as heart attacks and preventing subsequent mortality or disability. Observations indicate that the KOS-Zawał program reduces the risk of death in the first year after myocardial infarction by almost 30% compared to standard care. By promoting a patient-centric and technologically advanced approach, this paper contributes to the evolving field of BPM in chronic disease management, underscoring the critical role of telemedicine and BPM in transforming healthcare delivery and improving patient outcomes.

Keywords: diagnostic and therapeutic processes · business process management (BPM) · telemedicine · chronic disease therapy · illustrative case study

C. Di Ciccio et al. (Eds.): BPM 2024, LNBIP 527, pp. 205–219, 2024.
https://doi.org/10.1007/978-3-031-70445-1_13

1 Introduction

The healthcare industry grapples with delivering personalised, high-quality care while facing resource limitations. This underscores the need to reimagine diagnostic and therapeutic approaches, embracing a comprehensive patient-centred paradigm. Business Process Management (BPM) and telemedicine technologies emerge as promising solutions, having successfully optimised workflows across various sectors [13, 25]. In healthcare, BPM holds the potential for streamlining diagnostic and therapeutic processes and facilitating automated, continuous monitoring of key health indicators without constant medical oversight [21, 28]. However, broader acceptance within the clinical community and robust research are imperative to delineating its optimal applications and overcoming scepticism [10].

Chronic diseases present a significant and escalating burden, affecting a substantial portion of the European population [24]. Arterial hypertension is one of the most common chronic diseases. It currently affects approximately 40% of the population, which is expected to increase [27]. Managing these conditions, which often require long-term and costly therapies, necessitates coordinated interdisciplinary efforts. The evolving healthcare landscape underscores the importance of involving patients and caregivers, aligning with the increasing adoption of telemedicine for remote, real-time health monitoring. The active engagement of patients and caregivers is pivotal for success, as they play active roles in care rather than passive recipients [5]. However, effective utilisation of telemedicine demands a departure from traditional models, where physicians dictate treatment and patients passively comply. This shift is particularly critical for chronic conditions, where sustained collaboration between patients, caregivers, and healthcare providers is essential. Professionals must empower patients with knowledge, clearly define roles, and continuously adjust treatments based on real-time data and evolving medical insights. This underscores the need for individualised care plans (ICPs), dynamically tailored through ongoing assessment of vital signs. Adopting a diagnostic and therapeutic processes approach promotes collaboration among interdisciplinary teams with shared objectives, roles, and communication channels guided by evidence-based practices [26].

This study explores the transformative potential of integrating telemedicine and BPM methodologies in cardiologic rehabilitation. This article presents a case study on implementing the Comprehensive Post-Myocardial Infarction Care Program (KOS-Zawał). The program aims to improve the quality of treatment and reduce mortality, specifically focusing on the care of patients after a heart attack. The primary objective of the KOS-Zawał program model is to ensure continuity of care and provide access to necessary healthcare services according to the patient's health needs, within optimal timeframes and in line with clinical guidelines and expert knowledge [1]. Medical research confirms the significant positive impact of rehabilitation on the treatment course of cardiological diseases. Participation in cardiac rehabilitation reduces the risk of death in the first year after myocardial infarction (MI) by 29% and the risk of rapid re-hospitalization due to cardiovascular diseases by nearly 20% [19]. Rehabilitation and health monitoring significantly lower the risk of acute cardiovascular events, delay disease progression, and shorten the treatment duration following acute cardiac incidents and exacerbations.

This research sheds light on the evolving healthcare landscape, emphasising the importance of a holistic view of diagnostic and therapeutic processes and collaborative efforts among healthcare professionals, patients, and caregivers to address challenges associated with chronic diseases. Thus, the paper aims to demonstrate and evaluate the reconfiguration of diagnostic and therapeutic processes, facilitating a patient-centric approach to therapy by incorporating innovative telemedicine technologies.

The paper begins with an extensive review of pertinent literature in the field. Following this, the authors delineate the methodology utilised in the study. Subsequently, the case study results are presented. The final section encapsulates conclusions and practical recommendations drawn from the research.

2 Methodology

This article aims to demonstrate and evaluate the reconfiguration of diagnostic and therapeutic processes, emphasising a patient-centric approach by incorporating innovative telemedicine techniques. Specifically, we illustrate how BPM methods can be utilised to develop comprehensive diagnostic and therapeutic processes for cardiologic rehabilitation. The authors employed an illustrative case study method to achieve these objectives, utilising a descriptive, in-depth, context-rich approach to provide detailed insights crucial for supporting the research process and understanding its outcomes [15]. The case study focused on a single instance implemented KOS-Zawał program to comprehensively explore organising a complex care system. This approach is a guideline for interdisciplinary teams, including health professionals, patients, and caregivers, leveraging the latest telemedicine technology to optimise treatment effectiveness and enhance healthcare system efficiency. The complexity and deviation from routine diagnostic and therapeutic processes make this case particularly intriguing for exploration [31]. The study illustrates the development of a KOS-Zawał program as a process and highlights the benefits of applying BPM and telemedicine technologies. These technologies enable the ongoing monitoring of various patient characteristics and parameters, ultimately improving patient outcomes and care management.

3 Related Work

3.1 BPM in Chronic Disease Management

The implementation of BPM in healthcare is driven by a focus on meeting recipients' needs, delivering quality services, ensuring accountability and transparency, and fostering continuous improvement [22]. Traditional BPM approaches fall short in dynamic healthcare environments like cardiological rehabilitation after strokes due to the diagnostic and therapeutic processes' complexity and context-specific nature [29]. BPM implementation in healthcare emphasises several key considerations:

- Patient-centric approach – treatment revolves around the patient, emphasising their needs and involvement in the treatment process.
- Engagement of patients and caregivers – active participation of patients and their caregivers is crucial for effective treatment outcomes.

- Cohesive system – diagnostic and therapeutic processes and various supportive processes must align to form a cohesive care system.
- Accessibility of knowledge – knowledge embedded in diagnostic and therapeutic processes should be accessible to all stakeholders through information technology and telemedicine systems.

Effective execution of therapeutic processes relies on empowering process executors to make informed decisions based on their expertise and the specific context of the clinical process [29]. As diagnostic and therapeutic processes are dynamic and inherently knowledge-intensive, success depends on efficiently utilising knowledge and engaging all stakeholders [4]. Modelling such processes requires flexibility to adapt to unforeseen events or changes in patient conditions. Business Process Modeling Notation (BPMN), with features like the "ad hoc subprocess," enables dynamic process flow, allowing process executors, including patients, to make real-time decisions during the execution of tasks [2, 22]. This adaptability is essential, particularly in cardiological rehabilitation after strokes, where patient conditions can vary widely. Analysing executed diagnostic and therapeutic processes facilitates the assessment of therapy effectiveness and the identification of deviations and opportunities for improvement [30]. Therapeutic processes serve as knowledge repositories, enabling continuous learning and refinement of clinical practices based on real-world data and experiences [22].

Despite Poland having one of the best systems for saving lives in cases of myocardial infarction, it also faces one of the highest mortality rates in the first years post-heart attack in Europe. This underscores the importance of holistic diagnostic and therapeutic processes that consider not only medical interventions but also patient responsibility and everyday engagement. In cardiological rehabilitation after strokes, integrating BPM methodologies can enhance patient-centred care, optimise treatment processes, and improve overall health outcomes. This approach emphasises adaptability, stakeholder engagement, and continuous learning to address the complexities of post-stroke rehabilitation effectively.

3.2 Cardiovascular Diseases (CVD) and Cardiac Rehabilitation (CR)

Cardiovascular diseases (CVD) are the leading cause of mortality globally, responsible for approximately 30% of all deaths worldwide, with a worrying trend towards increasing prevalence in the future. The global burden of CVDs has risen dramatically, with cases nearly doubling from 271 million in 1990 to 523 million in 2019. This rise is primarily driven by population growth and ageing, highlighting the urgent need for effective prevention and control strategies.

Cardiac rehabilitation (CR) is a comprehensive, medically supervised program designed to improve cardiovascular health following a cardiac event. It includes baseline patient assessments, nutritional counselling, risk factor modification (such as managing lipids, blood pressure, weight, and diabetes and promoting smoking cessation), psychosocial management, physical activity counselling, and exercise training. CR has been proven to expedite recovery post-cardiac events, reduce cardiovascular mortality and hospitalisation rates, and enhance patients' health-related quality of life. Despite its benefits, CR remains significantly underutilised worldwide.

To address this underutilisation, innovative delivery strategies such as home-based cardiac rehabilitation (HBCR) are being explored. Unlike traditional centre-based CR (CBCR), HBCR utilises remote monitoring and coaching with indirect exercise supervision. Technological advancements facilitate the implementation of HBCR by enabling real-time assistance and monitoring in the patient's home environment [3]. Studies have shown that cardiovascular telerehabilitation (CTR), a form of HBCR, can be an effective alternative to traditional CBCR.

CTR delivers CR treatment modules at home using wearable devices and remote communication between healthcare professionals and patients. This approach can increase CR utilisation and completion rates by overcoming barriers associated with traditional CBCR. For patients, CTR offers a more flexible and convenient option, which can lead to higher referral rates and improved enrollment and completion of CR programs. At the system level, CTR requires fewer personnel and training facilities per patient, making it a cost-effective solution that allows for the expansion of CR services without significant additional resources. Systematic reviews and meta-analyses have confirmed that CTR provides clinical and economic benefits similar to CBCR for patients with CAD and chronic heart failure (CHF).

The implementation of CTR faces several significant barriers, including uncertainty regarding ideal candidates for CTR, the lack of personalised CTR programs, insufficient reimbursement, concerns over investment, and the need for new clinical workflows. Furthermore, there is a notable absence of implementation research, which exacerbates these challenges. BPM offers a robust solution to address these issues and facilitate the broader adoption of CTR:

- Personalization and candidate identification—by leveraging data analytics and decision–support systems, BPM can streamline the identification and categorisation of patients suitable for CTR. Through BPM, healthcare providers can develop dynamic and personalised CTR programs tailored to individual patient needs, thus addressing the uncertainty regarding ideal candidates for CTR.
- Reimbursement and investment—by providing a clear and efficient framework for managing CTR processes, BPM can help demonstrate the cost-effectiveness and improved outcomes associated with CTR. This evidence can be used to advocate for better reimbursement policies and justify investments in CTR infrastructure. BPM's ability to provide detailed process documentation and performance metrics is crucial in building a compelling case for financial support.
- Creation of new workflows—one of the most significant barriers to CTR is the creation of new clinical workflows. BPM excels in designing, modelling, and optimising workflows, ensuring that all steps in the CTR process are efficient and well-coordinated. This includes integrating telemedicine technologies into existing healthcare processes and reducing the disruption caused by the transition to digital health solutions.
- Implementation research and stakeholder engagement – BPM encourages the co-creation of new processes with input from all relevant stakeholders, including patients. This collaborative approach enhances the commitment to new clinical workflows and ensures that the needs and preferences of end-users are met. Assessing and incorporating the digital readiness and health literacy levels of both patients and healthcare

professionals into BPM processes are crucial for the successful implementation of CTR.
- Continuous improvement and adaptation – BPM supports continuous improvement through ongoing monitoring and analysis of process performance. This capability is essential for adapting CTR programs based on real-time feedback and evolving medical knowledge. By continuously refining CTR processes, BPM ensures that they remain effective and aligned with best practices.

As presented above, BPM provides a comprehensive framework to address the key barriers to CTR implementation. By enhancing personalisation, securing financial support, optimising workflows, engaging stakeholders, and supporting continuous improvement, BPM can significantly advance the development and adoption of CTR, ultimately improving patient outcomes and the efficiency of cardiac rehabilitation programs. Moreover, the increasing incidence of hypertension, with projections indicating a worsening scenario, exacerbates the burden of CVD. Although Poland boasts one of the most effective systems for emergency cardiac care, it simultaneously suffers from one of the highest post-heart attack mortality rates in Europe. This paradox highlights the critical need for effective CR interventions.

3.3 Telemedicine, Data Significance, and AI Capabilities

Digital and telecommunication technologies have fundamentally transformed medical services. Early studies on telemedicine indicated its potential as a viable alternative to in-person care [8]. However, conclusive evidence of its positive impact on clinical outcomes was initially sparse [14], with benefits apparent in only a few healthcare areas [16]. Subsequent research has yielded more promising results. Of the reviewed articles [12], 72.5% indicated that telemedicine services delivered acceptable clinical outcomes, 67.7% observed cost-effectiveness, and 83.9% reported provider and patient satisfaction. Today, the advantages of telemedicine, such as enhanced access to care and continuity of care, have significantly improved patient satisfaction by providing more convenient and personalised care, better communication between patients and providers, and reduced frequency of in-person visits through daily monitoring of patient vital signs.

In a 2017 study [20], 95% of respondents identified clinical data (i.e., data related to regular patient care) as the most useful healthcare data sources, with only a slight predicted decline in their significance over the next five years. Current treatment models have evolved, largely based on data [7]. Data-driven patient care significantly facilitates and reduces the cost of treating any disease at an early stage, allows real-time notifications, and ensures prompt responses to concerning results. This improves diagnostic and therapeutic processes and enhances patient care.

Numerous studies, such as Eze et al. [11], have demonstrated that telemedicine does not compromise clinical care effectiveness compared to traditional healthcare delivery methods. One of the most comprehensive analyses evaluating telemedicine's efficacy in heart failure is the Telemedical Interventional Management in Heart Failure II (TIM-HF2) study [17]. Conducted on 1,571 participants (765 in the remote patient care group), the study showed that telemedicine significantly reduced the number of days lost due to unplanned cardiovascular events and lowered all-cause mortality through the patient diagnosis and treatment process. The expansion of available data volume and new data sources (including through telemedicine) is crucial for the development and acceleration of medical knowledge. This involves two stages of development: (1) Automation enabled by ICT, and (2) Generation of new knowledge through computational methods.

Automation, involving both hardware and software, has made it possible to address tasks previously unmanageable due to time constraints. This has expedited information exchange, patient management, and data analysis. Automated analysis of current and historical patient data and other healthcare-related datasets provides physicians with critical insights, enhancing decision-making processes.

The integration of data processing, modelling, and healthcare is reflected in areas like computational medicine and computational healthcare. According to [18], "there is a remarkable opportunity to promote medicine as an information science and strengthen the foundations of a learning healthcare system…". This implies that comprehensive data, computational patient models, and disease models can be utilised to discover new symptoms or risk indicators, improve diagnosis and treatment, and predict disease progression. For example, computational pharmacology methods predict how new drugs will interact with the human body. Data analysis at this level has achieved new quality—health data analytics involves analysing large datasets using quantitative and qualitative methods to identify trends, correlations, and patterns. Such analytics not only save money and time but also save patients' lives. However, leveraging all these possibilities requires not only new data sources but also new diagnostic and therapeutic processes [9], new thinking from medical authorities [7], and new methods and tools.

4 Changing Diagnostic and Therapeutic Processes in Cardiac Rehabilitation: A Case Study

A holistic diagnostic and therapeutic process cannot begin with treatment or even initial patient diagnostics. Its first stages should be health education and prevention. A fundamental element of prevention is information about current values of significant health parameters and the directions of their changes.

4.1 Monitoring the Condition of Cardiac Patients

Prior to the introduction of telemedicine devices, prevention, diagnosis, and management of cardiac diseases relied primarily on periodic check-ups and subsequent recommendations or temporary hospitalisations for specialised diagnostics or procedures when necessary. Health assessments such as ECGs were conducted during these check-ups, typically at monthly or longer intervals. Continuous data collection on critical health parameters in a patient's daily environment was limited to Holter monitoring, which had several significant limitations:

- Limited availability due to the restricted number of devices patients could borrow from healthcare facilities.
- Monitoring duration typically restricted to 24 or 48 h.
- Access to collected data only upon the completion of monitoring and device return.

- No capability for real-time response to critical health parameter deviations, posing potential risks to patient health and safety.

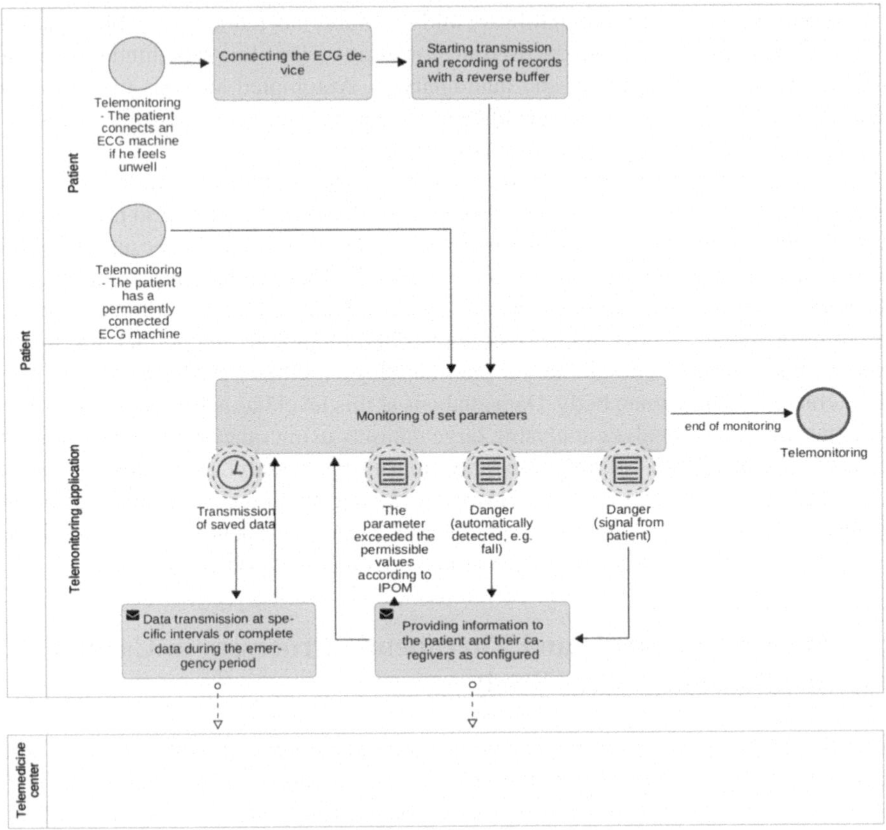

Fig. 1. Telemedicine Monitoring Diagram.

Cardiac telemonitoring, on the other hand, overcomes these limitations by enabling indefinite monitoring periods using increasingly accessible devices that ensure a high quality of measurements. These devices, often classified as "medical devices," can continuously collect and analyse measurement data and provide real-time alerts to patients and their caregivers regarding any health threats, such as exceeding permissible health parameter values. Additionally, they can be used by patients to signal distress to caregivers or automatically detect incidents like falls or loss of consciousness. As illustrated in Fig. 1, this monitoring can be conducted in collaboration with a Telemedicine Center.

The advantage of telemonitoring lies in its ability to continuously measure critical health parameters in a patient's daily living environment. This makes it an ideal solution for preventive care, not only for cardiovascular diseases but across various medical fields. Within cardiology, telemonitoring allows for the collection and analysis of data that would be unattainable with Holter monitors, especially in the early stages of the disease or in the case of rare heart rhythm disturbances that a Holter monitor might miss due to its limited monitoring duration.

Consequently, telemonitoring enables the collection of patient data even before acute conditions arise, eliminating the need for frequent hospitalisations or consultations. This data facilitates the analysis of trends in the patient's health status and tracks responses to implemented drug therapies.

4.2 The Course of the Therapeutic Process for Cardiovascular Diseases, with a Focus on Cardiac Telerehabilitation

For cardiovascular diseases, a crucial aspect of therapy aimed at preventing the deterioration of a patient's health after identified disorders or acute events (such as myocardial infarction) is cardiac rehabilitation. Before the introduction of telemedicine devices, cardiac rehabilitation was conducted either in stationary or outpatient settings, both of which were outside the patient's daily, natural environment. The main reasons for the low percentage of patients participating in cardiac rehabilitation included:

- The need to sacrifice daily family and work responsibilities.
- Inaccessibility of rehabilitation centres due to distance from the patient's residence (limited patient mobility, travel costs, commuting time).
- Lack of tailored rehabilitation programs for specific patient groups (e.g. elderly individuals, working individuals).
- Long waiting times to start rehabilitation (early initiation of rehabilitation is medically significant).
- Limited availability of hospital and outpatient facilities due to budget constraints.
- Financial constraints lead to limited available slots for rehabilitation courses.

These factors contributed to the low percentage of patients participating in cardiac rehabilitation. Additionally, conducting rehabilitation outside the patient's daily environment reinforced the mistaken belief that it was a "temporary" and "exceptional" activity, overseen solely by rehabilitation staff and concluding with the end of the rehabilitation course, without requiring further continuation between courses, which typically lasted only 2–3 weeks per year in practice.

This situation contradicts medical knowledge confirming the significant positive impact of rehabilitation on the course of cardiovascular disease therapy. Participation in cardiac rehabilitation reduces the risk of death by over 25% and significantly decreases the risk of acute cardiovascular events, delays disease progression, and shortens the treatment duration following acute cardiac incidents and exacerbations. To address this situation, the Comprehensive Post-Myocardial Infarction Care Program (KOS-Zawał) was established [23]. Its fundamental premise regarding cardiac rehabilitation was to shift rehabilitation from rehabilitation centres to the patient's home, ensuring their safety. The program's implementation was realised as a process consisting of three consecutive stages (Fig. 2).

Stage I: The initial Phase conducted in the Cardiac Rehabilitation Ward or Outpatient Rehabilitation Center.

During the initial two-week period of Stage I, the following activities take place:

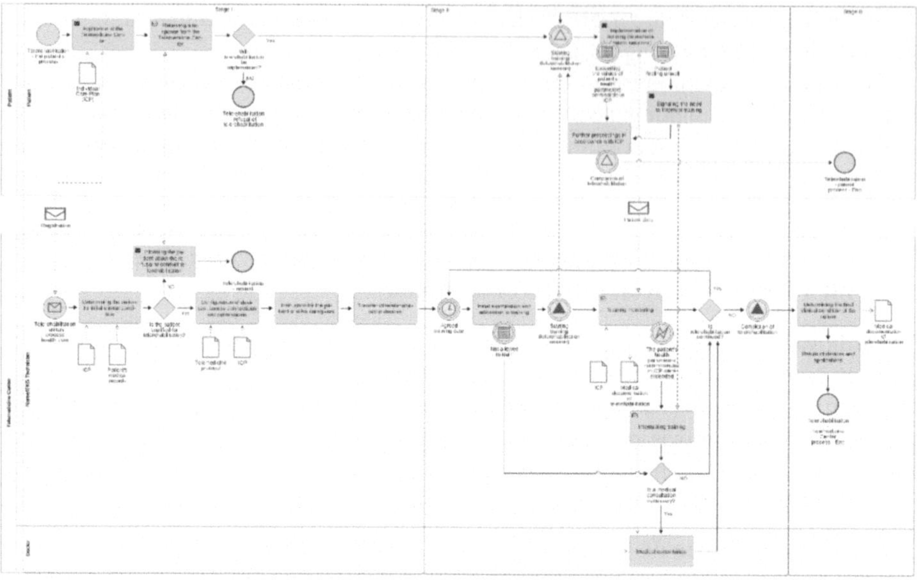

Fig. 2. The course of the process of human resources rehabilitation, including the use of telemedicine.

- Assessment of the patient's initial clinical condition (review of medical documentation, subjective and objective examination, ECG).
- Evaluation of physical fitness, including exercise stress testing/6-min walk test, and ergo spirometry to determine heart rate range.
- Assessment of the patient's psychological state.
- Development of an Individual Care Plan (ICP) tailored to the patient's current health status and recommendations.
- Patient health education, including familiarisation with the ICP and the planned course of rehabilitation.
- Training on device operation and transmission, including instruction on proper measurement execution.
- Provision of recommended responses if health parameter values exceed permissible limits defined in the ICP.
- Planning and execution of several training sessions.

Stage II. The core rehabilitation phase is conducted by the patient and potentially their caregivers in the patient's everyday living environment – simply put - at home. Within the KOS-zawał program, the patient is cleared for training (telerehabilitation session) if their health permits. They undergo training sessions under the supervision of medical staff, such as a nurse/ECG technician or physiotherapist. At the current stage of the KOS-zawał program's development, the activities conducted by the nurse/ECG technician hold particular significance, including:

- Remote measurement of patient health parameters.
- Clearing the patient for training.

- Remote monitoring of the training session (40 min – 1 h).
- Assessment of training using the Borg Scale.
- Reporting issues and consultation needs to other staff.
- Adjusting apparatus parameters.
- Remote monitoring and intervention in emergencies, if necessary.
- Maintenance of electronic rehabilitation documentation.
- Periodic follow-up consultations are also conducted during Stage II to determine the patient's current clinical status and assess the effectiveness of telerehabilitation. Based on these assessments, adjustments may be made to the frequency, duration, and scope of training activities.

In the case of post-myocardial infarction rehabilitation, the rehabilitation supervisor may consider concluding Stage II of the telerehabilitation process and initiating Stage III, which focuses on promoting the patient's health-conscious lifestyle. The aim is to maintain the effects of treatment and reduce the risk of disease recurrence. In this stage, telemonitoring becomes a natural continuation of telerehabilitation, enabling real-time patient health status monitoring.

4.3 Benefits of Implementing the KOS-Zawał Program

The KOS-Zawał program exemplifies a structured approach to post-myocardial infarction care, integrating initial in-hospital assessments with extended home-based rehabilitation and continuous telemonitoring. This three-stage process ensures a seamless transition from acute care to long-term health maintenance, promoting sustained health improvements and reducing the risk of recurrence. The initial in-hospital phase allows for comprehensive assessments and personalised care planning, while the home-based phase empowers patients to take an active role in their recovery, supported by remote monitoring and periodic consultations.

Cardiac telerehabilitation implemented within the KOS-zawał program has resulted in a reduction in the risk of cardiac diseases and re-hospitalisation compared to cardiac rehabilitation in stationery and outpatient settings, as evidenced by:

- A threefold increase in the number of rehabilitated patients.
- Significant improvement in the accessibility of cardiac rehabilitation (no limits on services provided within the KOS-zawał program).
- Increased access to services for patients living far from outpatient rehabilitation centres.
- Long-term telemonitoring capability and detailed patient data analysis not only during training sessions.
- High patient acceptance (less than 1% of patients interrupt the telerehabilitation cycle).
- Significantly accelerated adoption of health-promoting behaviours in the patient's living environment.
- Accelerated return to full social and professional activity for patients.

Beyond the significant increase in rehabilitation accessibility, the healthcare system benefits from much higher cost-effectiveness. Since the patient undergoes the procedure at home, the rehabilitation centre, and thus the healthcare system, does not incur the costs of the patient's stay at the facility, including providing medication, meals, medical care, or cleaning. There is also reduced utilisation of specialist working time, reducing the cost of service provision and additionally shortening the waiting time for specialist consultations, thereby reducing, for example, the waiting time to start rehabilitation.

Cardiac telerehabilitation naturally progresses to telemonitoring. It allows for real-time monitoring of the patient's health status practically without interfering with their living environment. It enables the preemptive detection of changes in the patient's health threatening the recurrence of acute conditions (e.g., another heart attack), leading to hospitalisation, loss of health preventing the previous social functions (e.g., leaving work and transitioning to disability), or other social costs (e.g., the need for continuous care). However, not only after suffering a heart attack and rehabilitation. Telemonitoring can also prevent heart attacks, informing the patient and his caregivers in advance about an upcoming heart attack or generally about deteriorating values of selected parameters characterising the health condition.

The high acceptance rate of telerehabilitation, with less than 1% of patients discontinuing the program, underscores its viability as a preferred care method. Integrating rehabilitation into the patient's daily environment fosters greater adherence to health-promoting behaviours, facilitating a quicker return to normal social and professional activities. Patients benefit from the continuous support and real-time feedback telemonitoring provides, reinforcing positive lifestyle changes and encouraging long-term health maintenance.

5 Conclusions

CVD remains the main cause of death worldwide, reaching about 30% of all deaths. One of the examples is also Poland, which, despite the existence of one of the most effective systems for emergency care for heart attacks, has one of the highest post-heart attack mortality rates. In Europe, the first year of post-myocardial infarction shows that it is crucial to have a holistic view of the patient's treatment process. One which accounts for preventive care and rehabilitation following severe cases or medical procedures. Such a holistic view is provided by BPM, which enables the joint management and coordination of the diagnostic-therapeutic process. Furthermore, BPM accounts in improved processes for the possibilities provided by the constantly developed ICT solutions.

The article presented the transformative potential of integrating telemedicine with BPM in treating one of the most prevalent chronic conditions on the example of cardiac telerehabilitation within the Comprehensive Post-Myocardial Infarction Care Program (KOS-Zawał). The example demonstrates how the BPM approach, supported by the use of telemedicine devices, improves the effectiveness and quality of treatment and, at the same time, allows for shortening wait times for specialist care and reduces resource consumption or costs of care. This is made possible on the one hand thanks to ensuring the use of personalised telemedicine devices, remote care and rehabilitation, along with the previously unavailable ongoing control of the patient's health parameters. On the other, the KOS-Zawał programme raises the effectiveness of treatment thanks to a holistic approach to therapy. Cardiac telerehabilitation at the foundation of this programme offers considerably larger accessibility than traditional rehabilitation and preventative care benefits by lowering the frequency of serious conditions and preventing later patient morbidity or impairment. The implementation of the KOS-zawał programme lowers the risk of death in the first year following myocardial infarction by almost 30% in comparison to standard care, thanks to the use of a proactive approach in place of a reactive one. The extension in accordance with BPM of the use of telemedicine solutions along with processes from the area of preventive care and rehabilitation will allow for the prevention of acute conditions thanks to the ongoing control of the patient's health parameter, including the prevention of first or later heart attacks thanks to the execution of preventative interventions, and not just reactive measures.

Both of the described services also stimulate the development of medicine. The solutions developed in the KOS-Zawał program can generally be used to prevent and treat chronic diseases. By allowing the collection and analysis of data collected before acute conditions, they allow for identifying changes in the patient's health parameters before exacerbations. This allows for developing detection algorithms for upcoming exacerbations and effective prevention procedures. The quality of these algorithms will improve with the growth in the quantity of available data. This will ensure the safe functioning of people at risk of chronic diseases in society without the danger of overloading or a drastic increase in the burden and costs of the healthcare sector.

While this case study provides valuable insights into the benefits of telemedicine in cardiac rehabilitation, it is important to acknowledge several limitations that could impact the generalizability and applicability of the findings. The study focuses primarily on implementing the KOS-Zawał program within a specific healthcare context. Consequently, the results may not directly apply to other regions or healthcare systems with different infrastructures, resources, or patient populations. Future studies should consider diverse settings to validate the findings and ensure broader applicability. Additionally, the study predominantly assesses telemedicine interventions' immediate and short-term outcomes in cardiac rehabilitation. Long-term effects, such as sustained patient adherence, long-term cost savings, and enduring health outcomes, require further investigation. A longitudinal study design would be necessary to capture these longer-term impacts comprehensively. Finally, although the study highlights the cost-effectiveness of telemedicine interventions, a comprehensive economic analysis that includes all associated costs and savings was not conducted. Future research should incorporate detailed cost-benefit analyses to provide a more thorough understanding of the economic implications of widespread telemedicine adoption in cardiac rehabilitation.

Disclosure of Interests. The authors have no competing interests to declare that are relevant to the content of this article.

References

1. Agency for Health Technology Assessment and Tariff System, (2016) Report No: AOTMiT-WT-553-13/2015 Comprehensive Care Post-Myocardial Infarction (2016)

2. Aissaoui, N.O., Ben Mbarek, H., Layeb, S.B., Hadj-Alouane, B.A.: A BPMN-VSM based process analysis to improve the efficiency of multidisciplinary outpatient clinics. Prod. Plan. Control **35**(5), 461–491 (2022). https://doi.org/10.1080/09537287.2022.2098199
3. Amantea, I.A., Sulis, E., Boella, G., Marinello, R., Grosso, M., Crespo, A.: A modeling framework for an innovative e-health service: the hospital at home. In: Obaidat, M.S., Oren, T., Rango, F.D. (eds) Simulation and Modeling Methodologies, Technologies and Applications. SIMULTECH 2020. Lecture Notes in Networks and Systems, vol. 306. Springer, Cham (2022). https://doi.org/10.1007/978-3-030-84811-8_6
4. Baptista dos Santos França, J., Manhães Netto, J., Carvalho, E.S., Santoro, F.M., Araujo Baião, F., Pimentel, M.: KIPO: The knowledge-intensive process ontology. Softw. Syst. Model. **14**, 1127–1157 (2015)
5. Becker, J., Fischer, R., Janiesch, C.: Optimizing U.S. health care processes - a case study in business process management. In: AMCIS 2007 Proceedings. Paper 504 (2007). https://aisel.aisnet.org/amcis2007/504. Accessed on 31 Oct 2021
6. Calzon, B.: 24 Examples of big data analytics in healthcare that can save people. Bus. Intell. (2024). https://www.datapine.com/blog/category/business-intelligence/
7. Compton-Phillips, A.: Care Redesign Survey: What Data Can Really Do for Health Care. NEJM Catalyst (2017)
8. Currell, R., Urquhart, C., Wainwright, P., Lewis, R.: Telemedicine versus face to face patient care: effects on professional practice and health care outcomes. Cochrane Database of Syst. Rev. (2) (2000). https://doi.org/10.1002/14651858.CD002098
9. Davenport, T., Bean, R.: Five key trends in AI and data science for 2024. MIT Sloan Manag. Rev. (2024). https://sloanreview.mit.edu/article/five-key-trends-in-ai-and-data-science-for-2024/
10. De Ramón, F.A., Ruiz Fernández, D., Sabuco, G.Y.: Business process management for optimizing clinical processes: a systematic literature review. Health Inf. J. **26**(2), 1305–1320 (2020). https://doi.org/10.1177/1460458219877092
11. Eze, N.D., Mateus, C., Cravo Oliveira Hashiguchi, T.: Telemedicine in the OECD: an umbrella review of clinical and cost-effectiveness, patient experience and implementation. PLoS One **15**(8), e0237585 (2020). https://doi.org/10.1371/journal.pone.0237585
12. Goharinejad, S., Hajesmaeel-Gohari, S., Jannati, N., Goharinejad, S., Bahaadinbeigy, K.: Review of systematic reviews in the field of telemedicine. Med. J. Islam Repub. Iran **29**(35), 184 (2021). https://doi.org/10.47176/mjiri.35.184
13. Gómez-Valiente, P., Pérez, J., Lillo-Castellano, J., Marina-Breysse, M.: Smart-IoT business process management: a case study on remote digital early cardiac arrhythmia detection and diagnosis. IEEE Internet Things J. 10/19 (2023). https://doi.org/10.1109/JIOT.2023.3269820
14. Hailey, D., Roine, R., Ohinmaa, A.: Systematic review of evidence for the benefits of telemedicine. J. Telemed. Telecare **8**(Suppl 1), 1–30 (2002)
15. Hayes, R., Kyer, B., Weber, E.: The Case Study Cookbook; Worcester, Polytechnic Institute: Worcester. MA, USA (2015)
16. Hersh, W.R., et al.: Clinical outcomes resulting from telemedicine interventions: a systematic review. BMC Med. Inform. Decis. Mak. **1**, 5 (2001). https://doi.org/10.1186/1472-6947-1-5
17. Koehler, F., Koehler, K., Deckwart, O., Prescher, S., Wegscheider, K., et al.: Efficacy of telemedical interventional management in patients with heart failure (TIM-HF2): a randomised, controlled, parallel-group, unmasked trial. Lancet **392**(10152), P1047-1057 (2018). https://doi.org/10.1016/S0140-6736(18)31880-4
18. Krumholz, H.M.: Big data and new knowledge in medicine: the thinking, training, and tools needed for a learning health system. Health Aff (Millwood). **33**(7), 1163–1170 (2014)
19. Kubielas, G., Diakowska, D., Uchmanowicz, I.: Survival analysis of patients with acute coronary syndrome receiving comprehensive coordinated care after myocardial infarction (KOS-Zawał). Polish Heart J. (Kardiologia Polska) **80**(3), 315–321 (2022)

20. Lin, M., Yuan, W., Huang, T., Zhang, H., Mai, J., Wang, J.: Clinical effectiveness of telemedicine for chronic heart failure: a systematic review and meta-analysis. J. Investig. Med. **65**(5), 899–911 (2017). https://doi.org/10.1136/jim-2016-000199
21. Mans, R., Aalst, W., Vanwersch, R.: Process Mining in Healthcare Evaluating and Exploiting Operational Healthcare Processes. Springer, Berlin, Germany (2015)
22. Marrella, A., Mecella, M., Sharf, M., Catarci, T.: The TESTMED Project Experience. Process-aware Enactment of Clinical Guidelines through Multimodal Interfaces. arXiv 2019, arXiv:abs/1807.02022
23. NHL (National Health Fund). Communication for service providers: Communication on the implementation of services after myocardial infarction (KOS-Zawał). NFZ, Warszawa 2017
24. OECD/EU. Health at a Glance: Europe 2016–State of Health in the EU Cycle; OECD Publishing: Paris, France (2016)
25. Otto, L., Schlieter, H., Harst, L., Whitehouse, D., Maeder, A.: The telemedicine community readiness model—successful telemedicine implementation and scale-up. Front. Digit. Health **5**, 1057347 (2023). https://doi.org/10.3389/fdgth.2023.1057347
26. Salvi, S., et al.: The burden of chronic respiratory diseases and their heterogeneity across the states of India: the global burden of disease study 1990–2016. Lancet Glob. Health **6**, e1363–e1374 (2018)
27. Scaratti, C., Leonardi, M., Silvaggi, F., Ávila, C., Muñoz-Murillo, A., Ferraina, S.: Mapping European welfare models: state of the art of strategies for professional integration and reintegration of persons with chronic diseases. Int. J. Environ. Res. Public Health **15**, 781 (2018)
28. Schuster, D., Benevento, E., Aloini, D., et al.: Analyzing healthcare processes with incremental process discovery: practical insights from a real-world application. J Healthc Inform Res (2024). https://doi.org/10.1007/s41666-024-00165-6
29. Szelagowski, M., Berniak-Wozny, J.: A process-centered approach to the description of clinical pathways -forms and determinants. Int. J. Environ. Res. Public Health **2019**(16), 2638 (2019)
30. Williams, S., Radnor, Z.: An integrative approach to improving patient care pathways. Int. J. Health Care Qual. Assur. **31**, 810–821 (2017)
31. Yin, R.L.: Case Study Research: Design and Methods, 4th edn. SAGE Publications, Thousand Oaks, CA, USA (2009)

Adapting to the Dynamic Nature of Business Processes in the Digital Age

Marek Szelągowski[1(✉)], Natalia Potoczek[2], Piotr Sliż[3], Justyna Berniak-Woźny[3], Aneta Napieraj[4], Marta Podobińska-Staniec[4], Arkadiusz Kozłowski[3], and Piotr Senkus[5]

[1] Systems Research Institute of the Polish Academy of Sciences, Warsaw, Poland
marek.szelagowski@dbpm.pl
[2] Institute of Economics of the Polish Academy of Sciences, Warsaw, Poland
[3] Faculty of Management, University of Gdańsk, Sopot, Poland
[4] AGH University of Kraków, Kraków, Poland
[5] Faculty of Economic Sciences, University of Warsaw, Warsaw, Poland

Abstract. Every organization's changing and increasingly complex environment poses increasingly serious challenges for BPM. Many processes are difficult to design, which reduces the chance of their predictability. Investigating the nature of processes and increasing efforts to support unstructured and knowledge-intensive processes becomes necessary. Therefore, the paper aims to explore the intricate nature and diversity of business processes in a multifaceted way, recognizing their dynamic adaptation to the evolving economic landscape. This study utilizes established classifications in order to categorize processes into four types: structured, structured with ad hoc exceptions, unstructured with pre-defined fragments, and completely unstructured. It takes a survey-based opinion research approach, employing CAWI for data collection, and utilizes statistical analysis. The analysis reveals that traditional, structured processes still constitute a substantial portion of organizational processes. Nevertheless, unstructured processes and unstructured processes with pre-defined fragments are gaining increasing relevance. The study also identifies significant correlations between the implementation of IT tools and the prevalence of unstructured processes. The research results can serve as a crucial knowledge resource for managers implementing and using BPM and software suppliers developing IT tools supporting BPM.

Keywords: BPM · digital technologies · process nature · process diversity

1 Introduction

Studying the nature and diversity of business processes (BPs) is a constant challenge and a necessity due to the dynamics of the economic environment. Consequently, the impact of external and internal factors in individual groups of BPs in the enterprise is not the same and unchanged. The current predictability or knowledge intensity of individual groups of processes or BPs may change very quickly. In response to this scientific and practical imperative, this study aims to illuminate the relationships between the nature

C. Di Ciccio et al. (Eds.): BPM 2024, LNBIP 527, pp. 220–234, 2024.
https://doi.org/10.1007/978-3-031-70445-1_14

of business processes and the support of digital technologies in process management. Researchers point out the diverse nature of the structuring of business processes, which is most often related to the dynamics of the business environment and the predictability of activities and results obtained [31]. The mechanisms of economies of scale constantly motivate companies to automate and robotize processes that work best in conditions of relative predictability and, consequently, the detailed structuring of processes [40]. However, recent global changes have increased the motivation to look for technological solutions that will equally support processes with a low level of structure [14]. By the approach adopted in research at the interface of knowledge management (KM) and business process management (BPM), the last two groups of processes were classified as knowledge-intensive business processes (kiBPs), in line with the approach of Gronau and Weber [16], that is, those in which the created value depends on the use of knowledge of process participants. As emphasized by Di Ciccio et al. [13], methods of the execution knowledge-intensive BPs are closely dependent on using the knowledge of process executors and they require significant flexibility during design and execution. Therefore, they are usually unpredictable. Due to the above, it can be assumed that the importance of processes in the organization perceived by those implementing them and the type of IT support in managing and implementing processes will be different for structured (predictable) and unstructured (unpredictable) processes. To examine the nature of BPs, the criterion of their unpredictability was chosen, as it was more general, more intuitively understandable and possible to objectively explore based on the criterion of having or not having full knowledge determining the likely course of BPs. Since the competitiveness of structured and unstructured (knowledge-intensive) processes is also conditioned by the development of information technologies, unstructured and knowledge-intensive processes currently constitute the most significant challenge for managers and software developers. The presented study referred to the most popular IT tools: Enterprise Resource Planning (ERP), Customer Relationship Management (CRM), Supply Chain Management (SCM), Web Portal, Workflow, Business Process Management System (BPMS), Robotic Process Automation (RPA), and Artificial Intelligence (AI). Due to various external and internal factors, the same processes may become more or less structured in different use cases. For this study, 3 research questions were formulated: RQ1: What is the share of structured and unstructured BPs in organizations? RQ2: What is the relationship between the importance of BPs in organizations and their nature? RQ3: What is the relationship between the IT tools used in the BPM and the nature of BPs?

2 Theoretical Background

This study was preceded by the analysis of many process classifications, including Gronau and Weber [16], Bahrs and Müller [4], Dalmaris et al. [11], Pesic et al. [30], Kemsley [19], van der Aalst [41], Marjanovic and Freeze [9, 10], Di Ciccio et al. [13], Işık et al. [17], Olding and Rozwell [28], or Berniak-Woźny and Szelągowski [8]. Ultimately, the Kemsley classification was adopted for research purposes. With reference to the predictability of the execution of BPs, the study distinguished four categories of business processes: structured (static) – entirely predictable, structured with ad hoc exceptions – processes in which most of the course is predictable and structured, but in which there are activities whose implementation may be unpredictable, unstructured with the predefined fragments – mostly not predictable, but with parts where it is possible to impose the way of implementing activities and unstructured – impossible to design in advance and therefore unpredictable. These classifications are crucial to answering RQ1.Based on the research conducted, the Beverungen team formulated seven paradoxes of BPM in the modern hyper-connected world [9]. They emphasize, among other things, that as a result of changes taking place in the company's environment, we are increasingly faced with the need to create scenarios for redesigning and implementing new variants of processes during their implementation, which, of course, contradicts the standardization of processes understood as imposing a unified, exclusively correct course. On the one hand, companies are faced with the need to standardize most processes to take advantage of economies of scale and reduce process costs, but on the other hand, we are increasingly faced with new needs or market problems. At the same time, the authors draw attention to the personalization of products in individualized service processes, which is, in fact, contrary to efforts to force the same course of business processes [9]. IT artefacts in BPM are becoming increasingly complex, and their assessment requires hard, data-driven field evidence. Process performance and data may vary depending on the scenario, so the context of process implementation will force actions that will be difficult to standardize or even design [9]. Many researchers have undertaken the contextual approach to process management and IT tool development, and recent years have demonstrated the increasing importance of this research issue [7, 9, 21]. Nowadays, context awareness in BPM seems obvious, so there is a need to create new methods supporting context-sensitive BPM. These paradoxes highlight the dynamic nature of BPs, addressing RQ2. Vom Brocke et al. [10] recognize this problem and, with regard to the design sciences paradigm, propose the Cluster-Aware Multiagent System (CAMAS), which is intended to help practitioners select BPM methods by assessing the context of process implementation. The proposed Classification Framework is a meta-model referring to three dimensions: process life cycle, process goal, and process context (use case). The context of the process itself is created by: value contribution, repetitiveness, knowledge intensity, creativity, interdependence, and variability, which directly refers to the need to identify the nature of the BPs. As a result of recent global experiences, an increasing number of researchers refer to external factors affecting management in enterprises [22, 23]. For many years, attention has been paid to the fact that external shocks usually lead to unintended radical changes in processes [15, 19], which is a challenge for managers, process implementers, and IT tool providers. Röglinger et al. [31] emphasize that this problem in the area of BPM has been neglected. Actions should be taken to research processes and generate ideas that increase safety and ensure process efficiency in unpredictable conditions. These postulates undoubtedly correspond to the developing trend of research on the nature of business processes, which was initially related to the possibilities and consequences of the knowledge-based economy [16, 20, 23, 27, 28], followed by the development of digital technologies [5, 6, 14, 19, 35] and more recently, the dynamics of processes determined by external factors [20, 33]. The significant role of humans in unpredictable (unframed) BPs emphasized van der Aalst [41]. Research on the role of executors in BPs with various levels of structure was also conducted by the team of Di Ciccio et al. [13]. Using earlier works by Gronau and

Weber [16] and Seethamraju and Marjanovic [34], researchers defined kiBPs as those processes that require a unique human role in implementation. The less rigid structure of these processes is most often associated with the impact on their implementation of hidden knowledge, which is difficult to codify and often unconscious, and the dynamism (including energy, willingness to act, and commitment) of the employees using them [35]. In this case, IT tools should support process implementers in making quick and accurate decisions, which means that they should facilitate access to data and their swift processing. Difficulties related to the implementation of structured and unstructured processes simultaneously are indicated, among others, in research by Antunes et al. [1]. They propose an approach that integrates formal and informal process structures. The composite approach adopts a socio-material conceptual lens in which material and human factors are supported. Modern AI in these machine learning (ML) technologies creates new opportunities to support and even replace humans in business processes. Still, in many areas, as van der Aalst [39] emphasizes, there remain tasks that require common sense, contextual knowledge, creativity, adaptability, culture, and empathy, which only humans can provide. According to van der Aalst, combining these two spheres of competence will require people to develop hybrid intelligence, which combines human and machine intelligence. The synergy effect achieved from this cooperation will probably become one of the most important research trends in many scientific disciplines. To examine the impact of these technological advancements on the management of structured and unstructured business processes, RQ3 explores the relationship between the IT tools used in BPM and the nature of BPs.

3 Material and Methods

As a result of the theoretical study, the stages of the scientific and research process were designed to facilitate the achievement of the formulated goal. The research was conducted using the diagnostic survey method among managers and specialists actively engaged in business processes. The gathered empirical material provided clear answers to the three research questions. This section outlines the study design and characterizes the study sample.

3.1 Research Design

The pilot study and the main research phase were conducted in Poland in 2023. The design of the research questionnaire was a pivotal step, with meticulous attention to the clarity and comprehensibility of the questions to ensure effective capture of the desired information. Incorporating the Process Classification Framework (PCF) developed by the American Productivity & Quality Center (APQC) [2, 3], the questionnaire was tailored to the declared operational profile (production, commercial, service) of the surveyed organizations. The research questionnaire consisted of three sections. The first one was the background information section. The second section required respondents to assess the diversity of business processes aligned with their operational profile, following APQC's framework. In the third section, participants were asked to evaluate the significance of these business processes within the organization. Before full-scale data collection, a pilot study was conducted to evaluate the comprehensibility of the questionnaire, addressing any potential ambiguities or issues. This pilot study included 18 surveys. Subsequently, the main research phase commenced, employing a survey-based opinion research approach using the Computer Assisted Web Interview (CAWI) techniqe [18]. Post-data collection, an exhaustive evaluation of data quality was undertaken to ascertain the reliability and validity of the acquired dataset. The ensuing data analysis phase involved the application of statistical techniques and data visualization methods to derive meaningful insights and provide a comprehensive description of the research findings. Measures of ordinal association (Somers' D, Stuart-Kendall τ_c, and Spearman's rank correlation coefficient) were used to investigate the correlation between variables on ordinal scales. For an in-depth analysis and visualization of the relationship observed in the contingency table, correspondence analysis was used.

3.2 Sample Selection

The research was conducted using a survey-based opinion research approach with non-probabilistic sampling. The criterion for selecting organizations primarily stemmed from the invitation of respondents who predominantly hold managerial or expert positions. The research questionnaire was administered to respondents using the CAWI technique. On a total of 466 completed research questionnaires collected, an assessment of the quality of the empirical data was performed. Initially, 14 records were excluded from the dataset, as they corresponded to organizations that had not defined a single process according to the presented framework. Following an extensive data quality assessment phase, 220 records were removed from the dataset, significantly improving the reliability and accuracy of the results. A total of 232 questionnaires underwent further analysis, involving a partial assessment through the creation of a matrix to verify the presence or absence of defined processes. The empirical material collected in the quantitative study was verified for the quality of the obtained results, which influenced the final number of questionnaires included in the study. For this purpose, a verification of variables concerning the number of identified processes was conducted, along with the evaluation of processes based on their significance in each category. In case of discrepancies, the questionnaire was discarded. During this stage, an additional 21 records were excluded to enhance data quality and accuracy. This meticulous data curation process resulted in a final dataset comprising 211 questionnaires. By rigorously scrutinizing and selecting only the most relevant and reliable data, we ensured the high quality of the dataset, establishing a robust foundation for subsequent research analysis and interpretation. Frequency distributions for variables describing the final sample are presented in Table 1. The respondents represented organizations of various sizes, scales of operations, stages in the organizational life cycle, and levels of BPM maturity and IT support.

Table 1. Statistical summary of the research sample.

Variable/category	Count	Percent
Size of the organization		
Micro organization	46	21.8%
Small organization	53	25.1%
Medium organization	29	13.7%
Large organization	83	39.3%
Scale of operations		
Local	62	29.4%
Regional	9	4.3%
National	46	21.8%
International	94	44.5%
Stage in the organizational life cycle		
Stage 1. Initial level	19	9.0%
Stage 2. Growth stage	69	32.7%
Stage 3. Maturity stage	119	56.4%
Stage 4. Decline stage	4	1.9%
Level of BPM maturity		
Level 1. Initial level	25	11.8%
Level 2. Repeatable level	16	7.6%
Level 3. Defined level	77	36.5%
Level 4. Managed level	51	24.2%
Level 5. Optimizing level	42	19.9%
IT support		
Website	128	60.7%
CRM	127	60.2%
ERP	86	40.8%
Workflow	81	38.4%
SCM	64	30.3%
BPMS	55	26.1%
RPA	17	8.1%
AI	15	7.1%

4 Results

First, an attempt was made to answer the first research question (RQ1) concerning the identification of structured and unstructured processes within the studied group of 211 units. Subsequently, efforts were directed towards addressing RQ2 and establishing the relationship between the significance of business processes in organizations and their diversity. Finally, the analysis focused on RQ3, aiming to define the relationship between the IT tools used in Business Process Management (BPM) and the nature of the processes.

4.1 Assessment of Process Diversity

44 types of processes were distinguished based on the APQC's framework [2, 3]. For each type of process, respondents indicated if the process is defined in their organizations and, if so, assessed its nature (structured/ unstructured) and importance. In this way, a total of 6,506 processes were evaluated by 211 respondents. It is essential to emphasize that, as a result of the rigorous data quality analysis, only those questionnaires were included in the subsequent phases of the study that demonstrated 100% response consistency. This stringent criterion serves to ensure data integrity, thereby bolstering the reliability of the findings. The obtained results depicting the percentage distribution of process diversity are presented in Table 2.

Table 2. Distribution of specific business process categories.

Business process category	Count	Percentage
Structured (static) processes	1,665	25.6%
Structured processes with ad-hoc exceptions	2,441	37.5%
Unstructured processes with pre-defined fragments	1,722	26.5%
Unstructured processes	678	10.4%
Total	6,506	100.0%

Based on the analysis of the obtained results, it is evident that the nature of business processes in contemporary organizations is indeed diverse, with a majority exhibiting a non-traditional, dynamic character. In this respect, the obtained results confirm previous studies [28]. Surprisingly, traditional, structured (static) processes still constitute a significant portion, accounting for nearly 26% in the studied organizations. This finding is intriguing, especially in the context of earlier predictions that suggested a decline in the share of traditional, structured business processes. The study highlights the coexistence of various process types, emphasizing the ongoing need for organizations to manage both structured and unstructured processes effectively. In the context of these results, it is evident that semi-structured processes, encompassing both structured processes with ad-hoc exceptions as well as unstructured processes with pre-defined fragments, collectively represent a majority, accounting for 64% of the total. This underscores the significant importance of these process types in the analyzed context, warranting further consideration and in-depth analysis. In summary, by answering RQ1, it should be stated that structured business processes are a minority among business processes in surveyed organizations. The findings underscore the diverse nature of business processes in contemporary organizations, indicating that the landscape includes both traditional and dynamic processes, each requiring tailored management approaches and IT support.

4.2 The Relationship Between the Significance of Business Processes in Organizations and Their Nature

The assessment of the significance of business processes involved evaluating their impact on the organization using a 5-point scale (from 1, meaning very low or no impact, to 5, meaning very high impact). Table 3 illustrates the percentage distribution of the significance of various categories of business processes within the surveyed organizations.

Table 3. Joint relative frequency distribution of the significance and category of BPs.

	Business process (BP) significance/impact				
Business process category	very low	low	moderate	high	very high
Structured (static) processes	3.9%	2.9%	5.1%	8.5%	5.2%
Structured processes with ad-hoc exceptions	1.6%	5.0%	14.5%	10.9%	5.5%
Unstructured processes with pre-defined fragments	0.4%	2.0%	4.0%	11.4%	8.7%
Unstructured processes	0.2%	0.5%	2.0%	3.9%	3.8%
Total	6.2%	10.4%	25.5%	34.7%	23.2%

In total, the combined data from Table 3 indicates that business processes within the studied organizations vary intensively in terms of their significance. These results collectively illustrate the broad spectrum of impact levels across different categories of business processes, underlining the need for a nuanced approach to their management and support. The nature of business process and its significance for the organization are two ordinal variables. Their categories can be ordered from structured to unstructured processes and from very low to very high impact. Based on the contingency table (with the same structure as Table 3 but with absolute frequencies), various measures of ordinal association were computed to assess the direction and strength of the relationship between these variables. The results are as follows: Somers' D = 0.21, Stuart-Kendall $\tau\_c = 0.20$, Spearman $r\_S = 0.24$. All three measures take values between -1 and 1, where an absolute value near 0 indicates no correlation and the more the actual value deviates from 0, the stronger the agreement (for positive values) or disagreement (for negative values) of ranks. The values of these measures show the same picture of the relationship between the nature of the business process and its impact on the organization, namely, the variables are positively correlated, and the strength of the correlation is moderate. It means that the more unstructured a process is the higher significance it has on the organization. To investigate this relationship further, the correspondence analysis was performed [6]. The principal inertia of the first two singular vectors accounted for as much as 98.6% of the total inertia of the contingency table (62.2% for the first and 36.4% for the second singular vector). A two-dimensional plot covering rows and columns categories of the contingency table is presented in Fig. 1 (the mapping is symmetric, i.e. both rows and columns are in principal coordinates).

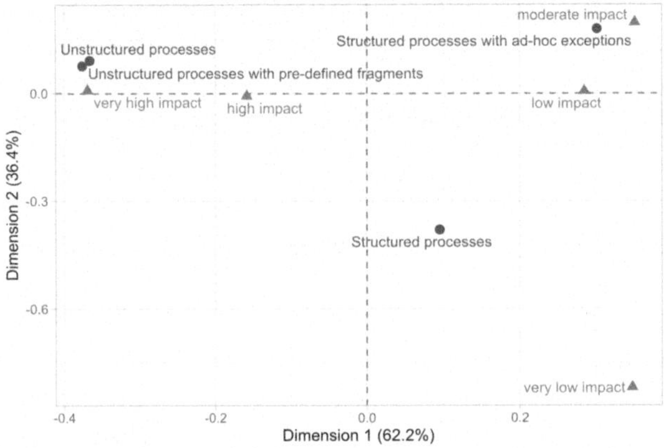

Fig. 1. Two-dimensional representation of the relationship between the nature of a process (filled circles) and its assessed importance (triangles).

The two categories for unstructured processes (fully unstructured and semi-unstructured) are situated on the plot near the two categories for the most impactful processes (very high impact and high impact) and are clearly separated from the rest of the categories on the first dimension. This is in line with the previous observation of a positive ordinal correlation between the business process category and its significance. Unstructured processes are generally associated with a greater impact on the organization. The pattern for the two categories of structured processes is less clear. Structured processes with ad-hoc exceptions are relatively more often of moderate or low impact on the organization. Fully structured processes are the most separated from other categories on both dimensions, which means that some structured processes were important, some not, with a distinctive pattern, which to some extent disrupts the ordinal association between the nature of the business process and its impact on the organization. For the data obtained in Table 3, the authors have performed another analysis by differentiating sources of data by their impact on management, and indirectly on the responsibility for the organization. To achieve this, the results were segregated based on the criterion of the groups of employees participating in the study. In accordance with the information in the questionnaire, the respondents were divided into two groups. The first group represents the managerial cadre, while the second comprises operational employees, specialists, and experts (non-managerial employees). The first group (Group

I), representing the managerial cadre, consisted of 77 employees who evaluated 2.386 processes. In contrast, the second group (Group II) consisted of 134 respondents who evaluated 4.120 processes. It is noteworthy that from the perspective of the positions held, a significant difference was observed. The share of structured business processes in Group I evaluations is notably lower and constitutes 18.36%, while for Group II, it is 29.78%. In the case of structured processes with ad-hoc exceptions, a similar result was recorded for both Group I and Group II, with percentages of 37.51% and 37.52%, respectively. Differences, though not as pronounced as for structured (static) business processes, were observed for the groups of unstructured processes with pre-defined fragments and unstructured processes. For Group I, the respective shares were 31.77% and 12.36%, whereas for Group II, they were 23.40% and 9.3%. In the above analysis, the statistical unit was the business process, with one respondent evaluating multiple processes. An analysis was also carried out using the organization/respondent as the unit of the study. Respondents indicated on average 31 processes in their organization. For each of the 211 respondents separately, relative frequency distributions of their business process categories and importance were calculated. Spearman rank correlation coefficients between the response proportions for each category were then calculated, which were visualized in Fig. 2. High positive correlation coefficients appear on the main diagonal of the matrix (from the upper left to the lower right), i.e. in accordance with the previously observed relationship – a higher proportion of structured processes is associated with a higher proportion of insignificant processes, and a higher proportion of unstructured processes is associated with a higher proportion of significant processes. Consequently, the correlation coefficients outside the main diagonal (lower left and upper right corners) are negative, indicating that the more unstructured processes, the fewer insignificant ones, and the more structured processes, the less significant ones.

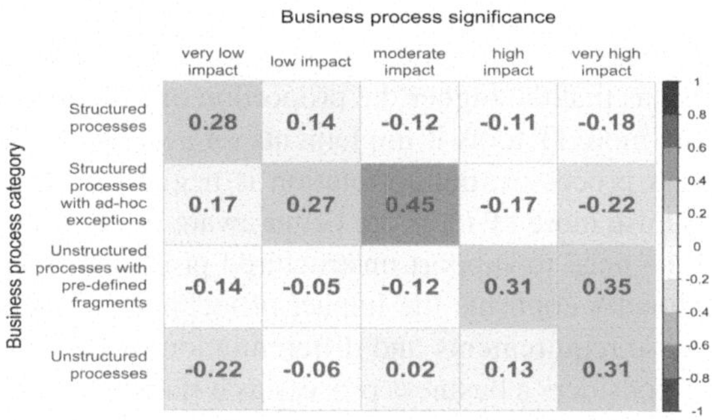

Fig. 2. Spearman's correlation coefficients between the proportions of processes assessed as structured/unstructured (rows) and as having given impact on an organization (columns).

In summary, the research findings provide clear evidence of a relationship between the significance or level of importance of processes and their nature within the studied business processes. Answering RQ2, it should be stated that most processes important for the surveyed organizations (having a high or very high impact) are not traditional business processes. Semi-structured processes constitute 63% of the most important processes in the surveyed organizations. These insights contribute to a deeper understanding of how the nature of processes influences their significance and, consequently, informs organizational decision-making and management practices.

4.3 The Relationship Between the IT Tools Used in the BPM and the Nature of the Processes

The analysis of the collected empirical data prompted the authors to attempt to identify correlations between the stated implementation of IT tools in BPM and diversity within the studied business processes. This analysis directly addresses RQ3, which seeks to uncover whether a correlation exists between the adoption of IT tools and the nature of BPM processes. In the first step, it was checked whether the sheer number/variety of IT tools used is related to the occurrence of business processes of a given category. For this purpose, Spearman's rank correlation coefficients between the number of implemented IT tools and the proportion of structured and unstructured processes within the organization were calculated. The results are presented in Table 4.

Table 4. Spearman's correlation coefficients between the number of implemented IT tools and the proportion of each business process category within the organization.

Proportion of:	Correlation with the number of IT tools
Structured processes	−0.002
Structured processes with ad-hoc exceptions	−0.096
Unstructured processes with pre-defined fragments	0.121
Unstructured processes	0.316

The correlation coefficient is substantial only in the case of the proportion of unstructured processes. It means that the higher the proportion of fully unstructured processes in an organization, the more IT tools it implements on average. For the proportions of other types of business processes, the correlation is negligible. These results suggest that organizations that use more IT tools are better aware of the occurrence of unstructured processes and the need to support unstructured processes with various IT tools. This finding also indirectly confirms the impact of technology adoption on increased awareness of the diverse requirements and differentiation of BPs within the organization. Further analysis considers a business process as a statistical unit and examines the specific classes of IT tools. Deviations from the overall relative frequency distribution of business process categories by the IT tool implemented are depicted in Fig. 3. The tools on the graph are ordered according to the combined share of unstructured processes and unstructured processes with pre-defined fragments. Naturally, one organization might use more than one tool, so the groups overlap to some extent. Nevertheless, the graph is useful in discovering the association between IT tools used and the incidence of business processes. The overall frequency distribution of business process categories is presented in Table 2. The percentage points in Fig. 3 show how implementing a given IT tool changes this distribution. For example, organizations that use BPMS had 20.2% of processes assessed as unstructured, which is 9.8 percentage points higher than the overall share of unstructured processes; these organizations had also a relatively higher share of unstructured processes with pre-defined fragments and a lower share of structured and semi-structured processes. The IT tools that make the share of fully structured and unstructured processes the most different from the overall distribution is AI (a considerably higher proportion of unstructured processes and lower proportion of structured processes). Other IT tools that are associated with a higher proportion of unstructured processes are SCM, BPMS, and ERP. A peculiar situation occurs in the case of Workflow and RPA, where there is a higher relative frequency of both fully structured and fully unstructured processes and a lower frequency of semi-structured ones. Implementation of Websites and CRM are the least changers of the overall distribution of business process categories. In general, implementing any IT tools for BPM is associated with an increased proportion of unstructured processes (deviations of unstructured processes in Fig. 3 are always positive).

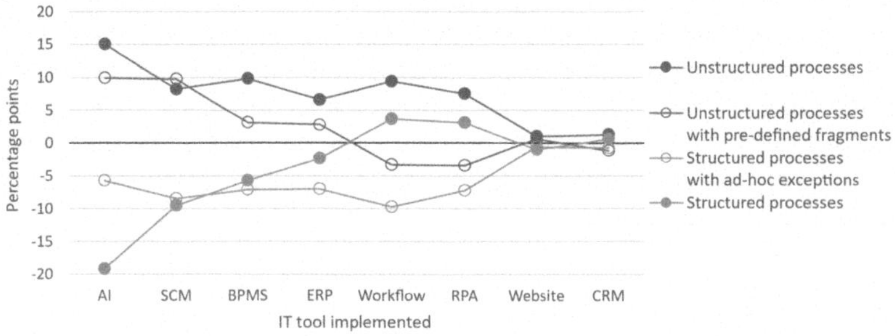

Fig. 3. Deviations from the overall distribution of business process categories by IT tools implementation.

In summary, the study reveals a significant positive correlation between the number of implemented IT tools and the prevalence of unstructured processes, suggesting that organizations with more IT tools tend to have a higher occurrence of unstructured processes. This can be interpreted as these organizations having already identified and being aware of the existence of unstructured business processes, while their competitors may not have done so yet. The specific IT tools is also a factor differentiating the distribution of business processes. Implementing AI, SCM, BPMS, and ERP is most likely to follow the increase of unstructured processes and decrease of structured ones. These findings underscore the influence of technology adoption to awareness nature of the business processes within organizations, contributing valuable insights for strategic decision-making and management practices.

5 Conclusions

The results obtained in the group of 211 surveyed enterprises confirm the fact that business processes are differentiated due to their dynamic. In the surveyed group of organizations, the largest share (64%) is semi-structured processes, i.e., structured processes with ad-hoc exceptions and unstructured processes with pre-defined fragments.

The share of traditional structured processes covers almost 26% of all examined processes. This finding addresses RQ1: What is the share of structured and unstructured BPs in organizations? The research also answered the question (RQ2) about the correlation between the importance of processes and their nature. It turned out that unstructured processes are more often of high or very high importance for the organization than traditional structured processes are. The observation suggests that very important processes often coexist with an increased preponderance of unstructured processes. The study also showed that there is a relationship between the information technologies used in processes and the nature of the processes, answering RQ3: Is there a relationship between the IT tools used in the BPM and the nature of BPs?. The sheer number of different IT tools used is positively correlated with the prevalence of unstructured processes. The IT tools that correlate with the higher incidence of unstructured or semi-unstructured processes are AI, SCM, BPMS, and ERP. According to the authors, a more significant share of semi-structured or completely unstructured processes is caused by the need to adapt to the requirements of customers expecting individualization of the delivered products or services and the changing economic environment of enterprises dynamically. The conducted research provides valuable knowledge about the importance of processes of various nature and their relationships with information technologies. However, attention should be paid to the limitations of the presented research, which result from the fact that the sample was not a probability sample. The formulated conclusions concern 211 organizations located in Poland.

References

1. Antunes, P., Piono, A.J., Nkhoma, M., Thuan, N.H.: The composite approach as a hybrid approach to business process modeling: proposition and empirical evaluation. Bus. Process Manag. J. ahead-of-print (2023). https://doi.org/10.1108/BPMJ-11-2022-0569
2. APQC (n.d.), Process Classification Framework. https://www.apqc.org/process-frameworks. Accessed 30 Oct 2023
3. Auksztol, J., Chomuszko, M.: Model APQC – Uniwersalny Model Klasyfikacji Procesów (APQC Model – Universal Process Classification Model), In Auksztol, J., Chomuszko, M. (Eds.) Modelowanie organizacji procesowej (Modeling of process organization), pp. 73–90 (2013). Wydawnictwo Naukowe PWN, Warszawa
4. Bahrs, J., Müller, C.: Modelling and analysis of knowledge intensive business processes. In: Althoff, K.-D. et al. (eds.) Professional Knowledge Management, pp. 243–247 Springer, Berlin, Heidelberg (2005). https://doi.org/10.1007/11590019_28
5. Becker, J., vom Brocke, J., Heddier, M., Seidel, S, S.: In search of information systems (grand) challenges. Bus. Inf. Syst. Eng. **57**(6), 377–390 (2015). https://doi.org/10.1007/s12599-015-0394-0
6. Beh, E.J., Lombardo, R.: Correspondence Analysis: Theory, Practice and New Strategies (2014). https://doi.org/10.1002/9781118762875
7. Berger, S., Denner, M.S., Röglinger, M.: The Nature of Digital Technologies – Development of a Multi-layer Taxonomy (2018)
8. Berniak-Woźny, J., Szelągowski, M.: Towards the assessment of business process knowledge intensity – a systematic literature review. Bus. Process. Manag. J. **28**(1), 40–61 (2021). https://doi.org/10.1108/BPMJ-01-2021-0012

9. Beverungen, D., Buijs, J.C.A.M., Becker, J., Di Ciccio, C., van der Aalst, W.M.P.: Seven paradoxes of business process management in a hyper-connected world. Bus. Inf. Syst. Eng. **63**(2), 145–156 (2021). https://doi.org/10.1007/s12599-020-00646-z
10. vom Brocke, J., Baier, M.S., Schiedel, T., Stelzl, K., Röglinger, M., Wehking, C.: Context-aware business process management. Bus. Inf. Syst. Eng. **63**(5), 533–550 (2021). https://doi.org/10.1007/s12599-021-00685-0
11. Dalmaris, P., Tsui, E., Hall, B., Smith, B.: A framework for the improvement of knowledge-intensive business processes. Bus. Process. Manag. J. **13**(2), 279–305 (2007). https://doi.org/10.1108/14637150710740509
12. Denner, M.S., Röglinger, M., Schmiedel, T., Stelzl, K., Wehking, C.: How context-aware are extant BPM methods? - development of an assessment scheme. In: Weske, M. et al. (eds.) Business Process Management, pp. 480–495 Springer International Publishing, Cham (2018). https://doi.org/10.1007/978-3-319-98648-7_28
13. Di Ciccio, C., Marrella, A., Russo, A.: Knowledge-intensive processes: characteristics, requirements and analysis of contemporary approaches. J. Data Semant. **4**(1), 29–57 (2015). https://doi.org/10.1007/s13740-014-0038-4
14. Goni, J.I.C., Van Looy, A.: Process innovation capability in less-structured business processes: a systematic literature review. Bus. Process. Manag. J. **28**(3), 557–584 (2022). https://doi.org/10.1108/BPMJ-07-2021-0487
15. Grisold, T., Groß, S., Stelzl, K., Mendling, J., Röglinger, M., Rosemann, M.: The five diamond method for explorative business process management. Bus. Inf. Syst. Eng. **64**(2), 149–166 (2022). https://doi.org/10.1007/s12599-021-00703-1
16. Gronau, N., Weber, E.: Management of knowledge intensive business processes. In: Desel, J. et al. (eds.) Business Process Management, pp. 163–178 Springer, Berlin, Heidelberg (2004). https://doi.org/10.1007/978-3-540-25970-1_11
17. Işik, Ö., Jones, M.C., Sidorova, A.: Practices of knowledge intensive process management: quantitative insights. Bus. Process. Manag. J. **19**(3), 515–534 (2013). https://doi.org/10.1108/14637151311319932
18. Kagerbauer, M., Manz, W., Zumkeller, D.: Analysis of PAPI, CATI, and CAWI methods for a multiday household travel survey. In: Zmud, J., Lee-Gosselin, M., Munizaga, M., Carrasco, J.A. (Ed.) Transport Survey Methods, Emerald Group Publishing Limited, Leeds, pp. 289–304 (2013). https://doi.org/10.1108/9781781902882-015
19. Kemsley, S.: The changing nature of work: from structured to unstructured, from controlled to social. In: Rinderle-Ma, S. et al. (eds.) Business Process Management, p. 2 Springer, Berlin, Heidelberg (2011). https://doi.org/10.1007/978-3-642-23059-2_2
20. Kregel, I., Koch, J., Plattfaut, R.: Beyond the hype: robotic process automation's public perception over time. J. Organ. Comput. Electron. Commer. **31**(2), 130–150 (2021). https://doi.org/10.1080/10919392.2021.1911586
21. Legner, C., et al.: Digitalization: opportunity and challenge for the business and information systems engineering community. Bus. Inf. Syst. Eng. **59**(4), 301–308 (2017). https://doi.org/10.1007/s12599-017-0484-2
22. Li, J., Sun, W., Jiang, W., Zhang, L.: How the nature of exogenous shocks and crises impact company performance?: the effects of industry characteristics. IJRCM **6**(4), 40–55 (2017). https://doi.org/10.4018/IJRCM.2017100103
23. Marjanovic, O.: A novel mechanism for business analytics value creation: improvement of knowledge-intensive business processes. J. Knowl. Manag. **26**(1), 17–44 (2021). https://doi.org/10.1108/JKM-09-2020-0669
24. Marjanovic, O., Freeze, R.: Knowledge intensive business processes: theoretical foundations and research challenges. In: 2011 44th Hawaii International Conference on System Sciences, pp. 1–10 (2011). https://doi.org/10.1109/HICSS.2011.271

25. Marjanovic, O., Freeze, R.: Knowledge-intensive business process: deriving a sustainable competitive advantage through business process management and knowledge management integration. Knowl. Process. Manag. **19**(4), 180–188 (2012). https://doi.org/10.1002/kpm.1397
26. Martín-Navarro, A., Lechuga Sancho, M., Medina-Garrido, J.: Determinants of BPMS use for knowledge management. J. Knowl. Manag. ahead-of-print, (2023). https://doi.org/10.1108/JKM-07-2022-0537
27. McCormack, K., et al.: A global investigation of key turning points in business process maturity. Bus. Process. Manag. J. **15**(5), 792–815 (2009). https://doi.org/10.1108/14637150910987946
28. Olding, E., Rozwell, C.: Expand your bpm horizons by exploring unstructured processes. Gartner Technical Report G00172387, Published: 10 December 2009, Refreshed: 22 May 2015
29. Papavassiliou, G., Mentzas, G.: Knowledge modelling in weakly-structured business processes. J. Knowl. Manag. **7**(2), 18–33 (2003). https://doi.org/10.1108/13673270310477261
30. Pesic, M., Schonenberg, H., van der Aalst, W.M.P.: DECLARE: full support for loosely-structured processes. In: 11th IEEE International Enterprise Distributed Object Computing Conference (EDOC 2007), p. 287 (2007). https://doi.org/10.1109/EDOC.2007.14
31. Röglinger, M., et al.: Exogenous shocks and business process management. Bus. Inf. Syst. Eng. **64**(5), 669–687 (2022). https://doi.org/10.1007/s12599-021-00740-w
32. Rosemann, M., Recker, J., Flender, C.: Contextualisation of business processes. Int. J. Bus. Process. Integr. Manag. **3**(1), 47–60 (2008). https://doi.org/10.1504/IJBPIM.2008.019347
33. Sarnikar, S., Deokar, A.V.: A design approach for process-based knowledge management systems. J. Knowl. Manag. **21**(4), 693–717 (2017). https://doi.org/10.1108/JKM-09-2016-0376
34. Seethamraju, R., Marjanovic, O.: Role of process knowledge in business process improvement methodology: a case study. Bus. Process. Manag. J. **15**(6), 920–936 (2009). https://doi.org/10.1108/14637150911003784
35. Szelągowski, M.: The knowledge and process dimensions. VINE J. Inf. Knowl. Manag. Syst. **51**(2), 271–287 (2020). https://doi.org/10.1108/VJIKMS-09-2019-0150
36. Szelągowski, M., Berniak-Woźny, J.: The knowledge and process continuum. Knowl. Process. Manag. **26**(4), 308–320 (2019). https://doi.org/10.1002/kpm.1611
37. Thomas, O., et al.: Global crises and the role of BISE. Bus. Inf. Syst. Eng. **62**(4), 385–396 (2020). https://doi.org/10.1007/s12599-020-00657-w
38. Van Der Aalst, W.M.P., Hinz, O., Weinhardt, C.: Impact of COVID-19 on BISE research and education. Bus. Inf. Syst. Eng. **62**(6), 463–466 (2020). https://doi.org/10.1007/s12599-020-00666-9
39. Van Der Aalst, W.M.P.: Hybrid Intelligence: to automate or not to automate, that is the question. IJISPM **9**(2), 5–20 (2021). https://doi.org/10.12821/ijispm090201
40. Van Der Aalst, W.M.P., Bichler, M., Heinzl, A.: Robotic process automation. Bus. Inf. Syst. Eng. **60**(4), 269–272 (2018). https://doi.org/10.1007/s12599-018-0542-4
41. Van Der Aalst, W.M.P.: Business process management: a comprehensive survey. Int. Schol. Res. Not. **2013**, e507984 (2013). https://doi.org/10.1155/2013/507984
42. Yang, C.G.: A study on the changes in the ICT industry after the COVID-19 pandemic. Ind. Manag. Data Syst. **123**(1), 64–78 (2023). https://doi.org/10.1108/IMDS-03-2022-0165
43. Zelt, S., Schmiedel, T., vom Brocke, J.: Understanding the nature of processes: an information-processing perspective. Bus. Process. Manag. J. **24**(1), 67–88 (2018). https://doi.org/10.1108/BPMJ-05-2016-0102

Enhancing Business Process Management Through Nature Assessment: Development and Deployment of the Business Process Nature Assessment Tool

Marzena Grzesiak[1], Marek Szelągowski[2](✉), Marek Moszyński[3], and Justyna Berniak-Woźny[2]

[1] Faculty of Management and Economics, Gdańsk University of Technology, Gdańsk, Poland
marek.szelagowski@dbpm.pl

[2] Systems Research Institute of the Polish Academy of Sciences, Warsaw, Poland
marek.szelagowski@dbpm.pl

[3] Faculty of Electronics, Telecommunications and Informatics, Gdańsk University of Technology, Gdańsk, Poland

Abstract. In contemporary business landscapes, the diverse nature of processes necessitates a thorough assessment before implementing Business Process Management (BPM). However, conducting such assessments can be arduous, particularly for small and medium-sized enterprises. This study aims to address this challenge by developing and showcasing the utility of an online tool based on the Business Process Nature Assessment Framework (BPNAF). The BPNAF facilitates integrated assessments of business processes, aligning with established BPM methodologies and lifecycles. Through an illustrative case study, this research examines the efficacy of this tool in assessing the processes/nature. The BPNA tool enables swift and effective evaluations of process unpredictability and knowledge intensity, aiding in selecting appropriate BPM methodologies, Knowledge Management (KM) and Human Resource Management (HRM) strategies, and supporting IT systems. Visualised results offer clear recommendations tailored to the organisation's unique process landscape, accommodating standard and customised recommendations. By contributing insights into practical BPNAF implementation, this study plays a pivotal role in advancing BPM assessment methodologies and enriching the understanding and practice of BP management. This research underscores the critical role of BP assessment in organisational success, advocating for informed decision-making and best practices in BPM.

Keywords: business process · business process management (BPM) · process nature · business process nature assessment framework (BPNAF) · illustrative case study

C. Di Ciccio et al. (Eds.): BPM 2024, LNBIP 527, pp. 235–249, 2024.
https://doi.org/10.1007/978-3-031-70445-1_15

1 Introduction

For nearly 20 years, management practitioners and researchers have observed the diverse nature of business processes (BPs) [5, 17]. Business processes with different execution dynamics require the application of various methods for identification, design, and implementation. Unstructured, unpredictable BPs cannot be managed by designing detailed models and then always imposing the same way of execution. Simultaneously, case management methods are unsuitable for traditional, structured, and repetitive BPs. A study on the nature of business processes conducted by an interdisciplinary group of Polish researchers in business process management (BPM) in 2023 showed that only about 35% of the processes in the surveyed organisations have a traditional, repetitive, and predictable nature [14]. Additionally, the study revealed that unstructured processes were significantly more important for the surveyed organisations. Therefore, before implementing BPM, it is essential to assess the nature of the organisation's processes and select implementation methodologies and tools according to the actual nature of the processes being implemented, including human resource management (HRM) principles, IT systems supporting BPM, and knowledge management (KM) methods and tools.

Consistent with previous proposals, including those by Zelt et al. [19, 20], such an assessment is facilitated by the Business Process Nature Assessment Framework (BPNAF) [1, 13]. This comprehensive framework enables a holistic evaluation of the nature of BPs, offering standardised recommendations for various BPM areas such as implementation methodologies, human resource management (HRM), knowledge management (KM), and IT systems. The development of a supporting tool facilitates the practical application of BPNAF, significantly reducing the effort required to ensure its effectiveness from a business user perspective. The simplicity and speed of the assessment allow the use of BPNAF in organisations without experience in BPM, including medium and small companies. Developed on its basis the Business Process Nature Assessment tool (BPNA tool) not only maintains the advantage of easy and quick but also allows for the seamless integration of the BPNA with other implementation tools, enhancing its usability and value. The BPNA tool presented in this study enables the integrated assessment of BPs, aligning with established BPM implementation methodologies and BPM lifecycles dedicated to both traditional BPs [3] and dynamically managed BPs [11]. Additionally, by permitting repeated assessments of all or selected processes within the organisation, the tool enables tracking changes in BPs due to factors such as shifts in organisational culture following BPM implementation or adopting new technologies and IT tools. Consequently, when used with BPM maturity assessment (BPM MA), this tool provides a means to evaluate changes in BPM maturity and the effectiveness of achieving pre-implementation goals.

The paper is structured as follows: Sect. 2 presents the research methodology. Section 3 discusses the BPNAF, including the assumptions for the tool implementing the assessment of BPs' nature. The subsequent section describes the BPNA tool, enabling the assessment of BPs' nature and additional possibilities arising from IT technologies. Section 5 presents the results of the assessment of BPs' nature conducted for a medium-sized Polish company in the production industry. The article ends with Conclusions presenting, among other things, directions for further research.

2 Methodology

This research employs an illustrative case study methodology to investigate the effectiveness of the BPNA tool for assessing the nature of BPs. The choice of a case study approach aligns with Yin's [18] assertion that this method is particularly suited for exploring complex phenomena in real-life settings, especially when understanding the intricate boundaries between the phenomenon and its context. This approach ensures that our findings are theoretical and have practical implications for real-world business process management.

Utilising a comprehensive qualitative approach, primarily adopting an interpretive stance [6], this research design incorporates a mixed-method approach for data collection, which includes:

- Document Analysis: This involves thoroughly examining organisational documents such as process manuals, procedural documents, and previous assessments of business processes. This analysis provides insights into the existing documentation and understanding of BPs within the organisation.
- Observations: Real-time observations of selected business processes to understand their dynamics, variability, and interactions within the organisational environment. These observations contribute to a richer understanding of how BPs are executed and managed in practice.
- Semi-Structured Interviews: The conduct of semi-structured interviews with key stakeholders in BPM, including managers who oversee the implementation of BPs, process owners who are responsible for the day-to-day execution of BPs, and IT personnel who support the technological aspects of BPs. These interviews elicit nuanced insights into stakeholders' perceptions of the nature of business processes, the challenges encountered, and their expectations from the assessment framework.

Through this research design, the study aims to significantly contribute to the body of knowledge on BPM and assessment methodologies. By shedding light on the practical application and utility of the BPNAF in real-world organisational contexts, this research strives to inform and advance best practices in BP assessment and management, emphasising our findings' crucial importance.

3 Business Process Nature Assessment Framework

The BPNAF was developed in response to the need for a tool to assess the nature of BPs [1, 13]. To some extent, this function was fulfilled by Business Process Maturity Models (BPM MM) methods and to [4, 8, 12]. However, in most cases, their complexity and lack of integration with BPM implementation methodologies resulted in the costs associated with their use being significantly higher than the benefits obtained [12, 15]. The BPNAF is a simple and flexible solution that allows for its application regardless of the organisation's size. Even in large organisations, conducting a study, according to the BPNAF, does not take longer than 5 to 7 days. The assessment process for small and medium-sized enterprises does not exceed 2–3 days. The process of assessing the nature of BPs in the organisation has been synchronised with generally accepted BPM

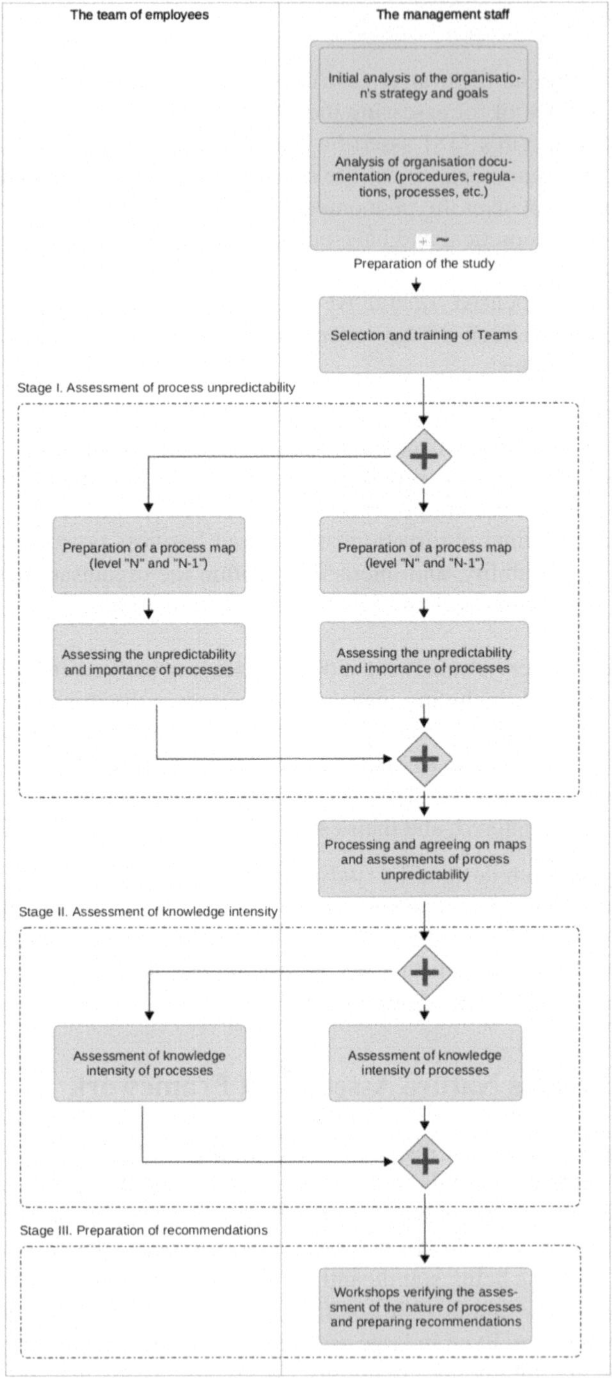

Fig. 1. Diagram of the Business Process Nature Assessment Process in BPMN.

implementation methodologies and divided into three stages carried out after initial preparation and initiation of the BPM implementation project (Fig. 1):

I. In Stage I, the preparation of process maps for levels N and N-1 is carried out according to the categorisation proposed by A. Mahal [7], along with an assessment of the unpredictability of N-1 level processes according to the categorisation proposed by S. Kemsley [5]. These tasks are performed in separate workshop sessions with top management and the Working Team, typically consisting of middle-level managers and independent employees. After this stage, the results of the teams' work are discussed, and a common final model is agreed upon.

II. In Stage II, each team assesses the intensity of knowledge required to execute the N-1 level processes.

III. In Stage III, the teams' work results are discussed, and a common final model for assessing the nature of BPs is agreed upon. Additionally, recommendations concerning BPM implementation methods, HRM principles, and the selection of IT systems supporting the organisation's operations are prepared.

The outcomes of implementing BPNAF include:

- awareness of the necessity for a holistic view of organisational BPs,
- experience in teamwork involving top management and middle management in the identification and recommendation of BPM improvements,
- agreed process maps for levels N and N-1, typically containing organisational environment data, assigned to specific groups of processes, key performance indicators (KPI), or major IT systems supporting the organisation's operations,
- data on unpredictability, knowledge intensity, significance, and projected changes in the significance of individual process groups, presented along with standard recommendations in a clear format such as The Business Process Nature Assessment Matrix (BPNAM),
- agreed recommendations for BPM improvements, addressing individual business process groups, HRM principles, and the systemic infrastructure supporting the organisation's operations.

Depending on the organisation's needs, the process of assessing the nature of BPs:

- may be repeated for all processes or selected groups of N-1 level processes,
- may be deepened through the preparation of maps or models and the assessment of N-2 level processes.

4 The Business Process Nature Assessment Tool

The practical use of BPNAF required the development of a tool to minimise the workload and automate the comparison of the results of subsequent assessment processes.

4.1 Tool Requirements

The main functions of the BPNA tool should include:

- Creation of research questionnaires that allow for: input of necessary identification data and a brief organisation profile; if such research has been conducted in the organisation previously, to minimise the workload, it should be possible to copy and edit the results; input of data (maps) of processes at levels N and N-1 or copying and updating data from a previous study; verification of the correctness of created process maps.
- Input of process dynamics assessments with the following rules: qualification of individual groups of N-1 level processes into one of four process categories: structured, structured with ad hoc exceptions, unstructured with pre-defined fragments, or unstructured; assessment of process significance (weight) for the organisation (on a scale from 1 to 6, where 1 – the least important, 6 – the most important); assessment of the direction of change in process significance for the organisation (on a scale from -2 to 2, where -2 indicates a significant decrease in significance, 0 indicates no change, and + 2 indicates a significant increase in significance).
- Input of knowledge intensity assessments for each group of N-1 level processes with the following rules: the overall assessment of the process group consists of 10 assessments of various criteria characterising knowledge on a four-point scale: strongly disagree, disagree, agree, strongly agree; the criteria should be presented in a separate questionnaire for each group of N-1 level processes, with the overall assessment of the process group expressed numerically on a scale from 0 (the process group does not require any knowledge) to 10 (intensive knowledge is required according to each of the 10 criteria); visualisation of results based on input assessments with basic statistics and creation of the BPNAM, along with standard recommendations for groups of N-1-level processes based on assessments presented in the matrix.
- The tool should also allow users to input individual recommendations on BPM, KM, HRM, or ICT tools tailored to the assessed organisation.

4.2 IT Architecture of BPNAF Tool

The BPNA tool has been designed as a simple but classical client-server program as this architecture meets all requirements and allows for future extensions. In this model, the server hosts the application's software code and resources while the client uses a web browser to request and execute the code and view the data in the browser's user-friendly interface (Fig. 2). The code was designed using open-source technologies what guarantees achieving a modern user interface on variety of devices.

The tool can be installed on any web server that executes PHP code. The server-side code is executed only when the user data needs to be changed, which happens when generating a questionnaire (create operation) and filling it (write operation). Such a solution allows for completing the questionnaire without storing and retrieving files locally, which speeds up the process of filling. However, there is also an option for storing the questionnaire locally, working with it without an Internet connection and inserting the questionnaire when online.

When used by various clients, one of the important requirements of the distributed system is the separation of client resources. This requirement is typically met by applying an annoying login/password approach and – for larger solutions – with virtual machines. The tool uses a link access approach instead. The link code is generated when finalising

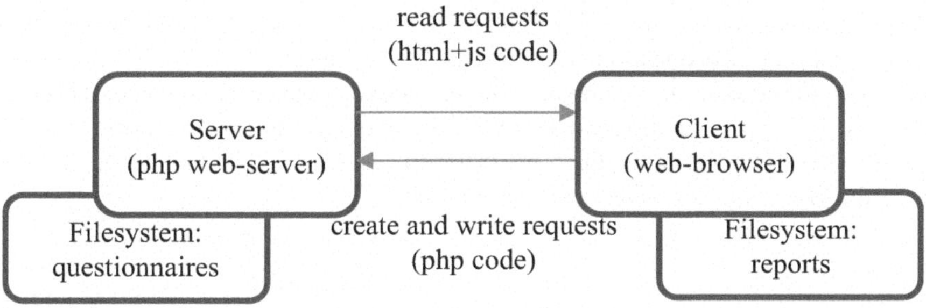

Fig. 2. IT architecture of BPNAF tool.

the definition of the questionnaire. It uses two items from the questionnaire content, namely the company name and evaluator name, to create a unique hex-coded part of the link. This solution, being more practical in operation, also allows access to and modification of the questionnaire structure when knowing only these two items. Moreover, the case of entering the company name with an empty evaluator name could be used to generate the questionnaire structure for the examined company and then for each evaluator – just by duplicating it based on previous generic definitions. Finally, the creation procedure results in different links that could be delivered by e-mail to different evaluators. Using a company name as a primary search key, there is also an option to simultaneously present all questionnaires for one company, which is useful when creating a final (averaged) questionnaire.

Each questionnaire-oriented calculation is executed on a client machine, which reduces the server's workload. The calculation and visualisation of the BPNAM matrix are performed dynamically when an evaluator (system client) fills out the questionnaire. There is also an option to generate HTML-sized reports to document the results. All questionnaires are stored in the server filesystem and could be accessed by a privileged person for further analysis. Internally, they constitute tab-separated files that could be imported to various applications, including Excel-like spreadsheets. It is worth mentioning that the copy-paste operation is implemented in both directions, which allows for exchanging all questionnaire data between the BPNA tool and spreadsheet applications, provided the spreadsheet fields match the tool fields.

4.3 BPNA Tool User Interface

The starting page of the BPNA tool consists of five sections:

1. Introduction – short description of the tool.
2. Getting started – user instruction.
3. Examples – three examples showing the assessment and enabling assessment to be carried out for a test organisation.
4. Questionnaire creator – the form for entering data about the analysed organisation, its process map, and generating access links for evaluators.
5. Maintenance – the area for the tool administrators and advanced users.

The questionnaire creator is essential to the starting page (Fig. 3). It enables the creation of one or two questionnaires for one organisation. The process map can be created after introducing the company's name, country, industry, and size. At level N, the processes category can be created within core, management, and support areas. After specifying the category's name, the process group's name (level N-1) must be filled. The name of the process group is automatically numbered.

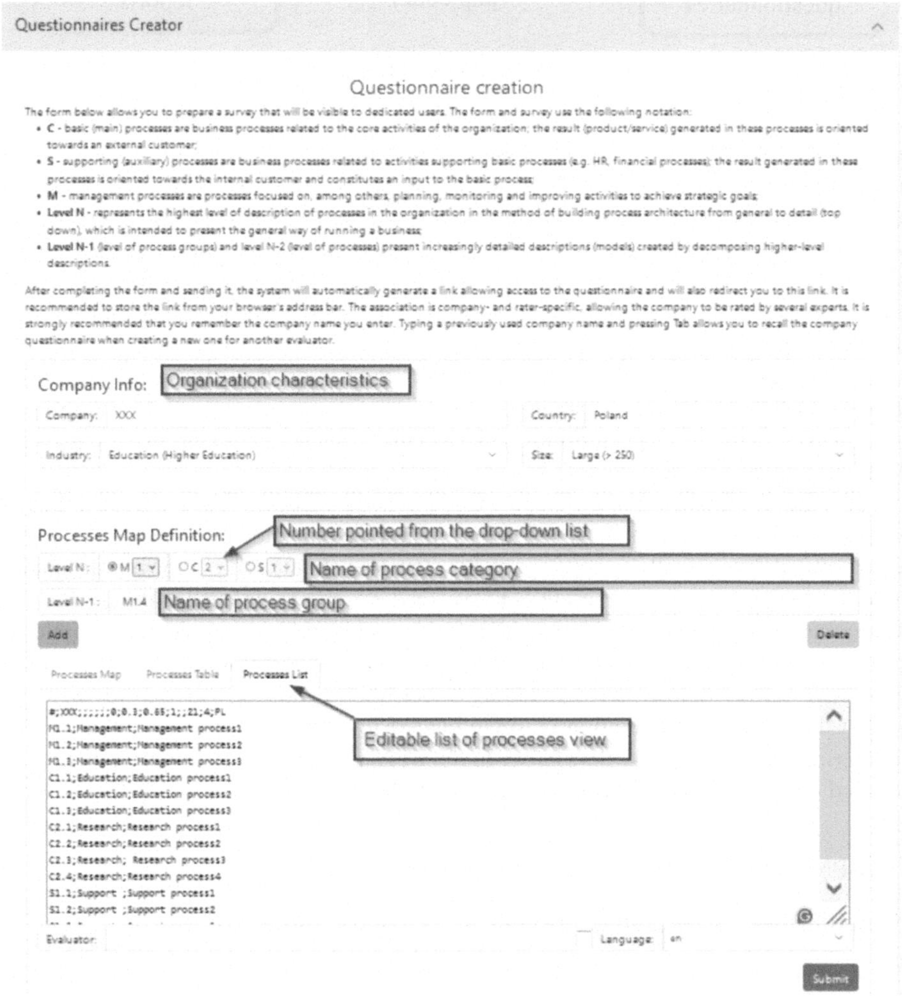

Fig. 3. Questionnaire creator view.

After completing the organisation's data and process maps, it is necessary to assess the predictability, weight, and direction of change in the weight of process groups at the N-1 level (Fig. 4). The data in the *Knowledge intensity* column are automatically filled in for each group of N-1 level processes, each time after completing its dedicated questionnaire. The overall level of knowledge intensity for the process group is calculated as the sum of the assessment values of each of the 10 criteria. The final result is presented in Fig. 5.

The method of filling out the questionnaires is flexible:

- the evaluator can first complete the questionnaire assessing the structuring for all processes and then fill in the level of knowledge intensity for each of them or
- the evaluator can sequentially assess the structure of processes, determine their knowledge intensity, and then repeat the same procedure for all processes.

The buttons in the toolbar above the questionnaire facilitate navigation between the process structure assessment questionnaires and knowledge intensity questionnaires. The buttons below the questionnaire (Prev and Next buttons) facilitate navigation between the questionnaires when assessing the intensity of knowledge for subsequent processes.

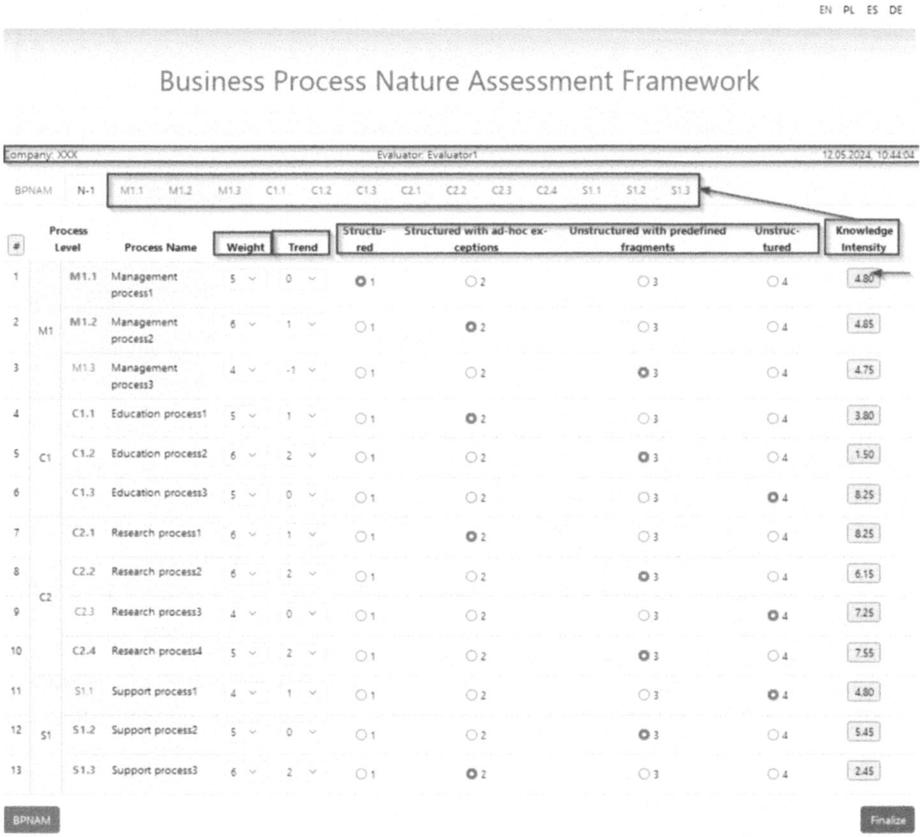

Fig. 4. Evaluation of the scope of the process structuring questionnaire.

After completing the questionnaires, the generated assessment matrix (BPNAM) can be analysed. The process categories are depicted as colour-coded dots: green for

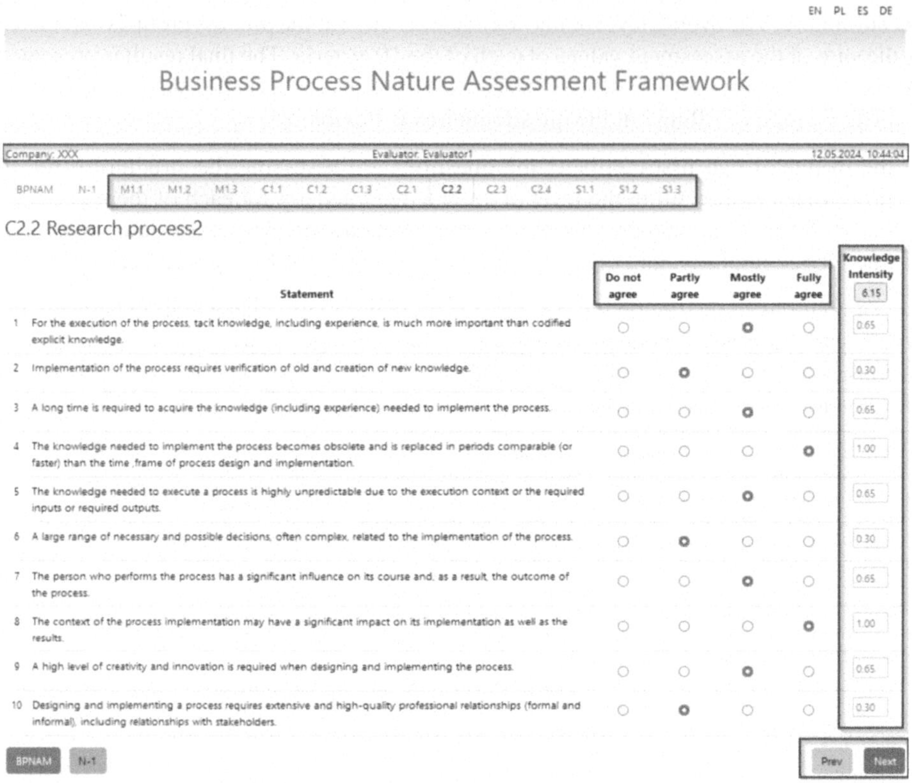

Fig. 5. Evaluation of the knowledge intensity for the process – single questionnaire.

management processes, blue for core processes, and yellow for support processes. The basic statistics for the processes are also presented next to the assessment matrix. It includes each matrix quadrant's share of process categories (Fig. 6).

For a given organisation, it is also possible to assess processes by two evaluators (e.g., representing the Management Board and the middle management level, such as department managers), compare the assessments, and reach a consensus. To do this, individual links need to be generated for each evaluator, and after assessing the processes, another link for the final evaluation can be generated. The combined link presents both assessments in preview mode and allows for marking the final, agreed-upon assessment.

After each study's completion, a summary report containing recommendations can be generated.

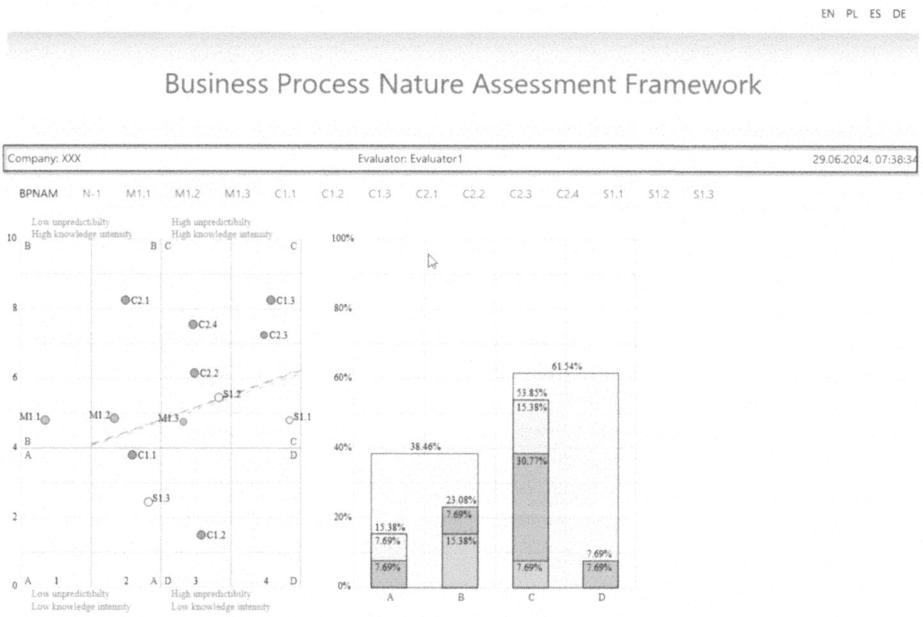

Fig. 6. BPNAM matrix and processes statistics.

5 Illustrative Case Study

The use of the BPNA tool for assessing the nature of BPs is illustrated through a case study that demonstrates the assessment process, results, and subsequent recommendations [2]. The studied company has extensive experience in the production and supply of steel structures and the execution of power network projects, including various machine and equipment components. The execution of steel structure supply projects involves several stages: material procurement, technological processes such as preparing elements for welding, welding, mechanical processing, painting, and functional testing. The company manufactures steel structures and components to construct industrial halls, production facilities, and warehouses. Another key area of the company's operations is constructing and reconstructing overhead and underground power lines, transformer stations, and street lighting.

Figure 7 presents the numerical results of the BP predictability and knowledge intensity study, while Fig. 8 presents these results using the BPNAM and a bar chart of process statistics. In both figures, green denotes managerial processes, blue represents core processes, and yellow signifies support processes.

The distribution of process nature assessments aligns with van der Aalst's [17] predictions, spanning from traditional processes (low knowledge intensity and predictability) to necessitating dynamic management (high knowledge intensity and unpredictability). In the organisation under study, the assessment indicates the potential to support business management and support processes, as well as most core processes, through implementing workflow or Document Management Systems (DMS). This would ensure that all participants in the selected processes are continuously informed about their current

#	Process Level N		Process Level N-1		Weight	Trend	Unpredictability Level				Knowledge Intensity
1	M1	Supervision of processes	M1.1	Monitoring and decision making	4	1		x		2	4,50
2			**M1.2**	Signing commercial contracts	6	1		x		2	2,80
3	M2	Ensuring financial liquidity	**M2.1**	Business financing management	6	0		x		2	3,40
4			M2.2	Acceptance of purchase and sales invoices	4	0	x			1	0,90
5	C1	Construction of power networks	**C1.1**	Field site visit	5	0			x	3	6,50
6			**C1.2**	Preparation of work schedules	5	0		x		2	5,80
7			**C1.3**	Investment management on the construction site	6	1			x	3	6,50
8	C2	Production of Steel Structures	**C2.1**	Marketing and sales of steel structures	6	2		x		2	6,50
9			C2.2	Purchase of materials and services	4	1	x			1	3,75
10			**C2.3**	Production of steel structures	5	2		x		2	6,15
11	C3	Preparation of offers	**C3.1**	Searching the market for orders	5	1		x		2	4,80
12			**C3.2**	Offering	6	0	x			1	2,80
13			**C3.3**	Participation in tenders	6	1		x		2	1,85
14	S1	Human capital management	S1.1	Employee recruitment	4	0		x		2	2,15
15			**S1.2**	Shaping HR and payroll policy	5	1			x	3	2,80
16			S1.3	Calculation and payment of remuneration	4	0		x		2	1,60
17			**S1.4**	Staff development	5	1	x			1	1,20
18	S2	Supply	**S2.1**	Verification of material requirements for the construction of energy networks	5	0		x		2	4,50
19			**S2.2**	Market reconnaissance and purchase of materials	6	0		x		2	4,15
20			**S2.3**	Delivery logistics	5	0		x		2	1,50
21			S2.4	Complaints and warranty service	4	0		x		2	1,25
22	S3	IT	S3.1	Ensuring IT security	4	0	x			1	1,80
23			**S3.2**	IT systems support	5	1		x		2	3,05
24			S3.3	IT support for employees	4	0			x	3	1,85
25			S3.4	Purchase, maintenance and repair of IT equipment	4	0		x		2	2,10

Fig. 7. Screendump of BP nature assessment inside the BPNA tool.

status and progress. For example, all subprocesses of Financial Liquidity Assurance (M2), Offer Preparation (C3), and Procurement (S2) are positioned in quadrant A of the BPNAM. The BPNA tool typically recommends their automation using Robotic Process Automation (RPA), support through a knowledge repository providing quick access to historical data, and adopting recruitment and development employees methods emphasising diligence, accuracy, and systematic work. The high knowledge intensity and unpredictability of core processes, particularly in Power Network Construction (C1), justify implementing elements of Business Process Management Systems (BPMS) and even Intelligent BPMS (iBPMS). This includes Process Mining and Artificial Intelligence (AI) technologies, which enable information acquisition and learning during process execution. Given the high knowledge intensity, the company should consider supporting processes in quadrants B and C of the BPNAM, particularly process C1, with an online Knowledge Management System (KMS) integrated with.

BPMS/iBPMS. This support should extend beyond a standard knowledge repository [9]. It is crucial to support the functioning of a Community of Practice (CoP) within the organisation, which actively disseminates knowledge and integrates collaboration with external experts in the knowledge acquisition process.

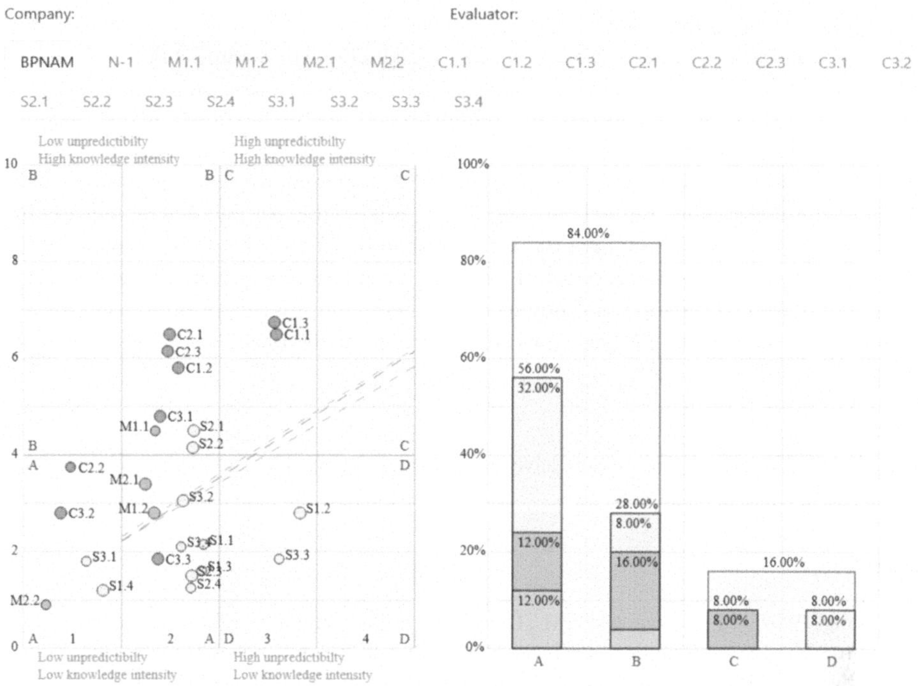

Fig. 8. The BPNAM and processes statistics as presented in the BPNA tool.

The individual recommendations proposed by participants in the study included:

- Improving selected processes within the domain of soft HR, particularly onboarding. Each contract represents a unique investment requiring high qualifications and experience. Given that the knowledge intensity assessments for core processes such as Power Network Construction (C1) and Steel Structure Production (C2) exceed 5 or even 6, new employees should receive substantial support.
- Integrating AI tools to facilitate decision-making through rapid and accurate analysis of existing business data.
- Ensuring the creation of professional databases that are fed by current business data, thanks to ensuring compatibility with the systems already in use.

6 Conclusions

The presented BPNA tool enables a quick and effective assessment of the nature of business processes in the organisation in terms of their unpredictability and the intensity of knowledge required to achieve the set BP goals. The simplicity and speed of the test allow the use of the BPNA tool in organisations without experience in BPM, including medium and small companies. Visualisation of the results allows for an understandable recommendation for selection, both for experts and non-specialists, including methodologies for BPM implementation, HRM strategies, or IT systems supporting BPM, according to the real nature of the BPs implemented in the organisation. The tool allows for the utilisation of standard recommendations and the introduction of recommendations tailored to the current developmental state of the specific, assessed organisation.

Further, BPNA tool development should primarily enable the utilisation of research data concerning assessments and recommendations as elements of the analysis of development paths for organisations implementing BPM. Agreeing to share anonymised individual recommendations should facilitate sharing practical experiences or benchmarking among organisations. Simultaneously, future research utilising anonymised data should allow for the preparation of increasingly accurate standard recommendations for selecting BPM implementation methodologies or managing employees for processes of varying natures or supporting systems for process groups with similar knowledge intensity and unpredictability.

Disclosure of Interests. The authors have no competing interests to declare that are relevant to the content of this article.

References

1. Berniak-Woźny, J., Szelągowski, M.: Business processes nature assessment matrix - a novel approach to the assessment of business process dynamism and knowledge intensity. Aslib J. Inf. Manag. **74**(2), 244–264 (2022). https://doi.org/10.1108/AJIM-04-2021-0110
2. Davey, L.: The application of case study evaluations. Pract. Assess. Res. Eval. **2**(1) (1991). https://doi.org/10.7275/02g8-bb93
3. Dumas, M., Rosa, L.M., Mendling, J., Reijers, A.H.: Fundamentals of Business Process Management. Springer-Verlag (2018)
4. Hammer, M.: The process audit. Harv. Bus. Rev. **85**(4), 111–123 (2007)
5. Kemsley, S.: The changing nature of work: from structured to unstructured, from controlled to social. In: Rinderle-Ma, S., Toumani, F., Wolf, K. (eds) Business Process Management. BPM 2011. Lecture Notes in Computer Science, vol. 6896. Springer, Berlin, Heidelberg (2011). https://doi.org/10.1007/978-3-642-23059-2_2
6. Klein, H.K., Myers, M.D.: A set of principles for conducting and evaluating interpretive field studies in information systems. MIS Quart. 67–93 (1999)
7. Mahal, A.: How work gets done: Business Process Management. Technics Publications, USA, Basics and Beyond (2010)
8. OMG. Business Process Maturity Model (BPMM) Version 1.0 (2008). http://www.omg.org/spec/BPMM/1.0/PDF. Accessed on 30 Jun 2024
9. Qin, H., Wang, H., Johnson, A.: Understanding the information needs and information-seeking behaviours of new-generation engineering designers for effective knowledge management. Aslib J. Inf. Manag. **72**(6), 853–868 (2020). https://doi.org/10.1108/AJIM-04-2020-0097
10. Szelągowski M., et al.: Exploring the diverse nature of business processes in organizations in industry 4.0/ 5.0. Fut. Bus. J. (ahead of print) (2024)
11. Szelągowski, M.: Evolution of the BPM lifecycle. In: M. Ganzha, L., Maciaszek, M. Paprzycki (Eds.), Communication Papers of the 2018 Federated Conference on Computer Science and Information Systems (FedCSIS), vol. 17, pp. 205–211. Springer (2018). https://doi.org/10.15439/978-83-952357-0-2
12. Szelągowski, M., Berniak-Woźny, J.: The adaptation of business process management maturity models to the context of the knowledge economy. Bus. Process. Manag. J. **26**(1), 212–238 (2019). https://doi.org/10.1108/BPMJ-11-2018-0328

13. Szelągowski, M.: Practical assessment of the nature of business processes. Inf. Syst. e-Bus. Manag. **19**(2), 541–566 (2021). https://doi.org/10.1007/s10257-021-00501-y
14. Sliż. P., Berniak-Woźny J., Siciński J., Babluza-Kudzian, M.: BPM maturity of organizations in Poland. Raport 2024, Uniwersytet Gdański, Gdańsk (2024). https://doi.org/10.13140/RG.2.2.33481.56169
15. Tarhan, A., Turetken, O., Ilisulu, F.: Business process maturity assessment: state of the art and key characteristics. In: 41st Euromicro Conference on Software Engineering and Advanced Applications, Madeira, Portugal, pp. 430–437 (2015). https://doi.org/10.1109/SEAA.2015.50
16. van der Aalst. Business process management: a comprehensive survey. ISRN Softw. Eng. **2013**, Article ID 507984 (2013). https://doi.org/10.1155/2013/507984
17. van der Aalst, W., Weske, M., Grunbauer, D.: Case handling: a new paradigm for business process support. Data Knowl. Eng. **53**(2), 129–162 (2005). https://doi.org/10.1016/j.datak.2004.07.003
18. Yin, R.K.: Case Study Research: Design and Methods. 4th Edition. SAGE Publications (2009)
19. Zelt, S., Schmiedel, T., Recker, J., vom Brocke, J.: A theory of contingent business process management. Bus. Process. Manag. J. **25**(6), 1291–1316 (2018). https://doi.org/10.1108/BPMJ-05-2018-0129
20. Zelt, S., Schmiedel, T., vom Brocke, J.: Understanding the nature of processes: an information processing perspective. Bus. Process. Manag. J. **1**(24), 67–88 (2018). https://doi.org/10.1108/BPMJ-05-2016-0102

Adapting to Change: Employees Ambidexterity as a Driver for Operational Adaptability and Organizational Development

Mariusz Hofman(✉), Grzegorz Grela, Paulina Orzelska, and Jarosław Banaś

Maria Curie-Sklodowska University, Pl. Marii Curie-Skłodowskiej 5, 20-031 Lublin, Poland
{mariusz.hofman,grzegorz.grela,paulina.orzelska,
jaroslaw.banas}@mail.umcs.pl

Abstract. The paper demonstrates that employee ambidexterity plays a critical role in achieving operational adaptability and the development of an organization in a dynamic environment. It advances theories of business process management by showing that process actors, who are capable of implementing change inherently and rapidly, can create flexible business processes. This allows for the effective meeting of customer needs and the achievement of business performance. The CATI method was utilized to examine 209 process actors from Polish enterprises. Given the survey's size and nature, the PLS-SEM method was employed. The research results confirmed that organizations operating within an adaptiveness paradigm rely on actors who perform their work appropriately to the emerging circumstances. This confirms the necessity of conducting in-depth research on the social implications of agile BPM.

Keywords: contextual ambidexterity · operational agility · agile BPM

1 Introduction

The currently widespread paradigm of adaptability requires enterprises to accommodate the changing business environment [48]. There are multiple solutions that an organization may use to continuously adapt to environmental changes in order to survive [47]. One of the ways is to search for operational adaptiveness which allows organizations to rethink, redesign and change business processes constantly [3]. Its achievement makes it possible to establish effective ways for organizations to create value for their customers [36, 71] and to reach the business performance [51, 66]. A central concern of adaptive organizations is the balance between the exploration of new possibilities and the exploitation of existing certainties [29]. Reaching operational agility depends on organizational actors who are able to simultaneously pursue exploitation and exploration [15]. They are prepared to unlearn the established ways of doing things in order to learn new ones so that their operations create products that meet customer expectations [27, 45].

C. Di Ciccio et al. (Eds.): BPM 2024, LNBIP 527, pp. 250–266, 2024.
https://doi.org/10.1007/978-3-031-70445-1_16

The approach by which employees ambidexterity facilitates operational adaptiveness, and thus the growth of their organization, seems most effective in a rapidly changing environment. It is based on the assumption that the capability of process actors to combine exploitation and exploration activities allows them to change operations inherently or rapidly [21, 30] and, consequently, to help their organization survive in a dynamic environment [66]. The approach is consistent with the need to rethink and redesign business processes unceasingly, as proposed by agile BPM advocates, so that continuous alignment between operations and environmental dynamics is maintained [3, 57, 65]. Therefore, from a cognitive standpoint, it is crucial empirically validate this approach. Thus, the following research questions have been formulated: If employees' ambidexterity facilitate operational agility, and does this support effective organizational operations in a volatile environment? To examine how contextual ambidexterity affects operations and business success, we developed a model linking employees' ambidexterity [15], operational adaptiveness [8, 37] and organizational effectiveness [32, 51].

The leading contribution of the paper is to show that human capabilities continue to be the key to attaining adaptability, which guarantees an organization's development in a dynamic environment. Process actors, who are able to implement change inherently and rapidly, can create flexible business processes. This allows them to establish effective ways to create value for customer and to reach the set business goals. This challenges the assumption of the key role of technology in reaching adaptability and market success of an organization in a dynamic environment [9, 12, 37, 44, 64]. This also confirms that organizations operating in the paradigm of adaptiveness are dependent on actors carrying out their work according to the occurring circumstances and can establish new ways of doing things [3, 61]. The second contribution is to advance theories of business process management by showing the importance of AI technologies, agile methods and organizational culture in reaching process actors' capabilities to pursue exploitation and exploration simultaneously [26, 54].

2 Research Framework and Hypothesis Development

The model links the following key aspects: individual behaviors (employee level), agile business processes (operational level), and business performance (organizational level). The model entails the assumption that process actors' ambidexterity [15] will be a condition for reaching operational agility [8, 37]. This means that process actors are ready to take risks and unlearn the established ways of doing things in order to create necessary changes in operations. Operational agility allows businesses to create products that meet customer expectations [27, 36] and reach the expected business goals [32, 66].

The model also includes identifying factors that affect process actors' ability to combine exploitation and exploration activities efficiently. These include the availability of AI-based solutions [4], the use of agile approaches [50] and agile organizational culture based on values and principles [26]. The links between the variables visible in the model define the research hypotheses adopted (Fig. 1). The accuracy of the prepared model was verified in the context of business processes.

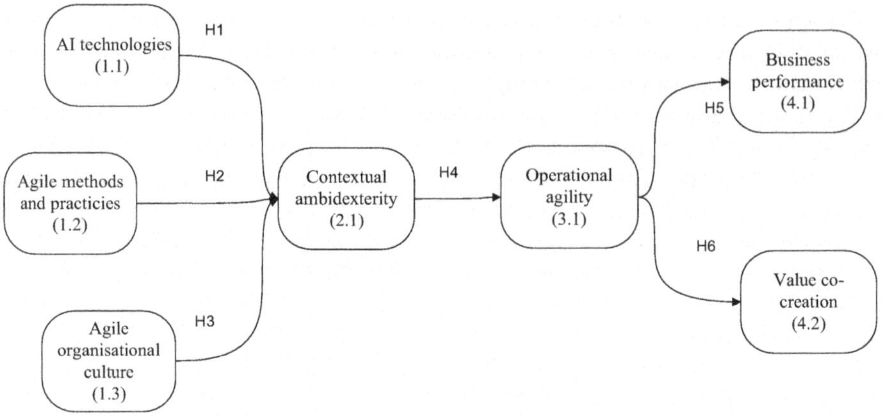

Fig. 1. Research model

2.1 Contextual Ambidexterity

Scholars demonstrate ambidexterity as the capability to pursue two dissimilar things simultaneously [13]. More specifically, ambidexterity is described as the ability to pursue exploitation and exploration activities at the same time [2, 48, 52]. Exploration includes the ability to take risks, search, discover and innovate in order to prepare the field for future exploitation activities [17]. Exploitation is associated with efficiency and productivity [23]. It involves the ability to improve, optimize, and execute tasks [13, 53]. Therefore, the more the environmental dynamism increases, the more an ambidextrous organization benefits from the dual capacity of exploration and exploitation [31, 41].

In this paper, contextual ambidexterity is understood as an individual's ability to pursue exploitation and exploration simultaneously [15, 48]. Contextual ambidexterity of process actors has the potential to significantly support business process change. However, this requires openness to change [3, 40], the ability to experiment [3, 57], and the capacity to implement change rapidly and inherently [50]. Developing such skills necessitates the development of unique tools, methods and organizational culture [7, 26, 50].

2.1.1 AI Technologies

AI technologies can significantly help organizations to identify and prioritize processes, to initiate necessary process changes, and to manage processes timely and accurately [4]. May suggest to employees some potential ways of acting more precisely, given technical and organizational problems [7, 74]. Their use allows process actors to effectively identify the necessary changes, to design new solutions and innovate solutions, and to assess their impact on business process performance. These technologies further permit process actors to be more ambidextrous and to develop new process models that might be implemented in a redesign or improvement formula, to compare them, and to verify their effectiveness [21, 69]. Machine learning algorithms also facilitate the identification of customers' behavioral patterns and their swift translation into changes to business processes [67, 69]. Therefore, it can be said that using AI technologies fosters the ability to combine exploitation and exploration activities, and to efficiently implement changes to business processes through improvement or innovation [53]. Given the above, the following research hypothesis can be formulated:

H1: The utilization of AI technologies has a positive impact on the contextual ambidexterity.

2.1.2 Agile Approaches and Practices

Many agile BPM approaches, i.e. "AGILIPO," have been developed to enhance the responsiveness of process actors, enabling effective implementation of changes to business processes [40, 43, 50]. These approaches are generally based on the Scrum framework [62] and allow the effective transformation of operations by way of continuous improvement [40] and process re-engineering [11, 50]. The specific nature of agile methods necessitates the need to apply agile practices, i.e., user stories, estimation or progress charts [50], as well as practices connected with stakeholder feedback [28], and customer requirements prioritization and their ordering in the product backlog [40, 50].

According to Lu and Ramamurthy [34], agility naturally encompasses the concept of ambidexterity. Agile methods and practices allow process actors to focus on customer needs [6, 65]. They stimulate their openness to change, the ability to experiment and to learn retrospectively [3, 40]. The use of agile BPM approaches translates into the readiness and ability of process actors to create change through process optimization and process re-engineering, which allows them to effectively pursue exploitation and exploration simultaneously [15]. Therefore, the following research hypothesis may be proposed:

H2: The utilization of agile methods and practices has a positive impact on the contextual ambidexterity.

2.1.3 Agile Organizational Culture

Organizations with a strong culture of agility tend to respond faster to changes and adapt to them more effectively [1]. Literature shows the usefulness of agile culture for organizations seeking continuous alignment between dynamically changing environments and operations [30, 72]. Agile culture shapes process actors' focus on continuously identifying customer needs and providing solutions that meet customer expectations [30, 36]. In this organizational culture, experimenting, collaboration and knowledge sharing are also allowed [60]. This boosts process actors' readiness to change and shapes their ability to implement changes in operations rapidly or inherently, depending on a given situation. It also allows them to be prepared to unlearn the established ways of doing things in order to learn new ones so that their operations create products that meet customer expectations [27, 45]. Thus, agile organizational culture elevates the process actors' ability to combine exploitation and exploration activities, which allows them to implement necessary changes in operations effectively [26, 30]. Thus, the following research hypothesis may be proposed:

H3: Agile organizational culture has a positive impact on contextual ambidexterity.

2.2 Operational Agility

Operational agility refers to an organization's capability to easily and quickly retool its operations to adapt to environmental changes [22, 33, 46, 63]. Operational agility manifests itself in responsiveness and flexibility, which enable efficient adaptation of operations to changing environmental conditions. Responsiveness emphasizes the readiness to follow evolving customer needs and introduce changes so that operations create products that meet customer expectations [34, 45]. Flexibility, on the other hand, is the ability to quickly and efficiently change operations, either through process innovation or by changing and following up with process improvements [3, 57, 65].

Organizations that wish to achieve operational agility depend on actors focused on customer needs, open to change and ready to establish different ways of doing things [61]. Contextual ambidexterity allows process actors to design responsive and adaptive business processes which facilitate the maintenance of continuous alignment between operations and environmental dynamics [21, 26, 42, 53]. Thus, the following research hypothesis may be proposed:

H4: Contextual ambidexterity of process actors positively influences operational agility.

2.3 Business Performance

Business performance is an indicator which measures how well an organization accomplishes its objectives [51]. Initially, this view primarily focused on the effectiveness of an organization [68]. However, with the increasing complexity and uncertainty of the business environment, a need emerged to adopt a broader perspective in measuring business performance [25]. It was made possible thanks to applying a balanced range of financial and non-financial measures and allowed organizations to appropriately and adequately assess business performance [19, 59]. Given the above, business performance may be defined as reaching multiple financial and non-financial objectives that contribute to the creation of competitive advantage and the long-term development of an organization.

Operational agility allows organizations to operate both flexibly and efficiently. This positively affects their ability to reach the set objectives and grow [48, 73]. In other words, an organization's business performance is associated with its capability for flexibility, responsiveness, and maintaining a balance between exploration and exploitation [24]. Thus, the following research hypothesis may be proposed:

H5: Operational agility positively influences organizational business performance.

2.4 Value Co-creation

Value is always co-created in an organization [70]. This means that customers play an active role in the value co-creation process [16, 35]. Therefore, an effective value co-creation process should be recognized as an element of business strategy and a key source of long-term competitive advantage [49].

Both customers and the organization might positively or negatively affect the value co-creation process. Customers must acquire the necessary skills and knowledge to participate effectively in the value co-creation process [16]. Therefore, in addition to manufacturing and delivering products, the key role of the enterprise is to facilitate customer learning so that customers can participate in the value co-creation process effectively [20]. This means that process actors and customers should collaborate on designing or modifying business processes to increase value co-creation efficiency [20, 26, 42]. Operational agility allows for the exploration of new possibilities and the exploitation of existing certainties so that business processes can create products that meet customer expectations [26, 39]. Thus, the following research hypothesis may be proposed:

H6: Operational agility positively influences value co-creation in organizations.

3 Method

3.1 Sample and Data

Given the specific nature of the research, purposive sampling was applied. The respondents included individuals engaged in business processes. The CATI method was used to collect data. Regarding the said method, respondents are interviewed by telephone, and the person conducting the survey reads out questions and records answers using a dedicated form. The qualifying question was aimed to verify whether a given respondent was taking part in any business process. Such an approach allowed the randomly select of a group of 209 process actors who could fully express their position on the statements in the survey questionnaire. They constituted the research sample, as described in Table 1. The analysis of the research sample structure confirmed that it included respondents engaged in various processes, serving diverse functions as part of the processes, employed in organizations of various sizes, and operating in different sectors. Therefore, it can be assumed that the research results aptly describe how values, methods and AI technologies affect the contextual ambidexterity of process actors. They also allow the assessment of employee ambidexterity and operational agility, business performance and value co-creation efficiency. The research was conducted in December 2023. The territorial reach of the study was limited to enterprises based in Poland.

3.2 Measurement

Due to the size of the survey sample and the character of the survey, the PLS-SEM method was used. This method makes it possible to verify the existence of a cause-and-effect relationship between the variables in the model. It is recommended when the survey sample is not large, the analyzed data are inconsistent with normal distribution, and the tested structural model is complex, meaning it contains vast constructs and measures [18]. The measurement scales used were prepared based on the relevant literature. The validity of the measurement model was verified with the confirmatory factor analysis [5]. Convergent and discriminant validity tests were carried out to assess the validity of construct distinction. The convergent validity of the constructs was measured using AVE and the discriminant validity of the constructs was measured using the HTMT ratio. The reliability of construct distinction was measured with Cronbach's alpha and Composite Reliability (CR) coefficients [10]. The adjustment of the structural model was assessed using the following measures: AVIF, AFVIF (examining the collinearity of variables), GoF (examining the predictive power of the model), SPR (occurrence of Simpson's paradox), and SSR (analysis of statistical suppression). The calculations were carried out using the WarpPLS 8.0.

Table 1. Sample characteristics

Role in the business process	Frequency	Percentage
Process owner	70	33.5
Process performer	139	66.5
Total	209	100
Experience in business process area	Frequency	Percentage
0–1 years	2	1.0
2–3 years	14	6.7
4–6 years	32	15.3
7–10 years	36	17.2
11–15 years	30	14.4
over 15 years	95	45.4
Total	209	100.0
Process in which the respondent participates	Frequency	Percentage
Basic	61	29.2
Auxiliary (support)	53	25.4
Management	95	45.4
Total	209	100.0
Size of the organization	Frequency	Percentage
Micro (1–9 employees)	57	27.3
Small (from 10 to 49 employees)	86	41.2
Medium-sized (from 50 to 249 employees)	45	21.5
Large (over 250 employees)	21	10.0
Total	209	100.0
Industry in which the organization operates	Frequency	Percentage
Services	80	38.3
Production	77	36.8
Trade	34	16.3
Mixed	18	8.6
Total	209	100.0

4 Results

Tables 2 and 3 show the results of the convergent and discriminant validity tests for the individual constructs. The analysis of the obtained results shows that the constructs have appropriate reliability and satisfactory convergent and discriminant validity.

Table 2. Analysis of reliability and convergent validity

	1.1 (AIT)	1.2 (AMP)	1.3 (AOC)	2.1 (CA)	3.1 (OA)	4.1 (BP)	4.2 (VC)
Adjusted R-squared coefficients	-	-	-	0.552	0.599	0.477	0.414
Composite reliability coefficients (CR)	0.930	0.897	0.864	0.910	0.956	0.917	0.923
Cronbach's alpha coefficients	0.895	0.847	0.790	0.868	0.947	0.879	0.888
Average variance extracted (AVE)	0.772	0.686	0.614	0.717	0.731	0.736	0.749

Table 3. Analysis of discriminant validity (HTMT ratio)

		1.1 (AIT)	1.2 (AMP)	1.3 (AOC)	2.1 (CA)	3.1 (OA)	4.1 (BP)	4.2 (VC)
1.1	AI technologies (AIT)	–	–	–	–	–	–	–
1.2	Agile methods and practices (AMP)	0.258	–	–	–	–	–	–
1.3	Agile organizational culture (AOC)	0.316	0.970	–	–	–	–	–
2.1	Contextual ambidexterity (CA)	0.312	0.760	0.861	–	–	–	–
3.1	Operational agility (OA)	0.313	0.773	0.847	0.857	–	–	–
4.1	Business performance (BP)	0.407	0.605	0.691	0.718	0.754	–	–
4.2	Value co-creation (VC)	0.365	0.642	0.703	0.703	0.697	0.600	–

The results of the confirmatory factor analysis verifying the validity of the measurement model are included in Appendix 1. The model shows a good fit to empirical data: APC = 0.489, $p < 0.001$, AARS = 0.511; $p < 0.001$, AVIF = 2.178, AFVIF = 2.681, GoF = 0.606, SPR = 1.000, SSR = 1.000. The results of an extended analysis of the goodness-of-fit indices of the model show that it is a good fit: SRMR = 0.084 and SMAR = 0.067. Figure 2 shows the structural model describing the direction and strength of the predicted relationships between the latent variables. The research hypotheses were verified on their basis.

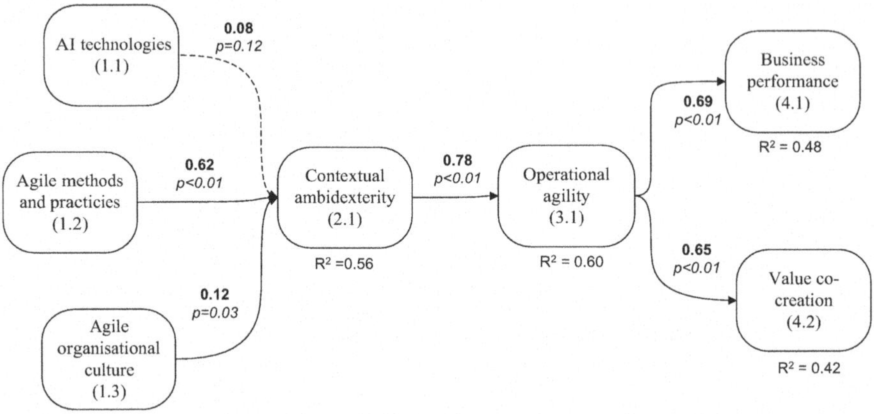

Fig. 2. Research model with respective informants and results ($n = 209$)

According to the adopted hypotheses, AI technologies (H1), agile methods and practices (H2), as well as agile organizational culture (H3) shape the contextual ambidexterity of process actors. The obtained results clearly indicate that AI technologies do not have a statistically significant impact on the contextual ambidexterity of process actors ($\beta = 0.081$, $p = 0.117$). Therefore, hypothesis H1 was not confirmed. Empirical data confirm that agile methods and practices have a positive and statistically significant influence on the contextual ambidexterity of process actors ($\beta = 0.616$, $p < 0.01$). Similarly, agile organizational culture has a weaker, but still statistically significant, impact on the contextual ambidexterity of process actors ($\beta = 0.123$, $p = 0.03$). Thus, hypotheses H2 and H3 were confirmed.

The adopted hypotheses also indicate that contextual ambidexterity has a favorable impact on operational agility (H4). The obtained data show that the ability of the process actors to pursue exploitation and exploration simultaneously has a strong, positive and statistically significant influence on the adaptiveness of business processes ($\beta = 0.775$, $p < 0.01$). Thus, hypothesis H4 was also confirmed.

The adopted hypotheses assumed that operational agility determines the achievement of business performance (H5) and value co-creation (H6). The analysis of empirical data confirms that operational agility has a positive, strong and statistically significant influence on business performance ($\beta = 0.692$, $p < 0.01$) and the value co-creation effectiveness ($\beta = 0.646$, $p < 0.01$). Thus, the research hypotheses assuming the existence of such cause-and-effect relationships, i.e., H5 and H6, were confirmed.

The analysis of the mediation effect [55] shows that AI technologies indirectly affect process actors' ambidexterity through agile methods and practices (AIT - > AMP, $\beta = 0.331$, $p < 0.01$ and AMP - > CA, $\beta = 0.616$, $p < 0.01$).

5 Discussion

The article advances an operations management theory in the business process management field [56]. It is consistent with the currently dominating paradigm of adaptability through implementing flexible and easy-to-change business processes in an organization [3, 57]. The research results show that the readiness and ability of process actors to combine exploration and exploitation activities is the key to achieving operational agility. Therefore, the capability of employees engaged in given processes to unlearn the established ways of doing things in order to learn new ones allows them to experiment and create new processes (or their variants) or to search for other ways to refine them. It enables them to develop a flexible process system that might be quickly modified on an ongoing basis in response to environmental changes. The research results confirm the effect of employees' capabilities to implement changes rapidly or inherently on the level of business process adaptability and the effectiveness of dynamic management, as was previously suggested in the literature on the subject [4, 26, 53]. The main goal of the BPM theory is to establish effective and efficient ways for organizations to create value for their customers and to contribute to business performance [36, 51, 71]. According to the conducted research, the achievement of operational agility allows organizations (with customers participating) to effectively modify processes so that products meeting customers' demands are created. Operational adaptability also facilitates fulfilling organizational goals, guaranteeing growth opportunities. This means that the conducted research confirmed the expected influence of operational adaptability on value creation efficiency [36] and on business performance [48, 73].

A central concern of organizational adaptive operations is the balance between the exploration of new possibilities and the exploitation of existing certainties [39, 53]. Thus, the development of process actors' ability to implement change rapidly and inherently requires the availability of appropriate tools and methods, and the proper organizational culture being put in place [14, 15, 58]. The results that we have obtained confirm that the use of agile BMP approaches based on the Scrum framework affects process actors' contextual ambidexterity to the greatest extent. This means that the flexibility that agile methods and practices offer provides an opportunity to effectively and efficiently implement changes (inherently or rapidly) whilst their application contributes to shaping employees' behavior patterns. The obtained results substantiate the influence of the methods and approaches being used on the way the contextual ambidexterity of the process actors is shaped, as previously suggested in the literature on the subject [15, 40, 50]. They might also constitute valuable guidance to enterprises wishing to reach operational adaptability. The research results also indicate the positive effect of agile organizational culture on process actors' ability to combine exploitation and exploration activities. Encouraging employees to be open to change and allowing them to experiment and search for new ways of doing things contributes to shaping their ability to pursue exploitation and exploration simultaneously. It seems that agile methods provide process actors with well-tested approach in the event of implementing process changes through improvement or innovations, whilst agile organizational culture creates conditions for applying them in practice. This supports the assumption proposed in the literature that the organizational culture which promotes agile principles and values influences the level of employees' contextual ambidexterity [26, 58]. Currently, the key issue is whether AI tools enhance employees' capabilities – in this case, their ambidexterity. Literature on the subject includes a call for such types of studies [38, 53]. The conducted research results show that AI technologies do not directly influence the contextual ambidexterity of process actors. This means that AI-based tools have not yet been made widely available to workers. Or they are not used by process actors to any significant extent to identify, model and implement changes to business processes. Therefore, the application of AI technologies in isolation from the appropriate methods and practices will not yield any expected results. This also means that AI tools are used by employees more intentionally, for example, to solve specified problems or work out solutions as part of pending process transformations.

The conducted research points to the critical role of human capabilities in creating operational adaptability and reaching organizational objectives in a volatile environment. This proves that collaboration, knowledge sharing and participation of people are significant factors in enabling the organization's agility [3, 61]. It also indicates how important and agile the BPM process is for the actors' ability to learn, unlearn and re-learn [27, 45]. This provides organizations with the opportunity to make changes in numerous ways. It also allows them to explore new possibilities and to exploit existing certainties. This challenges the assumption of the key role of technology in dynamic BPM [9, 21, 38, 44, 64] and corroborates the need to conduct in-depth research into the social implications of agile BPM [61]. Organizations should put in place mechanisms to facilitate communication, collaboration and fast decision-making, allowing process actors to experiment and search for new solutions.

6 Conclusions

The conducted research indicates that individuals' ability to respond to emerging changes is key to achieving operational adaptability and organizational development in a dynamic environment. This ability of process actors is crucial in agile BPM, providing organizations with the opportunity to explore new possibilities and exploit existing certainties. This corroborates the need for in-depth research into the social implications of agile BPM. Organizations should implement mechanisms to facilitate communication, collaboration, and fast decision-making, allowing process actors to experiment and search for new solutions.

The main limitations of this study concern the purposive sampling method used, which means that the sample does not meet the criteria of statistical representativeness. Nonetheless, the analysis of the sample structure indicates that the respondents represented different roles, processes, and organizations. Therefore, future research may focus on validating and scaling the model proposed in this article, including finding ways to measure aspects not included in this study but which play an important role in achieving agility at operational and strategic levels.

The considerations presented in this paper may also be valuable for practitioners, indicating that employees' capabilities could be an efficient path to attaining adaptability and development in a changing and dynamic environment.

Disclosure of Interests. The authors have no competing interests to declare that are relevant to the content of this article.

Specification of the measurement model (results of the confirmatory factor analysis)

1.1 AI technologies (AIT), CR = 0.930, Cronbach's α = 0.895, AVE = 0.772	
The use of artificial intelligence allows us to create new and innovative process solutions	0.665
Machine learning algorithms help us monitor process results	0.947
Machine learning algorithms help us choose the best method to streamline processes	0.952
AI-based technologies allow us to monitor changes to customer behavior on an ongoing basis	0.919
1.2 Agile methods and practices (AMP), CR = 0.897, Cronbach's α = 0.847, AVE = 0.686	
Agile methodologies help us implement changes to business processes	0.829
We focus on our customer expectations and requirements in the course of day-to-day meetings	0.763
The use of retrospective allows us to continuously perfect our abilities and learn from changes	0.866
The agile approach stimulates openness to change and experimentation	0.853
1.3 Agile organizational culture (AOC), CR = 0.864, Cronbach's α = 0.790, AVE = 0.614	
Thanks to organizational culture based on agility, our organization can quickly respond to changes	0.776
Our organizational culture promotes experimentation and learning from retrospection	0.785
In our organization, there is consent to the departure from the established operational methods and to the learning of new ones to meet our customer expectations better	0.777
The management allows us to combine adaptation- and development-oriented activities	0.795
2.1 Contextual ambidexterity (CA), CR = 0.910, Cronbach's α = 0.868, AVE = 0.717	
Our employees are able to innovate and refine processes simultaneously	0.841
As employees, we are open to change and able to implement it fast	0.823
We can explore new opportunities and exploit existing processes simultaneously	0.862
We are capable of conducting process-based and project-based operations simultaneously	0.860
3.1 Operational agility (OA), CR = 0.956, Cronbach's α = 0.947, AVE = 0.731	
Our operations are flexible and can be easily adapted to market changes	0.861
Our business processes are simple and easy to modify with a view to creating value	0.744
We are able to quickly adapt our operations to market changes	0.866

(continued)

(*continued*)

We adapt our business processes to meet the evolving market requirements	0.894
Our organization quickly adapts to changes in the business environment	0.888
The ability to respond fast to changes is the key to our adaptability	0.878
Our operations are flexible to meet the changing business demands	0.890
Our organization is constantly searching for new opportunities while relying on its existing experience	0.807
4.1 Business performance (BP), CR = 0.917, Cronbach's $\alpha = 0.879$, AVE = 0.736	
Our organization has better financial results than competing organizations	0.775
We fulfill the planned financial and non-financial objectives	0.877
Our operational methods contribute to strengthening our competitive advantage	0.895
Our operational methods support the long-term development of our organization	0.878
4.2 Value co-creation (VC), CR = 0.923, Cronbach's $\alpha = 0.888$, AVE = 0.749	
Our customers actively participate in the process of product and service value co-creation	0.868
Our processes are comprehensible to customers, which facilitates their engagement in value co-creation	0.887
We are open to changes which positively affect the value creation process for customers	0.811
We continuously cooperate with our customers on new methods to create value	0.893

References

1. Accenture: Business agility report 2020 (2020). https://www.accenture.com/us-en/insights/technology/business-agility-report-2020. Accessed 22 Jan 2024
2. Andriopoulos, C., Lewis, M.: Exploitation-exploration tensions and organizational ambidexterity: managing paradoxes of innovation. Organ. Sci. **20**(4), 696–717 (2009). https://doi.org/10.1287/orsc.1080.04062009
3. Badakhshan, P., Conboy, K., Grisold, T., vom Brocke, J.: Agile business process management. A ystematic literature review and an integrated framework. Bus. Process Manag. J. **26**(6), 1505–1523 (2020). https://doi.org/10.1108/BPMJ-12-2018-0347
4. Bazan, P., Estevez, E.: Industry 4.0 and business process management: state of the art and new challenges. Bus. Process Manag. J. **28**(1), 62–80 (2022)
5. Brown, T.A.: Confirmatory Factor Analysis for Applied Research, New York, London, The Guilford Press 34 (2006)
6. Bruno, G., et al.: Key challenges for enabling agile BPM with social software. J. Softw.: Evol. Process **23**(4), 297–326 (2011)
7. Caputo, A., Marzi, G., Pellegrini, M.M.: The internet of things in manufacturing innovation processes: development and application of a conceptual framework. Bus. Process. Manag. J. **22**(2), 341–358 (2016)
8. Chen, Y., Wang, Y., Nevo, S., Jin, J., Wang, L., Chow, W.S.: IT capability and organizational performance: the roles of business process agility and environmental factors, Eur. J. Inf. Syst. 1–17 (2013). https://doi.org/10.1057/ejis.2013.4

9. Chen, Y., Wang, Y., Nevo, S., Jin, J., Wang, L., Chow, W.S.: IT capability and organizational performance: the roles of business process agility and environmental factors. Eur. J. Inf. Syst. **23**(3), 326–342 (2014)
10. Cronbach, J.: Coefficient alpha and the internal structure of tests. Psychometrika **16**(3), 297–303 (1951)
11. Çulha, D., Dogru, A.: Towards an agile methodology for business process development, In: Nanopoulos, A., Schmid, W. (Eds), Proceedings of 6th International Conference, S-BPM ONE2014 in Eichst€att, pp. 133–142, Germany, Springer-Verlag, Cham (2014)
12. Di Ciccio, C., Marrella, A., Russo, A.: Knowledge-intensive processes: an overview of contemporary approaches. In: ter Hofstede A.H.M., Mecella M., Sardina S., Marrella A. (Eds.), 1st International Workshop on Knowledge-Intensive Business Processes (KiBPs), pp. 33–47 (2012)
13. Duncan, R.B.: The ambidextrous organization: designing dual structures for innovation. Manag. Organ. **1**(1), 167–188 (1976)
14. Ghoshal, S., Bartlett, C.: Linking organizational context and managerial action: the dimensions of quality in management. Strateg. Manag. J. **15**, 91–112 (1994)
15. Gibson, C.B., Birkinshaw, J.: The antecedents, consequences, and mediating role of organizational ambidexterity. Acad. Manag. J. **47**(2), 209–226 (2004)
16. Grönroos, C., Voima, P.: Critical service logic: making sense of value creation and co-creation. J. Acad. Mark. Sci. **41**(2), 133–150 (2013)
17. Guisado-González, M., González-Blanco, J., Coca-Pérez, J.L.: Analyzing the relationship between exploration, exploitation and organizational innovation. J. Knowl. Manag. **21**(5), 1142–1162 (2017)
18. Hair, J.F., Ringle, C.M., Sarstedt, M.: PLS-SEM: indeed a silver bullet. J. Market. Theory Pract. Taylor Francis **19**(2), 139–152 (2011)
19. Harrison, J.S., Wicks, A.C.: Stakeholder theory, value, and firm performance. Bus. Ethics Q. **23**(1), 97–124 (2013)
20. Hibbert, S., Winklhofer, H., Temerak, M.S.: Customers as resource integrators: toward a model of customer learning. J. Serv. Res. **15**(3), 247–261 (2012)
21. Houy, C., Fettke, P., Loos, P., Van Der Aalst, W.M., Krogstie, J.: Business process management in the large. Bus. Inf. Syst. Eng. **3**(6), 385–388 (2011)
22. Huang, P.-Y., Pan, S.L., Ouyang, T.H.: Developing information processing capability for operational agility: implications from a Chinese manufacturer. Eur. J. Inf. Syst. **23**, 462–480 (2014)
23. Hughes, M., Hughes, P., Morgan, R.E.: Exploitative learning and entrepreneurial orientation alignment in emerging young firms: implications for market and response performance. Br. J. Manag. **18**, 359–375 (2007). https://doi.org/10.1111/j.1467-8551
24. Junni, P., Sarala, R.M., Taras, V., Tarba, S.Y.: Organizational ambidexterity and performance: a meta-analysis. Acad. Manag. Perspect. **27**(4), 299–312 (2013)
25. Kaplan, R.S., Norton, D.P.: Transforming the balanced scorecard from performance measurement to strategic management: part I. Account. Horiz. **15**(1), 87–104 (2001)
26. Khan, S., J., Mir, A.A.: Ambidextrous culture, contextual ambidexterity and new product innovations: the role of organizational slack and environmental factors. Business Strat. Environ. **28**(4), 652–663 (2019)
27. Klammer, A., Gueldenberg, S.: Unlearning and forgetting in organizations: a systematic review of literature. J. Knowl. Manag. **23**(5), 860–888 (2019)
28. Kolar, J., Pitner, T.: Agile BPM in the age of cloud technologies. Scal. Comput.: Pract. Exper. **13**(4), 285–294 (2012)
29. Kuran, T.: The tenacious past: theories of personal and collective conservatism. J. Econ. (1988)

30. Lee, J.Y., Seo, Y., Jeung, W., Kim, J.H.: How ambidextrous organizational culture affects job performance: a multilevel study of the mediating effect of psychological capital,. J. Manag. Organ. **1**(16) (2017). https://doi.org/10.1017/jmo.2017.38
31. Lee, O.K., Sambamurthy, V., Lim, K.H., Wei, K.K.: How does IT ambidexterity impact organizational agility? Inf. Syst. Res. **26**(2), 398–417 (2015)
32. Leroi-Werelds, S.: An update on customer value: state of the art, revised typology, and research Agenda. J. Serv. Manag. **30**(5), 650–680 (2019). https://doi.org/10.1108/JOSM-03-2019-0074
33. Liang, H., Wang, N., Xue, Y., Ge, S.: Unraveling the alignment paradox: how does business-IT alignment shape organizational agility? Inf. Syst. Res. **28**(4), 863–879 (2017)
34. Lu, Y., Ramamurthy, K.: Understanding the link between information technology capability and organizational agility: an empirical examination. MIS Q. **35**(4), 931–954 (2011)
35. Lusch, R.F., Vargo, S.L.: An overview of service-dominant logic. In: Vargo, S.L., Lusch, R.F. (Eds), The Sage Handbook of Service-Dominant Logic, pp. 3–21, Sage Publications, London (2018)
36. Malinova, M., Mendling, J.: Identifying do's and don'ts using the integrated business process management framework. Bus. Process. Manag. J. **24**(4), 882–899 (2018)
37. Mao, H., Gong, Y., Titah, T.: Understanding the relationship between IT capabilities and operational agility: a multimethod approach. J. Enterp. Inf. Manag. **36**(2), 409–436 (2023). https://doi.org/10.1108/JEIM-12-2021-0521
38. Mao, H., Liu, S., Zhang, J., Zhang, Y., Gong, Y.: Information technology competency and organizational agility: roles of absorptive capacity and information intensity. Inf. Technol. People **34**(1), 421–451 (2021)
39. March, J.G.: Exploration and exploitation in organizational learning. Organ. Sci. **2**(1), 71–87 (1991). https://doi.org/10.1287/orsc.2.1.71
40. Martins, P.V., Zacarias, M.: An agile business process improvement methodology. Procedia Comput. Sci. **121**, 129–136 (2017)
41. Marzi, G., Zollo, L., Boccardi, A., Ciappei, C.: Additive manufacturing in SMEs: empirical evidences from Italy. Int. J. Innov. Technol. Manag. **15**(01) (2018)
42. Meyer, N., Schiffner, S.: Democratizing business process management: empowering process participants to contribute to the enactment of business processes. In: 2014 IEEE 16th Conference on Business Informatics, pp. 93–100 (2014)
43. Meziani, R., Magalhães, R.: Proposals for an agile business process management methodology. In: First International Workshop on Organizational Design and Engineering (2009)
44. Mikalef, P., Pateli, A.: Information technology-enabled dynamic capabilities and their indirect effect on competitive performance: findings from PLS-SEM and fsQCA. J. Bus. Res. **70**, 1–16 (2017)
45. Miller, K.D., Pentland, B.T., Choi, S.: Dynamics of performing and remembering organizational routines. J. Manage. Stud. **49**(8), 1536–1558 (2012)
46. Ngai, E.W.T., Chau, D.C.K., Chan, T.L.A.: Information technology, operational, and management competencies for supply chain agility: findings from case studies. J. Strateg. Inf. Syst. **20**, 232–249 (2011)
47. Niu, K.-H., Li, H.: Knowledge management and organizational adaptation effectiveness: an empirical study. Int. J. Bus. Manag. Commer. **7**(1), 10–26 (2022)
48. O'Reilly, C.A., Tushman, M.L.: Organizational ambidexterity: past, present, and future. Acad. Manag. Perspect. **27**(4), 324–338 (2013)
49. Osterwalder, A., Pigneur, Y.: Business Model Generation. Wiley, Hoboken, NJ (2010)
50. Petrillo, A., Di Bona, G., Forcina, A., Silvestri, A.: Building excellence through the agile reengineering performance model (ARPM): a strategic business model for organizations. Bus. Process. Manag. J. **24**(1), 128–157 (2018)

51. Prakash, A., Jha, S.K., Prasad, K.D., Singh, A.K.: Productivity, quality and business performance: an empirical study. Int. J. Product. Perform. Manag. **66**(1), 78–91 (2017). https://doi.org/10.1108/IJPPM-03-2015-0041
52. Raisch, S., Birkinshaw, J., Probst, G., Tushman, M.L.: Organizational ambidexterity: balancing exploitation and exploration for sustained performance. Organ. Sci. **20**(4), 685–695 (2009)
53. Rialti, R., Marzi, G., Silic, M., Ciappei, C.: Ambidextrous organization and agility in big data era. The role of business process management systems. Bus. Process Manag. J. **24**(5), 1091–1109 (2018). https://doi.org/10.1108/BPMJ-07-2017-0210
54. Rialti, R., Zollo, L., Pellegrini, M.M., Ciappei, C.: Exploring the antecedents of brand loyalty and electronic word of mouth in social-media-based brand communities: do gender differences matter? J. Glob. Mark. **30**(3), 147–160 (2017)
55. Sarstedt, M., Hair, J.F., Jr., Nitzl, C., Ringle, C.M., Howard, M.C.: Beyond a tandem analysis of SEM and PROCESS: Use of PLS-SEM for mediation analyses. Int. J. Mark. Res. **62**(3), 288–299 (2020)
56. Schmenner, R., Swink, M.: On theory in operations management. J. Oper. Manag. **17**(1), 97–113 (1998)
57. Schmitt, A., Hörner, S.: Systematic literature review – improving business processes by implementing agile. Bus. Process. Manag. J. **27**(3), 868–882 (2021)
58. Simsek, Z., Heavey, C., Veiga, J.F., Souder, D.: A typology for aligning organizational ambidexterity's conceptualizations, antecedents, and outcomes. J. Manage. Stud. **46**(5), 864–894 (2009). https://doi.org/10.1111/j.1467-6486.2009.00841.x
59. Sneyd, K.P., Rowley, J.: Linking strategic objectives and operational performance: an action research-based exploration. Meas. Bus. Excell. **8**(3), 42–51 (2004)
60. Spiegel, M., Schmiedel, T., vom Brocke, J.: What makes change harder – or easier. MIT Sloan Manag. Rev. **58**(3), 88–89 (2017)
61. Suša Vugec, D., Tomičić-Pupek, K., Bosilj Vukšić, V.: Social business process management in practice: overcoming the limitations of the traditional business process management. International Journal of Engineering Business Management **10**, 1–10 (2018)
62. Sutherland, J., Schwaber, K.: The scrum guide (2016). http://www.scrumguides.org. Accessed on 15 Jan 2024
63. Tallon, P.: Inside the adaptive enterprise: an information technology capabilities perspective on business process agility. Inf. Technol. Manage. **9**(1), 21–36 (2008)
64. Tan, F., Pan, S.L., Zuo, M.: Realising platform operational agility through information technology–enabled capabilities: a resource-interdependence perspective. Info Syst. J. **29**, 582–608 (2019). https://doi.org/10.1111/isj.12221
65. Thiemich, C., Puhlmann, F.: An agile BPM project methodology. In: Daniel, F. et al. (Eds), Business Process Management: 11th International Conference, BPM 2013, Proceedings, August 26–30, pp. 291–306, Springer, Beijing and Berlin Heidelberg (2013)
66. Thompson, G., Seymour, L.F., O'Donovan, B: Towards a BPM success model: an analysis in South African financial services organisations. In: Halpin, T., et al. (Eds), Enterprise, Business-Process and Information Systems Modeling. Lecture Notes in Business Information Processing, vol. 29, pp. 1-13, Springer, Heidelberg (2009).
67. Tian, X., Pavur, R., Han, H., Zhang, L.: A machine learning-based human resources recruitment system for business process management: using LSA BERT and SVM. Bus. Process Manag. J. **29**(1), 202–222 (2023)
68. Valmohammadi, C., Roshanzamir, S.: The guidelines of improvement: relations among organizational culture, TQM and performance. Int. J. Prod. Econ. **164**, 167–178 (2015). https://doi.org/10.1016/j.ijpe.2014.12.028
69. van der Aalst, W.M.P., et al.: Business process mining: an industrial application. Inf. Syst. **32**(5), 713–732 (2007)

70. Vargo, S.L., Lusch, R.F.: Evolving to a new dominant logic for marketing. J. Mark. **68**(1), 1–17 (2004)
71. vom Brocke, J., Rosemann, M.: Business Process Management, Wiley Encyclopedia of Management, Management Information Systems, vol. 7, pp. 1–9. Wiley (2014)
72. von Rosing, M., Gill, A.Q.: Applying agile principles to BPM. In: von Rosing, M., von Scheel, J., Scheer, A.-W. (Eds), The Complete Business Process Handbook - Body of Knowledge from Process Modeling to BPM, Vol. I, pp. 557–581 Elsevier – Morgan Kaufmann, Waltham, MA (2015)
73. Vrontis, D., Thrassou, A., Santoro, G., Papa, A.: Ambidexterity, external knowledge and performance in knowledge-intensive firms. J. Technol. Transf. **42**(2), 374–388 (2017)
74. Wamba-Taguimdje, S.-L., Fosso Wamba, S., Kala Kamdjoug, J.R., Tchatchouang Wanko, C.E.: Influence of artificial intelligence (AI) on firm performance: the business value of AI-based transformation projects. Bus. Process. Manag. J. **26**(7), 1893–1924 (2020)

Role and Application of OMG Notations in IT: Analysis and Survey on the Effectiveness of Business Notations Among IT Companies in Poland

Anna Suchenia[1(✉)], Karol Suchenia[1], Mateusz Ślażyński[2], and Leszek Szała[3]

[1] Cracow University of Technology, ul. Warszawska 24, 31-155, Kraków, Poland
{asuchenia,karol.suchenia}@pk.edu.pl

[2] AGH University of Krakow, al. A. Mickiewicza 30, 30-059 Krakow, Poland
mslaz@agh.edu.pl

[3] Department of Mathematics, Informatics and Cybernetics, Faculty of Chemical Engineering, University of Chemistry and Technology, Technicka 5, 166 28 Prague 6, Prague, Czech Republic
Leszek.Szala@vscht.cz

Abstract. This study analyzes OMG (Object Management Group) notations, their capabilities, and potential drawbacks. It includes a literature review and a survey focused on the purpose of using OMG models in the IT sector. Conducted via CAWI (Computer Assisted Web Interview) with 160 respondents from Polish IT companies, the survey aimed to identify which OMG notations are used, their practical applications, creation sequences, and the impact of job roles. The results confirmed the objective of validating the purposeful use of various notations. The survey established relationships and justified the use of specific models by Polish IT professionals with different job profiles. It highlighted the diverse perspectives needed in software development, depending on the role within the project. The survey also identified the most commonly used platforms for designing models with OMG notations and revealed that there is no universal modeling tool suitable for all stages of the process.

Keywords: BPMN · OMG · modeling

1 Introduction

In the digital era, where information technology and management systems form the foundation for effective business operations, OMG models play a crucial role in ensuring effective communication and knowledge exchange between various business entities. A detailed analysis of OMG notations, their capabilities, and potential drawbacks that may cause anomalies was the impetus for further research into their popularity and applicability.

Following this, an analysis of the literature, including studies on the application of specific OMG models, was conducted. Guided by the knowledge derived from these

C. Di Ciccio et al. (Eds.): BPM 2024, LNBIP 527, pp. 267–281, 2024.
https://doi.org/10.1007/978-3-031-70445-1_17

sources, a proprietary survey was developed, focused on the purpose of using OMG (Object Management Group) models, with a particular emphasis on the IT sector. The aim of this study was to understand which OMG notations are used, which models are utilized in practice, their sequence of creation, and how job roles influence the results.

This study provides a contribution to research on the purpose of using OMG models, offering valuable insights into their practical application in the local IT sector.

2 Analysis of the Use of Diagrams and Tools for Modeling

UML is a broad topic, providing a rich set of diagram notations divided into 14 different types, each with its own UML model serving various functions and addressing different aspects of software developers' needs [1]. The literature includes studies examining the scope of UML notations and the usage of specific UML diagrams, such as the works [2–6].

Scott Ambler in [7] points out that while the number of diagrams is constantly increasing, there are still missing elements, such as a user interface diagram or a diagram representing database structure. The new version of UML was also supposed to improve interoperability between modeling tools. However, the author believes that the intended goals might not have been fully achieved. Grady Booch, one of the key creators of the Unified Modeling Language, stated that "only 20% of UML is needed for 80% of all software" [8]. Moreover, most users, when learning UML notations, rely on books, courses, tutorials, or simply start using UML diagramming tools, which usually do not cover the entire UML specification. As a result, many UML users might never become aware of numerous specific constructs.

While UML notation is well-formalized, its extensive specification also becomes a disadvantage, making it difficult to thoroughly familiarize oneself with all aspects of the diagrams, leading to their poor utilization. Discussions about the pros and cons of UML are ongoing in various industry forums [7, 9], and new ideas for the further development of this notation continue to emerge. In the Software Engineering community, negative opinions are often encountered, with the complexity of UML frequently cited as one of its main drawbacks [7, 8]. The study [10] presents controversies regarding the significance, role, and usefulness of models, modeling, and modeling languages in the industry. While some consider UML the "lingua franca of software engineering," others claim they simply do not use UML notation [11]. The authors of [11] present UML as the most commonly used modeling language by academia and software companies. Research by Petre [12] critically approaches the universality of UML, yet confirms its significant role in model-based engineering, despite not all diagram types being utilized. This underscores a key assumption of modeling, that involves automatic code generation from corresponding models [13].

Another area of research concerns the application of UML notation in the industry, particularly studies [2, 14], which show that various UML components are used for different purposes by different users. These observations were confirmed and expanded in later analyses [4, 15]. Probably the only secondary study on UML and its practical application is [16], which dates back to 2010.

Another important group of scholarly works includes research related to industrial modeling [17–21]. These studies observed that modeling is used in the industry, with

UML being the dominant notation. They also concluded that the less formal the use of modeling, the more frequently it is applied in practice. The study [10] confirms the opinions presented in [11] and [12]. Further works [22, 23] also describe modeling practices in UML language used in the industry. The research presents practices related to modeling and a set of recommended practices aimed at increasing the efficiency of software modeling. The use of software modeling notations is seen as beneficial for software maintenance but must be tailored to the specific context.

The analysis of the study presented in [2] in 2006 suggests that although UML is often presented as a tool focused on use cases, class diagrams have become the most commonly used element, with 73% of respondents confirming their use in at least two-thirds of their projects. Use case descriptions ranked fourth (44%). Nevertheless, respondents were generally familiar with all elements, from only 3% claiming they never used class diagrams in their projects to 25% for collaboration diagrams. Respondents with the most experience working with UML indicated that their projects utilized a greater number of elements, suggesting that the level of usage may increase as users gain more experience. This result seems inflated. It can be assumed that it is related to the overall context of the study – individuals who are heavily involved in the topic of OMG notations were likely more inclined to participate in the survey due to their deep knowledge and interest in the subject.

The next article presents a study aimed at determining the extent of usage of various UML diagrams [15]. The presented results can be interpreted as follows:

- Widely Used: Diagrams used in over 60% of sources.
- Rarely Used: Diagrams used in less than 40% of sources.

The survey results indicate that the use of different UML diagrams is varied, and the entire UML notation is not fully utilized (Fig. 2). The article [15] also presents survey results indicating which UML elements are most frequently used. The study covered books, courses, tutorials, and tools related to UML. The most commonly used diagrams were class diagrams, activity diagrams, sequence diagrams, use case diagrams, and state machine diagrams. These diagrams are widely used without any doubt, whereas interaction overview diagrams, timing diagrams, and profile diagrams are rarely utilized. The dominant position of class diagrams is not surprising as they are fundamental to UML. On the other hand, the second place for activity diagrams is a significant result, reflecting their use in modeling business processes and SOA-based systems.

The continuation of these studies is the work [5], which, based on an individual survey conducted among 275 respondents from the industrial and academic sectors, presents results related to the usage of UML diagram types. It also includes an evaluation of which UML elements are considered and used in practice. A case study [24] from 2018, based on the analysis of UML notation application in the IT sector in Brazil, involved a survey completed by 222 people from 140 different IT companies. It addressed questions about experiences with UML, difficulties in adopting UML, and suggestions for improving UML usage in practice. The results show that only 28.2% of respondents used UML in their daily work, while 73.2% did not. Most (61.26%) stated that automatic creation of UML diagrams to present a comprehensive system view would help increase UML usage.

The study in [25] addressed different perspectives on modeling, providing interesting results. Although functional perspectives (96%) and informational perspectives (99%) are the most popular in modeling software architecture with UML, the operational perspective is the least popular, used by only 26% of specialists. The most commonly used modeling tool for UML is Enterprise Architect.

Another study examining the use of UML diagrams among respondents found that the most recognizable were use case and class diagrams, likely due to their specificity favoring simpler design of IT system modules [26]. Following these were activity and sequence diagrams, with package, component, and state diagrams being the least recognized. Table 1 presents the results of this study. Respondents indicated that the most frequently used diagrams in their IT projects were use case and class diagrams, noted by nearly half of the respondents. Activity, component, state, sequence, and package diagrams were chosen less frequently. The results presented in Table 1 confirm that familiarity with diagrams does not always equate to their universal application.

Table 1. UML Diagrams Used by Respondents. Source: Own elaboration based on [26].

Diagram Name	Respondents (%)
Use case	30.4%
Class	24.1%
Activity	20.3%
Component	11.4%
Sequence	8.9%
State	8.9%
Package	5.1%
Others	0.0%

Table 2 summarizes the results from [26] regarding the use of tools in the modeling process by respondents. Analyzing respondents' preferences for UML diagramming tools, it was noted that Visual Paradigm was the most popular. Its popularity may stem from its frequent use in educational institutions, translating into professional practice, and its high compliance with the UML standard issued by OMG. The second most frequently used editor was Microsoft Visio. This might seem surprising, considering that Visio is a diagramming tool rather than a dedicated UML diagramming tool. However, its widespread availability and ease of access could explain its popularity among respondents. The survey results also suggest that modeling functions integrated with programming tools, such as Eclipse or NetBeans, are not commonly used. This includes both built-in tools and plugins for creating UML diagrams directly within the development environment. The "Other" category included tools like StarUML and Enterprise Architect. Both these applications are dedicated UML modeling tools but have not gained as much popularity as Visual Paradigm or Microsoft Visio among respondents.

In summary, the choice of tools for creating UML diagrams may be strongly linked to the availability of tools, habits acquired from education, and compliance with the

Table 2. Tools Used for Modeling. Source: Own elaboration based on [26]

Diagram Name	Respondents (%)
Visual Paradigm	39.2%
Microsoft Studio	13.9%
PlantUML	5.1%
yEd	5.1%
UML plug-in for Eclipse	2.5%
UML plug-in for NetBeans	2.5%
UML plug-in for IntelliJ IDEA	2.5%
Draw.io	2.5%
Others	5.1%

UML standard. There is no single, versatile tool that meets all modeling needs. The IT market offers various tools that provide a wide range of functions enabling effective and professional model creation.

OMG, as the organization responsible for creating and maintaining the UML standard, provides an official list of supported tool vendors on its website [27]. Additionally, surveys among system design specialists have resulted in rankings of modeling tools for the years 2019 [28], 2020 [29], 2021 [30], and 2022 [31].

The study [32] includes a review of tools used for modeling in UML notation. These tools can be used for creating UML diagrams, model and schema exchange, model transformations, code generation, as well as reverse and bidirectional engineering. Many vendors offer their UML tools for free download from the internet.

With the growing demand for tools that enable online collaboration, a new category of modeling tools, online modeling tools, has emerged and is developing dynamically. While many of them have more features of diagramming tools than modeling tools, they allow for the creation and sharing of models online, including UML diagrams, database models, and BPMN process models. Some of these tools may also offer online versions.

3 Survey on the Purpose of Using Specific OMG Notation Models

The survey on the purpose of using specific OMG notation models was initiated based on the analysis of available literature and existing rankings of design tools. This served as the foundation for creating our own study on the applicability of various OMG models. The survey was conducted using the CAWI technique (Computer Assisted Web Interview) with 160 respondents from Polish IT companies. The prepared closed-ended questions (11 in total) were made available to respondents through the online tool Google Forms. The primary objective of the survey was to explore the justification for using certain diagrams and OMG notations in the IT sector. Below is the structure of the survey along with the available responses:

1. **What is your primary role in the IT industry?**
   Closed-ended question, allowing only one option to be selected from the list. If the respondent selected "I do not work in IT" the survey was automatically terminated.
2. **Do you create diagrams using OMG notations (UML, BPMN, CMMN, DMN)?**
   Closed-ended question with the option to select one answer. If the respondent chose "Yes," they were directed to question three. Otherwise, they were directed to question four.
3. **Which environment do you use most often for creating diagrams?**
   The question allows selecting up to 3 options. The list of tools was prepared according to rankings but included an option for a custom answer.
4. **Do you think using a single notation is sufficient?**
   Closed-ended question allowing only one answer. If the respondent selected "No," they were directed to question five. Otherwise, the next question was question six.
5. **What is the minimum number of notations that should be used to design a modern information system?**
6. **Which models, in your opinion, should be created first?**
7. **In your opinion, what should be the order of creation for the listed diagrams?**
   This question required each participant to assign a unique value from 1 to 5 to a specific type of diagram. Assigning a lower value indicated that the respondent believed that a particular diagram should be prioritized in the creation process.
8. **Which OMG notations do you use at work?**
9. **Which of the listed diagrams do you use most frequently? (You can select up to 5 models)**
10. **Does the company you work for use automatic code generation based on OMG notation models?**
    This detailed survey structure aims to gather comprehensive insights into the use and preferences for OMG notations in the IT industry.
11. **Please choose only one answer. In the company where you work:**

a. Too many OMG models are being used.
b. Too few OMG models are being used.
c. The OMG models being used are not the ones that I believe should be used.
d. The appropriate OMG models are being used in the right number.

3.1 Analysis of Survey Results

The survey was available in two public thematic groups related to the IT sector in Poland on the social networking site:

- Group "*Problems of the Polish IT Industry*"
- Group "*IT Jobs for Intern/Junior*"

The total number of members in both groups at the end of the survey was approximately 120,000, with the first group having around 37,000 members and the second group having around 83,000 members. The survey was completed by 160 people from Poland, of which about 9% declared that they are not related to the IT industry (Fig. 1). Programmers constituted the largest group among the respondents.

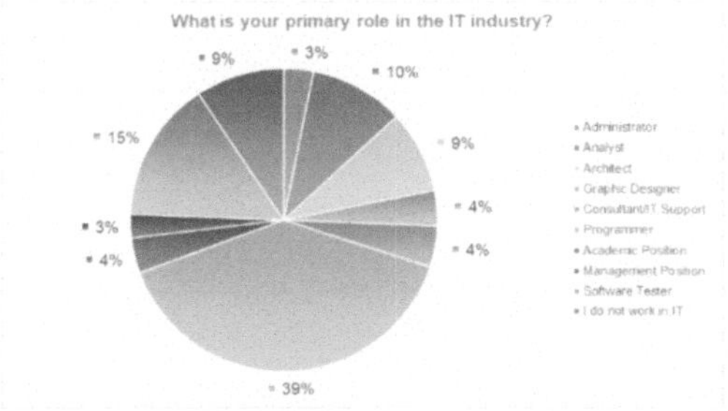

Fig. 1. Type of Role in the IT Sector. Source: own elaboration.

The study suggests that approximately 41% of IT sector employees are involved in creating diagrams using OMG notations (Fig. 2). This result appears to be inflated. It can be assumed that this is related to the overall context of the survey—individuals more strongly associated with the subject of OMG notations were likely more inclined to complete the survey due to their greater knowledge and interest in the topic.

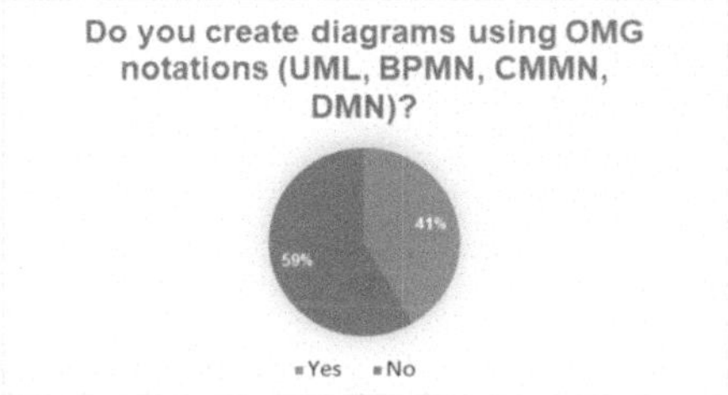

Fig. 2. Involvement in the Process of Creating OMG Notation Diagrams.

Respondents who confirmed their involvement in creating diagrams using OMG notations were asked to indicate the tools they most frequently use (Fig. 3). The most popular tool was Enterprise Architect. Other commonly used environments included Draw.io, Microsoft Visio, and Visual Paradigm.

An important aspect to consider is the potential influence of the role within the company on the choice of tool (related to its adequacy for the tasks performed). In this context, the following diagrams (Figs. 4, 5, 6, and 7) are presented. They illustrate the tool choices of those who are actively involved in the software production process, breaking down these choices according to their roles.

These diagrams provide a detailed view of the preferences for various tools based on the specific roles of the respondents within their organizations. This analysis helps to understand how role-related responsibilities might influence the selection of tools for creating OMG notation diagrams.

The analysis of the presented results shows that the distribution of tool usage is similar for each group. We can hypothesize that the choice of tools for creating diagrams is not strictly related to the role performed in the enterprise.

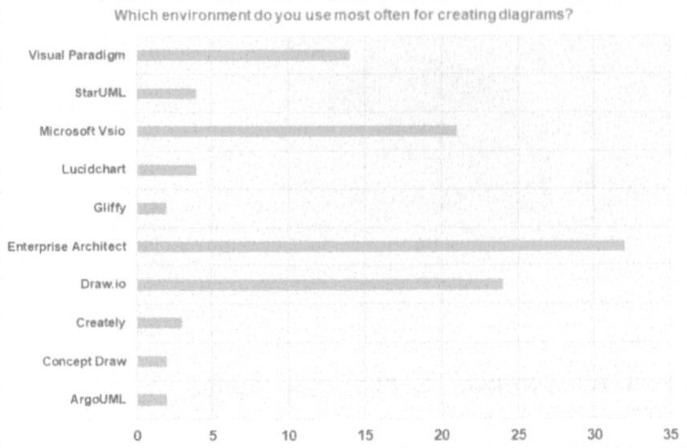

Fig. 3. Working Environment for Creating OMG Models. Source: Own elaboration

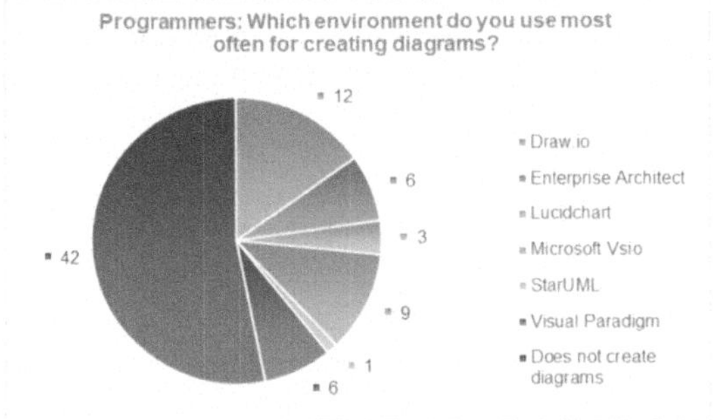

Fig. 4. Modeling Environment for Programmers. Source: Own elaboration.

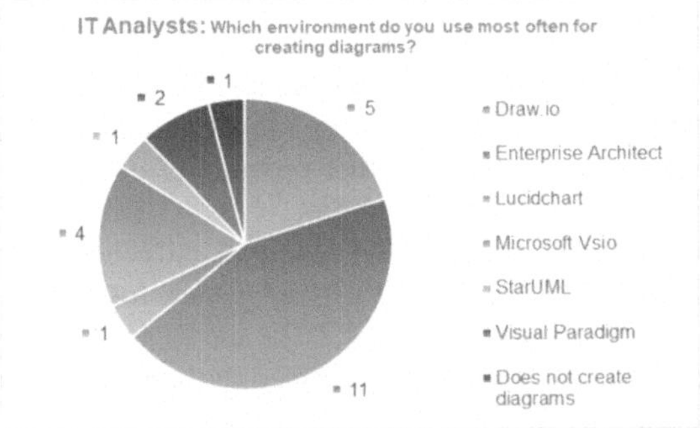

Fig. 5. Modeling Environment for IT Analysts. Source: Own elaboration.

The next element of the analysis was the assessment of the number of OMG notations required in the process of creating advanced software. Respondents who chose the option "No" in the previous question received an additional question about the minimum number of OMG notations that they believe should be involved in the process of building a modern

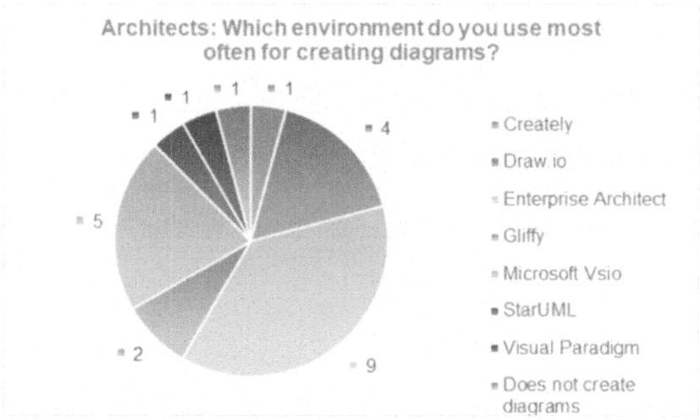

Fig. 6. Modeling Environment for IT Architects. Source: Own elaboration.

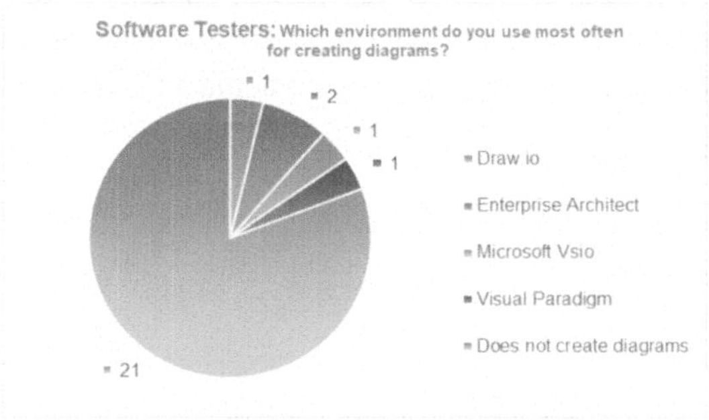

Fig. 7. Modeling Environment for Software Testers. Source: Own elaboration.

application. The majority of responses suggested the use of two or three notations, as presented in Fig. 8.

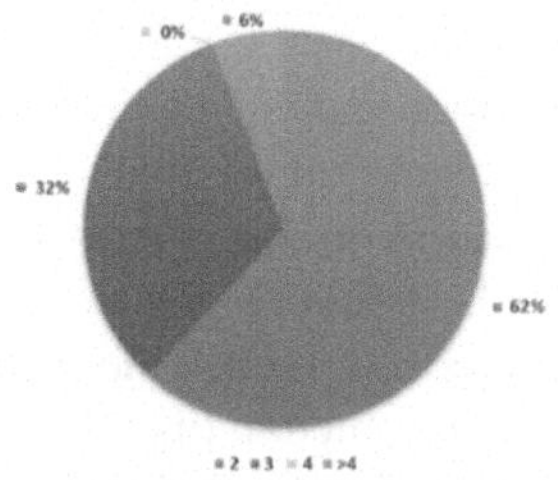

Fig. 8. Minimum Number of OMG Notations Required. Source: Own elaboration.

The chart in Fig. 9 shows that only 15% of respondents consider one notation to be always sufficient. About one-third of the participants believe that the number of notations should depend on the nature of the project. This can be interpreted as a recommendation to use more notations for more complex projects involving intricate business processes, whereas a single notation might be sufficient for relatively simple applications.

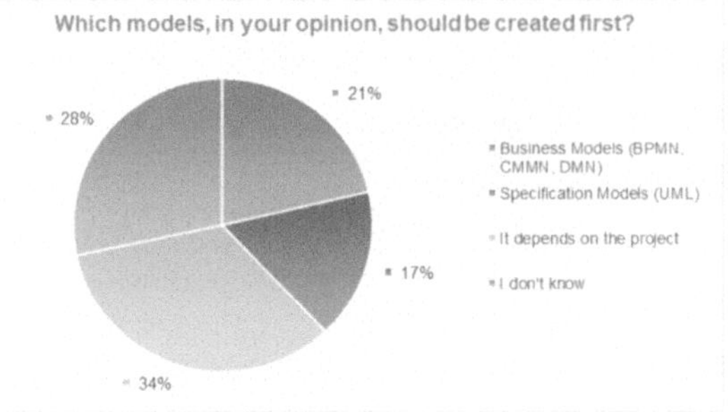

Fig. 9. Sufficiency of Single Notation. Source: Own elaboration.

The next survey question concerned the order of creating models using commonly used OMG notations. Participants were asked to determine the sequence in which various models should be created. The goal was to verify whether there is a standard and generally accepted design path for diagrams. The results of this question are presented in Fig. 10.

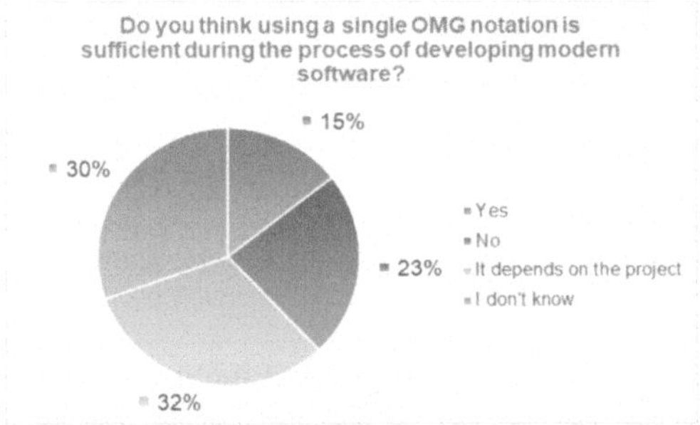

Fig. 10. Sequence of Creating OMG Models. Source: Own elaboration.

The analysis of the collected responses (Fig. 11) indicates that there is no defined path. Approximately one-third of respondents believe that the order of creating models should depend on the specific project. Based on the gathered results, we cannot definitively determine whether business models should be created first or specification models.

The following question attempted to establish the sequence of creating specific types of diagrams. Analyzing the charts presented in Figs. 12 and 13, it can be observed that a significant majority of respondents prefer to use BPMN as the first model. Additionally, there is a tendency to use the sequence diagram as the last model in most projects. Determining the order of creation for the remaining models based on the obtained results proves to be a complicated matter.

The analysis of the responses to the next question allowed us to verify the degree of use of OMG tools (Fig. 14). According to the obtained data, the most popular notation is UML, used by nearly 70% of respondents. The second most popular notation is BPMN, used by about 45% of respondents. Other notations, such as CMMN or DMN, are used by less than 10% of respondents.

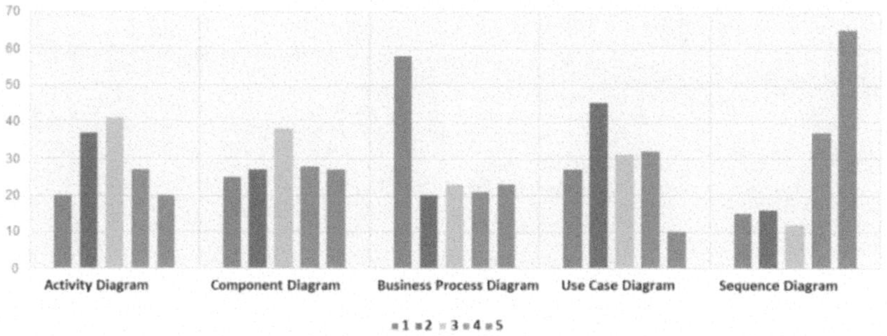

Fig. 11. Preference for Business or Specification Models. Source: Own elaboration.

In the "other" section, some respondents mentioned using other OMG tools, such as SysML. It is worth noting that about one-quarter of respondents in the "other" section indicated that they do not use OMG tools at all. This indicates that despite the popularity of these tools among some IT specialists, there is also a significant number of people in the industry who do not use them in their work.

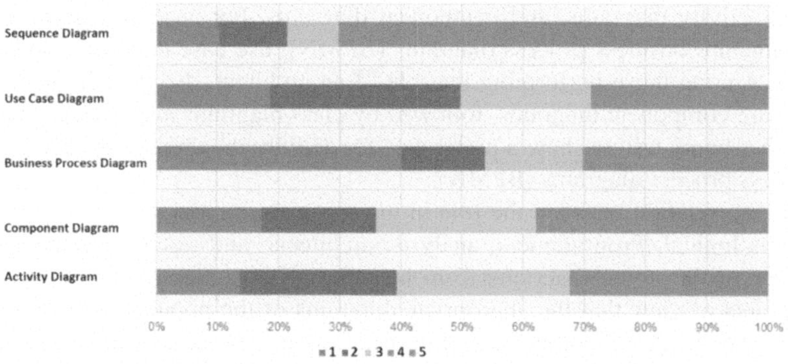

Fig. 12. Proposed Sequence of Creating OMG Models. Source: Own elaboration.

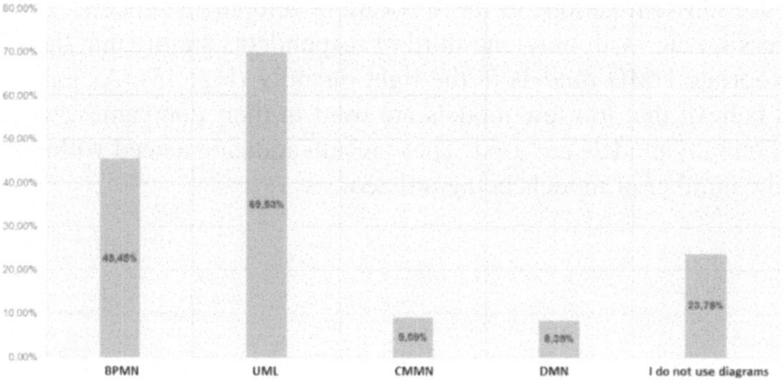

Fig. 13. Usage of OMG Notations. Source: Own elaboration.

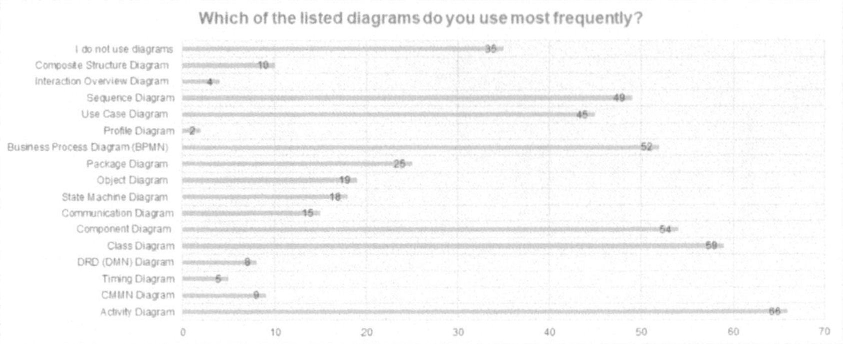

Fig. 14. Usage of OMG Tools. Source: Own elaboration

Considering whether the choice of specific diagrams might depend on the role an individual plays in an IT company, it raises the question of whether the popularity of certain diagrams is due to the predominance of certain roles among respondents.

Based on the presented charts, we can conclude that the types of diagrams used are related to the role performed in the IT company. Programmers most often use class diagrams, activity diagrams, and component diagrams. For analysts, the most popular diagrams are business process diagrams (BPMN), use case diagrams, and activity diagrams. Among those performing the role of an architect, the most frequently used diagrams are component diagrams, followed by class diagrams and package diagrams. On the other hand, testers show a preference for activity diagrams, use case diagrams, and business process diagrams (BPMN).

Such a correlation between the role in the organization and the types of diagrams used seems logical. Programmers, analysts, architects, and testers may have unique perspectives on the project that stems from their specialized duties. It is precisely for these diverse points of view that the appropriate diagrams of the popular OMG consortium notations are dedicated. Another interesting aspect to investigate is the issue of automatic code generation based on OMG notation models. Most companies do not use automatic code generation.

The final question aimed to investigate whether and how companies where respondents work use OMG notations in the software development process. The results are somewhat pessimistic, with only one-third of respondents stating that their companies use the appropriate OMG models in the right quantity (Fig. 15). As many as 42% of respondents believe that too few models are used in their companies, while only 8% report that too many models are used. These results indicate a need within the industry to increase the number of models being utilized.

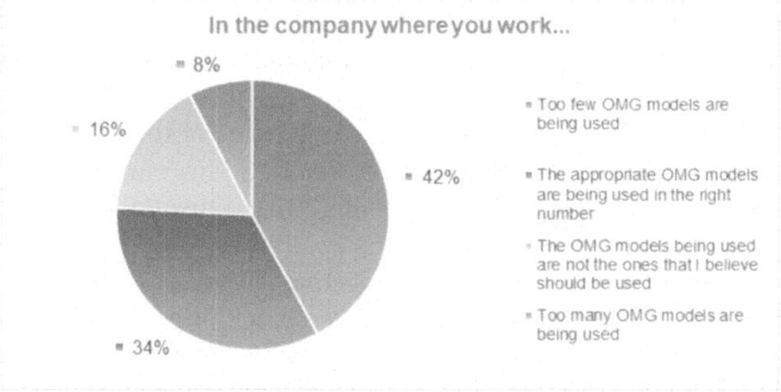

Fig. 15. Appropriateness of OMG Model Usage. Source: Own elaboration.

4 Summary

The survey confirmed the goal of identifying and confirming the use of specific notation models by IT professionals. Diverse roles in the process of creating modern software require a varied perspective on the project, often filtered through the tasks assigned to a particular position. The study also confirmed that the use of specific models is determined by the necessity to obtain a particular project cross-section. Architects, working at an abstract level, use such models as component diagrams or package diagrams compared to analysts who gather client requirements using activity diagrams or business process diagrams. Similar differences are observed for programmers and testers, who also play crucial roles in the software development process and use different models depending on specific needs and objectives.

Our survey identified the most commonly used platforms for OMG notations but found no direct link between tool choice and diagram creation. There is no single universal modeling tool applicable at all stages of the process. The choice of the appropriate tool requires users to understand the goals they intend to achieve in a given environment and then look for a tool that most effectively meets these requirements. Various tools available on the market offer diverse functions that enable professional and intuitive model creation.

The survey found no uniform sequence for creating diagrams, as each project has unique requirements. Every undertaking related to the creation of modern software requires analysis from multiple perspectives, such as complexity, functionality, and scale. Only after such a comprehensive analysis is it often possible to determine the most appropriate course of action, including the sequence of model creation, as well as the choice and number of models necessary to achieve a specific goal.

References

1. OMG. Unified Modeling Language ver. 2.5.1. https://www.omg.org/spec/UML/2.5.1/PDF, OMG Document Number: 2017-12-05 (2017)
2. Dobing, B., Parsons, J.: How UML is used. Commun. ACM **49**(5), 109–113 (2006)
3. Grossman, M., Aronson, J.E., McCarthy, R.V.: Does UML make the grade? insights from the software development community. Inf. Softw. Technol. **47**(6), 383–397 (2005)

4. Langer, P., Mayerhofer, T., Wimmer, M., Kappel, G. On the usage of UML: initial results of analyzing open UML models. In: Fill, H., Ulrich Reimer, D.K. (eds.) Modellierung 2014, pp. 19–21 (2014)
5. Reggio, G., Leotta, M., Ricca, F.: Who knows/uses what of the UML: a personal opinion survey. In: Dingel, J., Schulte, W., Ramos, I., Abrahão, S., Insfran, E. (eds.) MODELS 2014. LNCS, vol. 8767, pp. 149–165. Springer, Cham (2014). https://doi.org/10.1007/978-3-319-11653-2_10
6. Scanniello, G., Gravino, C., Tortora, G.: Investigating the role of UML in the software modeling and maintenance - a preliminary industrial survey. In: Filipe, J., Cordeiro, J. (eds.) ICEIS 2010, pp. 141–148. SciTePress (2010)
7. Ambler, S.W.: UML 2.5: Do you even care? (2013). https://www.drdobbs.com/architecture-and-design/uml-25-do-you-even-care/240163702#disqus_thread
8. Cabot, J.: What do their creators think about UML now? Modeling Languages (2018). https://modeling-languages.com/uml-opinions-creators/
9. Cabot, J.: Grady Booch's comments on UML/MDA and software engineering. Modeling Languages (2009). https://modeling-languages.com/grady-boochs-comments-umlmda/
10. Störrle, H.: How are conceptual models used in industrial software development? A descriptive survey. In: Proceedings of the 21st International Conference on Evaluation and Assessment in Software Engineering, EASE 2017 (2017)
11. Petre, M. UML in practice. In Proceedings of the 2013 International Conference on Software Engineering, ICSE 2013, pp. 722–731. IEEE Press (2013)
12. Petre, M.: "No shit" or "Oh, shit!": responses to observations on the use of UML in professional practice. Softw. Syst. Model. **13**, 1225–1235 (2014)
13. Selic, B.: The pragmatics of model-driven development. IEEE Softw. **20**(5), 19–25 (2003)
14. Dobing, B., Parsons, J.: Dimensions of UML diagram use: a survey of practitioners. J. Database Manag. (JDM) **19**(1), 1–18 (2008)
15. Reggio, G., Leotta, M., Ricca, F., Clerissi, D.: What are the used activity diagram constructs? - a survey. In: Pires, L.F., Hammoudi, S., Filipe, J., das Neves, R.C. (eds.) MODELSWARD 2014, pp. 87–98. SciTePress (2014)
16. Budgen, D., Burn, A.J., Brereton, O.P., Kitchenham, B.A., Pretorius, R.: Empirical evidence about the UML: a systematic literature review. Softw.—Pract. Exp. **41**, 363–392 (2011)
17. Baltes, S., Diehl, S.: Sketches and diagrams in practice. In: Proceedings of the 22nd ACM SIGSOFT International Symposium on Foundations of Software Engineering, pp. 530–541 (2014)
18. Davies, I., Green, P., Rosemann, M., Indulska, M., Gallo, S.: How do practitioners use conceptual modeling in practice? Data Knowl. Eng. **58**(3), 358–380 (2006)
19. Forward, A., Lethbridge, T.C.: Problems and opportunities for model-centric versus code-centric software development: a survey of software professionals. In: MiSE 2008, pp. 27–32. Association for Computing Machinery, New York (2008)
20. Lange, C., Chaudron, M., Muskens, J.: UML software architecture and design description. IEEE Softw. **23**(2), 40–46 (2006)
21. Nugroho, A., Chaudron, M.R.V.: A survey into the rigor of UML use and its perceived impact on quality and productivity. In: Rombach, H.D., Elbaum, S.G., Münch, J. (eds.) ESEM 2008, pp. 90–99. ACM (2008)
22. Fernández-Sáez, A.M., Chaudron, M., Genero, M.: An industrial case study on the use of UML in software maintenance and its perceived benefits and hurdles. Empir. Softw. Eng. **23**, 3281–3345 (2018)
23. Fernández-Sáez, A.M., Genero, M., Chaudron, M.R.: Empirical studies concerning the maintenance of UML diagrams and their use in the maintenance of code: a systematic mapping study. Inf. Softw. Technol. **55**(7), 1119–1142 (2013)

24. Farias, K., Gonçales, L., Bischoff, V., Silva, B., Guimarães, E.T., Nogle, J.: On the UML use in the Brazilian industry: a state of the practice survey. In: SEKE (2018)
25. Ozkaya, M., Erata, F.: A survey on the practical use of UML for different software architecture viewpoints. Inf. Softw. Technol. **121** (2020)
26. Zyła, K., Ulidowski, A., Wrzos, J., Włodarczyk, B., Krocz, K., Drozd, P.: UML – a survey on technical university students in Lublin. J. Comput. Sci. Inst. **13**, 279–282 (2019)
27. OMG. UML vendor directory listing (2021). https://www.omg.org/uml-directory/vendor/list.htm
28. Modeling Languages. Top online UML modeling tools in 2019 (2019). https://modeling-languages.com/web-based-modeling-tools-uml-er-bpmn/
29. Cabot, J. Modeling, low-code tools (2019). https://twitter.com/i/lists/744723693801209856
30. Cabot, J., 25 best UML tools free UML diagram software in 2021 (2021). https://www.guru99.com/best-uml-tools.html
31. Modeling Languages.com. Curated list of UML tools – 2022 edition (2022). https://modeling-languages.com/uml-tools/
32. Thong, W.J., Ameedeen, M.A.: A survey of UML tools. In: Abawajy, J., Othman, M., Ghazali, R., Deris, M., Mahdin, H., Herawan, T. (eds.) DaEng-2015. Lecture Notes in Electrical Engineering, vol. 520, pp. 61–70. Springer, Singapore (2019). https://doi.org/10.1007/978-981-13-1799-6_7

Extending CMMN for Effective Management of Data in Knowledge-Intensive Processes

Mateja Bule(✉) and Gregor Polančič

Faculty of Electrical Engineering and Computer Science, University of Maribor, Koroška cesta 46, 2000 Maribor, Slovenia
{mateja.bule,gregor.polancic}@um.si

Abstract. Knowledge-intensive processes, such as product innovation, product development, and emergency management, are increasingly gaining importance in modern organisations and society. As such processes are characterised by a high degree of flexibility, creativity, and complexity, they usually rely on different techniques and technologies compared to traditional "well-structured" processes. Accordingly, the Case Management Model and Notation (CMMN) was proposed similarly to BPMN for the structured processes. Even though CMMN is certainly necessary in this field, our preliminary analysis revealed some shortcomings. In this article, we address some of these shortcomings related to the presentation and manipulation of data. More precisely, we are dealing with the extension of the element Case File Item and its associated elements, where metadata is poorly defined. With the help of the existing Dublin Core metadata element set, the element Case File Item was extended structurally and visually to achieve an improved understanding of the business process models.

Keywords: CMMN · shortcomings · extension · Case File Item · Dublin Core · metadata scheme

1 Introduction

In modern organisational environments, business processes can be broadly categorised into structured and unstructured. Structured processes follow a predefined sequence of activities with well-defined rules and predictable outcomes, making them suitable for routine and repetitive tasks. In contrast, unstructured processes have gained increasing importance in recent years. They are more dynamic and flexible, adapting to varying circumstances, requiring decision-making, and relying heavily on the expertise and experience of individuals. These knowledge-intensive business processes are covered by *case management* approach that involves overseeing the entire life-cycle of a *case*, from initiation to resolution. Case management aims to ensure that cases are handled efficiently

C. Di Ciccio et al. (Eds.): BPM 2024, LNBIP 527, pp. 282–296, 2024.
https://doi.org/10.1007/978-3-031-70445-1_18

and effectively, often integrating technology and methodologies to manage information, and improve communication. By leveraging case management, organisations can improve the quality of service delivery and optimise resource utilisation. This approach often integrates different technologies, to facilitate the dynamic and adaptable nature of the case, allowing for the accommodation of unpredictable and non-linear workflows. Effective case management enhances communication and collaboration among stakeholders. It ensures that all aspects of a case are tracked and addressed systematically, leading to better decision-making and more favourable results.

To address the need for effectively modelling and managing unstructured, knowledge-intensive processes, the Object Management Group (OMG) introduced the Case Management Model and Notation (CMMN) [18]. CMMN provides a framework for capturing the variability and adaptability inherent in these processes, offering a flexible approach to case management that supports dynamic decision-making and contextual actions.

Nevertheless, online evidence and findings from the literature suggest that CMMN has not gained acceptance as anticipated [9,24]. Several factors may contribute to this, and we managed to identify two reasons. Firstly, although BPMN is primarily a complement to CMMN, it can also be used to model unstructured business processes, positioning BPMN as a potential competitor to CMMN. BPMN's earlier publication date and wider adoption are evident in Google Trends [7] data over the past five years, which shows a 1:20 search ratio in favour of BPMN. While CMMN has a solid foundation for unstructured business processes, it faces significant competition from the well-established BPMN. And secondly, the current knowledge society promotes a more flexible and adaptable organisational approach. However, many organisations still operate on assembly line principles, as transitioning to a new way of functioning is gradual. Consequently, organisations may not yet be fully prepared to adopt new methodologies like CMMN, leading to its limited use.

Starting from the preliminary research, where we explored the potential shortcomings, we identified some potential areas for improvement. Consequently, this paper examines the possible extensions for CMMN to further enhance its capability in modelling unstructured business processes, particularly in data management. While CMMN is suitable for handling cases, it can be further improved to address data management better. By proposing enhancements to the data structures within CMMN, we aim to enhance its applicability and effectiveness in various real-world scenarios. Additionally, our goal is to extend CMMN's capabilities to accommodate the intricate needs of modern business environments, enabling more efficient, flexible, and data-driven solutions.

In our effort to extend CMMN's data capabilities, we integrated a widely recognised metadata standard known as Dublin Core [4], which offers a core set of 15 elements that are designed to provide a simple and standardised way to describe digital resources. By incorporating Dublin Core into the CMMN structure, we aim to enhance its ability to effectively describe the properties, types and states of data. This integration not only facilitates better data presentation but also ensures that the extended CMMN can leverage a globally accepted metadata standard to improve clarity, consistency, and usability.

The remainder of this article is structured as follows: Sect. 2 briefly reveals related works, Sect. 3 reviews the Case Management Model and Notation (CMMN). In Sect. 4, the Dublin Core Metadata element set is presented. The proposed extension of data structures of CMMN and the corresponding discussion are presented in Sects. 5 and 6. Finally, conclusions are given in Sect. 7.

2 Related Work

In our preliminary research, we conducted a comprehensive analysis of the available literature to identify existing work related to extending CMMN. We identified a pivotal study [19] that provided a succinct overview of the application of CMMN. The analysis in [14] indicates that CMMN aligns with most KIP characteristics related to case modelling, which is also relevant for our research. Some similar conclusions can also be found in [23]. In [21] authors presented an intriguing extension focused on the relationship between Case File Items and tasks. An innovative approach was presented, where the connections between these elements are annotated with the specific actions performed on the data.

3 Case Management Model and Notation (CMMN)

CMMN has its foundations in Case Management (or also Case Handling) [26], introduced in 2004 and presented as an alternative paradigm for business process support, addressing some shortcomings of the existing paradigm - Business Process Management (BPM). Meanwhile, Case Management was presented as less restrictive and more flexible [13,26], with numerous examples in medicine, law, employment, etc. Processes within these fields are presented with *a case*, where each instance differs. For example, cases from medicine demonstrate how each case instance can be carried out completely differently.

With Case Management as a paradigm, the need for a standard arose, and in 2014, consortium OMG[1] released specification for CMMN 1.0. [18], suitable for managing knowledge-intensive, flexible processes. The latest version 1.1, was released in 2016.

3.1 CMMN Metamodel

CMMN metamodel provides a basis for defining the various case components and relationships that constitute a case, specifying the structure and semantics of CMMN, and detailing how elements such as Case File Items, Tasks, Stages, and Milestones interact and contribute to the overall process. It outlines the roles and responsibilities of participants, the conditions for task activation, and

[1] The Object Management Group® Standards Development Organisation (OMG® SDO) is a global, open membership, non-profit consortium.

the dependencies between different parts of the case. By offering a well-defined metamodel, CMMN ensures consistency and clarity in representing complex, unstructured business processes. The CMMN metamodel is presented using a UML class diagram, which provides a clear and structured visualisation of the components and their relationships within the case [18]. CMMN has a built-in extension mechanism that allows practitioners to perform adaptions or extensions to meet specific requirements of their business processes. This is achieved through the use of extension points where additional elements and attributes can be integrated.

3.2 Visual Notation

On top of the metamodel, CMMN covers a set of graphical elements that enable one to model and manage a case. These elements are depicted through standard shapes and symbols, enabling process managers to map out the flow of activities, dependencies, and decision points in a way that is both accessible and precise. Some core elements are described in Table 1 and graphically presented in Fig. 1.

Table 1. Basic elements of CMMN

Element name	Description
Case Plan Model	is the outermost element, within which the whole case is presented and defined
Task	is a central element that defines a single action, a unit of work, that needs to be performed. There are different types of tasks
Case File Item	is an element that defines data (e.g. document, file) required for case processing
Stage	may be considered as an "episode" or a "phase" of one case. Elements, like Tasks with Connectors, can be united within one Stage element
Milestone	represents an achievable target, defined to enable evaluation of the case's progress
Event Listeners	is an element that waits for an event in a case, to trigger other elements, like Tasks or Stages
Sentry (Sentries)	combines an "event and condition." It is not a stand-alone element, it requires to be part of a Case Plan Model, Stage, Task or Milestone. It is graphically represented with a rhombus
Connectors	can be used to visualise dependencies between elements but do not have associated execution semantics

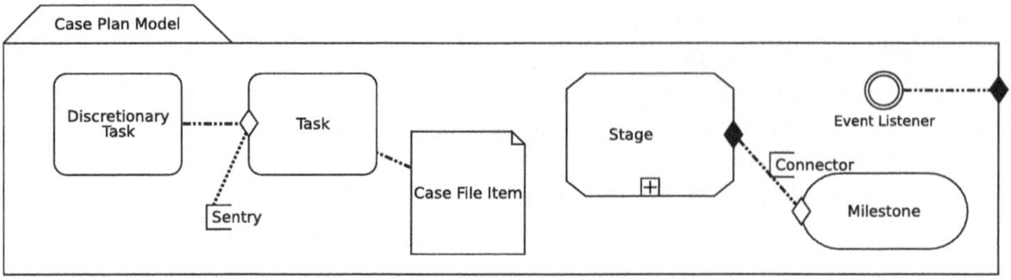

Fig. 1. Basic elements of CMMN

3.3 Applications

CMMN is applicable in industries and domains where processes are dynamic, non-linear, and require a high degree of flexibility, such as healthcare diagnosis and treatment processes, legal services, product innovation, emergency management, insurance, customer service management, etc. It is particularly effective in scenarios where predefined process models like BPMN are insufficient due to the unpredictable nature of the tasks involved. CMMN allows for the comprehensive modelling of such cases, providing a sufficient scope of structure and flexibility to support, both automated processes and manual interventions. By leveraging CMMN, organisations can achieve greater efficiency, agility, and accuracy in managing complex, knowledge-intensive cases.

For example, in the healthcare sector, CMMN can be effectively utilised to manage patient care processes, which are often complex and require a high degree of flexibility. For instance, consider the case of managing chronic disease care for patients with conditions such as diabetes. The patient care process involves multiple stages, including initial diagnosis, treatment planning, ongoing monitoring, and adjustments to the treatment plan based on the patient's response. Each of these stages can have various tasks such as scheduling appointments, conducting tests, reviewing results, and updating the patient's care plan.

Using CMMN, a healthcare provider can model this entire process to accommodate the dynamic and individualised nature of patient care. Tasks such as'Blood Sugar Monitoring,''Diet and Exercise Counselling,' and'Medication Adjustment' can be defined, along with events that trigger these tasks, like a change in the patient's condition. Milestones such as'Stabilised Blood Sugar Levels' can be set to track progress. Sentries can specify conditions under which certain tasks are activated, for example, initiating'Medication Adjustment' if blood sugar levels are outside the target range.

This approach ensures that all aspects of the patient's care are managed efficiently, with the ability to adapt to changes and new information as the case evolves. Thus, CMMN enhances the quality and responsiveness of chronic disease management in healthcare.

Based on the example above, we recognise that effective data management is crucial for successful case management. However, our preliminary analysis revealed that the existing support for data management within CMMN was inadequate. Therefore, our objective was to refine and augment the features related to data management, ensuring that they align closely with the needs and complexities of real-world case scenarios.

4 Dublin Core

Generally, there are several possibilities for describing digital resources, each providing a standard for metadata that helps in organising, managing, and retrieving these resources efficiently. Each of these standards or schemas offers specific advantages depending on the type of digital resource, the needs of the organisation, and the intended use of the metadata. By choosing the appropriate metadata standards, organisations can ensure effective management, discovery, and preservation of their digital assets. Some examples are: Dublin Core [3], Metadata Object Description Schema (MODS) [12], Machine-Readable Cataloging (MARC) [10], Metadata Encoding and Transmission Standard (METS) [11], etc.

According to the fact that Dublin Core is one of the most widely adopted metadata standards, used across various domains, including libraries, digital repositories, and educational institutions, especially due to its simplicity and flexibility [2,22], we recognised its potential within our work as well.

Dublin Core is a standard describing digital resources with specific metadata elements (or terms) [3]. The *core set* consists of 15 metadata elements. Additionally, there is also a *qualified set*, which extends the core set with some additional elements to describe any digital resource in even more detail.

Dublin Core was first introduced in the report of a workshop in 1995, in Dublin, Ohio, US [3]. After years of discussions and formalisation, Dublin Core became an international standard ISO 15836-1 [8], a US standard ANSI/NISO Z39.85 [16], and also IETF RFC 5013 [17]. The Dublin Core Metadata Initiative (DCMI) is responsible for formulating and maintaining the Dublin Core [3].

The aforementioned core set of 15 elements is also known as Dublin Core Metadata Element Set (DCMES) [4] and is presented alphabetically in Table 2 with the corresponding definitions attached. Among all core elements of DCMES, the most frequently used are typically *title*, *creator*, *description*, *type*, and *date*, as these provide essential information for identifying and understanding the resource. These elements are also fundamental in various other metadata schemes, like Metadata Object Description Schema (MODS) [12] or Machine-Readable Cataloging (MARC) [10].

For DCMES one of the most common examples of use is in digital libraries and institutional repositories. For example *Europeana* [6] is a digital platform that provides access to millions of books, music, artworks, and more from European archives, museums, and libraries. It uses the Dublin Core metadata standard to describe and manage its vast collection of digital resources [5]. Additionally, DCMES is also in use by some WordPress plugins and themes to enhance the metadata of posts, pages, and other content types [25]. This can improve search engine optimisation (SEO) and content discovery.

Table 2. 15 core elements of DCMES

Element (or Term)	Definition
Contributor	An entity responsible for making contributions to the resource
Coverage	The spatial or temporal topic of the resource, the spatial applicability of the resource, or the jurisdiction under which the resource is relevant
Creator	An entity primarily responsible for making the resource
Date	A point or time associated with an event in the life-cycle of the resource
Description	An account of the resource
Format	The file format, physical medium, or dimensions of the resource
Identifier	An unambiguous reference to the resource within a given context
Language	A language of the resource
Publisher	An entity responsible for making the resource available
Relation	A related resource
Rights	Information about rights held in and over the resource
Source	A related resource from which the described resource is derived
Subject	The topic of the resource
Title	A name given to the resource
Type	The nature or genre of the resource

5 Extension of CMMN with Dublin Core Support

Our preliminary analysis of the advantages and disadvantages of CMMN, derived from the available and relevant literature, revealed some interesting insights. One notable finding is the weak support for data modelling capabilities within CMMN, where the Case File Item serves as a crucial element and simultaneously carries the data within a case. However, despite its importance, CMMN offers a very limited view of data, and consequently also to the Case File Item element. We presume that the format and nature of the represented data are not sufficiently extensively defined, leading to challenges in achieving comprehensive and consistent data handling. Consequently, there is a need to address these shortcomings and enhance the data modelling capabilities of CMMN to ensure it meets the complex requirements of modern case management scenarios. We propose an extension or improvement of the Case File Item and its associated elements and by addressing these deficiencies, we seek to provide a greater robustness for data representation within CMMN.

5.1 Design Science Research Method

Within our research, the main research method was the Design Science Research Method (DSRM) [20], which is particularly appropriate for the development and creation of novel artifacts aimed at addressing specific problems. DSRM facilitates a structured approach to problem-solving by iterative designing, implementing, and evaluating artifacts within a real-world context. Based on the given description, this research method seemed particularly appropriate for our work.

The DSRM process sequence starts with *identifying problem and motivating.* After the initial step *defining the objectives of the solution* is required, followed by the *design and development* of the new artifacts. Later, the steps of *demonstration*, *evaluation* and *communication* follow each other. The process can be iterative, depending on current needs when researching. A simplified representation of the DSRM process model is presented in Fig. 2, obtained from [1].

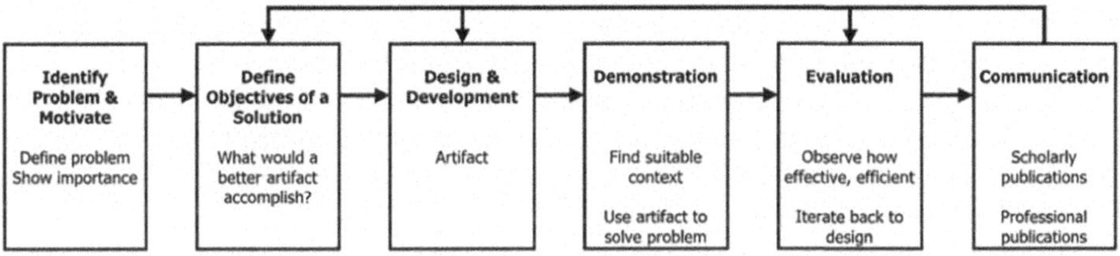

Fig. 2. DSRM process model (sequence) with iterations

Within our research, we performed the initial four steps of the DSRM, including *identification of the problem*, *defining objectives*, *designing and developing* the artifact and *demonstrating.* The following steps of extensive *evaluation* and *communication* of results are planned in our future work.

5.2 Extending the CMMN Structure

The CMMN specification details the structure and definition of a Case's information model. It "comprises of classes for the management of the data aspects of a Case" [18]. UML classes (or class diagrams) are a demonstration of the CMMN metamodel or its structure. Figure 3 is an excerpt from CMMN specification, where the definition of Case File is given. Here, the two classes *CaseFileItem* and *CaseFileItemDefinition* play crucial roles in organising and defining the data and involved documents. *CaseFileItem* represents an instance of data or a document that is used within a case. It can be linked to one or more *CaseFileItemDefinition* to specify the type of data that it represents. Meanwhile, the

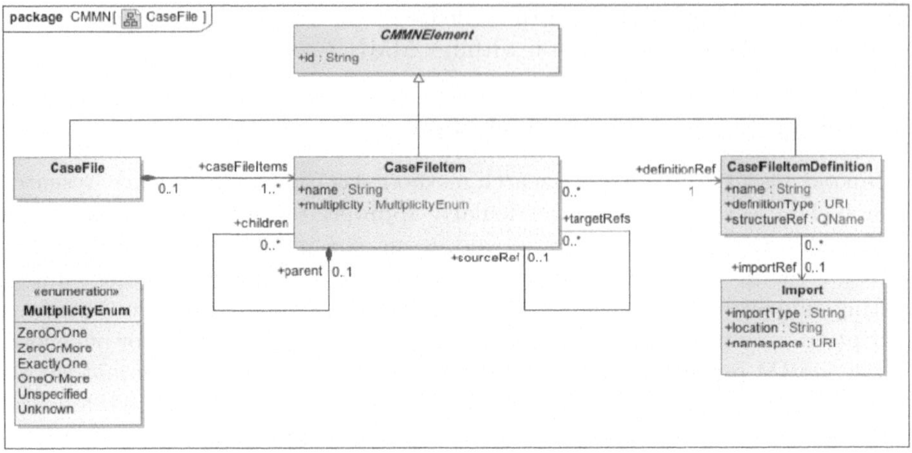

Fig. 3. Case File class diagram

CaseFileItemDefinition defines the structure and semantics of the data or a document that *CaseFileItem* instances can hold. For example, multiple instances of *CaseFileItem* can reference the same *CaseFileItemDefinition* [18].

Also in Fig. 3, attributes that are already defined for the *CaseFileItem* are: *name*, *multiplicity*, *definitionRef*, *children*, *parent*, *targetRef* and *sourceRef*. Already defined attributes for *CaseFileItemDefinition* are: *name*, *definitionRef*, *structureRef*, *importRef* and *properties*.

Although the existing attributes satisfy the essential requirements for effective data management, we believe that incorporating additional attributes is necessary to achieve a more comprehensive representation of the data within the case. These supplementary attributes will not only enhance the granularity and accuracy of the data but also provide a more holistic view, thereby improving the overall functionality and usability of the case model.

Further, we propose some extensions according to the representation of data, or more precisely according to the representation of *CaseFileItem* and *CaseFileItemDefinition*. We propose to extend the mentioned classes by adding additional attributes that enhance the definition's ability to describe and manage data within a case. To find the most appropriate attributes, we first analysed the Dublin Core metadata element set (see Sect. 4). Out of the set of 15 core elements, we identified six of them appropriate for our work. After a minor adjustment of attributes, we assigned them to classes, as can be seen in Fig. 4. Additional attributes and enumerations are coloured yellow.

CaseFileItem, which is an instance of data (or a document) used in a case, we redefined with five new attributes. The first one is *relation*, where the output or input relation of the data instance is indicated, including *RelationEnum*. The second one is *state*, defining the state or condition of the data currently. It has a certain stock of values, defined with *StateEnum*. The file location reference of the

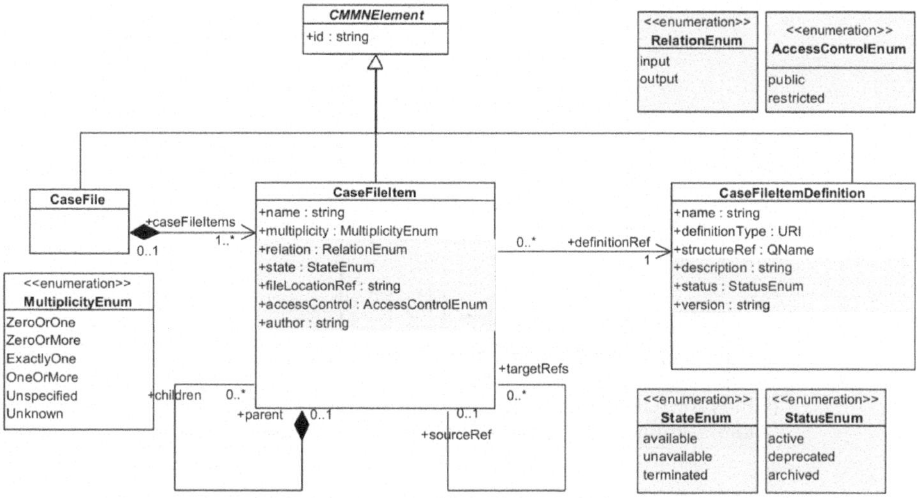

Fig. 4. Extended Case File class diagram

data (or document) instance is defined in *fileLocationRef*. Additional attributes *accessControl* and *author* are also defined.

CaseFileItemDefinition, which specifies the definition of data represented, we redefined with three additional attributes. The first new attribute is *description*, as a textual description of what the definition represents. The next is *status*, with *StatusEnum*, to indicate the different states of the Case File Item definition. The remaining new attribute is *version*.

5.3 Visual Extension

In our research, we aim to extend the Case File Item and its associated elements, not only structurally but also visually. While the *CaseFileItemDefinition* class is crucial for defining the structure and schema of Case File Items, a direct visual representation is not needed. Meanwhile, *CaseFileItem* class represents an instance of a data (or a document) in the CMMN model and it is also represented visually with a corresponding shape (see Case File Item in Fig. 1). Within our extension process, we aimed to improve this shape.

In Fig. 5, the proposed visual extension of the Case File Item element is presented. The main shape of the Case File Item elements remains intact, the same as the existing attribute *name* of *CaseFileItem* class (marked with the label *CFI#1 name*[2]). Below, we added an existing attribute *name* from *CaseFileItemDefinition* class (*CFID#1*[3]) to show it in textual form.

[2] CFI stands for *CaseFileItem* class.

[3] CFID stands for *CaseFileItemDefinition* class.

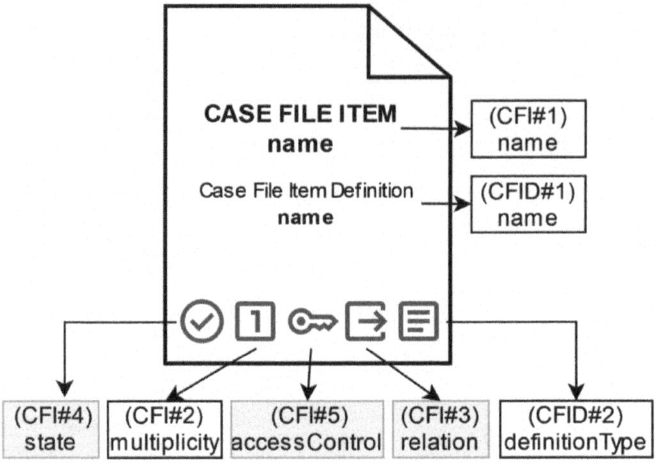

Fig. 5. Visual extension of Case File Item element

At the bottom of the Case File Item shape, attributes like *state*, *multiplicity*, *accessControl*, *relation* and *definitionType* are represented with corresponding icons and marked with the labels[4] (for example *CFI#4 state*). Attributes *multiplicity*, *definitionType* and *name* are existing attributes from the current specification of CMMN [18], but are also presented visually within this proposed extension. The presented icons for a particular attribute are chosen to best adjust to the concept of the attribute. Furthermore, there is potential for further research, according to the principles from the theory *The "Physics" of Notations* [15].

Some of the proposed attributes, like *fileLocationRef* and *author* from *CaseFileItem* class and some others from *CaseFileItemDefinition* class, do not have a visual representation within the proposed extension. We suggest that attributes like this can be covered within the tools for modelling and implementation, as additional properties. This is also a possible solution for all other proposed attributes, that were presented visually in Fig. 5.

Other possible variations of the extension exist. The proposed extension presented above is showing one of the possible variations of visually extending data structures in CMMN. There can be several other options. For example, icons shown on the shape of the Case File Item can be visible or hidden, depending on the existence of the definition of the attribute. Further, icons representing one exact attribute could have different appearances, depending on the value of the attribute.

By enhancing the visual aspects of Case File Item, we seek to improve the clarity and usability of data management in CMMN, making it easier for users to interact with and understand the data within cases.

---

[4] Labels are not part of the extension.

5.4 Demonstration

The demonstration of the extended Case File Item element is shown through a straightforward example (Fig. 6), referring to the given brief example of applicability in Sect. 3.3, where multiple tasks are linked to different documents. The symbol for Case File Item (a document with a folded corner) includes various icons that carry specific meanings. For instance, the first icon, depicted with a tick, serves to symbolise the document's current state, which can encompass values such as "ready," "error," or "corrupted." Meanwhile, the second icon, featuring a number, indicates the document's multiplicity, providing insight into the quantity of associated documents. Access control is denoted by the third icon, characterised by a key symbol, signifying the level of accessibility to the document. The fourth icon, represented by an arrow pointing inwards or outwards, distinguishes whether the document serves as an input or output in the given context. Finally, the last icon categories the type of document, offering clarity regarding its format or purpose within the system. Together, these icons streamline document management by pointing out crucial details to users.

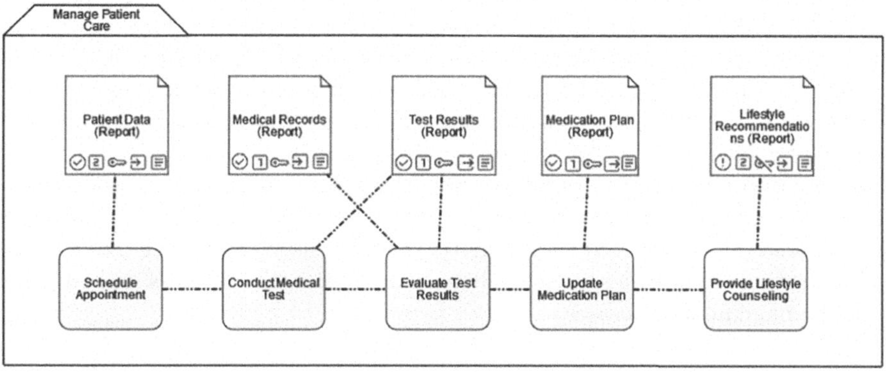

Fig. 6. Example of improved Case File Item element

This example demonstrates how additional attributes can be represented using icons, similar to how characteristics are indicated in Tasks (e.g., required, manual, discretionary, etc.). These icons provide visual cues that enhance the comprehensibility of the data associated with each task. For instance, attributes such as document status, multiplicity, access control, and type are depicted through distinct icons, making it easier for users to quickly understand these characteristics.

Moreover, modelling tools can contain some extra attributes, offering further customisation and detail in the representation of tasks and their associated data. This approach improves clarity and also ensures that all relevant information is readily accessible and easily interpreted. At the same time, there is also a significant opportunity for modelling tools to reduce complexity and enhance the comprehensibility of models. This can be achieved through innovative features such as mouse-over functionalities that display explanatory information, thereby providing immediate context and clarification without cluttering the visual interface. Additionally, introducing a new layer that reveals supplementary information can offer a deeper level of detail when needed, while maintaining a clean and user-friendly interface.

6 Discussion

As was already reported, the element Case File Item and consequently the entire management of data of CMMN has been identified as poorly defined. We aimed to address the encountered issue, by complementing some necessary attributes, that would contribute to the entire case.

In the following paragraph, we presented the discussion about the advantages of use, for all additional attributes within *CaseFileItem* and *CaseFileItemDefinition*.

Starting with core metadata set of attributes like *description*, *status*, *state*, *fileLocationRef* and *relation* can improve how data in general is presented to users. Appropriate attributes can enhance the user experience, making the case management system more intuitive and easier to use. Attributes like *version* and *author* can improve transparency and accountability within case management processes. Attribute *accessControl* can ensure that sensitive data is only accessible to authorised users, enhancing data security.

By extending the data capabilities of CMMN also visually, organisations can tailor their case management systems to better meet their specific requirements. This can lead to improved data integrity, security, user experience, compliance, and integration capabilities, ultimately enhancing the effectiveness and efficiency of case management processes.

6.1 Limitations and Future Work

This research is still in its early phases, which inherently brings certain limitations. One notable limitation is related to the presented icons (see Fig. 5) for a particular attribute. Currently, these icons have been chosen to best fit the concept of the attribute. However, in future stages, we intend to select icons according to the theory *"The Physics" of Notations* [15] to enhance accuracy and coherence. Additionally, there are several other limitations related to different variations for the visual presentation of additional attributes, which have not yet been fully explored in this phase. These limitations highlight the need for continued development and refinement in our ongoing research.

Although the paper does not explicitly tackle challenges specific to CEE countries, the suggested enhancements to CMMN might indirectly aid organisations in these regions. These improvements could enhance the flexibility, data integrity, and compliance of BPM systems.

7 Conclusion

In this article, we aimed to briefly present the current, still ongoing research, about potential extensions for data structures within the Case Management Model and Notation (CMMN). By examining the current classes of *CaseFileItem* and *CaseFileItemDefinition*, which are crucial components in data presentation in CMMN, we proposed several enhancements that could improve flexibility, data integrity, security, user interaction, and compliance. This proposed extension is intended to address specific business needs and enhance the overall effectiveness of case management systems. Our work lays the foundation for further discussion and development in this area.

References

1. Boné, J., Ferreira, J., Ribeiro, R., Cadete, G.: Disbot: a Portuguese disaster support dynamic knowledge chatbot. Appl. Sci. **10**, 9082 (2020). https://doi.org/10.3390/app10249082
2. Caplan, P.: Metadata fundamentals for all librarians. American Library Association (2003)
3. Dublin Core Metadata Initiative: DCMI Metadata Terms (2024). https://www.dublincore.org/. Accessed 8 May 2024
4. Dublin Core Metadata Initiative: Dublin™ Core Metadata Element Set, Version 1.1: Reference Description (2024). https://www.dublincore.org/specifications/dublin-core/dces/. Accessed 8 May 2024
5. EU: Europeana Semantic Elements Specification and Guidelines (2013). https://pro.europeana.eu
6. EU: Europeana (2024). https://www.europeana.eu
7. Google: Google Trends (2023). https://trends.google.com/
8. ISO: ISO 15836-1 (2024). https://www.iso.org/standard/71339.html. Accessed 8 May 2024
9. Kurz, M., Schmidt, W., Fleischmann, A., Lederer, M.: Leveraging CMMN for ACM: examining the applicability of a new OMG standard for adaptive case management. In: S-BPM ONE 2015, pp. 1–9 (2015). https://doi.org/10.1145/2723839.2723843
10. Library of Congress: Machine-Readable Cataloging (MARC) (2024). https://www.loc.gov/marc/umb/. Accessed 8 May 2024
11. Library of Congress: Metadata Encoding and Transmission Standard (METS) (2024). https://www.loc.gov/standards/mets/. Accessed 8 May 2024
12. Library of Congress: Metadata Object Description Schema (MODS) (2024). http://www.loc.gov/standards/mods/. Accessed 8 May 2024
13. Marin, M., Hull, R., Vaculín, R.: Data centric BPM and the emerging case management standard: a short survey. In: La Rosa, M., Soffer, P. (eds.) BPM 2012. LNBIP, vol. 132, pp. 24–30. Springer, Heidelberg (2013). https://doi.org/10.1007/978-3-642-36285-9_4
14. Marin, M.A., Hauder, M., Matthes, F.: Case management: an evaluation of existing approaches for knowledge-intensive processes. In: Reichert, M., Reijers, H.A. (eds.) BPM 2015. LNBIP, vol. 256, pp. 5–16. Springer, Cham (2016). https://doi.org/10.1007/978-3-319-42887-1_1

15. Moody, D.: The "physics" of notations: toward a scientific basis for constructing visual notations in software engineering. IEEE Trans. Software Eng. **35**, 756–779 (2009)
16. NISO: ANSI/NISO Z39.85 (2024). https://www.niso.org/publications/ansiniso-z3985-2012-dublin-core-metadata-element-set. Accessed 8 May 2024
17. NISO: IETF RFC 5013 (2024). https://www.ietf.org/rfc/rfc5013.txt. Accessed 8 May 2024
18. OMG (Object Management Group): Case Management Model and Notation 1.1 Specification (2016)
19. Pańkowska, M.: Process modeling paradigm change. Knowl. Process Manag. **30**, 163–175 (2023). https://doi.org/10.1002/kpm.1749
20. Peffers, K., Tuunanen, T., Rothenberge, M.A.: A design science research methodology for information systems research. J. Manag. Inf. Syst. **24**, 45–77 (2014)
21. Plebani, P., Marrella, A., Mecella, M., Mizmizi, M., Pernici, B.: Multi-party business process resilience by-design: a data-centric perspective. In: Dubois, E., Pohl, K. (eds.) CAiSE 2017. LNCS, vol. 10253, pp. 110–124. Springer, Cham (2017). https://doi.org/10.1007/978-3-319-59536-8_8
22. Riley, J.: Understanding Metadata: What is Metadata, and What is it For? NISO (2017)
23. Routis, I., Bardaki, C., Nikolaidou, M., Dede, G., Anagnostopoulos, D.: Exploring CMMN applicability to knowledge-intensive process modeling: an empirical evaluation by modelers. Knowl. Process Manag. **30**, 33–54 (2023). https://doi.org/10.1002/kpm.1738
24. Routis, I., Nikolaidou, M., Anagnostopoulos, D.: Modeling collaborative processes with CMMN: success or failure? An experience report. In: Gulden, J., Reinhartz-Berger, I., Schmidt, R., Guerreiro, S., Guédria, W., Bera, P. (eds.) BPMDS/EMMSAD -2018. LNBIP, vol. 318, pp. 199–210. Springer, Cham (2018). https://doi.org/10.1007/978-3-319-91704-7_13
25. Seolus: Dublin Core Metadata Generator (2020). https://wordpress.com/plugins/dublin-core-metadata-generator
26. van der Aalst, W., Weske, M., Grünbauer, D.: Case handling: a new paradigm for business process support. Data Knowl. Eng. **53**, 129–162 (2005). https://doi.org/10.1016/j.datak.2004.07.003

Educators Forum

Preface of the Educators Forum

The primary goal of the Educators Forum is to provide a platform for sharing experiences, ideas, and best practices in teaching BPM, as well as inspiring and facilitating discussions among educators involved in BPM education. This forum is crucial for exchanging insights and building a network of substantive support for BPM educators. In today's world, both the quality of course content and the form of its delivery are essential, and the challenges teachers face are similar regardless of their location. The diversity of papers submitted to the Educators Forum highlights the universality of these challenges and demonstrates that BPM teaching is a rich, multidimensional research area worthy of scientific exploration. This year, the forum received twenty-four submissions, of which, after rigorous reviews, twelve were selected for presentation at the Educators Forum and inclusion in the conference proceedings.

The article "Order Selection Methodology of OMG Diagrams in Teaching Designing Information Systems" introduces a methodology for selecting the optimal order of diagrams in information systems design education. It demonstrates that the *max_diff* strategy, focusing on maximizing new element introductions per diagram, achieves the quickest total coverage of elements. This approach aims to enhance students' comprehension and efficiency in system modeling by swiftly exposing them to diverse system components. Practical advice for educators includes starting with diagrams like class diagrams that introduce the most new elements and monitoring student progress to adapt diagram sequencing as required.

The article "Comprehension of (Business) Process Models via Tokens: an Eye-tracking Approach" examines how process models are understood using animated representations and eye-tracking, comparing static and token-animated models. Differences in eye-gazing patterns indicate distinct cognitive processes for animated versus static models, suggesting interactive animations could enhance BPMN education and instructional design. Future research may refine modeler interaction strategies and develop educational multimedia environments (EMMEs) based on animated model interactions to improve BPMN teaching methodologies.

In another study reported in "From Design to Evaluation: Students Experiencing the BPM Lifecycle Hands-on", teachers at the Hasso Plattner Institute describe a seminar offering practical BPM experiences for bachelor's students. The seminar guided students through a complete BPM lifecycle, including process elicitation, Camunda implementation, DMN decision-making, and process analysis with simulation and KPI evaluation. Students gained insights into BPM's practical applications and provided positive feedback, suggesting improvements such as clearer workshop objectives and increased engagement of domain experts during modeling sessions.

The authors of "Beyond Traditional Learning: The LLM Revolution in BPM Education at University" explore integrating Large Language Models (LLMs) in BPM education, highlighting their potential to supplement traditional teaching methods. Conducted at a technical university in Poland with 33 master's students, the study compares LLM-generated responses to human assessments in BPM tasks. While LLMs like ChatGPT-3.5 and ChatGPT-4o show promise in providing immediate feedback and diverse perspectives, they also reveal disparities in reliability compared to human grading. Educators are advised to cautiously integrate LLMs as complementary tools, recognizing their benefits while understanding their limitations in handling complex tasks and nuances in student responses.

The implementation and effectiveness of flipped learning in a Business Process Modelling course at Gdańsk University of Technology in Gdańsk, Poland are discussed in the article "Application of the Flipped Learning Methodology at a Business Process Modelling Course – a Case Study". Flipped learning, popularized since 2012, involves pre-class resources like pdfs and videos, followed by active in-class activities. Students' feedback indicates mixed perceptions, with some acknowledging its benefits while others express concerns about its effectiveness compared to traditional methods. The study concludes with recommendations for refining course design, including adopting blended learning approaches, shorter instructional videos, and integrating quizzes to enhance engagement and learning outcomes.

The article "Enabling BPM Competencies through Real-life Student Consulting Projects" discusses the effectiveness of integrating real-life student consulting projects to enhance BPM competencies based on the experience of teachers from Eastern Switzerland University of Applied Sciences in St. Gallen, Switzerland. These projects provide practical teaching experiences beyond theoretical foundations, benefiting both students and regional companies by applying BPM knowledge directly. The organization's methodology, developed over two decades of collaboration with local companies, emphasizes aligning student learning with the BPM lifecycle and gradually increasing project complexity to prepare students for professional challenges. This approach not only bridges the gap between research and practice in BPM but also delivers tangible, state-of-the-art solutions to client companies.

In the article "Fostering BPM Learning Through PBL: Development and Application of an Educational Framework" the creation and implementation of a Project-Based Learning (PBL) framework for BPM education at a Brazilian institution is presented. This framework integrates dynamic capabilities and holistic diagnosis, providing graduate students with a comprehensive approach to BPM. Implemented through a real project in a cosmetics company, the PBL-BPM framework enhanced students' ability to apply theoretical knowledge practically, develop essential BPM skills, and propose significant process improvements. The study underscores the framework's efficacy in bridging academic theory with industry demands, preparing students effectively for the job market while contributing to BPM education evolution.

The experience with the PBL implementations is presented also in the article "Project-Based Learning in BPM: Lessons from Flipped Classroom vs. Lecture-Based Instruction" based on two editions of a BPM course in a master's program, comparing flipped classroom with lecture-based instruction. Both approaches incorporated PBL where students engaged with real-world processes and stakeholders to enhance process improvement efforts. The study found that PBL combined with real-world applications and stakeholder interactions significantly enriched learning, fostering skills in communication and organizational management. The flipped classroom approach enhanced student engagement in collaborative learning and facilitated targeted feedback for individual comprehension gaps. These findings contribute valuable strategies for BPM educators seeking effective teaching methodologies and highlight the benefits of integrating practical, interactive learning experiences in BPM education.

Innovative teaching methods for business students in digitalization are the focus of the study "Relevance of Topics in Workflow-based Student Projects" which explores process automation using tools like Camunda, AI components, and decision models (DMN). It discusses criteria for selecting automation tasks based on difficulty, topic relevance (e.g., supplier selection, project evaluation), and alignment with DMN and AI, comparing preferences between Vietnamese and Swiss students. The study emphasizes a shift from traditional lecture-based teaching to case study-based and flipped classroom approaches, enhancing student engagement and understanding of business process and project management concepts through real-world applications. Integrating DMN and ChatGPT in coursework reflects industry relevance, highlighting the importance of preparing students for evolving business environments through innovative educational strategies.

The application of the Business Process Design Space (BPD-Space) in teaching explorative business process redesign was studied in the article "Can Business Process Design Space Help Students Generate Innovative Process Redesign Ideas?" Through three experiments involving BPM course students and workshops, including a case study on the Surabaya City Office for Population Administration and Civil Registry (COPACR), participants successfully utilized BPD-Space to generate a variety of innovative redesign ideas. Findings indicate that students with greater experience and domain knowledge excel in producing creative and novel process improvements, underscoring the effectiveness of BPD-Space in fostering innovative thinking among BPM learners.

A novel approach to educating on declarative process models (DPMs), distinguishing them from procedural models through a focus on constraint-based definitions of organizational behavior is presented in the article "Towards an E-Learning Approach for Declarative Process Modeling". By leveraging insights from scientific literature, the paper outlines a structured e-learning framework that progresses through various task categories, including basic concepts, complex constraint combinations, and handling model inconsistencies. This approach aims to facilitate comprehension and proficiency in DPMs among users with diverse backgrounds, paving the way for future interactive applications that automate task generation and evaluation. The ongoing development includes empirical validation within BPM courses and plans for an advanced, gamified e-learning tool to enhance both academic and professional training in declarative process modeling.

To address the intricate challenge of educating students on ERP systems and their associated business processes, such as order-to-cash, "Teaching Enterprise Resource Planning Systems and Business Process Management in Tandem using the Enterprise Online Guide" introduces the Enterprise Online Guide (EOG) as a navigational tool to simplify these systems' complexities. By illustrating the correlation between process models and database tables through EOG, the paper demonstrates how students can gain a comprehensive understanding of ERP systems. The approach highlights the efficacy of EOG in enhancing learning outcomes related to key concepts in business process management.

We express our warm gratitude to everyone who contributed to the success of the Educators Forum 2024. Our thanks go to the authors for their paper submissions, the Program Committee members for their meticulous reviews, the speakers for sharing their experiences, and the audience for their heated and lively discussions. We also thank the BPM 2024 chairs and organizers for their support in preparing the Educators Forum.

Organization

Program Chairs

Katarzyna Gdowska	AGH University of Krakow, Poland
Thomas Grisold	University of St. Gallen, Switzerland
Piotr Sliż	University of Gdańsk, Poland

Program Committee

Iris Beerepoot	Utrecht University, The Netherlands
Matthijs Berkhout	Hogeschool Utrecht, The Netherlands
Andrea Delgado	Universidad de la República, Uruguay
Mahendrawathi ER	Institut Teknologi Sepuluh Nopember, Indonesia
Daniela Grigori	University Paris-Dauphine, France
Marzena Grzesiak	Gdańsk University of Technology, Poland
Jennifer Haase	Humboldt University, Germany
Niels Martin	Hasselt University, Belgium
Fredrik Milani	University of Tartu, Estonia
Jorge Munoz-Gama	Pontificia Universidad Católica de Chile, Chile
Marta Podobińska-Staniec	AGH University of Krakow, Poland
Manuel Resinas	University of Seville, Spain
Tijs Slaats	University of Copenhagen, Denmark
Koen Smit	Utrecht University, The Netherlands
Gergana Vladova	University of Potsdam, Germany
Jan Martijn van der Werf	Utrecht University, The Netherlands
Francesca Zerbato	Eindhoven University of Technology, The Netherlands

Application of the Flipped Learning Methodology at a Business Process Modelling Course – A Case Study

Marzena Grzesiak(✉) and Marek Moszyński

Gdańsk University of Technology, Narutowicza 11/12, 80-230 Gdańsk, Poland
Marzena.Grzesiak@pg.edu.pl

Abstract. Flipped learning has been known for a long time, but its modern use dates back to 2012, with the publication of Bergmann and Saams. In the last decade, it has become an increasingly popular learning method. Every year, the number of publications on implementing flipped learning experiments is growing, just as the amount of research on the effectiveness of this educational method. The aim of the article is to analyze the possibilities of using the flip blended methodology for teaching in the area of BPM. The pilot flipped learning course was implemented as a Business Process Modeling course addressed to Gdańsk University of Technology students and the feedback from the course participants has been collected. Among the pre-class resources, PDF files (containing descriptions and screenshots of the relevant issue) and videos were most often used. The aim of this research was to gather students' opinions on flipped learning and, based on these insights, to design well-founded changes. Many students did not point out the disadvantages of flipped learning. Still, some believe this method may be less effective than the traditional one. Based on the research results, a course redesign will be prepared according to the flipped learning framework.

Keywords: flipped learning · Business Process Modelling · case study · engineering course · study course design · Educators Forum

1 Introduction

Although the concept of flipped learning has received many definitions (e.g. [2,3,16–18]), its basis consists of the following four pillars [7,12]:

- **F**lexible environment – enables adapting the learning space according to the chosen active learning method, flexible management of learning time, as well as creating appropriate objective assessment methods;
- **L**earning culture – use of a learner/student-centered approach; the teacher becomes a guide, not a "sage on the stage" [13]; active engagement allows for exploration of the topic and sustained building of knowledge;

C. Di Ciccio et al. (Eds.): BPM 2024, LNBIP 527, pp. 303–312, 2024.
https://doi.org/10.1007/978-3-031-70445-1_19

- **I**ntentional content – the teacher decides what parts of the content to deliver directly and which ones the pupil/student can work with outside of class; the aim is to maximize the time allocated to active learning;
- **P**rofessional educator – the teacher's role, although they remain as if on the sidelines, is very important—observing their students, providing them with feedback, assessing their work, and deciding when and how to transfer direct instruction from the group to the individual space.

The popularity of this educational concept and its application potential may be indirectly evidenced by the frequency of searches for the terms "flipped learning" and"flipped classroom" on the Internet [11]. They were first noted in 2004. The second wave of interest started in 2011 and has been increasing to this day. The apogee could be seen in the early months of the pandemic (May 2020). The increasing number of academic publications in this area also confirms the growing popularity of flipped learning. An analysis of publications in the Scopus database for the "flipped learning", "flipped classroom" and "inverted classroom" keywords confirms the increased interest in this area, especially after 2012. The number of publications is successively growing, and in the last three years, there have already been more than 500 of them per year (see Fig. 1). Although with some delay, there are also more and more publications on the effectiveness of flipped methods ("effectiveness" has been added to earlier keywords). As a result of a theoretical study based on resources from the Scopus database, a knowledge gap was identified, indicating a lack of publications describing the flipped classroom method in the context of BPM teaching (only 11 publications

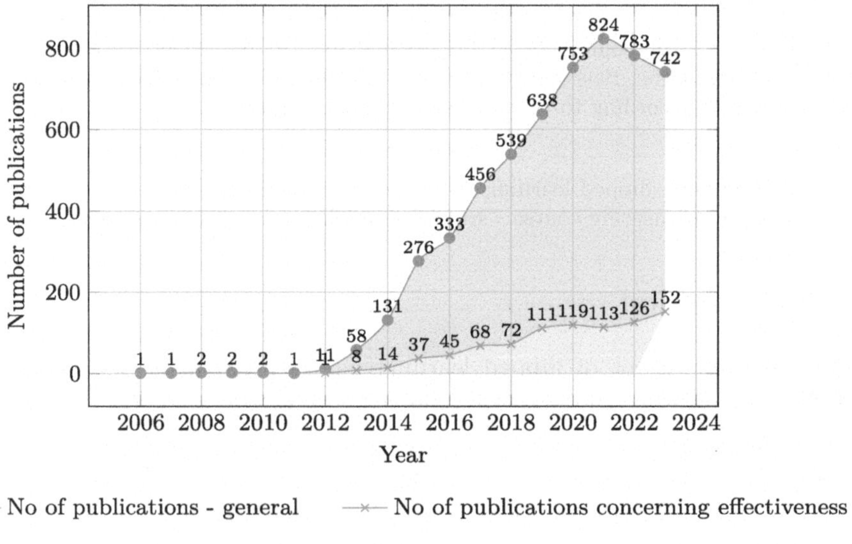

Fig. 1. The number of publications concerning the "flipped learning" or "flipped classroom" or "inverted classroom" (and "effectiveness") - Scopus database, extracted on May 2024.

were found from 2019–2023 regarding flipped methodology and Business Process Management).

When seeking ways to achieve desired learning outcomes and acquire knowledge and skills in business process modelling, leveraging the flipped learning experience is beneficial. Various methods and tools can be employed to design activities conducted before and during class meetings. Before classes, students can engage with videos or texts. For Generation Z and younger students, using methods familiar to them from YouTube or social media will likely be more appealing. During university classes, active learning techniques (e.g., case analysis, teamwork, or problem-solving) can stimulate cognitive processes and enhance understanding of the material. The inclusion of gamification also yields positive results. Active techniques support achieving higher learning goals as defined in Bloom's taxonomy [1]. The effectiveness of active methods, including flipped learning, in engineering education is supported by research on their impact [16] and especially in BPM courses e.g. [5]. However, it must be emphasised that designing practical classes and preparing materials using flipped learning requires significant effort from the teacher. External support, such as methodological training or insights from other educators' experiences, is often necessary [4,15]. Applying management methods based on the process life cycle [6] in the educational process is also valuable. Therefore, developing and implementing a flipped course and continuously monitoring and improving it based on personal observations and student feedback is crucial. The paper is structured as follows: Sect. 2 presents the pilot implementation of flipped learning approach for BPM course, Sect. 3 discusses analysis of research outputs and Sect. 4 reports conclusions observed.

2 Pilot Flipped Learning Implementation

The methods used in the educational process described below were not initially called flipped learning.

Many of the methods and tools were naturally used in activities conducted with students. Studying is a time for actively pursuing knowledge and skills; it is a time for establishing solid social relationships that often last for years. Nevertheless, an attempt was made to describe and evaluate the way the Business Process Modelling course was implemented within the framework of "a pilot flipped learning implementation".

2.1 Case Study: Flipped Learning at the Business Process Modelling Course

<u>What is the content of the course? What is it for?</u>

The Business Process Modelling course in question is dedicated to the Gdańsk University of Technology students. This course takes place during the fourth semester of the undergraduate Data Engineering studies. The information about the field of study [9] and the list of subject with the syllabus [10] are available at the ECTS catalog of Gdańsk University of Technology.

The level of qualification according to the European and national qualifications frameworks is full qualification at the 6th level of the Polish Qualifications Framework.

The Business Process Modelling aims are gaining knowledge and skills in the field of process modelling in the organization, acquiring the ability to use IT tools for process modelling, as well as getting acquainted with the BPMN notation. The learning outcomes are also included in the syllabus [10]

The subject learning outcomes are correlated to the field of study learning outcomes.

The main topics of the lectures included the definitions related to the area of business processes modelling, the evolution of BPM, changing the orientation in the organizations from functions to processes, the business process life cycle, the identification and classification of processes, process mapping, modelling and simulation tools, process metrics, as well as the fundamentals of the Business Process Model and Notation. The laboratories are strictly practical. Within a step-by-step approach, the students perform the exercises in order to get acquainted with the relevant software. The exercises strengthen the skills related to defining the parameters of models and diagrams, running simulations, analysing reports, building the scenarios and proposing process improvements. The development of individual projects takes 40% of the labs time. The course precedes the Business Process Analysis and Optimization course realized on 5th semester for the Big Data in Management stream.

Each year this course is attended by approximately by 50 students.

<u>What was the rationale for implementing the flipped learning approach?</u>

When teaching similar subjects (e.g., Process Management, Business Process Modelling, or Simulation Process Modelling) during different courses, it was easy to miss or forget some parts of the content when discussing the same issues. The first video materials were developed eight years ago—these were recordings of the computer screen with a commentary on the stage of a task being carried out, e.g. drawing a process diagram or recording a lecture for a prepared presentation. The recordings were also helpful in the case of teacher illness/dismissal, when it was impossible to organize a substitute or make up the class at another time. Over time, more materials were developed, and the pandemic further accelerated the refinement and development of new materials and the implementation of new methods and tools. Sometimes these were methods and tools that had been known before but found a new application. Gradually, the auxiliary materials were created and integrated more thoughtfully and systemically into the educational process. The Business Process Modelling field of study was implemented five years ago (the first Data Engineering course was organised in 2018). It was natural to incorporate methods implemented in other subjects.

The course consists of 15 h of lectures and 30 h of laboratory. The lecture is carried out in the first half of the semester (before students start to carry out an individual project in the laboratory):

- first lecture in stationary/synchronous form—introduction to the subject; other lectures in the form of e-learning (asynchronous; lessons on the Moodle platform and video recordings)
- laboratories (synchronous; at the laboratory), in the remote form as an exception.

Laboratories consist of three parts:

1. (a) introductory exercises allowing the student to become familiar with the process modelling environment (creating a process diagram, defining parameters, simulation, basic analysis); the exercise is prepared in the form of a file with commands and screenshots that allow the students to complete the tasks at their own pace;(b) a recording explaining what has been done in the exercise, what is important, and where mistakes are most likely to occur.
2. seven exercises to be carried out in the following weeks of classes. Each exercise allows the students to define specific parameters of the model and/or diagram and in each subsequent exercise, skills from the previous exercise are consolidated and new ones are introduced;
3. implementation of an individual project - students are assigned a project topic - a description of the process for which they create a diagram representing its flow, as well as parameters necessary to carry out the simulation, analyse the report results, create a new scenario, generate a report for it and compare and comment on the reported results.

What is the role of the students?

Before the in-lab classes students obtain the above-mentioned input with which they will work with during the subsequent classes in the laboratory. The e-learning lessons contain a literature review for a given topic and end with review questions that can be answered multiple times—the students can return to the correct part of the lesson if necessary or if there are questions that need to be answered. Video recordings, on the other hand, show practical issues that will be practiced in the laboratory.

During the laboratory, the students will learn about the tool used for modelling and simulating processes by carrying out the subsequent exercises. During classes, the students can use descriptive materials containing information of similar scope as in video materials, e.g. presentations. Performing a project in the last part of the semester allows them to verify and complete the previously acquired skills.

What is the role of the teacher?

The task to be performed is explained at the beginning of the class. The key issues are highlighted in the form of a note on the board. It is also a time for the students to ask questions if something is unclear to them. During the implementation of student exercises and projects, the teacher monitors their progress. In case any problems appear, the teacher indicates the right way to solve them or the resources where hints can be found. The students receive feedback and points for each completed exercise and project.

3 Research Methodology and Analysis of Outputs

The publications devoted to flipped learning motivated an attempt to revise the methodological correctness and examine students' opinions on the methods and tools used in the educational process. The collected data will be used to develop guidelines for the course re-design. The following two research methods were used:

1. Opinion polls were conducted in May 2023 using a Google forms. The form was completed by 33 out of 51 students. Although the course is ongoing, it is so advanced that students can already evaluate the methods, tools, and materials used during the classes.
2. Document analysis method—determining the correctness of the formulation of learning objectives and learning outcomes based on the Bloom's taxonomy [1].

Some of the research results are presented below as regards the learning outcomes, allowing for a starting point for the redesign of the course.

Students assessed the preferences for the different forms of auxiliary materials used to prepare for the classes within pre-class activities (see Fig. 2).

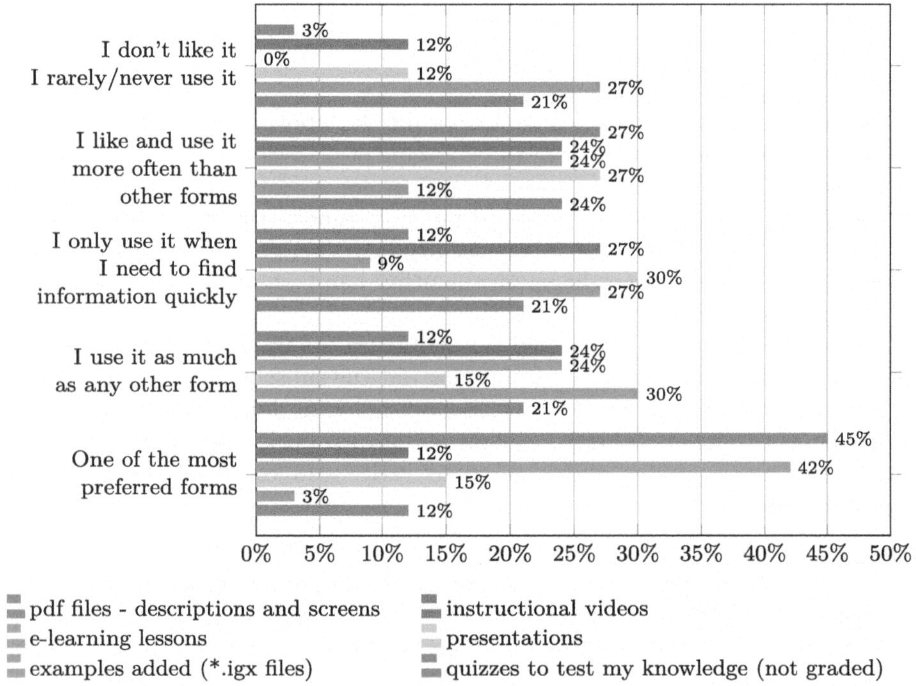

Fig. 2. Students' preferences for using the auxiliary materials.

The analysis of the research results shows that the preferred forms of materials are descriptive files with screenshots (45%) and instructional videos (42%). In turn, e-learning lessons, although indicated as disliked (27%), were used by 30% of students, so as often as other forms. The frequency of using various forms of materials increases during and after classes (see Fig. 2).

Students provided their opinions concerning the pros and cons of flipped learning. 45% of them did not point out any disadvantages. As the students indicated limited time as one of the barriers to using the auxiliary materials (30%), they were asked about the length of the lecture video and the instructional video.

As far as the lecture is concerned, 55% of the students pointed out that it should be no longer than 30 min, and 33% of them suggested that 15 min is the maximum desired length. The instructional video's length should be from 11 to 20 min. Some students prefer the shorter videos (18% prefer 5–10 minutes) or the longest ones (12% accept 21–30 min). Other reference values can be found in the literature (e.g. [2]), but the investigated students were able to connect this parameter with the topics learned. The above-mentioned lengths are acceptable for the students.

Another disadvantage, pointed out by 18% of students, was the need for face-to-face contact with the teacher. Nevertheless, they did not use the available consultation hours (79% of students have yet to attend them).

The advantages of flipped learning specified by the students included the possibility of using the auxiliary materials multiple times (82% of answers), their availability at any time (82% of answers), or the ability to learn at their own pace (55% of answers).

From the point of view of achieving the subject objectives and learning goals, the students' opinion regarding attaining these objectives in pre-class and in-class activities is essential. Most students (75%) declared that watching video lessons before the class has allowed them to achieve the learning objectives.

According to the results presented in Table 1 and 2, it could be stated that the aim of the subject (given in the previous chapter) and learning outcomes were achieved in cognitive and behavioural domains. However, the analysis of them, according to Bloom's taxonomy and its revision [1,8,14], points to the need to modify their formulation.

Table 1. Students' opinions concerning linking the pre-lab activities with the learning outcomes.

Pre-lab activities in this course (e.g. reading, lecture videos, quizzes etc.)	I strongly disagree	I do not agree	I neither agree nor disagree	I agree	I strongly agree
Were helpful to my learning	6%	3%	15%	58%	18%
Motivated me to learn more	3%	24%	33%	27%	12%
Enabled me to learn at my own pace	3%	9%	18%	33%	36%
Prepared me for in-lab activities	3%	3%	24%	39%	30%

Table 2. Students' opinions concerning linking the in-lab activities with the learning outcomes.

The in-lab sessions helped me:	I strongly disagree	I do not agree	I neither agree nor disagree	I agree	I strongly agree
Clarify what I had learned in pre-class activities	3%	3%	21%	42%	30%
Apply what I had learned in pre-class activities	6%	0%	18%	42%	33%
Develop problem-solving skills	0%	0%	18%	48%	30%
Improve my group work skills	9%	21%	12%	33%	24%
Develop better learning and study skills	9%	6%	39%	24%	21%
Improve my communication skills	6%	21%	33%	15%	24%

The aim of the subject was reformulated into (1) gaining knowledge and skills in the field of process modelling in the organisation and (2) using IT tools for modelling in the BPMN notation.

The proposal of the learning course's outcomes fitted to the Blooms' taxonomy levels and linked to the field of study learning outcomes is presented in Table 3.

Table 3. The new Business Process Modelling learning outcomes.

The learning outcome of the course	Bloom' taxonomy level	Realized field of study's learning outcome *
Applies the selected notation to create a business process model.	1	[K6_W08]
Interprets and compares observed process measures.	2 and 4	[K6_K03]
Identifies and classifies processes in the organization.	2	[K6_W08]
Justifies the selection of business process parameters.	4	[K6_K03]
Creates a business process model using IT tools.	3	[K6_U03]
Tests possible scenarios of the process implementation.	5	[K6_K03]
Distinguishes the elements of the model in the given notation.	1	[K6_W08]

* The internal Gdańsk Tech coding of the learning outcomes for the cognitive, behavioural, and competencies domains as used in the table: K6_U03 - analyses problems and creates appropriate models, data structures and algorithms (including heuristic and numerical ones), assesses their computational complexity, estimates errors of the received solutions; K6_K06 - knows how to cooperate or work in a project team and take managerial or executive functions; K6_W08 - knows the models and structure of the data mining process and their multidimensional analysis and can assess the results of such analyses.

4 Conclusions

The analysis of publications on flipped learning referred to in the introduction, the review of the didactic process implemented for Business Process Modelling, as well as students' opinions made it possible to formulate general conclusions and recommendations regarding improvement. Flipped learning is a good direction for implementing the educational process for the described course. In addition, the students also confirmed earlier observations regarding pre-class and in-class activities.

The starting point for a good flipped learning class design consists in the correct formulation of the subject objectives and learning outcomes. It is also worth preparing class scenarios, choosing appropriate forms of pre-class materials (e.g., videos, documents, podcasts), as well as active learning methods (e.g., brainstorming, problem solving, cases, teamwork). Scenarios can also be helpful in the case of other expectations of students in subsequent years. Equally important are the ways of verifying the achievement of learning outcomes.

Examples of planned changes in the implementation of the subject include:

- change of lecture method - instead of e-learning, blended learning (e.g., pre-class: video or scientific paper, in-class: group work or discussion);
- recording new instructional videos—dividing the material into smaller parts and shortening the videos;
- including quizzes/tests in pre-class activities (students will be stronger motivated to perform the pre-class activities; conversely, it could help test the students' knowledge).

Although students prefer voluntary pre-class activities, some incentives can effectively mobilize them to perform these, e.g. the possibility of obtaining a credit waiver, or the opportunity to earn extra points at an admission test.

Business Process Modelling is part of Business Process Management. Business Process Management educators could adopt experiences gained through the analysed education process. The main limitation of the presented study is the single group of students. Despite the small number of students surveyed, it is worth noting that such studies have not been conducted previously. These limitations also outline the directions for further research related to the continuation of work with the flipped classroom in BPM courses, as well as at other universities (covering various fields and specializations).

References

1. Anderson, L.W., et al.: A Taxonomy for Learning, Teaching, and Assessing : A Revision of Bloom's Taxonomy of Educational Objectives. Longman, New York (2001)
2. Bergmann, J., Saams, A.: Flip Your Classroom: Reach Every Student in Every Class Every Day. International Society for Technology in Education, Washington, D.C. (2012)

3. Brame, C.: Flipping the classroom. Vanderbilt University Center for Teaching (2013). http://cft.vanderbilt.edu/guides--sub--pages/flipping--the--classroom/. http://cft.vanderbilt.edu/guides-sub-pages/flipping-the-classroom/. Accessed 19 May 2023
4. Brewer, R., Movahedazarhouligh, S.: Successful stories and conflicts: a literature review on the effectiveness of flipped learning in higher education. J. Comput. Assist. Learn. **34**(4), 409–416 (2018). https://doi.org/10.1111/jcal.12250. https://onlinelibrary.wiley.com/doi/abs/10.1111/jcal.12250
5. Chow, W.: Teaching business process management with a flipped-classroom and problem-based learning approach with the use of apromore and other bpm software in graduate information systems courses. In: 2021 IEEE International Conference on Engineering, Technology & Education (TALE), pp. 1–8 (2021). https://doi.org/10.1109/TALE52509.2021.9678885
6. Dumas, M., Rosa, M., Mendling, J., Reijers, H.: Fundamentals of Business Process Management. Springer, Heidelberg (2018). https://doi.org/10.1007/978-3-662-56509-4
7. Flipped Learning Network (FLN): The four pillars of f-l-i-p^{TM} (2014). https://flippedlearning.org/definition-of-flipped-learning/
8. Forehand, M.: Bloom's taxonomy. Emerg. Perspect. Learn. Teach. Technol. **41**(4), 47–56 (2010)
9. Gdańsk University of Technology: ECTS Information Catalog. Data Engineering - subject list. https://ects.pg.edu.pl/en/courses/14508. Accessed 20 May 2023
10. Gdańsk University of Technology: ECTS Information Catalog. Data Engineering - sylabus. https://ects.pg.edu.pl/en/courses/14905/subjects. Accessed 20 May 2023
11. Google: Google trends 2023 (2023). https://trends.google.pl/trends/explore?date=2004-01-01%202023-05-22&q="flipped%20learning","flipped%20classroom". Accessed 20 May 2023
12. Hamdan, N., McKnight, P., McKnight, K., Arfstrom, K.M.: A review of flipped learning: flipped learning network. J. Comput. Assist. Learn. (2013). http://www.flippedlearning.org/cms/lib07/VA01923112/Centricity/Domain/41/LitReview_FlippedLearning.pdf
13. King, A.: From sage on the stage to guide on the side. Coll. Teach. **41**(1), 30–35 (1993). http://www.jstor.org/stable/27558571
14. Krathwohl, D.R.: A revision of bloom's taxonomy: an overview. Theory Into Pract. **41**(4), 212–218 (2002). https://doi.org/10.1207/s15430421tip4104_2
15. Long, T., Cummins, J., Waugh, M.: Use of the flipped classroom instructional model in higher education: instructors' perspectives. J. Comput. High. Educ. **29**(2), 179–200 (2017). https://doi.org/10.1007/s12528-016-9119-8
16. Prevalla, B., Uzunboylu, H.: Flipped learning in engineering education. TEM J. **8**(2), 656–661 (2019). https://doi.org/10.18421/TEM82-46
17. Talbert, R.: Flipped Learning: A Guide for Higher Education Faculty. Stylus Publishing, LLC, Washington, D.C. (2017)
18. Tse, W.S., Choi, L.Y.A., Tang, W.S.: Effects of video-based flipped class instruction on subject reading motivation. Br. J. Edu. Technol. **50**(1), 385–398 (2019). https://doi.org/10.1111/bjet.12569. https://bera-journals.onlinelibrary.wiley.com/doi/abs/10.1111/bjet.12569

Project-Based Learning in BPM: Lessons from Flipped Classroom vs. Lecture-Based Instruction

Kate Revoredo(✉) and Saimir Bala

Humboldt-Universität zu Berlin, Berlin, Germany
{kate.revoredo,saimir.bala}@hu-berlin.de

Abstract. This paper reports the experience and lessons learned from the teaching of two editions of the business process management (BPM) course in a masters program, following two different designs based on active-learning. In one course we used flipped classroom and Project-Based Learning (PBL) whereas in the second course we adopted lecture-based instruction with in-class assessments and PBL. In both courses the goal of the project was to improve a real-world process, which included various interactions with the stakeholders for elicitation, insights, redesign and validation of the to-be process. We observed that PBL applied to real-world processes and interaction with process stakeholders had a number of benefits on learning. Students could learn from the daily experience of the process participants, realize the complexity of BPM efforts in real world, and increase their communication and organization soft-skills. When flipped-classroom was applied, we observed more engagement in active and collaborative learning. From a teacher's point of view, flipped-classroom enabled us to better assess individual performance, perceive which BPM concepts were not fully understood, and provide objective feedback according to the specific needs. Our lessons learned serve to enrich the BPM educators community with effective teaching strategies and valuable insights.

Keywords: project-based learning · flipped classroom · active learning · business process management

1 Introduction

Business Process Management (BPM) [8] is a dynamic field that necessitates not only theoretical understanding but also practical application and interaction with real-world processes and stakeholders. Effective teaching strategies are essential to equip students with the skills and knowledge required to navigate the complexities of a BPM initiative and prepare them for the job market [12]. The design of a BPM course that combines theoretical and practical elements allows students to acquire not only conceptual knowledge on BPM topics but also develop skills that are highly valued by the industry [11]. Therefore, it is desirable that the course includes learning tasks that promote the development

C. Di Ciccio et al. (Eds.): BPM 2024, LNBIP 527, pp. 313–323, 2024.
https://doi.org/10.1007/978-3-031-70445-1_20

of soft skills such as communication and presentation. Indeed, it is argued that many business school undergraduates lack these skills [5].

Active learning [2] methodologies, such as the flipped classroom [10] and PBL [9] have been widely used in higher education given their potential to enhance student engagement, comprehension, and retention. The flipped classroom approach, which inverts the traditional teaching model by delivering instructional content outside of class and using in-class time for interactive activities, promotes a more student-centered learning environment. Conversely, PBL immerses students in real-world problems, providing them with a natural and valuable experience on the topics [13] while fostering critical thinking, collaboration, and practical problem-solving skills. For teaching BPM, these methodologies have already been considered [6,7], but not yet in a combination that includes the involvement of process participants and stakeholders.

This paper reports on the *experiences* and *lessons learned* from implementing two distinct designs for a BPM course in a master's program. The first design employed a combination of flipped classroom and PBL, while the second design utilized lecture-based instruction complemented with in-class assessments and PBL. In both course designs, students undertook projects aimed at improving real-world processes, involving significant interaction with stakeholders for elicitation, insights, redesign, and validation.

Through this comparative study, we aim to highlight the benefits and challenges of each teaching approach, providing valuable insights for educators in the BPM community. Our findings suggest that while both approaches offer unique advantages, the integration of real-world PBL and stakeholder interaction significantly enhances the learning experience. Additionally, the flipped classroom model appears to further engage students in active and collaborative learning, offering substantial benefits for both learners and instructors. By sharing our experiences and the lessons learned, we hope to contribute to the ongoing discourse on effective BPM education, offering practical strategies that can be adapted and applied by educators to enhance their teaching methodologies and improve student learning.

The rest of the paper is structured as follows. In Sect. 2 we describe the two designs we used for the BPM course. In Sect. 3 we discuss the outcomes and lessons learned. Finally, in Sect. 4 we conclude our work and provide some future considerations.

2 Courses Design

This experience-report concerns two editions of the BPM course taught at the master program at an Austrian university. We refer to them respectively as *Course Variant 1* and *Course Variant 2*. BPM is a 4 ECTS course with 22.5 h time in class, divided in five classes of 4h each and a sixth class of 2 h:30 m, usually reserved for the final exam. The two courses were designed for in-presence teaching and examination. However, due to a COVID-19 wave the classes and the exam of *Course Variant 1* were moved to online.

The learning objectives for both variants was to learn the fundamentals of BPM, with a focus on how to conduct a BPM initiative in an organization.

The students had a background knowledge on business related subjects and a tendency to look at problems with an analytical perspective. In the following, we describe the two course variants in detail.

2.1 Course Variant 1

Course Variant 1 was taught in the winter semester 2021–2022 for 61 students. Table 1 depicts the topics of the syllabus and how they were distributed over the classes (columns 1–2).

Table 1. Distribution of topics per class and assessments for Course Variant 1

Class	Agenda	Assessment			
		Quiz	**BPM Case**	**Project**	**Exam**
1	Introduction to BPM Process Identification Process Modeling (BPMN)				
2	Process Modeling (Advanced BPMN) Process Discovery				
3	Process Analysis Process Redesign				
4	Process Implementation Process Monitoring				
5	Project Presentation				
6	Exam				
Grade		20%		40%	40%

The methodology applied was *Lecture-Based Instruction* with *In-Class Assessments*. The material used for the lecture instructions were based on the support material for the book of Dumas et al. [8][1]. The *In-Class Assessments* were made by means of (self-test) Quizzes, which were provided in the beginning of each lecture. The goal was to recap the content of the previous lecture, allowing the students to identify learning gaps, thereby motivating them to engage in discussion afterwards to address these learning gaps. The quizzes were done individually and with a limited time for each question. The time was stipulated with the assumption that the students had studied before and therefore would know the answer immediately after finishing reading the question. It was not possible for the student to search for the answer while taking the quizzes.

Besides the in-class assessments, two other tasks were given: a *BPM Case* and a group *Project*. The goal of these tasks was to explore the practicability of BPM. As first task, the students were asked to analyze a real application scenario of BPM. For that the students were assigned to work on a case from the BPM Cases books [3,4]. This was an individual task, where the student had one month to deliver the review. The deadline was before the presentation of the project. Such schedule was designed in with the aim that students could transfer the knowledge acquired by analyzing the BPM Case directly to their group project. In particular,

[1] http://fundamentals-of-bpm.org .

the students were asked to (i) perform a general evaluation of the work, including the identification of which BPM lifecycle phases were covered and which techniques were applied in each of the phases, (ii) provide a critical discussion of pros and cons of the case, and (iii) recommendations for future research and possible alternatives for solving the BPM problem addressed in the case. As second task, the students were asked to develop a *Project* with the goal to improve a real-world process. For that, the students were teamed-up in groups of 4 to 5. Each group decided to work on a process from the back- and front-end administration offices of the university. In the beginning, they only had a hint at the situation faced in the domain, the related participants and stakeholders but no clearly defined process was given (i.e., no clear process boundaries, no defined activities, decision points and actions). Both teaming-up and selecting the target process were done in the beginning of the first class. To facilitate their collaboration, the students were asked to create teams immediately after the presentation of the course design. A doodle was used to allow student teams to include their name in one of the offered real-processes. Students that were not present in the moment of the selection, were instructed to contact their peers at a later stage and join an existing team that had not yet reached the maximum capacity. The output of this task was a project presentation and a report. The presentations were held in the last class before the exam. The report, incorporating feedback received during the presentation, was due two weeks after. Following a PBL methodology, the students were stimulated to conduct teamwork during the whole course as a way to practice the concepts presented in class. An overview of the alignment of the topics presented in each class with the requirements of the project were presented to the students in the first class for their organization. Figure 1 depicts the dynamic of a unit period. A unit encompass a set of BPM topics. For example the second unit corresponds to the topics taught on class 2.

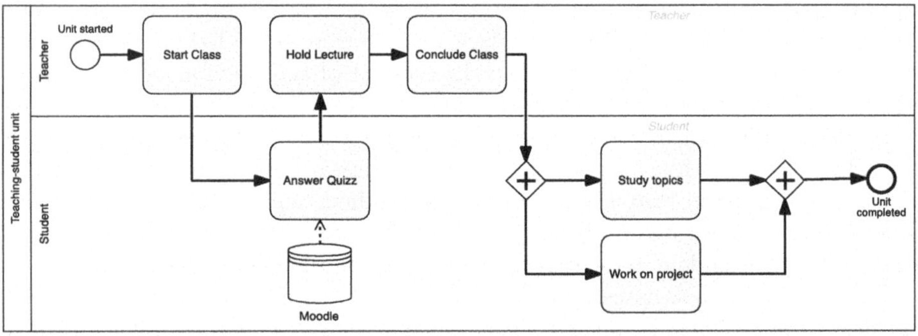

Fig. 1. Dynamic of a unit of the Course Variant 1

For the final grade, three components were considered. Assignments which included the quizzes and the analysis of the BPM case accounting for 20% of the grade. The project development and the exam accounted for 40% of the grade each. Table 1 summarizes the distribution of the points for the final grade (columns 3–6). Also, the period when each of the assignments took place is illustrated with the blue color.

2.2 Course Variant 2

Course Variant 2 was taught in the summer semester of 2022 for 6 students. *Flipped classroom* [1] with a PBL approach were used. With the flipped classroom method, students were expected to study the teaching material outside of the class hours and use class time for discussions and clarifications. The suggested material for the self study comprised the book [8] and the further material on the website[2]. The classroom sections were used to discuss the progress of the project and to reflect on the experience of applying the BPM concepts to the real-world scenario. Table 2 describes the agenda concerning the topics of the syllabus.

Table 2. Distribution of topics per class and assessments for Course Variant 2

Class	Agenda	Assessment		
		Progress Report	Formal Feedback	Final Report
1	Course Introduction Projects Organization BPM Overview			
2	Process Identification	Process profile		
3	Process Discovery Process Modeling	BPMN as-is model	Report on the model	
4	Process Analysis	Results on quantitative and qualitative analysis	Report on the analysis	
5	Process Redesign Process Monitoring	BPMN to-be model	Report on the model	
6	Project presentation			Presentation Report
Grade		20%	30%	50%

The project was developed in groups of two students. The groups acted as consultants who conducted a BPM initiative with the aim of improving the process. The groups were created in the first lecture. Each group selected to work on a specific process from the front or back-end administrative offices of the university (the processes were different from the *Course Variant 1*). For each process, a domain expert, who was also the customer of the process, was interviewed by the students. The groups had to interact with the domain experts in different stages including when presenting their redesign proposal. The domain experts could attend the final presentation and provided feedback about the project and the soft skills they observed.

The development of the project included four stages (rows 2–5 in Table 2) according to the BPM phases of identification, design, analysis, and redesign.

[2] cf. footnote 1.

Each stage had an expected outcome: a report on the weekly development of the topic and the main artifact of the corresponding BPM phase. The report and the generated artifact would be used as input for the next stage of the project. Rows 2–5 in Table 2 indicate the artifact generated.

The first stage covered the introduction to BPM and Process Identification, with the process profile being the main output artifact. The second stage covered Process Discovery and Process Modeling, with the as-is process model in BPMN being the output artifact. Process Analysis, including quantitative and qualitative analysis, was covered by the third stage. The output was a report describing the issues identified and a justification for the chosen process analysis techniques.

The last stage covered Process Redesign, where the output was the to-be process model in BPMN and a report describing the redesign technique used with a justification, as well as the feedback received from stakeholders. The artifacts generated as output of each stage were reviewed by another team and by the lecturer (formal feedback). Informal feedback was also provided by colleagues from other groups and the lecturer during the class sessions. An exchange of experiences among the groups was also stimulated, which contributed to learning. Figure 2 depicts the dynamic of each stage of the project.

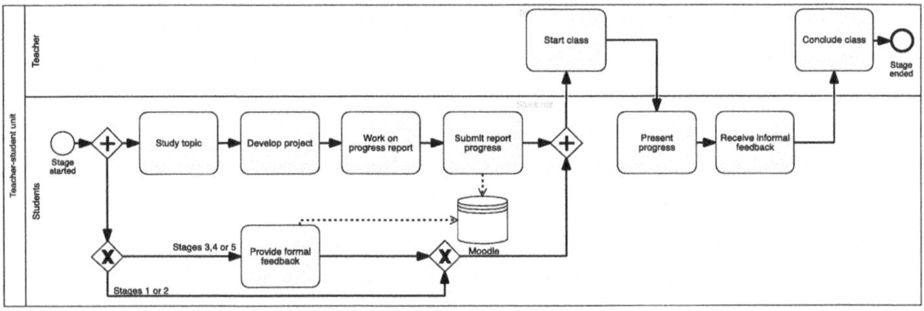

Fig. 2. Dynamic of a stage of the development of the project

For the students, each stage started with the studying of the corresponding topics, followed by the development of the project, and preparing the progress report. The stage ended with a class section where each team presented the results of that stage, commented on the review received (when applicable) and raised questions for discussions. At this moment, the teacher moderated the discussions, enriched and complemented them with points not mentioned by the students. Once all the topics were covered, the students were asked to finalize their projects and present them. After the presentation, they were given one week time to improve the final written document (i.e., report) incorporating the feedback received in the final presentation.

The course assessment was based on partial and final results of the project. Four components were used. The first was based on the *progress* of the project which was measured at each class based on the outputs of the stages and accounting for 20% of the grade. The *formal feedback* given to their peers' partial reports accounted for 30% of the grade (10% each time they provided a peer review, up to 3 times). The final *presentation* and the final *report* accounted for 10% and 40% of the grade.

Students were encouraged to continuously reflect on their learning and the processes used to complete each assignment. During the classes, we asked questions promoting reflection on what they had learned, what worked well, and what they might do differently next time. Students received feedback on their work and were encouraged to use the feedback to improve their work.

3 Outcome and Lessons Learned

In this section, we informally describe the impressions collected from students and our own observations of the methodologies applied. It is not our intention to decide on one approach over the other. The goal is to share our experiences, discuss the benefits and challenges of each teaching approach used, and to encourage BPM educators to try and improve them.

After having adopted PBL with different levels of engagement in the two BPM courses, it is possible to state that it does support the teaching of BPM. Next we present some lessons learned from the use of PBL.

Lessons Learned 1: *PBL improves the learning experience.* When asked about PBL, teams from both course variants acknowledged a better learning experience. The *Course Variant 2* showed more enthusiasm on the experience, adding that having feedback and discussion during class, after exploring the topics on their own, allowed them to increase their understanding on the topic. From the perspective of the lecturer, students from *Course Variant 2* showed more enthusiasm about the topics, and they engaged more in the activities. The comments and questions during the discussions showed more interest on the topics and engagement with the domain, providing quality suggestions for improvement of the process. Different then the students from *Course Variant 1*, where the reports were not providing a deeper analysis of the processes. In some occasions it seemed that the students were only doing the comments or questions to collect the points associated to the assignments and therefore to pass in the course (i.e., driven by points and not by knowledge).

Lesson Learned 2: *partial assignments with grade for the project development increase the engagement of students.* For the *Course Variant 1*, although the students were presented in the first day of the course with the alignment of the topics of each upcoming lecture and the requirement of the project and stimulated to conduct the project alongside the course as a way to practice the subject's concepts, the groups left the whole development of the project to the end of the course after the presentation of all the topics. As a consequence, there was not enough time for them to follow a proper BPM initiative. For example, only a few groups validated the as-is model with the process participant and none of the groups presented to the process participant their suggestion for improvement. Some of the students acknowledged that they underestimated the time required to develop the project. *Course Variant 2* on the other hand, made the students deliver intermediate artifacts that could be graded. Only one group did not do the first assignment (reasons for that discussed on *Lessons Learned 6*). All the other students acknowledged the benefit for the final quality of the project to have intermediate results, feedback and immediately grades. Also, the dependency among the artifacts generated in each stage of the project required the students to master all the concepts taught. For example, good knowledge of the as-is process model was necessary to generate the quantitative and qualitative analysis. In turn, good understanding of analyses methods was necessary to justify the redesign proposed and generate a to-be model.

Lessons Learned 3: *feedback on partial results of the project improves the quality of the outcome.* While the teams on *Course Variant 2* were receiving informal and formal feedback weekly from colleagues and the lecturer, teams from *Course Variant 1* only received it at the end when presenting the final results and mainly from the lecturer. Also, the feedback was not exhaustive, given time restrictions. Concerning the topics, one point to notice is that the modeling of project's scenario needs moderation from the lecturer to lead the student to use more advance BPMN elements. Students tend to focus only on basic elements of BPMN, which in some cases were not enough to fully represent the scenario. A deep feedback in this regard was only possible for *Course Variant 2* on the specific class about the process modeling. A suggestion for future editions where project-based is used in a similar way to *Course Variant 1* is to have a discussion with feedback session specifically for the modeling task.

Lesson Learned 4: *interactions with stockholders develop practical knowledge and soft skills.* The students from *Course Variant 2* reported that interacting with the process participants (stakeholders) was unexpectedly hard. That they had to develop ways of interviewing, including strategies to keep the participant focused on the process at hand without discussing issues in other processes. They also acknowledged that process identification, while apparently easy in theory, turned out challenging in practice, especially what concerns the identification of initial and final activities. They experienced in practice the connection among different processes and the importance of setting the scope of the analyzed process. During the discussion sections the teams were also able to identify connections between their processes and make joint decision of which process should cover certain identified aspects and therefore which group was responsible for detailing the information. The discussions also lead to the experience of different levels of abstraction of processes, with one group being responsible to detailing the process in a lower level while the other group would model the same process in a higher level, given that it was secondary to them. Another benefit of the involvement of process participant concerns the redesign phase. In this phase, the student, as the BPM analysts, tend to propose creative solutions which in many situations are not suitable and sometimes even unfeasible for the client. Having the validation of their proposal with the process participants allowed them to revise their proposal and find a solution that it was feasible and acceptable to the client.

Concerning the flipped-classroom, we observe that it is a suitable methodology, especially if combined with practical collaborative activities such as the development of a project. When the class has less students, the use of flipped-class room as designed in our course is more feasible. The *Course Variant 2* having only six students allowed the lecturer to more comprehensively assess the level of understanding of the topics and engagement of the students. Students participated actively in the sessions, contributing with comments and suggestions on the other team projects. In every session, they were asked to share not only the content of the topic of the week, but also their struggles on the corresponding stage, and lessons learned. For example, in the section about process identification one of the group acknowledged that the meeting with the *client* for the identification of the process took longer than expected and required effort to keep the focus. Hence, they needed to structure better the upcoming meetings and improve their communication skills. With more students, a different dynamic will need to take place. For example, it would not be possible to have all the groups present their project progress in every class. A lesson learned concerning the use of flipped-class room are the following.

Lesson Learned 5: *it takes time for the students to engage with a different course setting.* The students from one group in *Course Variant 2* reported that the beginning of the course was hard for them. It took them long to change their mindset towards active learning, as opposed to passive learning and the usual course setting. They also mentioned that one aspect of active learning is that they should be available during the whole period of the course, studying and generating artifacts guided by the task related to the correspondent stage. Conversely, in passive learning they were used to just have to be available mainly in the end for the grading assignments. Therefore, they also had to work on their agendas to make it possible to fully attend the course. For the *Course Variant 1*, the students reported that it was challenging to finish the quizzes in the short period of time provided. At first, they thought they would be able to catch up on the topic while doing the quizz. They evaluated their performance on the first quizz as not satisfactory given the lack of time to finish properly. They acknowledged that after the first quizz when they noticed the short period of time provided to do the quizz, they prepared for each class in advance, studying the concepts presented in the previous class. A suggestion for future editions is to grade less the first assignments or have a possibility to disregard the lower grade understanding that the students need time to adjust to the new design.

4 Conclusion

In this paper we reported on the experience of teaching BPM using two different courses design. In the first one we combined lecture-based instruction with in-class assessments while in the second flipped classroom was used. In both designs PBL was used where the students worked as consultants to improve a real-process having interactions with process participants.

Our experience showed that given the nature of the BPM discipline, PBL combined with a real-process and the interaction with practitioners promotes a better learning and it is feasible to be implemented. We describe 5 lessons learned from our experience and feedback from the students. As future work, we want to explore a similar design as *Course Variant 2* focusing on process implementation and monitoring.

Acknowledgements. Kate Revoredo is funded by the Berliner Chancengleichheitsprogramm (BCP) as part of the DiGiTal Graduate Program and Saimir Bala is supported by the Einstein Foundation Berlin under grant number EPP-2019-524.

References

1. Bishop, J., Verleger, M.A.: The flipped classroom: a survey of the research. In: ASEE Annual Conference & Exposition 2013, pp. 23–1200 (2013)
2. Bonwell, C.C., Eison, J.A.: Active Learning: Creating Excitement in the Classroom. 1991 ASHE-ERIC Higher Education Reports. ERIC (1991)
3. vom Brocke, J., Mendling, J. (eds.): Business Process Management Cases, Digital Innovation and Business Transformation in Practice. Management for Professionals. Springer, Heidelberg (2018). https://doi.org/10.1007/978-3-319-58307-5
4. vom Brocke, J., Mendling, J., Rosemann, M. (eds.): Business Process Management Cases Vol. 2, Digital Transformation - Strategy, Processes and Execution. Springer, Heidelberg (2021). https://doi.org/10.1007/978-3-662-63047-1
5. Chakabuda, T.C., Seymour, L.F., Merwe, F.I.V.D.: Uncovering the competency gap of students employed in business process analyst roles - an employer perspective. In: IST-Africa 2014, pp. 1–9 (2014)
6. Chow, W.: Teaching business process management with a flipped-classroom and problem-based learning approach with the use of apromore and other BPM software in graduate information systems courses. In: TALE 2021, pp. 1–8. IEEE (2021)
7. Delgado, A., Calegari, D.: Teaching BPM fundamentals: a project-based hands-on course for process-driven systems modeling and development. In: Köpke, J., et al. (eds.) BPM 2023. LNBIP, vol. 491, pp. 211–221. Springer, Cham (2023). https://doi.org/10.1007/978-3-031-43433-4_14
8. Dumas, M., Rosa, M.L., Mendling, J., Reijers, H.A.: Fundamentals of Business Process Management, 2nd edn. Springer, Heidelberg (2018). https://doi.org/10.1007/978-3-662-56509-4
9. Guo, P., Saab, N., Post, L.S., Admiraal, W.: A review of project-based learning in higher education: student outcomes and measures. Int. J. Educ. Res. **102**, 101586 (2020)
10. O'Flaherty, J., Phillips, C.: The use of flipped classrooms in higher education: a scoping review. Internet High. Educ. **25**, 85–95 (2015)
11. Ravesteyn, P., Versendaal, J.: Design and implementation of business process management curriculum: a case in Dutch higher education. In: Reynolds, N., Turcsányi-Szabó, M. (eds.) KCKS 2010. IAICT, vol. 324, pp. 310–321. Springer, Heidelberg (2010). https://doi.org/10.1007/978-3-642-15378-5_30

12. Senkus, P., et al.: Bridging the gap: an evaluation of business process management education and industry expectations - the case of Poland. In: Köpke, J., et al. (eds.) BPM 2023. LNBIP, vol. 491, pp. 250–262. Springer, Cham (2023). https://doi.org/10.1007/978-3-031-43433-4_18
13. Witt, C.M., Sandoe, K., Dunlap, J.C., Leon, K.: Exploring MBA student perceptions of their preparation and readiness for the profession after completing real-world industry projects. Int. J. Manag. Educ. **17**(2), 214–225 (2019)

Relevance of Topics in Workflow-Based Student Projects

A. Dannecker(✉) and L. Meyer

University of Applied Sciences and Arts Northwestern Switzerland FHNW, Windisch, Switzerland
achim.dannecker@fhnw.ch

Abstract. One of the significant challenges today involves developing innovative teaching concepts for business students in the field of digitalization. At the beginning of their studies, students usually only know the basics of IT, primarily related to their personal (mobile) environments, rather than real business scenarios such as business process management or workflow management.

This paper explores an innovative teaching concept in the digitalization field for business students. It focuses on how students perceive process automation in workflows using tools like Camunda, especially when tasks involve artificial intelligence (AI) components and decision models (DMN).

In this paper, we discuss the decision criteria relevant to the selection of an automation task for business students, examining the criteria students use to choose topics—whether based on difficulty, topic selection (such as supplier selection, project evaluation, product development, etc.), or the relevance to DMN and AI. Additionally, we compare whether there are differences in topic selection between Vietnamese and Swiss students.

Keywords: Flipped Classroom · Case Study · AI in Education · Workflow · Business Process Management · AI · Natural Language Processing · ChatGPT · DMN

1 Introduction

One of the big challenges today are innovative teaching concepts for business students in the area of business process and project management.

At a university in Switzerland most of the teaching for business students is done in a classical ex cathedra set-up. In the past the topics of business process management and project management were taught in such a setup and without any linkage between the topics. At the end, students were able to solve problems not more than Bloom's Taxonomy level "Understand". Because in Switzerland both topics are highly relevant in job applications. It is expected that business students can manage these topics after the university so the way the topic should be taught had to change.

On the one hand, the methodology of the course is based on case study-teaching and on the other hand on the concept of a flipped classroom. For this purpose, a case study

C. Di Ciccio et al. (Eds.): BPM 2024, LNBIP 527, pp. 324–332, 2024.
https://doi.org/10.1007/978-3-031-70445-1_21

was developed in which a start-up company is described that wants to develop a new crowdsourcing process within a project to develop new community driven products. The case study was developed from concrete experiences of a start-up company.

For twelve weeks the students had to familiarize themselves with the theory on their own according to the topics of business process management and project management. Within these sessions the students had to complete exercises in groups based on the preparation tasks and exercises that were developed from business process and project management professionals. The lecturers coached the different groups during these exercises.

During the semester, the students were able to work on two different assignments. These assignments related to the development of a process based on BPMN 2.0 (Business Process Model and Notation, an internationally used modelling language for business process modelling) and an implementation of parts of their process within a workflow engine. Workflow describes the steps in a business process, through which a piece of work passes from initiation to completion; and how these steps can be executed and automated according to a set of procedural rules with the help of a workflow engine. Organizations use workflow to coordinate tasks, improve organizational efficiency, add responsiveness, and increase profitability. Workflow may be sequential, with each step contingent upon completion of the previous one, or parallel, with multiple steps occurring simultaneously. Due to the final exam, the assignments counted for 30% of their overall grade.

The designed course meets the criteria of good teaching from Ulrich [1] very well [2–4], with an initial change in student perception taking place post-COVID19 [3].

To elaborate evidence-based criteria for good teaching, we consulted the syntheses of Schneider and Preckel [5], of Ulrich [1], and of Schneider and Mustafic [6] and use the reported criteria of good teaching. These were compiled by the respective authors from meta-analyses.

During five weeks, fixed groups work on individual topics on process modelling and workflow modelling. This work is evaluated and counts for 30% of the final grade. At the end of the 12 weeks, there is a written exam that counts 70% towards the final grade. We have found out that in general the quality in the first seven weeks decreases from semester to semester [3]. In contrast, the quality of the group work remains stable or improves. In the final exam, on the other hand, the performance decreases progressively [3]. This applies to both full- and part-time students.

Furthermore, students prefer to invest their time in relevant and challenging group work rather than in an exam. Even though the exam counts for more, we invest significantly more time in the group work [7]. For this reason, we always strive to make the group work as relevant and interesting as possible.

In the past, DMN was an integral part of group work. DMN, which stands for Decision Model and Notation, is a standard in process management developed by the Object Management Group (OMG). It's specifically designed for business decision management and is used to model and execute decisions that are determined by business rules. [8] DMN enhances the effectiveness and efficiency of decision-making processes in organizations by providing a clear, standardized, and collaborative approach to modeling and managing business decisions [9].

New this year was the opportunity to incorporate AI components into the group work. Since AI, as mentioned earlier, is a relevant and current topic, we created tasks that include these components. Our hypothesis is now that group works containing AI components are perceived as more relevant and exciting. This is, because we are increasingly seeing that ChatGPT, for example, is an integral part of students' research. [10–15].

In [16], it was shown that a correlation can be established between the perception of relevance and enjoyment factor, and the thematic areas in workflows that involve DMN and ChatGPT. Accordingly, topics related to ChatGPT are seen as more hedonistic and more relevant for the students' future. However, this perspective was captured in retrospect. This paper explores the motivations behind the choice of topics, rather than how the work was perceived in the end.

2 Methodology

2.1 Teaching Setup

Empirical studies, which use different designs to determine what "good" university teaching is, are now available in many publications [5, 6, 17–19].

In the following, the criteria with positive effects on the assessment of the course by means of teaching evaluations are presented first (cf. Table 1). Since only Ulrich [1] reports meta-analyses with the dependent variable teaching evaluation, we only include these criteria in our considerations (Table 1).

We are aware that the assignment of the individual criteria to the clusters can also be discussed in detail. However, since we are only concerned with substantiating the significance of the individual clusters, this allocation shall suffice here.

Among the criteria that have positive impact on teaching evaluation, the most valuable criteria can be found in the clusters "Good relationship building" and "Stimulating meaningful teaching".

As shown in [17] and [4], it is of importance that the lecturer as well as the dean are convinced of and dedicated to the course format to achieve a positive perception of the course. Therefore, in the evaluation of the students' feedback only 20 courses of 24 were taken into consideration (Table 2).

Thus, the abovementioned business process management and project management courses fulfill the evidence-based criteria for good teaching.

The group work, which accounts for 30% of the final grade, deals with two aspects. On the one hand, a process that is modeled and documented and a workflow that is created. Both results must be presented. When presenting the workflow, one person is randomly selected from the group. This ensures that everyone in the group deals with the content and, above all, with the workflows.

Overall, two categories of workflow tasks were distributed to the groups. Firstly, tasks that as part of automation involved the integration of a decision model. Such decision models (DMN, Decision Model and Notation) are an integral part of many processes in companies and are important for automating decision-making processes [9, 24, 25]. Specifically, in our tasks, examples of this include 'determining a product price

Table 1. Criteria with positive effects on teaching evaluation (criteria compiled from Ulrich [1], only criteria with d > .8 and criteria that can be directly influenced by teachers were considered).

Stimulation of meaningful learning	Social interaction
• Good preparation (teachers) and structure of the course [20] • Transparent learning objectives and performance requirements [21] • Relevance and usefulness of course content [20] • Type and usefulness of course materials [20]	• Encouragement of student questions and discussions, including critical questions [20] • Intellectually challenging teaching and encouragement of independent student ideas [20]
Assessment - learning support	Presentation
• Attention to student learning levels and progress [20] • Type, quality, and frequency of feedback from the teacher to the students [20]	• Rhetorical skills of the teacher [20] • Clarity and comprehensibility [20]
Use of technology	Good relationship building between teacher and students
	• Friendliness, respect, and interest of the teacher towards the learners [20] • Helpfulness and accessibility of teachers [20] • Reduced social distance to learners [22] • Professional, humanistic relationship building [23]
Good teacher	Not clustered
• Enthusiasm of the teacher for the subject and for teaching [20] • Expert knowledge of the teacher's course topic [20]	• Fair and objective assignment of grades [20]

based on a pricing strategy', 'selecting customer loyalty measures based on business rules', or 'making a decision regarding an application for further training in a company based on predefined rules'. The second part of the tasks involved integrating ChatGPT into the workflow. Here, the goal was to automate a business decision with the help of ChatGPT. Specific examples in our tasks include 'supplier evaluation and selection process', 'approval of expenses', or 'approval of project proposals'. Both categories of tasks, those with DMN and those with ChatGPT, aim to outsource complex decision-making tasks. We distributed the tasks evenly to all groups in all classes.

2.2 Student Questionnaire

First, we asked which topic they would choose. All topics are described in detail in the attachment. The 8 different topics available for selection were:

1. Product Pricing Strategy (DMN)

Table 2. Adoption of criteria due to the course format.

Stimulation of meaningful learning	Social interaction
• Structure of the course is well defined, received feedback over more than 60 times by students and taught from 12 different lecturers • All exercises are outlined by clear learning objectives and performance requirements • Relevance and usefulness are reflected by a real case study that has emerged from a real project • All content within exercises etc., is a one-to-one result (e. g. transcript of an interview) from the described project	• Direct interaction during the time the students work in groups on different challenging topics
Assessment - learning support	Presentation
• Give feedback during "class-time" and outside of the learning sessions • All uploaded solutions are provided with feedback • If groups and/or students need more time, feedback is also provided on the solutions • In-depth feedback to evaluated assignments during different coaching sessions	• During the "class-time" at the end, one solution is discussed in depth guided by the teacher
Use of technology	Good relationship building between teacher and students
• Use of professional tools that are relevant in the industry • Microsoft Teams is used for collaboration • Camunda is used for process and workflow modelling	• Open door for students via Microsoft Teams even outside usual time frames
Good teacher	Not clustered
• All teachers are enthusiastic about the course format and the topic • All teachers are professionals with more than ten years of experience in projects within the course's topics	• "Fair and objective assignment of grades" cannot be considered, because it takes place after the evaluation

2. B2C Customer Loyalty Program (DMN)
3. Reorder Process for Raw Materials (DMN)
4. Training Approval Process (DMN)
5. Supplier Evaluation and Selection Process (ChatGPT)
6. Expense Approval Process (ChatGPT)
7. Project Approval Process (ChatGPT)
8. Training Recommendation Process (ChatGPT)

In addition, we posed four open-ended questions and, in conclusion, asked whether the relevance of the topic influenced their decision, whether the topic involved DMN or ChatGPT, or whether both factors were decisive in their choice (Table 3).

Table 3. Questions related to relevance.

No	Question
Q1	What is the reason, please elaborate?
Q2	Do you think this topic is more relevant for the professional future than the others? Please explain
Q3	Do you think this topic will be more fun than the others? Please explain
Q4	Do you think this topic is easier than the others? Please explain
Q5	Was the integration of DMN/ChatGPT decisive for you, or was it the thematic orientation, the topic? (Multiple answers possible)

3 Results

During the spring semester of 2024, we launched the survey at the time when students had to choose their topics. The survey was conducted via MS-Forms. A class of business students at the Foreign Trade University in Vietnam and two classes at the University of Applied Sciences Northwestern Switzerland were surveyed.

A total of 23 students from Vietnam and 28 students from Switzerland participated.

3.1 Topic Related Results

Table 4 shows the distribution of students' decisions on which topics they would choose, divided into Vietnam and Switzerland.

Table 4. Number of topics selected

	1	*2*	*3*	*4*	*5*	*6*	*7*	*8*
Vietnam	3	8	2	3	1	1	2	3
Switzerland	0	4	3	3	3	6	5	4
Total	**3**	**12**	5	6	4	7	7	7

It can be seen that there is no major difference in the general selection of tasks in relation to DMN and ChatGPT. Slight differences can be seen in the selection between Vietnam and Swiss classes. In Vietnam DMN specific topics are more likely to be chosen and in the Swiss classes ChatGPT topics are more likely to be chosen. With regard to the topics, it should perhaps also be emphasized that the topic Product Pricing Strategy (DMN) is rarely chosen in contrast to B2C Customer Loyalty Program (DMN). In general, all topics appeal to students. This is also reflected in the first open question that was asked ("What is the reason, please elaborate?"). The students were mostly interested in the topic they chose, or already have experience in the subject area or want to work in this field later on.

3.2 Results as a Basis for Decision-Making

The following Table 5 shows how the students responded to questions 2, 3 and 4. If the students agreed with the question in their answer, this was marked with "y". If they disagreed, this was marked with "n". For question 5, the answers were differentiated, as multiple answers were possible. On the one hand, students could indicate whether the topic is relevant (marked with "T") or whether the embedding of DMN/ChatGPT is relevant (marked with DC). If both were selected, this is marked with "B" in the Table.

Table 5. Questions for decision-making

	Q2		Q3		Q4		Q5		
	y	n	y	n	y	n	T	DC	B
Vietnam	19	4	15	8	4	19	4	11	8
Switzerland	15	13	15	13	8	20	14	11	3
Total	34	17	30	21	12	39	18	22	11

In general, tasks are increasingly selected according to their relevance for the professional future (Q2). This is even more evident among students in Vietnam. A certain fun factor (Q3) also plays a role to a similar extent. The degree of difficulty, on the other hand (Q4), is not seen as an important aspect and plays a subordinate role. When making their selection, students usually focus either on the topic itself or whether DMN or ChatGPT was part of the task. In a few cases, both aspects played a role.

4 Conclusions

In conclusion, our study delved into the relevance of topics in workflow-based student projects, particularly focusing on business students' perceptions and choices regarding process automation using tools like Camunda, with emphasis on AI components and decision models (DMN). Despite limitations, including a relatively small sample size and the absence of data quality assessment, our findings shed light on several crucial aspects.

Firstly, we observed a shift in teaching methodology towards more interactive and case-study-based approaches, fostering students' engagement and understanding of complex topics like business process management and project management. The incorporation of a flipped classroom model and real-world case studies not only enhanced students' comprehension but also stimulated meaningful learning experiences.

Furthermore, our study introduced students to cutting-edge concepts such as AI integration, particularly ChatGPT, within workflow tasks. This addition aimed to make the projects more relevant, engaging, and aligned with current industry trends. We noted a correlation between the perception of relevance and enjoyment factor, particularly in tasks involving DMN and ChatGPT components.

Despite the absence of explicit statements regarding the correlation between the specific questions Q2, Q3, and Q4, our results suggest that students tend to prioritize the relevance of topics for their professional future and the enjoyment factor over perceived difficulty. Additionally, the integration of DMN and ChatGPT elements influenced topic selection, indicating the significance of aligning coursework with emerging technologies and industry demands.

In essence, our study underscores the importance of innovative teaching designs that bridge theoretical concepts with practical applications, such as workflow-based projects. By incorporating elements like DMN and ChatGPT, educators can not only enhance students' learning experiences but also prepare them for the dynamic and technology-driven business landscape. As we continue to refine and expand such teaching methodologies, we strive to ensure that students' investments in their education lead to high-quality work that is both fulfilling and relevant to their future careers.

References

1. Ulrich, I.: Gute Lehre in der Hochschule: Praxistipps zur Planung und Gestaltung von Lehrveranstaltungen, 2nd edn. Springer Fachmedien, Wiesbaden (2020). https://doi.org/10.1007/978-3-658-31070-7
2. Dannecker, A., Hanke, U.: Evidence-based good teaching succeeds face-2-face and online. In: ICERI2022 Proceedings, pp. 5650–5656 (2022)
3. Dannecker, A., Hanke, U.: Students struggle in coming back to face-2-face teaching in evidence based good teaching settings. In: INTED2023 Proceedings, pp. 4323–4329 (2023). https://doi.org/10.21125/inted.2023.1146
4. Dannecker, A.: Teaching each semester 12 classes with 7 lectures in a case study and flipped classroom setting. In: ICERI2019 Proceedings, pp. 8376–8382 (2019). https://doi.org/10.21125/iceri.2019.1996
5. Schneider, M., Preckel, F.: Variables associated with achievement in higher education: a systematic review of meta-analyses. Psychol. Bull. **143**, 565–600 (2017). https://doi.org/10.1037/bul0000098
6. Schneider, M., Mustafić, M. (eds.): Gute Hochschullehre: Eine evidenzbasierte Orientierungshilfe: Wie man Vorlesungen, Seminare und Projekte effektiv gestaltet, 1st edn. Springer, Heidelberg (2015). https://doi.org/10.1007/978-3-662-45062-8
7. Dannecker, A.: Business students invest their time in complex and challenging process digitalization tasks instead of preparing for the exam. In: EDULEARN23 Proceedings, pp. 8461–8467 (2023). https://doi.org/10.21125/edulearn.2023.2219
8. Decision Model and Notation Specification Version 1.3. https://www.omg.org/spec/DMN/1.3/. Accessed 17 Jan 2024
9. Janssens, L., Bazhenova, E., De Smedt, J., Vanthienen, J., Denecker, M.: Consistent integration of decision (DMN) and process (BPMN) models. In: Proceedings of the CAiSE'16 Forum, at the 28th International Conference on Advanced Information Systems Engineering (CAiSE 2016), pp. 121–128. CEUR-WS.org (2016)

10. Dempere, J., Modugu, K., Hesham, A., Ramasamy, L.K.: The impact of ChatGPT on higher education. Front. Educ. **8**, 1206936 (2023). https://doi.org/10.3389/feduc.2023.1206936
11. Firaina, R., Sulisworo, D.: Exploring the usage of ChatGPT in higher education: frequency and impact on productivity. BEI **2**, 39–46 (2023). https://doi.org/10.56741/bei.v2i01.310
12. Lo, C.K.: What is the impact of ChatGPT on education? a rapid review of the literature. Educ. Sci. **13**, 410 (2023). https://doi.org/10.3390/educsci13040410
13. Ngo, T.T.A.: The perception by university students of the use of ChatGPT in education. Int. J. Emerg. Technol. Learn. **18**, 4–19 (2023). https://doi.org/10.3991/ijet.v18i17.39019
14. Rahman, M.M., Watanobe, Y.: ChatGPT for education and research: opportunities, threats, and strategies. Appl. Sci. **13**, 5783 (2023). https://doi.org/10.3390/app13095783
15. Rawas, S.: ChatGPT: empowering lifelong learning in the digital age of higher education. Educ. Inf. Technol. (2023). https://doi.org/10.1007/s10639-023-12114-8
16. Dannecker, A., Meyer, L.: Relevance of artificial intelligence components in workflow-based student projects. In: INTED2024 Proceedings, pp. 3442–3450 (2024). https://doi.org/10.21125/inted.2024.0912
17. Dannecker, A., Hanke, U.: "Die gute Lehrveranstaltung" - eine Feldstudie (2021)
18. Ulrich, I.: Gute Lehre in der Hochschule: Praxistipps zur Planung und Gestaltung von Lehrveranstaltungen, 1st edn. Springer (2016). https://doi.org/10.1007/978-3-658-11922-5
19. Hattie, J.: Lernen sichtbar machen. - Hattie-Studie: Visible Learning (2013). http://www.visiblelearning.de/hattie-j-2013-lernen-sichtbar-machen/. Accessed 26 Sept 2019
20. Feldman, K.A.: The association between student ratings of specific instructional dimensions and student achievement: refining and extending the synthesis of data from multisection validity studies. Res. High. Educ. **30**, 583–645 (1989). https://doi.org/10.1007/BF00992392
21. Hattie, J.: The applicability of visible learning to higher education. Scholarsh. Teach. Learn. Psychol. **1**, 79–91 (2015). https://doi.org/10.1037/stl0000021
22. Allen, M., Witt, P.L., Wheeless, L.R.: The role of teacher immediacy as a motivational factor in student learning: using meta-analysis to test a causal model. Commun. Educ. **55**, 21–31 (2006). https://doi.org/10.1080/03634520500343368
23. Cornelius-White, J.: Learner-centered teacher-student relationships are effective: a meta-analysis. Rev. Educ. Res. **77**, 113–143 (2007). https://doi.org/10.3102/003465430298563
24. Mertens, S., Gailly, F., Poels, G.: Enhancing declarative process models with DMN decision logic. In: Gaaloul, K., Schmidt, R., Nurcan, S., Guerreiro, S., Ma, Q. (eds.) BPMDS EMMSAD 2015 2015. Lecture Notes in Business Information Processing, vol. 214, pp. 151–165. Springer, Cham (2015). https://doi.org/10.1007/978-3-319-19237-6_10
25. Song, R., Vanthienen, J., Cui, W., Wang, Y., Huang, L.: A DMN-based method for context-aware business process modeling towards process variability. In: Abramowicz, W., Corchuelo, R. (eds.) BIS 2019. Lecture Notes in Business Information Processing, vol. 353, pp. 176–188. Springer, Cham (2019). https://doi.org/10.1007/978-3-030-20485-3_14

From Design to Evaluation: Students Experiencing the BPM Lifecycle Hands-on

Maximilian König(✉), Maximilian Völker, and Mathias Weske

Hasso Plattner Institute, University of Potsdam, Potsdam, Germany
{maximilian.koenig,maximilian.voelker,mathias.weske}@hpi.de

Abstract. Business process management (BPM) has gained significant traction in both industry and academia, leading to its inclusion in many universities' curricula. Due to the broad scope of BPM and its intersections with many other disciplines, there is a considerable amount of content to cover when teaching the basics to students. Hence, courses mostly focus on teaching the theoretical foundations. However, to successfully apply BPM in practice, i.e., to design, analyze, implement, and evaluate real-world business processes, hands-on experience is necessary. In this paper, we present a course structure guiding students through one iteration of the whole business process lifecycle for a specific scenario, encouraging the students to explore the various methods available in each phase. In addition to the course structure, we share our lessons learned from the first iteration and the students' feedback.

Keywords: BPM · Teaching · Hands-on Project Seminar · Process-oriented Information Systems

1 Introduction

In today's rapidly evolving business environment, the discipline of business process management (BPM) emerges as a cornerstone of organizational agility and operational efficiency. Business processes support the majority of organizational activities and therefore play a key role in how effectively an organization can adapt to changes and optimize operational workflows. Thus, expertise in BPM becomes an essential skill for future professionals, highlighting the need to reevaluate educational strategies to incorporate and emphasize these competencies.

Comprehensive textbooks (e.g., [2,12]) provide a strong theoretical foundation and introduce frameworks such as the business process lifecycle, emphasizing the dynamic nature of business processes. However, that dynamic nature highlights the need for experiential learning opportunities to reach a profound understanding of the involved methods and their applicability to real-world problems.

To that extent, other course structures have been presented in related work, concentrating on practical experiences in different phases of the lifecycle. Delgado and Calegari reported on their hands-on course focusing on the modeling and

C. Di Ciccio et al. (Eds.): BPM 2024, LNBIP 527, pp. 333–343, 2024.
https://doi.org/10.1007/978-3-031-70445-1_22

automation of business processes [1]. Another practical course by Grisold et al. emphasizes the design and analysis phase and compares BPM to routine dynamics [3].

In this paper, we present our experiences and learnings from a new seminar encompassing a full iteration of the business process lifecycle. It was offered to bachelor computer science students of the Hasso Plattner Institute (HPI) at the University of Potsdam in Germany. The goal of the seminar titled "process-oriented information systems in practice" was to provide the students with more practical hands-on experiences in business process management, especially in regard to process elicitation, implementation, and analysis. It was designed as a follow-up course to a more theoretical lecture we offer for bachelor students that teaches the foundations of BPM with focus on modeling (BPMN + DMN), process analysis using formal methods, and the basics of process mining. Based on fictitious scenarios, four groups of students conducted workshops to elicit processes, modeled and implemented them in a business process management system, and analyzed them by various means.

In the following, we report on the vision and design of the seminar in more detail (Sect. 2). Subsequently, in Sect. 3, we provide a description of the seminar structure that was inspired by phases of the business process lifecycle. Before concluding the paper in Sect. 5, we elaborate on our lessons learned and the students' feedback in Sect. 4.

2 Course Design

We envisioned the seminar as a follow-up course for a more theoretical lecture on BPM foundations to provide the students with hands-on experiences on BPM, especially on process elicitation, implementation, automation, and analysis. For this, the students should act out the phases of the business process lifecycle on artificial use cases, i.e., (i) design and analysis, (ii) configuration, (iii) enactment, and (iv) evaluation [12].

The course's timespan was one semester with 15 weeks of lecture period, and it was rewarded with six ECTS, corresponding to roughly 150 h of work expected in total for each student. As illustrated in Table 1, the fifteen weeks of the semester were partitioned into three parts, with the enactment and evaluation phases combined, and interspersed with presentations for reporting on the progress and for grading purposes. Each of the phases was introduced by a theory unit of 90 min, in which the required knowledge was refreshed and the scope and tasks were defined. Besides the introductory lectures and the presentations, there were no fixed regular meetings. However, the students were encouraged to schedule meetings with their supervisors if needed.

As indicated in italics in Table 1, there were three artifacts used to grade the seminar. Two-thirds of the final grade was based on the two presentations, a midterm presentation and a final presentation at the end of the semester. We asked the groups to ensure that each student is actively involved in at least one of the presentations. The remaining third of the grade was determined by

Table 1. Overview of the seminar schedule

Weeks	Phase	Contents
1		Introduction & BPM recap
2–5	1	Elicitation with t.BPM, digitalization, validation
6–10	2	Modeling and implementation in Camunda
11		*Mid-term presentations*
12–14	3 & 4	Enactment, simulation, analysis, process mining
15		*Final presentations*
(17)		*Final report submission*

a project documentation describing the artifacts of all phases as well as the challenges encountered along the way. As a guide, we provided a template for the documentation, including the requirements for each section, such as describing the major adjustments made to the model, and how certain aspects, like services and the cross-team process communication, were implemented.

The seminar was offered to all bachelor students at the Hasso Plattner Institute who had previously attended the lecture on business process management, which focuses on modeling and formal analysis. Since there is only one bachelor program offered at HPI, IT-Systems Engineering, all students have a strong computer science background. Therefore, the focus of the seminar was more on technical than on management aspects. We admitted 20 students, all of whom were at least in their fifth semester, i.e., at the end of their bachelor studies.

The students were divided into four groups of five students each, to fit the two case studies we had prepared beforehand. We decided to use fictitious case studies as opposed to real-world customers to be able to better scope and control the extent and outcome of the seminar, especially considering that it was a bachelors course and not a masters.

The first case study evolved around a fictive electronics retailer that experienced rapid growth in recent years and faces several obstacles that should be relieved by process digitization and automation. One group of students focused on processes performed at branch offices, and another group focused on processes performed at the headquarters. This ensured different challenges for each group, while also providing points of contact between the teams and their processes.

For the second case study, we interviewed domain experts to fabricate a realistic scenario in a fictive hospital setting. Opposed to the first case, where each group had three smaller processes to elaborate, here, both groups were assigned one extensive process. The first one was assigned a general process of admissions in the emergency department, and the second was assigned a process as it could be performed in a radiology department. As with the first case, there are points of contact between the two processes for the teams to discover and discuss at a later point in time.

With this course design, we aimed to adhere to the recommendations for a successful application of project-based learning described by Kokotsaki et al. [4], as summarized in the following:

Student support/time management: While students were free to allot their time, dividing the semester according to the phases (cf. Table 1) with their distinctive artifacts helped to ensure consistent progress. Next to individual meetings that could be scheduled at any point, students could ask for help via email, the learning platform Moodle, and an institute-internal messaging service at any point throughout the semester.

Effective group work: The seminar was not compulsory and required successful completion of the theoretical lecture, which was intended to promote self-motivated group work at eye level.

Balancing didactic instructions with independent inquiries: The theory units ensured that all groups were equipped with the necessary knowledge to solve the tasks. However, the units did not provide a sample solution to be adapted, but rather a set of requirements along with references to further sources to be studied and applied by the participants themselves.

Assessment emphasis on reflection, self and peer evaluation: In addition to the meetings with the teams, emphasis was also placed on reflecting on the (team) work in the two presentations. As presentations were made in front of the other groups, the teams were able to discuss their progress.

Student choice and autonomy: The requirements for the individual phases only provided a rough direction and ensured that all necessary aspects were at least fundamentally addressed. However, the students were free to decide, for example, which aspects they would like to deal with in greater depth in their scenario/process. For example, some groups focused on implemented services, while other groups concentrated more on decision logic and data forms.

3 Course Structure and Content

The students were to experience the business process lifecycle in practice, so the seminar was organized according to the lifecycle as described above and illustrated in Fig. 1.

The first week of the semester was spent on refreshing the students' BPM knowledge with focus on modeling aspects. The lecture included a recap of the Business Process Model and Notation (BPMN) standard [8], some common workflow patterns [11], modeling guidelines [7], and the Decision Model and Notation (DMN) standard [9].

3.1 Design and Analysis

At the beginning of the project work, the student groups were handed a briefing of their scenario, which included an introduction to their "company", i.e., the

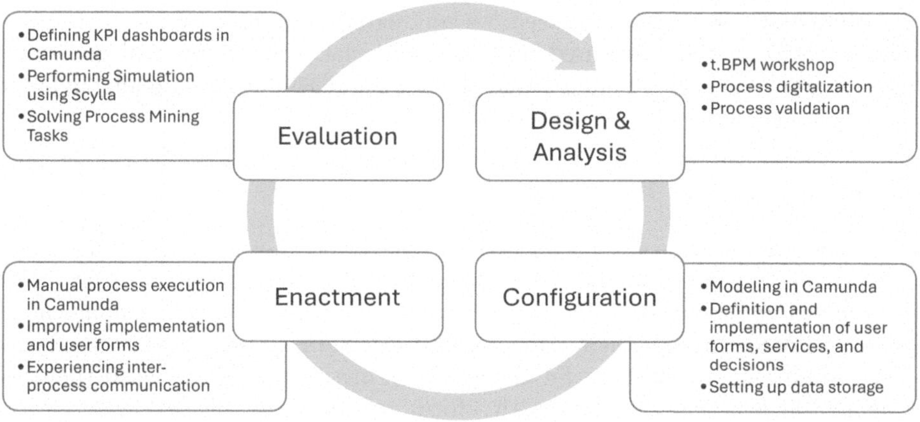

Fig. 1. Overview of course content and structure based on the business process lifecycle [12]

electronics retailer or the hospital, as well as a brief description of the process(es) they will deal with[1].

For the process elicitation, we decided to use a workshop method. Compared to interviews or questionnaires, a workshop requires less time to conduct and allows discussing and aligning different viewpoints of the participants immediately [5]. Specifically, we employed the "tangible business process modeling" (t.BPM) technique [6]. This method uses inscribable, blank shapes of basic BPMN elements and empowers the domain experts to model the process themselves by arranging and labeling the shapes under the guidance of a moderator.

To get to know t.BPM, we provided the necessary information, and, in accordance with the recommendations for t.BPM (cf. [6]), conducted warm-up exercises for the students to experience the technique themselves.

In the actual workshop sessions, the respective team was responsible for moderating and documenting the session, and the domain experts were played by volunteer students from the other scenario and by members of the chair. As a briefing for them, we prepared role and process descriptions with slight variations and small discrepancies to encourage discussion and better simulate a real workshop session where different views on the same process need to be reconciled.

After the workshop, the students' task was to digitize the collected process models based on their notes and the pictures taken of the results of the t.BPM workshop. The challenge here was to differentiate between necessary changes to the collected hand-drawn version with its shortcomings and possible inaccuracies, which were due to the time pressure and the workshop situation, and changes that seemed reasonable but would change the "as-is" process too much, and to evaluate them accordingly.

[1] The briefings for the four groups are available here: https://drive.google.com/file/d/1GZO23ipEZ7ATHmJtJre2RRx9BJ46rnvu/view.

After a first version was digitized, we conducted feedback meetings with each group to validate their captured processes and clarify problems and questions that arose during the modeling process. This was to ensure that no team deviated too far from the envisaged version(s) of the process(es).

3.2 Configuration

Before the configuration was performed, we conducted another round of meetings to discuss opportunities for process optimization the students came up with. For the process implementation, we used Camunda as business process management system (BPMS) through their Educational License Program[2].

As a minimal requirement for the students in this phase, we asked them to implement at least one user task with a form, one business rule task with DMN, one automated XOR-gateway, one message task (e.g., sending an email), and one service task. Additionally, they were expected to implement inter-process communication with the other group working on a different but interrelated process in the same scenario, e.g., if a patient is referred from the emergency department to the radiology for an X-ray examination. To do so, the teams had to communicate to agree on a common data exchange format.

While user forms and DMN can be defined directly in Camunda, calling external services requires additional setup. To that end, we provided a virtual machine to the students to use for deploying a database and hosting their service runners to be invoked by running process instances. With this shared hosting, the groups working on different processes in the same scenario were also enabled to work on the same data. This simplified the overall setup, but the groups individually had to agree on the data structure and the access to it.

The complexity of implemented models varied between the two scenarios. In the hospital setting, both teams implemented a larger process with 80 to 90 activities, gateways, and events, while the teams in the retailer scenario worked on multiple process models with 10 to 40 elements each, amounting to a similar overall complexity.

3.3 Enactment and Evaluation

In the final weeks of the semester, the students were tasked to generate execution data for their previously implemented processes. Based thereon, they should then monitor process metrics, assess the processes' performance, and propose potential improvements. We decided to merge the enactment and evaluation phases of the lifecycle to enable the student teams to distribute the work items more flexibly.

Enactment. For the enactment phase, all processes had to be executed at least 20 times, manually or automatically, with meaningful data for all required attributes. Based on these executions, a number of key performance indicators

[2] https://camunda.com/educationee/.

(KPIs) and additional statistics had to be compiled based on process variables, waiting times, and executing resources, among other things. For example, in the radiology scenario, the average treatment duration and the ratio of CT, MRT, and X-ray examinations should be measured. We provided each group with a small set of mandatory metrics, but also encouraged the students to define additional performance measures they deemed useful in the context of their scenarios. All metrics should be implemented in Camunda's Optimize dashboard for automated evaluation on the previously executed process instances.

Evaluation. In the final phase, we wanted to enable the students to explore different methods to generate additional insights into the performance of their processes using process mining, derive process improvement ideas, and measure the impact of such process changes.

First, to create a larger set of execution data, we introduced business process simulation. Specifically, we introduced the students to the BPMN process simulator Scylla[3] [10]. Since the configuration of simulation models is a challenging task, we provided each team with distributions for activity durations, event arrival rates, and branching probabilities for every splitting gateway, as well as staffing and work schedules. Additionally, due to the size of some teams' models, we decided to reduce the scope of these processes to about 12 activities by ignoring edge cases and abstracting from interactions with other processes. Once the simulation model had been configured by the students, Scylla enabled them to generate event logs for process mining.

To analyze the logs, we further introduced the tool Disco[4]. Using the tool's process mining and analysis capabilities, the students were tasked to discover bottlenecks and subsequently propose process improvements. In addition, we prepared for each team a number of potential changes to process and staffing, which should be evaluated in terms of their impact. Therefore, they had to adapt their simulation models to mirror the suggested changes and compare the resulting logs' metrics to the baseline. Example questions were: "How many additional employees have to be hired to maintain the current average examination time if radiological examinations always have to be prepared and conducted by two employees?" and "How does the average fulfillment time change if warehousing is expanded so that on average 95% of orders can be fulfilled without additional deliveries from the central warehouse?"

4 Feedback and Lessons Learned

Throughout the seminar, we continuously encouraged our students to provide feedback. In addition, students were asked to complete an anonymous general survey at the end of the semester via the Institute's course feedback platform, which 12 out of 20 students completed. The survey included a series of general questions asked for each course, as well as specific questions related to project

[3] https://github.com/bptlab/scylla.

[4] https://fluxicon.com/disco/ (using the academic license).

seminars. All questions were to be answered using a Likert scale with five levels, ranging from "Strongly agree" (1) to "Strongly disagree" (5), and invited for optional free-text responses, all of which were collected anonymously. In this section, we want to share that feedback together with our lessons learned from a teaching perspective.

Overall Reception. In total, the course was rated with 1.2 by the students, indicating that it was well received. The students were very satisfied with their increase in knowledge they gained from the course (1.6) and also found it very enjoyable (1.6). Additionally, it motivated students to further deepen their knowledge in this field (1.7) and they considered it relevant for their studies (1.6).

In addition to some difficulties that were to be expected with working in groups, some students noted that the tasks were not always clearly specified (2.1). We intentionally provided only a small set of mandatory requirements for the artifacts of each phase to encourage individual exploration, e.g., regarding the definition of process metrics, process mining results, or how the process implementation should be approached. However, this requires a high degree of self-organization and motivation on the part of the students.

Learning: The requirements definition for the outcome of each phase should be more comprehensive. To achieve that, we intend to provide checklists and include tasks such as "Come up with at least x sensible process metrics in the context of your scenario".

Design and Analysis. Using t.BPM for process elicitation was well received by the students, especially since they had to perform both as domain experts and BPM experts. The fact that different domain expert roles had access to different perspectives on the process led to intense discussions that required careful moderation, which students found very difficult. Furthermore, maintaining a consistent level of abstraction proved challenging for them, as domain experts quickly digressed into small details of the process. As points for improvement, students mentioned a comparison to alternative process elicitation approaches such as interviews, and they wished for a more extensive example workshop to better learn how to encourage active participation of domain experts and how to prevent too detailed discussions. From a teaching perspective, it is important to note that this approach may lead to process models that differ greatly from the envisaged solution. In that case, the feedback meetings can be used to guide the teams in the intended direction and to reduce the scope, if necessary. For example, waiving edge cases that were incorporated by the students helps them to focus on the main process.

Learning: In the next iteration, we plan to extend the introductory t.BPM workshop by more best practices and pitfalls to better convey how to deal with digressing domain experts and how to encourage their active participation.

Configuration. During the transition from elicited, rather high-level process models to low-level, implementable process models, the student groups required some guidance to decide which parts of their processes should be automated, and which parts cannot be automated at all. While implementing the processes, including services and a database for their scenario, students reported a significant learning curve, since the implementation involved multiple technologies that not all students were familiar with, e.g., Docker, databases, and DMN's expression language FEEL. In addition to that, they had to deal with the design decisions and limitations of the chosen BPMS. For example, while data objects can be modeled, they have no execution semantics in Camunda, so that the data flow had to be implemented by other means. Implementing a BPMN process model in a BPMS also forced students to think about the exact semantics of the modeling elements used. Thereby, a profound understanding of BPMN aspects such as task types, error handling, or the difference between exclusive and event-based gateways was achieved.

Learning: Depending on the group size and familiarity with the technologies used, students should be provided with a database containing mock data or even services that should be connected to the implemented process, to focus the attention on the process and its implementation, rather than on auxiliary technologies.

Enactment and Evaluation. One of the most frequently mentioned points for improvement is the lack of real-world data in the enactment phase. Students had to come up with artificial data sets, and the manual execution of process instances lacked dimensions such as time and executing resources.

Another difficulty arose from the complexity of the simulation models. Manually preparing them for the students' processes is laborious and involves many artificial values that may lead to unexpected patterns in the resulting logs. While it is part of the challenge for the students to identify patterns and points for process improvement, the outcome of this analysis was difficult to anticipate.

Learning: In the next iteration, we plan to provide data sets for process enactment to have more control over derivable process metrics. Additionally, the simulation and analysis should be conducted in a more controlled setting, where it is easier to predetermine what the students might discover.

5 Conclusion

In this paper, we reported on a hands-on seminar we conducted for bachelor students with previous knowledge of business process management and modeling. Based on two fictitious scenarios, the students experienced the different phases of the BPM lifecycle in groups of five persons. They conducted t.BPM workshops for process elicitation and implemented the processes in the BPMS Camunda, including automated decision-making with DMN, user forms, and custom services, and worked out various suggestions for process optimization. Finally, based on key performance indicators and simulation, they analyzed the processes to derive further insights and recommendations.

Key learnings for the students were the different uses of process models as means of communication versus implementation, a profound understanding of the semantics behind BPMN's modeling elements, and the wide range of application of BPM and its methods in general.

While the students' feedback was mostly very positive, we plan several improvements for the next iteration. Foremost, we plan to define the requirements and goals to reach for each phase in a little more detail. However, finding the right balance between providing guidance and allowing for a freer exploration with more open results will remain a challenge. Additionally, the t.BPM workshop introduction will be redesigned to focus more on dealing with digressing domain experts and encouraging experts' active participation in modeling. Further, we intend to provide data sets that can be used in the enactment phase to execute process instances to have more control over potentially interesting KPIs and process metrics.

Acknowledgments. We thank Camunda for providing access to the Camunda platform through the Educational License Program.

References

1. Delgado, A., Calegari, D.: Teaching BPM fundamentals: a project-based hands-on course for process-driven systems modeling and development. In: Köpke, J., et al. (eds.) BPM 2023. LNBIP, vol. 491, pp. 211–221. Springer, Cham (2023). https://doi.org/10.1007/978-3-031-43433-4_14
2. Dumas, M., Rosa, M.L., Mendling, J., Reijers, H.A.: Fundamentals of Business Process Management, 2nd edn. Springer, Heidelberg (2018). https://doi.org/10.1007/978-3-662-56509-4
3. Grisold, T., Wurm, B., vom Brocke, J., Kremser, W., Mendling, J., Recker, J.: Managing process dynamics in a digital world: integrating business process management and routine dynamics in IS curricula. Commun. Assoc. Inf. Syst. **51**, 5 (2022). https://aisel.aisnet.org/cais/vol51/iss1/5
4. Kokotsaki, D., Menzies, V., Wiggins, A.: Project-based learning: a review of the literature. Improv. Sch. **19**(3), 267–277 (2016). https://doi.org/10.1177/1365480216659733
5. Leyh, C., Bley, K., Seek, S.: Elicitation of processes in business process management in the era of digitization – the same techniques as decades ago? In: Piazolo, F., Geist, V., Brehm, L., Schmidt, R. (eds.) ERP Future 2016. LNBIP, vol. 285, pp. 42–56. Springer, Cham (2017). https://doi.org/10.1007/978-3-319-58801-8_4
6. Lübbe, A.: Tangible business process modeling: design and evaluation of a process model elicitation technique. Ph.D. thesis, University of Potsdam, Potsdam, Germany (2011)
7. Mendling, J., Reijers, H.A., van der Aalst, W.M.P.: Seven process modeling guidelines (7PMG). Inf. Softw. Technol. **52**(2), 127–136 (2010). https://doi.org/10.1016/J.INFSOF.2009.08.004
8. Object Management Group (OMG): Business Process Model and Notation (BPMN) Specification, Version 2.0.2 (2013). https://www.omg.org/spec/BPMN/2.0.2/PDF
9. Object Management Group (OMG): Decision Model and Notation (DMN) Specification, Version 1.4 (2023). https://www.omg.org/spec/DMN/1.4/PDF

10. Pufahl, L., Weske, M.: Extensible BPMN process simulator. In: Clarisó, R., et al. (eds.) BPM Demo Track, 2017. CEUR Workshop Proceedings, vol. 1920. CEUR-WS.org (2017)
11. Russell, N., Hofstede, ter, A., Aalst, van der, W., Mulyar, N.: Workflow control-flow patterns: a revised view. BPM Rep. (BPMcenter.org) **0622** (2006)
12. Weske, M.: Business Process Management - Concepts, Languages, Architectures, 3rd edn. Springer, Heidelberg (2019). https://doi.org/10.1007/978-3-662-59432-2

Enabling BPM Competencies Through Real-Life Student Consulting Projects

Sonia Lippe(✉), Franziska Bärtsch, Claudia Brönimann, and Clemens Mader

Eastern Switzerland University of Applied Sciences, St. Gallen, Switzerland
sonia.lippe@ost.ch

Abstract. Real-life student consulting projects represent a highly effective approach to teaching, enabling students to develop BPM competencies beyond the theoretical foundations taught in classrooms. In this context, we discuss and demonstrate the way our organisation manages these projects, providing students with practice-oriented teaching. Furthermore, the programme benefits regional companies, providing them with expert knowledge to enhance their BPM maturity. We conclude the discussion with a summary of the key insights for lecturers who wish to implement a similar approach for their students. These include setting clear goals with respect to the BPM lifecycle, implementing a professional project management approach for all projects, and defining practical BPM competencies to be enabled by the projects.

Keywords: BPM competencies · real-life teaching experiences · student consulting projects · local companies

1 Introduction

Business Process Management (BPM) plays an important role in all companies and implementing and systematically improving business processes enables adaptation to changing market demands and technological advancements [1]. The adoption of BPM as an enterprise-wide management paradigm can mitigate operational inefficiencies and foster innovative practices crucial for a company's success [2, 3]. However, the absence of specialized knowledge within the organization often leads to a deficiency in the necessary mindset and skills for effective BPM implementation [4]. As a result, education and training in BPM become essential in guiding this transformational process. Instructional formats that allow learners to apply theoretical principles learned in the classroom to practical business scenarios are crucial in cultivating a process-oriented mindset and developing abilities for systematic process transformation and optimization [5].

Teaching cases represent a valuable educational tool, as they facilitate the integration of real-world scenarios into the classroom environment [6, 7]. A growing number of BPM teaching cases are now accessible [8], with many focusing on specific sectors such as finance [9] or the public sector [10, 11]. While teaching cases offer valuable insights and a practical learning experience, there are also certain shortcomings. These include the

C. Di Ciccio et al. (Eds.): BPM 2024, LNBIP 527, pp. 344–353, 2024.
https://doi.org/10.1007/978-3-031-70445-1_23

simplification of complex real-world situations, an idealised view of problems, limited contextual complexity, hindsight bias and inadequate ambiguity [12].

Practice-driven projects represents another educational approach that provides a more comprehensive and realistic learning experience [13, 14]. Within this paper we introduce and discuss real-life student consulting projects as a particularly successful way of building up practical competencies in students and building a knowledge bridge between research and practice along the BPM journey. Here, our organization has developed a detailed concept carried out within local companies over the past two decades. This concept has matured to a point where we can outline the current methodology (Sect. 2), showcase an example project (Sect. 3), and share a collection of lessons learned for the benefit of fellow educators (Sect. 4).

2 Real-Life Student Consulting Projects

2.1 Our Approach

As part of our Business Information Technology curriculum, bachelor students are required to complete two real-life consulting projects over the course of their degree, accounting for a total of 10% of their ECTS. The projects are based on specific questions posed by companies and organisations, most of them being local SMEs. The initial project is designed to apply fundamental concepts in business process awareness and modelling. Aiming at the start of the BPM lifecycle, these projects are directed at process identification and essential process modelling to systematically select the business processes for digitalisation and to model the as-is situation. This is followed by topics advancing in the lifecycle, such as process redesign and re-engineering.

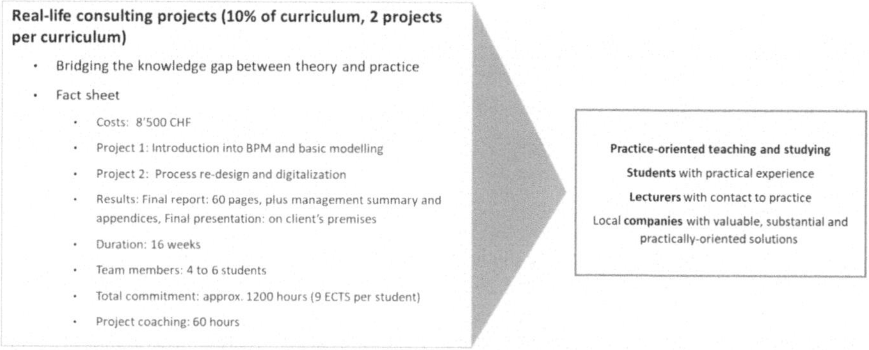

Fig. 1. Core concept and benefits for stakeholders of real-life consulting projects

The project teams comprise five to six students, who collectively invest approximately 1,200 h over the course of 16 weeks in the completion of their respective projects. Each project is coached by a faculty member for a minimum of 60 h and may also benefit from the input of other experts as required. Upon completion of the project, the client receives a comprehensive final report comprising 60 pages, a presentation of the results at the company's premises, and a public relations report for marketing purposes. The results are entirely owned by the client, who is also involved in grading the students' work to demonstrate its level of satisfaction with the project outcomes. The project expenses (consulting fee) of 8,500 Swiss francs paid by the client (and used to cover the coaching activities by our lecturers) validate the authenticity of the topics proposed and deter the creation of artificial questions. Moreover, the costs heighten the students' consciousness of the assignment's significance, which in turn stimulates the student team and the client to collaborate closely for optimal results and knowledge exchange.

Our approach benefits students, faculty, and project clients [15]. Student BPM consulting projects serve as a crucial bridge between academia, practical teaching, and students' professional careers. As consultants, students apply their academic knowledge to real-life projects, enhancing their professional and personal skills. They also develop solutions to specific BPM problems and apply critical thinking. For companies involved, the primary advantage of BPM projects is gaining access to expert knowledge. This is achieved by closely integrating projects and curricula and imparting theoretical foundations to students. Students, acting as third parties, can assess problems from an external perspective and provide neutral recommendations. Companies also benefit from a valuable recruiting platform through the BPM projects. For lecturers, the projects offer vast opportunities for aligning their research and teaching with practice and for understanding the practical implications of their subject areas. In addition, they get in direct contact with executives and other practitioners, potentially leading to further consulting or research projects. Figure 1 summarizes the core concept our student consulting projects and the discussed benefits for stakeholders.

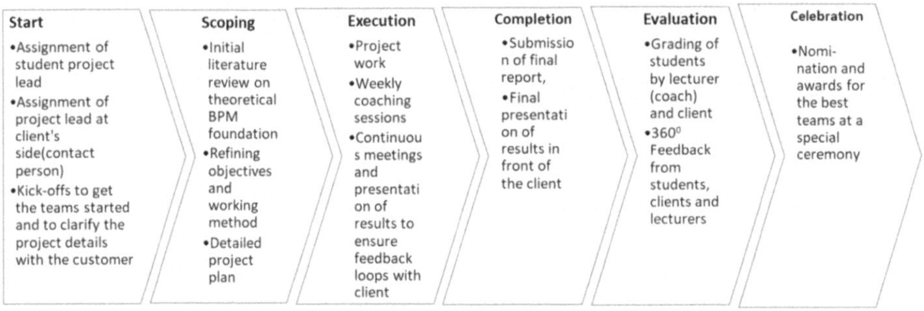

Fig. 2. Project life cycle of real-life consulting projects

All projects are procured, coordinated, and overseen by the Competence Center for Knowledge Transfer and Innovation, a division that was specifically stablished for this purpose 20 years ago and today incorporates this wealth of project experience into further teaching and research initiatives. During the *acquisition phase*, the center is tasked with ensuring an adequate number of projects, assessing the complexity level, and defining the objectives. Acquisition initially addresses companies and organizations from the university's large network. Every year, all previous customers as well as other partners and alumni are contacted. The network is also expanded at local events and through presentations at the university. Cold calling is used as a last resort to find project partners. The content of the project as well as the working methods and other legal aspects are set out in a project contract.

Throughout the entire project lifecycle, students follow a pre-established project management methodology [16]. This approach, grounded in universally applicable project management principles, has been instituted, standardized, and progressively refined over numerous years of hands-on experience with executing real-life consulting projects (see Fig. 2).

Start: Before the project's initiation, a student team member is designated as the project leader, responsible for overseeing the entire team and managing all client communications. Additionally, an individual from the client organization is assigned to the project. These two individuals, acting as the interface between the client and the project team, maintain a steady information exchange and are accountable for the project's successful completion. The projects begin with several official kick-offs, including a kick-off between the Competence Center and the coaching lecturers to establish common guidelines and a project framework, a kick-off between the Competence Center and the students to convey project expectations, rules, and guidelines, and a kick-off between the students and the client to discuss the assignment in detail.

Scoping: The subsequent scoping phase is a critical project period. Students must develop a comprehensive understanding of the company and its project motivation or problem to formulate project objectives and an appropriate approach. The project's specific scope is then formally outlined in a written agreement between the students and the client, building on the initial document created during the acquisition phase.

Execution: Once the assignment is clarified, the student project team begins the actual work, typically starting with a literature review for the theoretical foundation and additional meetings and interviews to understand the customer's specific context. The student project team is closely monitored by the coaches through weekly coaching sessions to ensure the results' quality and applicability. The first interim results are discussed with the client within a predetermined period. Based on the feedback received, the project undergoes further development.

Completion: The results are presented to the customer once more before being formally delivered as a final report and presented to a larger audience at the client's location.

Evaluation: The project concludes with a final evaluation by the coach and the client. A comprehensive debriefing (360^0 Feedback), obtained from the students, coaches, and the client's project lead, allows the Competence Center to continually refine the concept of real-life consulting projects for students and enables lecturers to adapt their BPM curriculum to industry needs.

Celebration: The six best projects are recognized each year by an external jury at an awards ceremony. The objective is not only to acknowledge the students' achievements but also to facilitate networking among clients and coaches and to identify potential candidates for future BPM projects.

2.2 Underlying Didactic Principle

Student consulting projects are recognized as effective tools for enhancing professional competence and cognitive complexity, thus involving higher levels of Bloom's taxonomy, such as levels 4, 5, and 6, respectively [14]. Participating in teams and taking on the role of a business consultant provides students an alternative perspective to real problems and thus promotes experiential learning and reflection of theoretical knowledge already acquired during the study degree [17]. This active-learning approaches allow students to extend their learning beyond traditional lecture-based classroom instruction [13, 18], while building on this this knowledge to work on the BPM challenges in a professional and methodical manner. The variety of content and methods used in real-life BPM projects give the students the required freedom possible for exploration, analysis and experimentation. While the students are on safe ground as far as the general process and the accompanying support measures are concerned, they are free to choose the required methods and tools in the project, based on the given context and project objectives. Due to the variety of topics, students are often confronted with content in which they first must find their own way for example by means of desk research. This requires a high degree of self-organisation, which can be achieved through self-chosen creativity, selection, and structuring techniques. The high degree of independence of the students is supported by the supportive environment in the sense of accompanying self-study. As a result, students in later projects are already practiced in gaining a fast overview of a subject area and concentrate more on the independent development of models and concepts. In summary, the student consulting projects aim to address the following learning goals:

Participants can:

- apply the previously learned theory and methods to solve a specific practical problem in BPM and implement the formal requirements of a holistic, scientific conceptual work in within a real-life project,
- meet the formal and scientific requirements of a rigorous consulting project involving understanding customer requirements, as well as secondary and primary research,
- plan and carry out a complex BPM project and document and present results in a manner appropriate to the target group,
- reflect and challenge their own judgement as well as the opinions of the team members, familiarize themselves with a new topic in a short time, demonstrate flexibility, independence, and responsibility, work under pressure, show endurance and be able to take criticism, and develop a sense for conflict potential and find constructive solutions, and
- demonstrate their project management, communication, team, and leadership skills.

This is evaluated based on the following assessment (same grade for the overall group), for which a grading sheet ensures the equal assessment of all coaches:

- The students provide a project report which shows their results (max 60 pages).
- The students give a poster presentation in one of the coaching sessions (grade integrated in the assessment of project work).
- The students present and discuss the project results with the client - duration 60' + 30' discussion (grade integrated in assessment of project work).

3 Project Example - Introducing Business Process Management at Glaroform AG

Company Description. GLAROFORM AG is active in the field of tool and mould making (High quality and thin wall moulds from Glaroform AG). It produces high-performance injection molds for both domestic and international customers, focussing on packaging of food and medical products. The company's market position is in a high-quality segment and its key differentiators are the highest achievable levels of efficiency and longevity of its products, in addition to an exemplary service delivery and technical support. GLAROFORM AG's activities include providing customers with technical and commercial advice as well as designing, manufacturing, and assembling the mould. Furthermore, the company offers a range of after-sales services such as spare parts, repairs and tool revisions, ensuring lifelong product support for its customers.

GLAROFORM AG was established in 1986 and currently employs 46 individuals, including six apprentices, and generates annual sales of approximately CHF 10 million. GLAROFORM AG is highly internationalised, with an export share of approximately 90%. Its principal markets are Europe, followed by Africa and Australia.

Project Goals. As a result of steady growth, GLAROFORM AG has reached a critical size, making it necessary to introduce a systematic approach to documenting, analysing, improving, and streamlining business processes to achieve organisational goals. This project constituted a first step in this journey with the aim to create an understanding of the existing processes and to document and optimise the steps taken in order fulfilment. Precisely, the primary objectives of the project were to

1. provide a complete overview of the existing processes at GLAROFORM AG in the form of a process map,
2. model the existing order fulfilment process at GLAROFORM AG, from enquiry to dispatch,
3. define and illustrate the responsibilities of the tasks in the order fulfilment process,
4. document the information flow, including data storage and exchange, within the process, and
5. to identify optimisation potential allowing for a later digitalisation.

Results. The project groups began with a systematic literature review to define the key terms in the BPM glossary, the purpose and scope of process maps, and basic modelling principles. The results were used to provide the team and the customer GLAROFORM AG with a theoretical basis for understanding and further developing BPM capabilities. A participatory approach (interviews with involved employees) was then chosen for creating the overall process map and documenting the processes, with the aim of actively involving and educating employees from the very beginning and minimising resistance to change. The order fulfilment process was then modelled using BPMN. Two further levels of detail were derived, and responsibilities were added for each task. Besides the process models, a detailed representation of information and process interfaces was created. When analysing the order processing process, the project group identified several areas for optimisation and derived two recommendations for action. As a first recommendation, the project group proposed a project management and collaboration tool that could increase efficiency through information exchange and centralised data storage. The second recommendation focussed on measures to improve data quality in the business software. *"In intensive cooperation with the customer, the project team solved the task professionally and laid the foundation for the successful introduction and use of process management at GLAROFORM AG. In addition, the project team went the extra mile and developed pragmatic recommendations for optimizing the processes, which have a high practical benefit."* (Sabrina Javaid, Lecturer and Coach)

4 Lessons Learned for Educators

From the vast experience we derive several best practices for designing and executing successful real-life student consulting projects, which we divide into topics related to the integration within the BPM curriculum, as well as topics related to project management.

Iterative Synchronization of Organizational BPM Curriculum and Real-Life Consulting Projects: Real-life BPM consulting projects allow students and lecturers to engage with complex issues. However, the inherent volatility and complexity of these projects often surpass the stability of the BPM curriculum. This necessitates a robust self-learning competence among students and mandates educators to perpetually update and adapt the curriculum to align with contemporary scientific discoveries and practical requirements. To effectively navigate these academic challenges, two strategies have been identified as beneficial. Firstly, at the conclusion of each BPM consulting project, an online survey is conducted to collect feedback from all participants - students, educators, and clients - on the applicability and potential improvements of the project outcomes. This feedback is then utilized to ensure the acquisition of projects that complement the theoretical lessons delivered in classrooms. Secondly, educators involved in regular teaching also undertake the role of project coaches, thereby gaining firsthand insights into the specific industry demands and incorporating them into the curriculum. Figure 3 illustrates the implementation of these feedback loops in relation to the BPM curriculum and each individual teaching unit (TU). This iterative approach facilitates the continuous realignment of BPM course content with practical necessities, thereby enhancing overall pedagogical effectiveness.

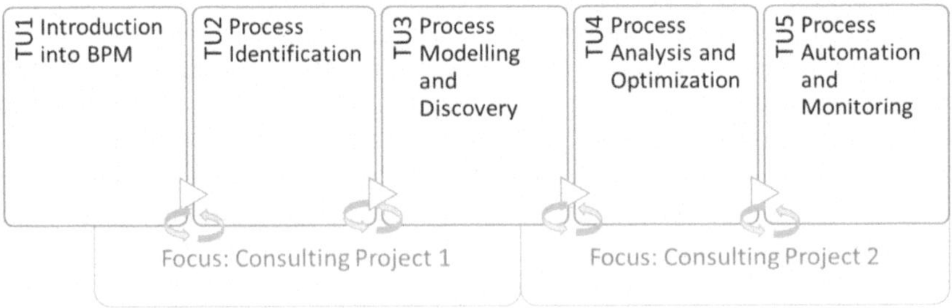

Fig. 3. Embedding real-life consulting projects within the BPM curriculum (teaching units) and BPM life cycle

Competency Development Along the BPM Lifecycle: An important mechanism to increase student's competencies is to align their development along the BPM lifecycle, which outlines the stages involved in managing a process from start to finish [1, 19]. Following the approach taken in the BPM curriculum of our organization, it has proven successful to also increase the complexity of the consulting projects accordingly. This leads to students learning successively about real-life processes, their design, the execution, and the optimization (see Fig. 3).

Establishment of Clear Objectives: Consultants typically engage in a social process of dialogue and debate with their clients to refine the problem definition and the purpose of the project. This initial phase should be accorded great importance in student projects and should include an examination of the company as well as the initial situation and the actual problem the project is intended to solve. It is recommended that a manageable focus is established for the project. This supports the approach of vom Brocke et al. (2021) to ensures that the students select the correct methods and approaches suitable for the given context and client [20].

Consensus on Scope Statement: It is imperative for students to submit a scope letter that delineates the objectives by a predetermined date and to secure a formal agreement from all stakeholders involved. This facilitates the completion of the project within the stipulated timeframe and to the anticipated quality, as well as accommodates changes requested by the customer during the project's duration.

Adoption of a Common High-Level Project Plan: A crucial determinant of successful BPM project implementation is the establishment of a temporal and content framework for the students. The provided schedule, inclusive of deadlines for the submission of various interim reports, aids in project orientation and progression, thereby mitigating significant delays. However, it is strongly recommended that the student team independently formulates a detailed, dynamic project plan, inclusive of team members' responsibilities, to guide them throughout the project's duration. This should be developed at the project's inception in collaboration with the coach and dynamically adjusted as necessary during the project. Regular project status checks should be conducted during coaching sessions, enabling early intervention if the project lags schedule.

5 Conclusion

In this paper, we showcased the effectiveness of real-life student consulting projects to enhance practical BPM competencies and bridge the gap between research and practice within the BPM domain. Our organization's well-established methodology, refined through decades of local company collaborations, has yielded valuable insights. By aligning student development with the BPM life cycle, we empower learners with relevant skills for process management and consulting. As educators, we recognize the importance of gradually increasing project complexity to prepare students for the challenges they'll face in their professional careers. And the customers benefit from directly applicable solutions that are based on state-of-the-art BPM-knowledge.

Disclosure of Interests. The authors have no competing interests to declare that are relevant to the content of this article.

References

1. Dumas, M., La Rosa, M., Mendling, J., Reijers, H.A.: Fundamentals of Business Process Management, 2nd edn. Springer, Heidelberg (2018). https://doi.org/10.1007/978-3-662-56509-4
2. Hammer, M., Champy, J.: Reengineering the Corporation: A Manifesto for Business Revolution. Harper Business, New York (1993)
3. Kerpedzhiev, G.D., König, U.M., Röglinger, M., Rosemann, M.: An exploration into future business process management capabilities in view of digitalization. Bus. Inf. Syst. Eng. **63**(2), 83–96 (2021)
4. Lückmann, P., Feldmann, C.: Success factors for business process improvement projects in small and medium sized enterprises – empirical evidence. Procedia Comput. Sci. **121**, 439–445 (2017)
5. Bandara, W., et al.: Business process management education in academia: status, challenges, and recommendations. Commun. Assoc. Inf. Syst. **27**(41), 743–776 (2010)
6. Chang, M.C.: Historical and theoretical foundations of case-based learning: situated cognition and collaborative learning theory. In: Quek, C.L.G., Wang, Q. (eds.) Designing Technology-Mediated Case Learning in Higher Education, pp. 49–68. Springer, Singapore (2022). https://doi.org/10.1007/978-981-19-5135-0_4
7. Orr, L., Weekley, L.: Teaching with case studies in higher education. In: Baron, A., McNea, K. (eds.) Case Study Methodology in Higher Education, pp. 180–208. IGI Global, Hershey (2019)
8. Bandara, W., Opsahl, H.: Developing organization-wide BPM capabilities in an SME: the approaches used, challenges and outcomes. J. Inf. Technol. Teach. Cases **7**(2), 92–113 (2017)
9. Bandara, W., Abbott, C., Mathiesen, P., Meyers, L., Nagra, M.: Developing enterprise-wide business process management capability: a teaching case from the financial sector. J. Inf. Technol. Teach. Cases **8**(2), 192–208 (2018)
10. Bandara, W., Bailey, S., Mathiesen, P., McCarthy, J., Jones, C.: Enterprise business process management in the public sector: the case of the department of human services (DHS) Australia. J. Inf. Technol. Teach. Cases **8**(2), 217–231 (2018)
11. Bandara, W., Thennakoon, D., Syed, R., Mathiesen, P., Ranaweera, K.D.S.: Improving the clinical-care pathway of an ayurvedic hospital: a teaching case for developing process improvement capabilities. J. Inf. Technol. Teach. Cases **6**(2), 111–120 (2016)
12. Baron, A., McNea, K. (eds.): Case Study Methodology in Higher Education. IGI Global, Hershey (2019)
13. Mebert, L., et al.: Fostering student engagement through a real-world, collaborative project across disciplines and institutions. High. Educ. Pedagog. **5**(1), 30–51 (2020)
14. Canziani, B., Tullar, W.L.: Developing critical thinking through student consulting projects. J. Educ. Bus. **92**(6), 271–279 (2017)
15. Bechter, M., Brönimann, C., Ivancic, R., Mueller, P.: Applied sciences at FHS St. Gallen University of Applied Sciences. Real-life student consulting projects as the link between academic, practice-oriented education and economic value. Magazin Erwachsenenbildung.at: das Fachmedium für Forschung, Praxis und Diskurs (2019)
16. Brönimann, C., Bechter, M., Ivancic, R.: Einfachheit in didaktischer Form – Erkenntnis durch komplexe Herausforderungen. Wissenstransfer in studentischen Beratungsprojekten. In: Moll, G., Schürz, Julia (eds.) Wissenstransfer - Komplexitätsreduktion - Design, pp. 131–148. wbv Media GmbH (2022)

17. Lucey, S.: Using student consultancy projects in enterprise and marketing programmes: some reflections. In: Santos, E., Ribeiro, N., Eugénio, T. (eds.) Rethinking Management and Economics in the New 20's. Springer Proceedings in Business and Economics, pp. 439–455. Springer, Singapore (2023). https://doi.org/10.1007/978-981-19-8485-3_19
18. Krain, M.: The effects of different types of case learning on student engagement. Int. Stud. Perspect. **11**(3), 291–308 (2010)
19. Macedo de Morais, R., Kazan, S., Inês Dallavalle de Pádua, S., Lucirton Costa, A.: An analysis of BPM lifecycles: from a literature review to a framework proposal. Bus. Process Manag. J. **20**(3), 412–432 (2014)
20. vom Brocke, J., Baier, M.-S., Schmiedel, T., Stelzl, K., Röglinger, M., Wehking, C.: Context-aware business process management. Bus. Inf. Syst. Eng. **63**(5), 533–550 (2021)

Towards an E-Learning Approach for Declarative Process Modeling

Sabine Nagel(✉) and Patrick Delfmann

University of Koblenz, Koblenz, Germany
{snagel,delfmann}@uni-koblenz.de

Abstract. In contrast to procedural process models, where every valid execution trace is explicitly modeled, declarative process models (DPMs) implicitly define allowed company behavior using a set of constraints. Related works have identified many challenges humans face when trying to make sense of DPMs, including combinations of constraints, inconsistencies, and the (graphical) notation of declarative constraints themselves. In this work, we provide the foundation for an e-learning approach designed to gradually familiarize users with the modeling language *Declare*. More specifically, we introduce a comprehensive collection of different types of tasks with increasing levels of difficulty. These tasks cover basic concepts, individual constraints, constraint combinations of different complexity, model behavior, as well as redundancy and inconsistency within DPMs. With this work, we aim to lay the foundation for future interactive applications to support not only teaching but also improving both comprehension and (consistent) declarative process modeling.

Keywords: Declarative Process Models · Process Modeling · Declare · Comprehension · E-Learning

1 Introduction

Declarative process models (DPMs) implicitly define allowed company behavior using a set of constraints [1], which differs from procedural process models, where every valid execution trace is explicitly modeled. The general idea of DPMs is that every process execution is allowed, as long as it does not violate any constraint [2]. While this approach allows for more flexible process definitions, many works have identified challenges humans face when trying to make sense of DPMs, which is mainly due to their implicit and circumstantial nature [3]. This stresses the need for resources to aid humans in becoming familiar with and learning how to read and actively model DPMs.

Thus, this paper aims to provide the first step towards an e-learning approach for declarative process modeling, based on insights from scientific literature. The idea is to identify different categories and types of tasks that not only cover basic concepts of DPMs but also specifically target understanding challenges identified in related studies. To support users with different backgrounds and experience levels, we focus on flexibility and extensibility and design tasks in a way that allows automatic evaluation and task

C. Di Ciccio et al. (Eds.): BPM 2024, LNBIP 527, pp. 354–364, 2024.
https://doi.org/10.1007/978-3-031-70445-1_24

generation based on user-defined settings and complexity levels. Thereby, we aim to lay the foundation for future interactive applications to support not only BPM education but also improving declarative process modeling and comprehension.

In Sect. 2 we provide preliminaries and related work on DPMs and DPM understanding. We introduce our approach in Sect. 3, followed by a discussion of possible task categories and scenarios in Sect. 4. We conclude with a summary and outlook.

2 Preliminaries and Related Work

2.1 Declarative Process Models

A declarative process model is a tuple $\boldsymbol{DPM} = (\boldsymbol{T}, \boldsymbol{A}, \boldsymbol{C})$, where.

- $\boldsymbol{T}$ is a finite, non-empty set of templates with each template being a unary $P(x_1)$ or binary predicate $P(x_1, x_2)$ representing existence and relation templates.
- $\boldsymbol{A}$ is a finite, non-empty set of activities
- $\boldsymbol{C}$ is a finite set of constraints with a constraint $c \in \boldsymbol{C}$ being an instantiation of a template $t \in \boldsymbol{T}$ by mapping an activity $a \in \boldsymbol{A}$ to each variable x.

In this paper, we consider the modeling language *Declare*, which provides a set of pre-defined templates that can be used to model constraints [4]. *Existence* templates describe properties of single activities while *relation* templates describe the relationship between two activities with one activity (= *activation*) requiring or prohibiting a target activity. Furthermore, templates follow a subsumption hierarchy. For example, all traces that satisfy *ChainResponse(a, b)* ("if *a* occurs in a trace, it must *immediately* be followed by *b*") also satisfy *Response(a, b)* ("if *a* occurs in a trace, it must *eventually* be followed by *b*"), as the latter constraint has a weaker condition and is, therefore, *redundant*. For a full overview of templates, including a detailed description and formalization, we refer to the paper by Di Ciccio and Montali [2]. *Declare* constraints can be represented in different ways, with the two main forms of representation being textual and visual representations. Table 1 provides different representations for two exemplary constraints, including a textual description of the condition in natural language and a deterministic finite automaton visualizing the allowed behavior of the constraint.

Table 1. Overview of Different Forms of Constraint Representation

	Existence	**Relation**
Description	Each trace must start with a.	If a occurs in a trace, it must eventually be followed by b.
Textual Constraint	*Init(a)*	*Response(a,b)*
Visual Constraint	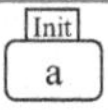	
Deterministic Finite Automaton (DFA)	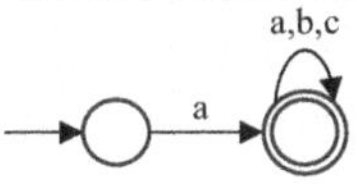	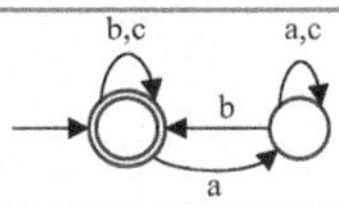

If a DPM does not accept any finite execution trace (e.g., *{Init(a), Response(a, b), NotResponse(a, b)}*), we refer to the model as unsatisfiable, as it is inconsistent in the classic-logical sense [4]. Additionally, recent works [5] have introduced so-called potential inconsistencies, i.e., constraint sets that are consistent in the classic-logical sense but become inconsistent upon the occurrence of the activating activity in a trace (e.g., *{Response(a, b), NotResponse(a, b)}*, with *a* as the activating activity). Furthermore, inconsistencies are considered *minimal* if removing a single constraint automatically resolves the problem.

2.2 Existing Studies on DPM Understanding

In previous work, we have conducted a systematic literature review to identify research models and empirical studies investigating DPM understanding [3]. As a result, we were able to derive seven understanding challenge categories that each comprise one or more challenges humans face when trying to make sense of DPMs. For a detailed discussion of the identified categories and related challenges, we refer to [3]. As we aim to base our approach on insights from these works, we extended this body of literature to also include works published since then, which led to the identification of two recent studies [6, 7]. Table 2 provides an overview of all relevant papers, including the considered language, artifacts, and focus. For artifacts, we distinguish between textual (T) or visual (V) models, generic (NL_G) or specific (NL_S) natural language descriptions, means for simulation (S) and dependency graphs (D). In the following, we will discuss the tasks applied in these studies, as these provide the basis for our task development.

Table 2. Overview of Papers on DPM Understanding

Paper		Language	Artifacts	Focus
Andaloussi et al. (2019)	[8]	DCR-HR	V, NL_S, S	Hybrid Representations
Andaloussi et al. (2019)	[9]	DCR-HR	V, NL_S, S	Hybrid Representations
Andaloussi et al. (2020)	[10]	DCR	V	Modularization
Andaloussi et al. (2021)	[11]	DCR	V, NL_G	Complexity
Andaloussi et al. (2021)	[12]	DCR-HR	V, NL_S, S	Hybrid Representations
Andaloussi et al. (2023)	[6]	DCR	V, NL_G	Complexity
De Smedt et al. (2016)	[13]	Declare	V + T, D	Hidden Dependencies
De Smedt et al. (2018)	[14]	Declare	V + T, D	Hidden Dependencies
Figl et al. (2020)	[1]	Declare	V, NL_S	Visual Representation
Haisjackl et al. (2013)	[15]	Declare	V	Explore Understanding
Haisjackl et al. (2014)	[16]	Declare	V, T	Visual vs. Textual
Haisjackl et al. (2016)	[17]	Declare	V	Hierarchical vs. Flat Models
Nagel & Delfmann (2024)	[7]	Declare	V, T, NL_G	Inconsistencies
Pichler et al. (2012)	[18]	ConDec	V	Declarative vs. Imperative
Zugal et al. (2011)	[19]	ConDec	V, S	Maintainability
Zugal et al. (2015)	[20]	ConDec	V	Hierarchy

Andaloussi et al. [8, 9, 12] investigated hybrid DCR graphs and confronted participants with dichotomous understandability questions: constraint tasks that target the relationship between two activities, decision tasks focusing on contextual information participants must identify, and scenario tasks that provide participants with a case history and ask them about allowed behavior based on the provided history. In another study, Andaloussi et al. [10] examined the impact of modularization on understanding DPMs by asking participants to solve dichotomous local and global tasks, addressing "the presence and absence of constraints in the model, order of activities and validity of process executions". Next, Andaloussi et al. [6, 11] studied the effects of DPM complexity on model comprehension by asking participants to provide a valid execution path containing each activity only once. According to the authors, this choice was made to keep the chances of guessing rather low and to be able to make the entire model relevant for answering the question, as order, dependencies, and mutual influence must be checked.

De Smedt et al. [13, 14] focused on understanding hidden dependencies in *Declare* models by asking participants about activities being enabled after firing a certain sequence and about reaching provided firing sequences in general. Figl et al. [1] examined the effect of visual model representation on DPM comprehension. They applied Wason selection tasks to investigate whether participants understood the concept of constraint activation and target. To study comprehension of entire models, participants were asked to classify process runs as valid or invalid. Haisjackl et al. [15, 17] conducted a study asking participants for valid, invalid, and minimal traces and included questions on hidden dependencies, pairs of constraints, combinations of constraints, and existence constraints. Using the same tasks, Haisjackl et al. [16] performed another study, focusing on differences in understanding the graphical and textual representation of models.

Nagel and Delfmann [7] conducted an exploratory study on inconsistency understanding in textual and visual *Declare* models. They provided participants with inconsistent models and asked them to either provide a valid trace or – if there is no valid trace due to the model being unsatisfiable – to explain the problem in their own words.

Pichler et al. [18] compared declarative and procedural process models in their study and covered both sequential (i.e., order, concurrency, exclusiveness, and repetition) and circumstantial tasks (i.e., existence, relation, and negation constraints) to cover all aspects of process model understandability. Zugal et al. [19] investigated the maintainability of DPMs by providing participants with visual *Declare* models, respective test cases, and so-called change assignments (e.g., "Remove precedence between A and B"). These assignments comprised of requirements "that hold for the initial model and must not be violated by the changes conducted". Lastly, Zugal et al. [20] studied the expressiveness and understandability of hierarchy by confronting participants with questions to cover all types of constraints (i.e., existence, negation, and ordering). Additionally, participants had to determine if a provided trace is valid for the given model.

3 Approach Overview and Basic Concepts

To develop the foundation for a flexible e-learning approach, we analyzed the existing literature on DPM understanding (cf. Sect. 2.2) to determine general requirements for an e-learning approach, based on the identified challenges (see Table 3). Next, we specifically considered all papers that have planned and/or executed DPM comprehension studies and extracted further details on the task they provided the participants with (cf. Sect. 2.2) as a basis for designing appropriate tasks. Additionally, we considered all possible relations between artifacts and analyzed their potential to be represented and/or covered in different tasks, which we will explain in detail in Sect. 4.

Table 3. Mapping of Understanding Challenge Categories and E-Learning Approach

Challenge Category		Sources	E-Learning Approach
UC1	Complexity	[6, 7, 17]	- increasing difficulty of tasks (e.g., scenario complexity)
UC2	Individual Constraints	[14, 16, 17]	- template definitions - constraint behavior and basic inference - hierarchy/redundancy
UC3	Constraint Combinations	[7, 14, 17]	- implicit behavior (e.g., valid/invalid traces, redundancy) - inconsistency
UC4	Representation	[7, 16]	- matching textual & visual representation - cover both forms of representation (when applicable)
UC5	Reading Order	[9, 17]	- model behavior (e.g., valid/invalid traces)
UC6	Background & Experience	[1, 17]	- customizable terminology of activities - generation of custom tasks with specific parameters - set learning objective for tasks (e.g., focus on backward constraints)
UC7	Task Type	[1, 18]	- provide different task types to cover both comprehension (= recognition) and execution/modeling (= recall)

As we aim for an approach that enables a flexible extension and automatic task evaluation and generation, we only consider task types that provide selectable options to users or request users to input simple artifacts (e.g., activities, traces, constraints). Additionally, we cover the comprehension and execution of provided models, as well as active modeling, following the concepts of recognition and recall [21]. As a result, and based on the different types of tasks identified in related studies (cf. Sect. 2.2), we consider the following types of tasks, with the possibility for extension in future iterations: single choice questions (SC), where only one of the provided options is correct, multiple choice questions (MC) where any number of provided options is correct, true/false questions (T/F) where a correct or incorrect statement is provided, and text input questions (IN) where users must actively provide an answer. Generally, a task consists of a *question* (Q), an underlying *scenario* (S) that provides the background information needed to answer the question (e.g., a *declarative model* or a *trace*), and a set of correct and incorrect answers as *options* (O). Table 4 shows four tasks based on a shared scenario. While selection-based questions can be evaluated by comparing the selection with the correct answers, input tasks must be evaluated on the fly after a user has provided a solution, which emphasizes the need for automatic task evaluation.

Table 4. Exemplary Tasks of Each Type

	SC	MC	IN	T/F
Q	*Which trace satisfies the given model?*	*Which traces satisfy the given model?*	*Please provide a valid trace for the given model.*	*The trace <a,b,c,a> satisfies the given model.*
S			{a,b,c} {Init(a), Response(a,b)}	
O	*<a,b,c,a>* (x) *<a,b,c,b>* (✓) *<b,c,a>* (x) *<a,c,a>* (x)	*<a,b,c,a>* (x) *<a,b,c>* (✓) *<a,b,a,b>* (✓) *<a,a,b>* (✓)	-	True (x) False (✓)

To provide appropriate tasks for users of all experience levels and to support a user's development from beginner to expert, we require means for increasing the difficulty of a task. To this aim, we distinguish between three different factors that can be adjusted.

- *Task Type*: As previously described, recognition tasks are generally associated with lower cognitive load and mental effort compared to recall tasks.
- *Scenario Complexity*: DPM complexity is one of the main factors that affect DPM comprehension, which also affects the perceived difficulty of solving related tasks. For example, we can increase the number of constraints, distinct templates, or the connectivity to increase the overall complexity of a DPM.
- *Plausibility*: If selectable options are obviously unrelated or incorrect, elimination will be easier compared to options that are very similar to the correct answer, as the latter requires careful consideration to distinguish incorrect from correct answers.

4 Artifact Relations and Resulting Scenarios

As already explained in Sect. 3, we aimed to identify relevant artifact combinations by considering all relationships between artifacts (Table 5). In addition to simple "*contains*" relations, we were able to identify several combinations where two artifacts directly influence each other. We derived the following types of relationships: (1) *satisfaction* (a trace can satisfy a constraint/model), (2) *inference* (an activity can activate a constraint/model, which then requires or prohibits at least one target activity), (3) *redundancy* (a constraint/model can subsume another constraint/model), (4) *inconsistency* (a model is unsatisfiable and does not allow a single valid trace) and (5) *matching* (comparing two models in different representations). In the following, we explain these relationships in more detail, discuss their potential for being represented and/or covered in tasks, and provide exemplary questions. This also includes a discussion of the required characteristics of the involved artifacts, and the possibilities for automatic evaluation and task creation, which is a requirement for selecting appropriate tasks for an e-learning approach.

Table 5. Artifact Relationships

	Activity	Trace	Constraint	Model
Activity		contained in	activates required/prohibited by	activates required/prohibited by
Trace	contains		satisfies	satisfies
Constr.	activated by requires/prohibits	(not) satisfied by	equals/subsumes contradicts	contained in/subsumed by contradicts
Model	contains requires/prohibits	(not) satisfied by	contains/subsumes contradicts	equals subsumes

4.1 Model Satisfaction

When evaluating an execution trace against a model, we can determine whether the trace is valid (i.e., satisfies all constraint in the model) or invalid (i.e., it violates at least one constraint). Additionally, traces can be minimal, meaning that they only contain required activities and nothing else. In this category, we distinguish between two different directions of a satisfaction relationship: either the user is provided with a model and is asked for valid/invalid/minimal traces or vice versa, i.e., a trace is provided, and the user must enter a constraint that is (not) satisfied by the trace. To accept at least one valid trace, a DPM must be consistent in the classic-logical sense. If an inconsistent model is provided, there are only invalid traces. Thus, if a valid trace is requested for inconsistent models, the user must have the option to select that there is no valid trace. Furthermore, a minimal trace can only be constructed for models containing at least one existence constraint requiring the presence of a specific activity in a trace (e.g., *AtLeastOne(a)*), as otherwise, the empty trace would equal the minimal trace.

4.2 Inference, (Hidden) Dependencies and Redundancy

If an activity activates a constraint/model, it requires or prohibits one or more target activities. As a result, we can map the activity-model relationship to tasks where the user is provided with a constraint/model and must enter or select an activity that can be inferred from a condition specified in the question. Here, the questions are phrased analogous to textual constraint descriptions and the answer can be evaluated by checking whether the constraint implementing the condition is already explicitly or implicitly included in the model (i.e., conducting a redundancy check). If a single constraint is provided, the focus is on understanding constraint/template definitions themselves, while providing more complex models shifts the focus to understanding hidden dependencies between activities. In the following, we provide exemplary questions.

- *If activity X occurs in a trace, what activity must it eventually/immediately be followed by?*
- *If activity X occurs in a trace, what activity must also occur somewhere in that trace?*
- *If activity X occurs in a trace, what activities are (no longer) enabled?*
- *What activity must occur [at least|at most|exactly] [once|twice|three times] in a trace?*
- *What activity must all traces [start|end] with?*

In addition to such inference-related questions, we can also directly cover redundancy. This applies for subsumption relations between two constraints/models, i.e., the user is provided with a constraint/model and asked to provide a constraint that is already implicitly included in the provided model (e.g., "*Which constraint can be added/removed from the model, without changing the model's behavior?*"). Similarly to inference-related tasks, the focus when providing a single constraint is on understanding the subsumption hierarchy between templates, while more complex models rather require the user to understand the model as a whole and identify hidden dependencies between activities. The underlying models must still allow for or contain redundancy and be consistent in the classic-logical sense, as otherwise, the accepted language is already empty, which cannot be changed by adding constraints. Also, removing any constraints that are not involved in an inconsistency will not lead to any changes to the (empty) language, even though the removed constraint might not be redundant.

4.3 Inconsistency

Generally, inconsistency-related tasks can have different goals: (1) causing inconsistency, (2) resolving inconsistency to restore consistency or (3) maintaining consistency. Table 6 provides an overview of exemplary questions for each of these goals and any prerequisites for the underlying scenario. More specifically, based on the definition of (minimal) inconsistencies, certain conditions must hold for a single constraint to be able to cause or resolve inconsistency. In addition to inconsistencies in the classic-logical sense that already occur at design time, *potential inconsistencies* only occur during run-time, i.e., when an activity triggers an unsatisfiable set of constraints. This can be represented by tasks that provide a consistent model containing one or more potential inconsistencies and asking the user for an activity that activates them ("*Which activities must not occur in a trace, as they activate run-time inconsistencies?*").

Table 6. Exemplary Questions and Model Prerequisites for Inconsistency-Related Tasks

Goal	Exemplary Question	Model Prerequisites
(1) Cause Inconsistency	*Which constraint can be added to the given model, to make the model inconsistent?*	• consistent model • contains at least one potential inconsistency *or* an existence constraint
(2) Resolve Inc./Restore Consistency	*Which constraint can be removed from the model to restore consistency?*	• non-minimally inconsistent model • if multiple minimally inconsistent subsets: must have overlapping constraint
(3) Maintain Consistency	*Which constraint can be added to the given model, without making the model inconsistent?*	• consistent model

4.4 Matching Representations

As a last category, we consider different ways of matching constraints represented in different forms. Here we consider both textual and visual constraint representations, natural language descriptions, and deterministic finite automata (DFA), for examples of these forms of representation, please see Table 1. While all relevant combinations can be implemented as true/false questions *("Do the provided representations display the same constraint?")*, other combinations might, for instance, be limited to SC (as only one representation matches exactly). Furthermore, as textual input (IN) questions only allow for simple inputs, they are only applicable when textual constraints must be entered. While representation-related tasks would technically also be applicable for entire models, we decided to only consider individual constraints in the first version, as displaying entire models in different forms of representation might be too large to display in a meaningful way, especially if they are provided as selectable options.

5 Summary and Conclusion

In this work, we have taken a first step towards developing an e-learning approach for declarative process modeling by developing a collection of task categories and scenarios based on scientific literature, covering basic concepts, individual constraints, constraint combinations of different complexity, model behavior, as well as redundancy and inconsistency within DPMs. Currently, we are implementing this approach to develop an application for manual and automatic task creation, filtering tasks based on different characteristics (e.g., complexity) and evaluating the selected or provided answer using several evaluation algorithms (e.g., redundancy and inconsistency checks or validating traces against models). As a next step, we will empirically evaluate the developed approach in the BPM course at our university and work towards developing an advanced and gamified e-learning application for declarative process modeling. That way, we aim to support BPM education in both universities and companies. However, our approach is not only targeted at students attempting to learn *Declare* in a stepwise and interactive manner but also at educators themselves, as our approach provides flexible means for automatic task creation and evaluation. Additionally, insights gained during the development, evaluation, and use of our approach will contribute to research on DPM comprehension and (manual) declarative process modeling.

Acknowledgments. This paper was funded by the Deutsche Forschungsgemeinschaft (grant number DE 1983/9-3).

Disclosure of Interests. The authors have no competing interests to declare.

References

1. Figl, K., Di Ciccio, C., Reijers, H.A.: Do declarative process models help to reduce cognitive biases related to business rules?. In: Dobbie, G., Frank, U., Kappel, G., Liddle, S.W., Mayr, H.C. (eds.) ER 2020. LNCS, vol. 12400, pp. 119–133. Springer, Cham (2020). https://doi.org/10.1007/978-3-030-62522-1_9
2. Di Ciccio, C., Montali, M.: Declarative process specifications: reasoning, discovery, monitoring. In: van der Aalst, W.M.P., Carmona, J. (eds.) Process Mining Handbook. Lecture Notes in Business Information Processing, vol. 448, pp. 108–152. Springer, Cham (2022). https://doi.org/10.1007/978-3-031-08848-3_4
3. Nagel, S., Delfmann, P.: Investigating inconsistency understanding to support interactive inconsistency resolution in declarative process models. In: ECIS 2022 Research-in-Progress Papers (2022)
4. Di Ciccio, C., Maggi, F.M., Montali, M., Mendling, J.: Resolving inconsistencies and redundancies in declarative process models. Inf. Syst. **64**, 425–446 (2017)
5. Corea, C., Delfmann, P.: Quasi-inconsistency in declarative process models. In: Hildebrandt, T., van Dongen, B., Röglinger, M., Mendling, J. (eds.) BPM 2019. Lecture Notes in Business Information Processing, vol. 360, pp. 20–35. Springer, Cham (2019). https://doi.org/10.1007/978-3-030-26643-1_2
6. Andaloussi, A.A., Burattin, A., Slaats, T., Kindler, E., Weber, B.: Complexity in declarative process models: Metrics and multi-modal assessment of cognitive load. Expert Syst. Appl. **233**, 120924 (2023)
7. Nagel, S., Delfmann, P.: Exploring cognitive effects of inconsistency characteristics on understanding inconsistencies in declarative process models. In: Proceedings of the 57th Hawaii International Conference on System Sciences (HICSS) (2024)
8. Abbad Andaloussi, A., Slaats, T., Burattin, A., Hildebrandt, T.T., Weber, B.: Evaluating the understandability of hybrid process model representations using eye tracking: first insights. In: Daniel, F., Sheng, Q.Z., Motahari, H. (eds.) BPM 2018. LNBIP, vol. 342, pp. 475–481. Springer, Cham (2019). https://doi.org/10.1007/978-3-030-11641-5_37
9. Abbad Andaloussi, A., Burattin, A., Slaats, T., Petersen, A.C.M., Hildebrandt, T.T., Weber, B.: Exploring the understandability of a hybrid process design artifact based on DCR graphs. In: Reinhartz-Berger, I., Zdravkovic, J., Gulden, J., Schmidt, R. (eds.) BPMDS/EMMSAD -2019. LNBIP, vol. 352, pp. 69–84. Springer, Cham (2019). https://doi.org/10.1007/978-3-030-20618-5_5

10. Abbad Andaloussi, A., Soffer, P., Slaats, T., Burattin, A., Weber, B.: The impact of modularization on the understandability of declarative process models: a research model. In: Davis, F.D., Riedl, R., vom Brocke, J., Léger, P.-M., Randolph, A.B., Fischer, T. (eds.) NeuroIS 2020. LNISO, vol. 43, pp. 133–144. Springer, Cham (2020). https://doi.org/10.1007/978-3-030-60073-0_15
11. Andaloussi, A.A., Burattin, A., Slaats, T., Kindler, E., Weber, B.: Assessing the complexity of declarative process models using model-based metrics. In: A Framework for Enhancing the Modeling and Comprehension of Declarative Process Models, pp. 179–204 (2021)
12. Abbad Andaloussi, A., Zerbato, F., Burattin, A., Slaats, T., Hildebrandt, T.T., Weber, B.: Exploring how users engage with hybrid process artifacts based on declarative process models: a behavioral analysis based on eye-tracking and think-aloud. Softw. Syst. Model. **20**(5), 1437–1464 (2020)
13. De Smedt, J., De Weerdt, J., Serral, E., Vanthienen, J.: Improving understandability of declarative process models by revealing hidden dependencies. In: Nurcan, S., Soffer, P., Bajec, M., Eder, J. (eds.) CAiSE 2016. LNCS, vol. 9694, pp. 83–98. Springer, Cham (2016). https://doi.org/10.1007/978-3-319-39696-5_6
14. De Smedt, J., De Weerdt, J., Serral, E., Vanthienen, J.: Discovering hidden dependencies in constraint-based declarative process models for improving understandability. Inf. Syst. **74**, 40–52 (2018)
15. Haisjackl, C., et al.: Making sense of declarative process models: common strategies and typical pitfalls. In: Nurcan, S., et al. (eds.) BPMDS EMMSAD 2013 2013. Lecture Notes in Business Information Processing, vol. 147, pp. 2–17. Springer, Heidelberg (2013). https://doi.org/10.1007/978-3-642-38484-4_2
16. Haisjackl, C., Zugal, S.: Investigating differences between graphical and textual declarative process models. In: Iliadis, L., Papazoglou, M., Pohl, K. (eds.) CAiSE 2014. Lecture Notes in Business Information Processing, vol. 178, pp. 194–206. Springer, Cham (2014). https://doi.org/10.1007/978-3-319-07869-4_17
17. Haisjackl, C., et al.: Understanding Declare models: strategies, pitfalls, empirical results. Softw. Syst. Model. **15**, 325–352 (2016)
18. Pichler, P., Weber, B., Zugal, S., Pinggera, J., Mendling, J., Reijers, H.A.: Imperative versus declarative process modeling languages: an empirical investigation. In: Daniel, F., Barkaoui, K., Dustdar, S. (eds.) BPM 2011. Lecture Notes in Business Information Processing, vol. 99, pp. 383–394. Springer, Heidelberg (2012). https://doi.org/10.1007/978-3-642-28108-2_37
19. Zugal, S., Pinggera, J., Weber, B.: The impact of testcases on the maintainability of declarative process models. In: Halpin, T., et al. (eds.) BPMDS EMMSAD 2011 2011. Lecture Notes in Business Information Processing, vol. 81, pp. 163–177. Springer, Heidelberg (2011). https://doi.org/10.1007/978-3-642-21759-3_12
20. Zugal, S., Soffer, P., Haisjackl, C., Pinggera, J., Reichert, M., Weber, B.: Investigating expressiveness and understandability of hierarchy in declarative business process models. Softw. Syst. Model. **14**, 1081–1103 (2015)
21. Raaijmakers, J.G.W., Shiffrin, R.M.: Models for recall and recognition. Annu. Rev. Psychol. **43**, 205–234 (1992)

Order Selection Methodology of OMG Diagrams in Teaching Designing Information Systems

Anna Suchenia[1(✉)], Karol Suchenia[1], Marek Adrian[2], Krystian Jobczyk[2], Piotr Wiśniewski[3], and Antoni Ligęza[2]

[1] Cracow University of Technology, ul. Warszawska 24, 31-155 Kraków, Poland
{asuchenia,karol.suchenia}@pk.edu.pl

[2] AGH University of Krakow, al. A. Mickiewicza 30, 30-059 Krakow, Poland
{madrian,jobczyk,ligeza}@agh.edu.pl

[3] University of St. Gallen, 9000 St. Gallen, Switzerland
piotr.wisniewski@student.unisg.ch

Abstract. In learning how to design information systems, the specification of diagram order might speed up or slow down the work on the system design. Our study introduces and examines an order selection methodology employing different strategies using similarity or difference in the number of elements. The performance of these strategies was assessed by comparing their potential cumulative coverage of elements and diagrams. The results indicate that the strategy maximizing the number of new elements per selection provides the quickest total coverage of elements. The insights provided in this position paper might help educators select the most effective order of diagrams to teach students, thus enhancing their understanding and efficiency in system modeling.

Keywords: Teaching Strategies · Software Modeling Education · System Design Education · Modeling Process · Information Systems Design

1 Introduction

Modeling parts of the system with various diagrams is an integral part of the system development process, in which specific components are defined using visual models. Such models use various notations that facilitate understanding and analysis of the software design and implementation process. Frequently used notations [1] are those offered by *Object Management Group* (OMG), including *Unified Modeling Language* (UML), *Business Process Model and Notation* (BPMN), *Decision Model and Notation* (DMN), and *Case Management Model and Notation* (CMMN).

Understanding how to effectively teach these notations is crucial for educators in the field of process and system modeling. We discuss the choice of diagrams for teaching information systems design, focusing on optimizing the order of

C. Di Ciccio et al. (Eds.): BPM 2024, LNBIP 527, pp. 365–374, 2024.
https://doi.org/10.1007/978-3-031-70445-1_25

selecting diagrams for educational purposes. Our research covers many types of diagrams, such as various UML, BPMN, DMN, and CMMN diagrams. We believe that such decisions might influence the overall clarity, efficiency, and success of the learning process.

The paper is structured as follows: in Sect. 2 we give an overview of the existing works related to our research. Section 3 presents the formalization that will be further used in Sect. 4, where we introduce our algorithm for ordering the diagrams. Section 5 analyzes the results of our algorithm. The paper is concluded in Sect. 6 with recommendations for educators.

2 Related Works

The study [2] presents controversy over the meaning, role and usefulness of models and modeling languages in industry. While some consider UML to be the *lingua franca of software engineering*, others argue that it is not so commonly used [3]. Although the study [4] takes a critical approach to the versatility of UML, at the same time it confirms the significant role of UML in the context of model-based engineering, even though not all diagram types are used.

In practice, the usage of UML diagrams varies and the entire UML notation is not fully used. According to [5], the most commonly used diagrams turned out to be class, activity, sequence, use case and state machine diagrams.

These diagrams are widely used without question, while interaction overview models, timing diagrams and profile diagrams are rarely used. The dominant position of class diagrams is not surprising, since they are a basic element of UML. On the other hand, the second place for activity diagrams is an important result, which is due to the fact that they are also used to model business processes.

UML is used in education not only for software modeling, but also software development classes, as it might bridge documentation with code. Recent case study [6] analyzed the difference between groups of students that used and not used UML for their software project. The authors mention some differences like number of code churns when using UML and among various UML diagrams, class diagrams are highlighted for their importance in representing the structure and behavior of software applications, especially during the implementation phase.

According to [7], example models might improve learning modeling skills. Students who used the examples produced models that were better in quality and expressed greater confidence in their designs.

Although various best practices for teaching modeling and UML were identified [8], the suggested order of diagrams to teach does not cover all of them and various notations. Moreover, their approach is based on the artefacts that contributes in achieving the final goal.

There are many tools that might support teaching modeling diagrams [9], but to the best of our knowledge, non of them suggest the teacher the order in which the diagrams could be introduced in the learning process.

3 Preliminaries

For our algorithm formalization, let us define the following notation:

- *max* function that performs the binary element-wise OR operation on two vectors **X** and **Y**:

$$max(\mathbf{X}, \mathbf{Y}) = \begin{bmatrix} max(x_1, y_1) \\ max(x_2, y_2) \\ \vdots \\ max(x_k, y_k) \end{bmatrix},$$

  where $\mathbf{X} = [x_1, x_2, \ldots, x_k]^T$ and $\mathbf{Y} = [y_1, y_2, \ldots, y_k]^T$ are column vectors,
- the notion of a matrix **A** with removed specific columns or rows ($\bigcirc$ denotes that no column or row is removed, i.e. $\mathbf{A} = \mathbf{A}_{\bigcirc,\bigcirc}$):

$$\mathbf{A} = \mathbf{A}_{\bigcirc,\bigcirc} = \begin{bmatrix} a_{11} & a_{12} & \cdots & a_{1m} \\ a_{21} & a_{22} & \cdots & a_{2m} \\ \vdots & \vdots & \ddots & \vdots \\ a_{n1} & a_{n2} & \cdots & a_{nm} \end{bmatrix}$$

$$\mathbf{A}_{\bigcirc,\mathbf{-j}} = \begin{bmatrix} a_{11} & \cdots & a_{1(j-1)} & a_{1(j+1)} & \cdots & a_{1m} \\ a_{21} & \cdots & a_{2(j-1)} & a_{2(j+1)} & \cdots & a_{2m} \\ \vdots & \ddots & \vdots & \vdots & \ddots & \vdots \\ a_{n1} & \cdots & a_{n(j-1)} & a_{n(j+1)} & \cdots & a_{nm} \end{bmatrix}$$

$$\mathbf{A}_{\mathbf{-i},\bigcirc} = \begin{bmatrix} a_{11} & \cdots & a_{1m} \\ \vdots & \ddots & \vdots \\ a_{(i-1)1} & \cdots & a_{(i-1)m} \\ a_{(i+1)1} & \cdots & a_{(i+1)m} \\ \vdots & \ddots & \vdots \\ a_{n1} & \cdots & a_{nm} \end{bmatrix}$$

$$\mathbf{A}_{\mathbf{-i},\mathbf{-j}} = \begin{bmatrix} a_{11} & \cdots & a_{1(j-1)} & a_{1(j+1)} & \cdots & a_{1m} \\ \vdots & \ddots & \vdots & \vdots & \ddots & \vdots \\ a_{(i-1)1} & \cdots & a_{(i-1)(j-1)} & a_{(i-1)(j+1)} & \cdots & a_{(i-1)m} \\ a_{(i+1)1} & \cdots & a_{(i+1)(j-1)} & a_{(i+1)(j+1)} & \cdots & a_{(i+1)m} \\ \vdots & \ddots & \vdots & \vdots & \ddots & \vdots \\ a_{n1} & \cdots & a_{n(j-1)} & a_{n(j+1)} & \cdots & a_{nm} \end{bmatrix}$$

  Let $R \subset \{1, 2, \ldots, n\}$ be the subset of row indices and $C \subset \{1, 2, \ldots, m\}$ be the subset of column indices we want to remove from the matrix **A**.
  We denote by $\mathbf{A}_{-R,-C}$ the matrix obtained from **A** by removing the rows indexed by R and the columns indexed by C. Therefore:

$$\mathbf{A}_{-R,-C} = \{a_{ij} | i \in (\{1, 2, \ldots, n\} \setminus R), j \in (\{1, 2, \ldots, m\} \setminus T\}),$$

i.e. a_{ij} in the above expression denotes these elements of the original matrix $\mathbf{A}$ that are not in the rows indexed by R or the columns indexed by C. Note that if $R = \varnothing$ and $C = \varnothing$, then $\mathbf{A}_{-\varnothing,-\varnothing} = \mathbf{A}_{\bigcirc,\bigcirc} = \mathbf{A}$.

- The algorithm operates on a zero-one matrix $\mathbf{D} = [d_{ij}]$, $i = 1, 2, \ldots, n$, $j = 1, 2, \ldots, m$, where n is the number of entities (rows) and m is the number of diagrams (columns). Each element d_{ij} is defined as follows:
  - $d_{ij} = 1$ if entity i can be represented within diagram j.
  - $d_{ij} = 0$ if entity i cannot be represented within diagram j.

  Therefore, the matrix $\mathbf{D}$ can be represented as a set of vectors $\{\mathbf{D}_j : j = 1, 2, \ldots, m\}$, where each vector $\mathbf{D}_j \in \{0, 1\}^n$ represents a column in this matrix (the number of elements of the vector $\mathbf{D}_j$ equals $|\mathbf{D}_j| = n$).

$$\mathbf{D} = \begin{array}{r|ccccc} & \text{Diagram 1} & \text{Diagram 2} & \cdots & \text{Diagram } m \\ \hline \text{Entity 1} & d_{11} & d_{12} & \cdots & d_{1m} \\ \text{Entity 2} & d_{21} & d_{22} & \cdots & d_{2m} \\ \vdots & \vdots & \vdots & \ddots & \vdots \\ \text{Entity } n & d_{n1} & d_{n2} & \cdots & d_{nm} \end{array}$$

A row of the matrix denotes the potential existence of a particular entity within the available diagrams, and a column represents what elements a particular diagram consists of:

$$\mathbf{D}_j = \begin{bmatrix} d_{1j} \\ d_{2j} \\ \vdots \\ d_{nj} \end{bmatrix}$$

Therefore, the value of the element d_{ij} indicates whether the entity i can be represented within the diagram j. For example, if $d_{12} = 1$, it means entity 1 can be represented in diagram 2. If $d_{12} = 0$, it means that entity 1 cannot be represented in diagram 2.

4 Steps of the OMG Diagram Ordering Algorithm

The goal of the algorithm is to optimize the selection and ordering of diagrams for a given project based on specific criteria, such as maximizing overlap with entities already modeled, minimizing such overlap, or simply choosing randomly.

Input Data:

- $\mathbf{D} = [d_{ij}]^{n \times m}$, $d_{ij} \in \{0, 1\}$ – the matrix of diagrams, where n is the number of all possible entities and m is the number of all possible diagrams,
- p – a project for which the algorithm will be used (with the required diagrams and entities),
- s – a strategy for the algorithm.

Output Data:

- **R** – the result matrix that stores the order of the diagrams selected from the available diagrams **D** (as well as their coverage of elements and other parameters used when selecting the diagram) for the project p according to the strategy s.

Algorithm:

1. Select a subset of entities and diagrams suitable for the project p. In this way, we will determine the submatrix $\mathbf{D}^p$ of the matrix $\mathbf{D}$. Let $R^p \subset \{1, 2, \ldots, n\}$ represent the set of row indices representing entities and $C^p \subset \{1, 2, \ldots, m\}$ be the set of column indices representing diagrams selected for project p. Therefore:

$$\mathbf{D}^p = \mathbf{D}_{-R,-C},$$
$$dim(\mathbf{D}^p) = k \times l, \text{ where } k \leq n \text{ and } l \leq m,$$

and R and C represent row and column indices representing entities and diagrams that are not suitable or needed for the project p:

$$R = \{1, 2, \ldots, n\} \setminus R^p,$$
$$C = \{1, 2, \ldots, m\} \setminus C^p.$$

2. Initialize:
   (a) the cumulative vector $\mathbf{V} \in \{0, 1\}^k$ of the entities represented in the current step of the algorithm in the project:

$$\mathbf{V} = \mathbf{V}_0 = \begin{bmatrix} 0 \\ 0 \\ \vdots \\ 0 \end{bmatrix},$$

   (b) the set of the diagrams $\mathbf{D}^u$ that have not yet been used within the project p: $\mathbf{D}^u = \mathbf{D}^p$.
   (c) the empty result matrix $\mathbf{R}$, $dim(\mathbf{R}) = 0 \times 4$ for storing the order of the selected diagrams. The dimension of the matrix stems from the 4 properties stored in the matrix.
3. While $\mathbf{D}^u$ is not empty:
   (a) If $\mathbf{V} == \mathbf{V}_0$:

      i sort the columns of $\mathbf{D}^p$ in descending order based on the sum of each column, i.e. the number of the supported entities,
      ii let j be the index of the diagram with the maximum sum in $\mathbf{D}^p$, and $count(j) = \sum_{i=1}^{k} D_{ij}^p$.

Else:

i calculate cosine similarity between cumulative vector $\mathbf{V}$ and all remaining diagrams in $\mathbf{D}^u$: For each diagram $\mathbf{D}^u_i$ in $\mathbf{D}^u$ calculate: cosine_similarity$(\mathbf{D}^u_i, \mathbf{V}) = \frac{\mathbf{D}^u_i \cdot \mathbf{V}}{\|\mathbf{D}^u_i\| \|\mathbf{V}\|}$, where $\| \cdot \|$ is a vector norm.

ii according to the strategy s, select the diagram to use in the next iteration.

- If $s ==$ "*max_cosine_similarity*": Let j be the index of the column with the greatest cosine similarity to the cumulative vector of $\mathbf{V}$ diagrams.
- Else if $s ==$ "*min_cosine_similarity*": Let j be the index of the column with the least cosine similarity to the cumulative vector of $\mathbf{V}$ diagrams.
- Else if $s ==$ "*max_diff*": Let j be the index of the column with the maximum number of new elements compared to the cumulative vector of $\mathbf{V}$ diagrams.
- Else if $s ==$ "*min_diff*": Let j be the index of the column with the minimum number of new elements compared to the cumulative vector of $\mathbf{V}$ diagrams.
- Else $s ==$ "*random*":
  Let j be the index of the randomly chosen column.

(b) Use the selected diagram with the index j in the project. For this step, one can either model the diagram manually taking into account the existing diagrams in the project, or use one of the existing translation methods in order to remain consistent with existing diagrams.

(c) Update the cumulative vector with the selected diagram $\mathbf{V} = max(\mathbf{V}, \mathbf{D}^p_j)$,

(d) For the cumulative vector $\mathbf{V}$ calculate:

$$cumulative\_count = \sum_{i=1}^{k} V_i,$$

$$cumulative\_coverage = \frac{cumulative\_count}{n} \times 100,$$

$$individual\_coverage = \frac{count(j)}{k} \times 100.$$

(e) Add a new row to the result matrix $\mathbf{R}$:

$$\mathbf{R} = \begin{bmatrix} \mathbf{R} \\ [j, cumulative\_count, cumulative\_coverage, individual\_coverage] \end{bmatrix}$$

(f) Remove the column $\mathbf{D}^p_j$ representing the currently used diagram from $\mathbf{D}^u$, i.e.: $\mathbf{D}^u = \mathbf{D}^u_{\bigcirc,-c}$, where c is such that $\mathbf{D}^u_c == \mathbf{D}^p_j$ (the c-th column representing a specific diagram in $\mathbf{D}^u$ corresponds to the j-th column representing this diagram in $\mathbf{D}^p$).

4. Return the result matrix $\mathbf{R}$.

5 Algorithm Results Analysis

To compare the strategies, we prepared the dataset containing 17 diagrams and 120 entities that could be represented within some of these diagrams, as well as provided a prototype implementation[1] that calculates the order and the potential total coverage for all elements and diagrams according to the given strategy (see the screenshot in Fig. 1).

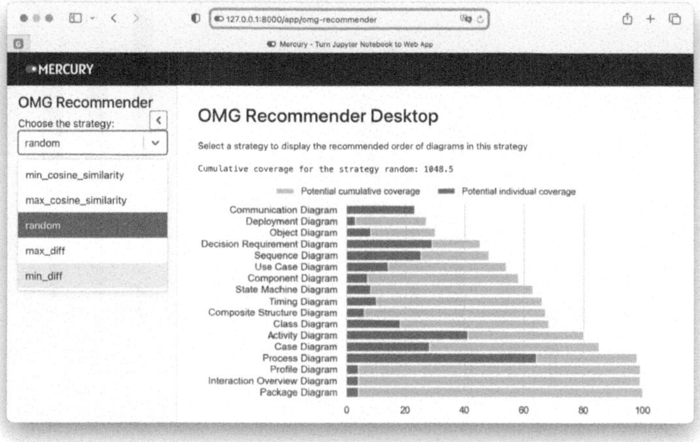

Fig. 1. Application graphical interface using the Mercury platform

As part of the analysis of the results of the developed strategies, the potential cumulative coverage is considered for all elements and diagrams (see Fig. 2). Coverage is potential, as in practice, people modeling the system will not use all the elements in the diagrams or all the diagrams. The case where all elements are not used could be computed in an analogous way after filtering the relevant entities and diagrams. The following strategies were developed and tested:

- *min_cosine_similarity* – selecting the next diagram with the least cosine similarity to the cumulative vector of diagrams,
- *max_cosine_similarity* – selecting the next diagram with the highest cosine similarity to the cumulative vector of diagrams,
- *random* – selection of the next diagram at random,
- *max_diff* – selection of the next diagram with the maximum number of new elements compared to the cumulative diagram vector,
- *min_diff* – selection of the next diagram with the minimum number of new elements compared to the cumulative diagram vector.

[1] https://github.com/KRaKEn-Research-Group/OMG-Recommender.

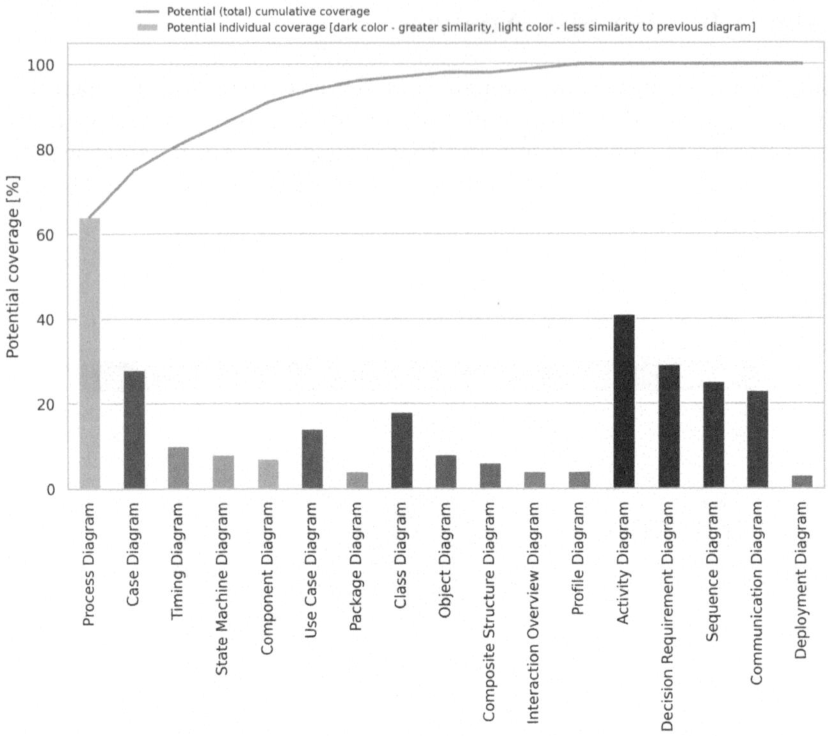

Fig. 2. The *max_diff* strategy result (order of diagrams: left to right)

These strategies can have different applications. In the *min_cosine_similarity* strategy, we choose the diagram that will be least related to existing diagrams, which can be useful, for example, when another person is modeling a part of the system or we do not want to have dependencies between diagrams at this stage of the project. For *max_cosine_similarity*, we choose the diagram that is most similar to existing diagrams, so that we can build on the knowledge already available or use the available translation methods [10]. The random strategy *random* serves as a reference point for the results of other strategies. The *max_diff* strategy is a strategy focused on maximizing the number of new elements relative to existing elements, so it can be useful when one wants to cover as much as possible at an early stage. The *min_diff* strategy, on the other hand, minimizes the number of new elements relative to existing ones, which can be useful if you want to monitor and compare models with each other.

In order to compare in numerical form all strategies presented in Fig. 3, we calculated the approximate area under the curve composed of the points of potential total (cumulative) coverage using the trapezoid method. Both based on the chart visualization and the result of the area under the curve for the *max_diff* strategy, it can be seen that it achieves the fastest total coverage. Taking the

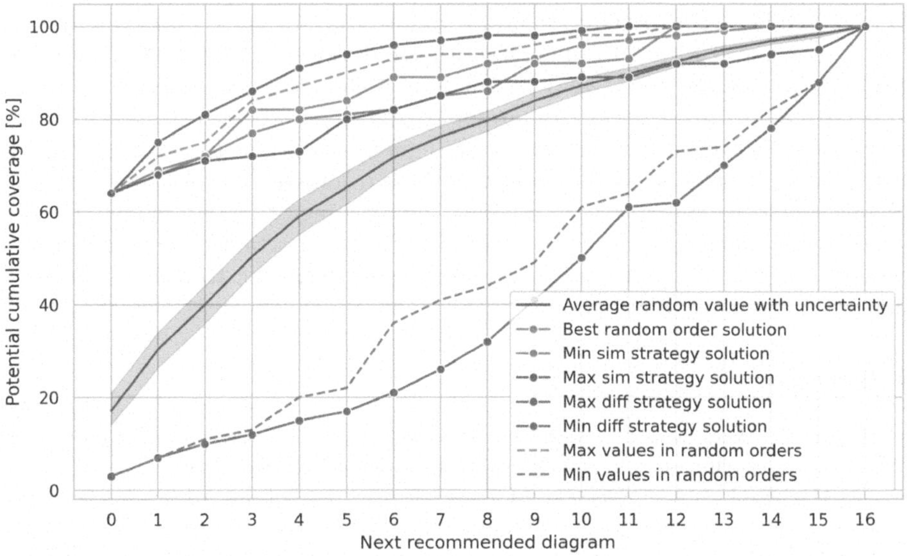

Fig. 3. Comparison of results of all strategies with random order of diagrams

average of 100 random results (by random strategy) as a benchmark, all strategies except *min_diff* achieve better results. Taking the Pareto principle (80:20) as a benchmark and determining the point when one achieves 80% coverage, one can see that only the *max_diff* strategy achieves 80% coverage of less than 20% of diagrams (i.e., 3 diagrams), the other strategies (not counting *min_diff*) reach 80% only after modeling from 4 (in the case of the best of a random 100 results) to 6 diagrams (in the case of the *max_cosine* strategy).

6 Conclusions

Our study reveals that the strategy *max_diff*, which focuses on maximizing the introduction of new elements in the next diagram selection process, achieves the most rapid total coverage of elements. This strategy is especially useful for covering as many aspects of the system as early as possible in the modeling process. On the other hand, the *min_diff* strategy lags behind other strategies in terms of the overall coverage. For educators, these insights highlight the importance of strategically ordering diagrams when teaching system modeling. By adopting effective strategies, educators can significantly enhance students' understanding and efficiency in mastering system modeling techniques.

Adopting the *max_diff* strategy for initial teaching might ensure students quickly grasp a wide range of system elements and their interrelations. Begin with diagrams that introduce the maximum number of new elements, and this approach will provide a broad understanding of the system's components early in the learning process. For example, start with class diagrams to cover a broad range of entities, followed by activity diagrams to introduce business processes. If your goal is for your students to understand dependencies and integration, use the *max_cosine_similarity* strategy to build on already familiar concepts. For varied perspectives, this strategy can be employed, e.g. for advanced classes focusing on system integration, you can start with sequence diagrams that build on previously learned diagrams. By incorporating these recommendations, educators can enhance the learning experience, making the teaching of system modeling more effective and engaging for students. However, it is important to regularly assess student performance and adapt the sequence of diagrams to address learning gaps or reinforce key concepts.

References

1. Kluza, K., Wiśniewski, P., Jobczyk, K., Ligęza, A., Suchenia, A.: Comparison of selected modeling notations for process, decision and system modeling. In: Ganzha, M., Maciaszek, L.A., Paprzycki, M. (eds.): Proceedings of the 2017 Federated Conference on Computer Science and Information Systems, FedCSIS 2017, Prague, Czech Republic, 3–6 September 2017. Annals of Computer Science and Information Systems, vol. 11, pp. 1095–1098 (2017)
2. Störrle, H.: How are conceptual models used in industrial software development? A descriptive survey. In: Proceedings of the 21st International Conference on Evaluation and Assessment in Software Engineering. EASE 2017, pp. 160–169. Association for Computing Machinery, New York (2017)
3. Petre, M.: UML in practice. In: Proceedings of the 2013 International Conference on Software Engineering, ICSE 2013, pp. 722–731. IEEE Press (2013)
4. Petre, M.: No shit or Oh, shit!: responses to observations on the use of UML in professional practice. Softw. Syst. Model. **13**, 1225–1235 (2014)
5. Reggio, G., Leotta, M., Ricca, F., Clerissi, D.: What are the used activity diagram constructs? - a survey. In Pires, L.F., Hammoudi, S., Filipe, J., das Neves, R.C. (eds.): MODELSWARD 2014 - Proceedings of the 2nd International Conference on Model-Driven Engineering and Software Development, Lisbon, Portugal, 7–9 January 2014, pp. 87–98. SciTePress (2014)
6. Apostol, D.C., Bogdan, R., Marcu, M.: UML diagrams in teaching software engineering classes. A case study in computer science class. In: 2024 IEEE 22nd World Symposium on Applied Machine Intelligence and Informatics (SAMI), pp. 000327–000332. IEEE (2024)
7. Karasneh, B., Jolak, R., Chaudron, M.R.: Using examples for teaching software design: an experiment using a repository of UML class diagrams. In: 2015 Asia-Pacific Software Engineering Conference (APSEC), pp. 261–268. IEEE (2015)
8. Kuzniarz, L., Staron, M.: Best practices for teaching UML based software development. In: Bruel, J.-M. (ed.) MODELS 2005. LNCS, vol. 3844, pp. 320–332. Springer, Heidelberg (2006). https://doi.org/10.1007/11663430_33
9. Huber, F., Hagel, G.: Tool-supported teaching of UML diagrams in software engineering education-a systematic literature review. In: 2022 45th Jubilee International Convention on Information, Communication and Electronic Technology (MIPRO), pp. 1404–1409. IEEE (2022)
10. Suchenia, A., Kluza, K., Wiśniewski, P., Jobczyk, K., Ligęza, A.: Towards knowledge interoperability between the UML, DMN, BPMN and CMMN models. In: MATEC Web of Conferences, vol. 252, p. 02011. EDP Sciences (2019)

Comprehension of (Business) Process Models via Tokens: An Eye-Tracking Approach

Ilia Maslov(✉) and Stephan Poelmans

KU Leuven, Warmoesberg 26, 1000 Brussels, Belgium
{ilia.maslov,stephan.poelmans}@kuleuven.be

Abstract. Process modeling involves creating diagrammatic representations of business processes, commonly employed for collaborative efforts, information exchange, and documentation. Considerable challenges hinder the training and learning of process modeling in comprehending process models, primarily due to the absence of systematic teaching approaches. Although various visualization and animation techniques, such as process tokens, exhibit significant potential in enhancing the guidance of modelers, this area remains underexplored. This research delves into an in-depth exploration of animated process model comprehension by novice modelers, using eye-tracking. The study compares and analyzes nine static with nine token-animated models. Utilizing statistical measures, visualized eye gaze data, and semi-structured interviews, the paper uncovers differences in the comprehension of both types of process models and suggests pedagogical implications.

Keywords: Process modelling · Token-animated process models · Process model comprehension · Process modeling education · Eye-tracking

1 Introduction

Process Modeling (PM) typically involves translating textual requirements into visual models, aiming to improve communication between business and IT, inform employees about their roles, enhance or re-engineer processes, or automate them using enterprise information systems. Business Process Model Notation (BPMN) is a *de facto* industry standard and needs PM experts [1, 2].

Comprehensibility of models is among the most essential qualities [3] and is impacted by the model reading behavior of the model reader [4]. For novices, it is often difficult to correctly understand process models as they don't know the correct procedures to analyze the models [4, 5], but it is also difficult for teachers to teach since they do not know what kind of reading patterns these students exhibit [2]. Thus, it is critical to research the factors contributing to the difficulty of process model comprehension via eye-tracking [6]. Eye-tracking is a fit approach for analyzing the comprehension of PM animations [7], as shown by [2, 6], since it can reveal the information-seeking behavior at a physiological level.

C. Di Ciccio et al. (Eds.): BPM 2024, LNBIP 527, pp. 375–385, 2024.
https://doi.org/10.1007/978-3-031-70445-1_26

One of the ways to achieve understandability is through animation and visualization, like tokens or highlighting techniques [8, 9]. With animations, the order of PM activities and the meaning or behavior of a combination of elements in a process can be better conveyed [5]. However, there is a lack of studies exploring the use of token animations in improving learning and teaching [10], though given strong empirical evidence for improving comprehension [9]. This study delineates our focus on reading and understanding process models [9]. The current paper is part of our broader research objective to advance the teaching of the BPMN 2.0 standard and to develop an animation-based PM education framework. To our best knowledge, no existing research has applied an eye-tracking analysis of PM animations.

This paper will combine the usage of eye tracking with reflective semi-structured interviews to assess the following research questions:

RQ1. *Are there any differences between static and animated process models regarding the model-reading patterns of inexperienced modelers?*

RQ2. *What can the differences imply for the PM education?*

In Sect. 2, we give an overview of the extant literature. In Sect. 3, we discuss our chosen eye-tracking methodology. In Sect. 4, we present the results of eye-tracking and semi-structured interviewing. Finally, in Sect. 5, we discuss the findings, limitations, and the threats to validity and conclude with future directions.

2 Extant Literature

Process Modeling (PM) involves making an internal mental model explicit through visual models to achieve shared understanding [7, 11]. Despite this, individuals may still perceive processes differently, making comprehension critical [1]. PM fundamentally involves reading and designing models, with comprehensibility being the most important quality, even more than syntactical or semantical correctness [3]. This experiment focuses on reading and comprehending process models.

Comprehension can be measured objectively by the number of correct answers and time taken to answer, and subjectively by perceived difficulty [4]. Factors affecting PM comprehension include personal and model-related features and their reading [4, 10, 12]. Internal cognitive factors crucial to comprehension include cognitive load capacity, influenced by an individual's working memory [4]. Higher education levels improve comprehension scores [13], familiarity with a domain reduces cognitive load [4], and age affects fixation count [6]. Cultural background also shapes reading preferences and expectations [14]. External cognitive resources, or model factors, also influence comprehension. These include the modeling language [4], model size [5], and semantic richness provided by features like task descriptions [4].

Complex logic, defined by numerous uncommon elements, can obstruct understanding [5]. Still, the explicit articulation of business rules within models boosts precision and diminishes cognitive load [15]. The arrangement of process tasks, including bending and modularization, is critical in their legibility and ease of interpretation [16]. Furthermore, visual cues, including task coloring, lower cognitive demands and support the assimilation of information [17]. Such cues are related to animations that can improve comprehension [5], which is the focus of this research. Different model reading strategies can influence comprehension, which is determined by the modeler's experience and the model's layout [16].

Reading strategies are often visualized via eye-tracking analysis. Eye Movement Modelling Examples (EMME) are proposed as patterns of typical eye-gazing behavior of modelers [2]. EMMEs include following the process path with occasional regressions, such as revisiting previous model elements and skimming and central-focused trial starts [14]. Participants may skip complex activities, noting congruence points, which are mental points to be revisited later [2]. Novices tend to read sequentially, with more fixations and longer reading times, while professionals focus on critical elements like decision points, reflecting their ability to handle higher-level constructs [4]. Reading intensity correlates with lower knowledge, implying increased re-reads and heightened intensity [17].

Comprehension of process models is improved by Visualization, Animation, and Simulation (VAS) features [4]. Visualization makes process model information visually accessible; animation makes it animate and float, and simulation generates instances based on provided conditions and contexts [10]. Visualized animations and participant-controlled simulations are necessary for better comprehension, as visual representations are easier to understand than abstract, mathematical concepts [5]. Animations enhance process model elements, making them more visual and easier to comprehend, and providing feedback on issues within the models [4, 5, 8]. Tokens, used for dynamic highlighting of process elements, lack comprehensive empirical analysis, with only recently research being conducted [9]. Multiple Instances Data Animator (MIDA) demonstrates the concurrent behavior of multiple tokens [18]. Emens et al. [8] designed a prototype with both types of animation and successfully tested their usefulness and ease of use in one case study. The Incipient Token-Based Process Modelling (TBPM) educational framework asserts that VAS features, particularly tokens, are foundational for better learning, enabling interaction with models at various levels and enhancing comprehension, thus improving modelers' knowledge and skills [10, 13].

3 Methodology

To answer the RQs and find differences in static and animated BPMN process model reading patterns, we conducted an eye-tracking analysis, completed with a semi-structured interview. Eye-tracking is suggested as a method to explore the places in the model that are protruding as cognitively overloading and thus challenging to comprehend and therefore can be improved by the designers of the models [16]. Building upon the work of [2], who advocated an exploratory approach integrating eye tracking to delineate gaze patterns, we adopt a similar methodology to analyze the effect of animations within process models. The following section describes the participants, experimental design, and materials.

Since eye-tracking research tends to be in-depth and cannot be conducted in groups, we invited a limited set of 7 participants (2 females and 5 males) from a business faculty at a European university with different educational levels via convenience sampling. The average age was 27 ($SD = 4.81$). 4 participants were PhD students, and 3 were enrolled in a Master program in economics. All participants knew BPMN; some knew other modeling languages, like UML. The business profile of the participants matches the typical process modeling stakeholders in the industry. Based on [5], 3 can be considered beginners with little BPMN practice. 4 mentioned some previous formal training. Except for the 2 novice modelers, participants knew the meaning of tokens in BPMN. We thus perceive the sample as mixed, with varied experience but limited knowledge of BPMN and tokens.

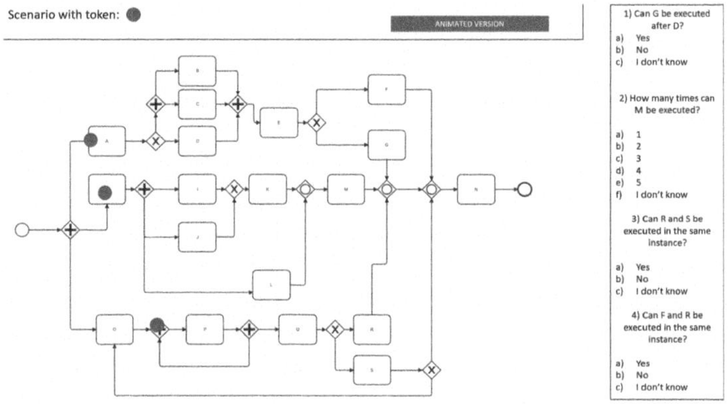

Fig. 1. Screenshot of an animated model with tokens (showing scenario in action).

Using the BPMN 2.0 standard in Camunda Modeler, we generated 4 abstract models that had 20 to 30 nodes, adapted from [5], and 4 models with 10 to 20 nodes, but semantically rich process models, based on our previous exam questions. This minimizes the effect of domain knowledge and analyzes if there is a difference between abstract and semantically rich notations [5]. There were one, two, or four multiple-choice questions per model. One final (abstract) model and its questions (Fig. 1) were constructed using past research (under revision), where we exploratorily tried to insert several BPMN patterns in one model (a deadlock, a livelock, and inclusive OR gateways) and which is hypothesized to produce the greatest differences in the eye-gazing patterns. Due to a lack of comprehensive animation tools, the animated videos were constructed in MS PowerPoint. Every model will be read from left to right, per the BPMN standard recommended by [16]. We added a process scenario (required for the answer) on top and multiple-choice question(s) on the right.

Participants were invited for an individual session. They filled in the pre-survey containing demographic and knowledge questions. After initial eye-tracking calibration, they were then shown a training model to make them comfortable with the study design and animation controls (replayability). For 9 models, the static process model and a set of multiple-choice questions (e.g., Fig. 1) were presented first, followed by the corresponding animated process model and the option to keep or to change the first answer.

Typical metrics in eye-tracking research on PM comprehension were used. The correct score to multiple choice questions is one of the most commonly used measures in measuring process model comprehension [4, 5]. Typical eye-tracking metrics include Area of Interest (AOI) on the screen, dwelling time, fixations (points where participants look), scan path (series of eye movements), and heatmaps [6, 16]. Higher dwelling times on the areas of interest (in heatmaps shown as warmer colors) typically indicate the perceived importance of the AOI and a higher demanded cognitive load.

The eye tracker ('SMI iView RED250mobile') recorded participants' eye-gazing data on a laptop with a 1920 mm × 1080 mm screen resolution. The distance between the participant's eyes and the computer monitor varied from 600 mm to 720 mm. Eye movement data were recorded at 60 Hz. The researchers were present with the participants in the same room and helped them in case of problems. Calibration (looking at 4 dots on the screen) and validation (confirming the distance between the recorded fixation and the center of the 4 dots) were acceptable: up to 0.1–0.5 mm deviation accuracy of fixation recordings. The setup has been used in similar eye-tracking research [6]. SMI Begaze software allows data to be exported for quantitative analysis in SPSS.

The post-experiment interview resulted in 96 answers from 7 participants to 15 semi-structured questions with the average time of interviewing about 23 min, ranging between 14 and 33 min. The questions were based on previous work [anonymized]. We asked participants to compare animated and static models and tell us about the benefits and disadvantages of tokens for comprehension, what they thought about visualization, animation, and simulation capabilities, and at what level (specific BPMN elements or recollection of elements) did participants focus when using tokens. Interviews were recorded, transcribed, and analyzed inductively to generate thematic codes [19].

For reliability and transparency reasons, available data can be accessed via: https://zenodo.org/records/12672574. In the following section, we present the general descriptives of the eye-tracking data and the analysis of the student's views on the use of animations and their advantages and drawbacks. Four typical animation interaction patterns described by the students provide us with a frame to better present and interpret the visual eye-tracking data.

4 Data Analysis Results

Table 1 presents descriptives in both conditions: static model first, followed immediately by an overlay of moving token instances, allowing participants to revise their first answers. Note that there were 2 questions in the Order model and 4 in the Final model. Using the Shapiro-Wilk test, we find that the data for several variables does not follow a normal distribution, likely attributed to the limited number of participants ($n = 7$).

The cumulative number of correct answers went from 3 to 33. The first answers given using static models did not improve (after the animation) in 2 models and decreased in 1 model (Abstract 2). In the static condition, there are longer dwell times and more and longer fixations for almost all the indicators across all models. We thus argue that in the animated condition, participants either (1) had a reduced cognitive load due to animations or (2) they remembered the questions and thus had to interact less with the model. We found that participants spent less time and had fewer fixations in the static abstract models (compared to the semantic models). However, we cannot see significant differences between semantically rich and abstract models in the animated condition. Participants, on average, replayed the animations 2 times.

In the post-experimental semi-structured interviews, tokens were named beneficial due to their visualization, simulation, replayability and interactivity, making it clearer and easier to understand the models in part by reducing complexity or by making BPMN rules explicit. Tokens were engaging and allowed exploration of the model with relevant feedback. However, students mentioned that animations occasionally introduced complexity, particularly in detailed multi-path scenarios. This led to increased reliance on animations to answer the questions (not providing true understanding). Furthermore, some students somewhat distrusted the information conveyed by the animations. Due to low interactivity, some respondents mentioned that the recorded speed was notably high and needed adjustments.

Table 1. Descriptive statistics comparing static and animated conditions.

	Static						Animated							
Model	Correct answers	Sum of Dwell Time [min]	Average of Dwell Time [min]	Sum of Fixation Count	Average of Fixation Count	Average of Average Fixation	Correct Answers	Sum of Dwell Time [min]	Average of Dwell Time [min]	Sum of Fixation Count	Average of Fixation Count	Average of Average Fixation	Sum of tries	Avg. Tries/Participant
Manager	3,0	6,2	0,8	846,0	105,8	424,3	3,0	1,4	0,2	242,0	34,6	320,3	27,0	3,9
Order	4, 2	7,9	1,1	1145,0	163,6	401,6	6, 2	2,8	0,2	467,0	38,9	346,5	12,0	1,7
Order tickets	2,0	11,6	1,7	1493,0	213,3	441,3	4,0	3,9	0,3	578,0	48,2	397,7	12,0	1,7
Goldcard	0,0	11,7	1,7	1439,0	205,6	476,1	7,0	6,7	0,6	826,0	75,1	479,8	11,0	1,6
Abstract 1	3,0	8,3	0,5	1223,0	67,9	363,8	6,0	3,0	0,3	453,0	41,2	362,8	11,0	1,6
Abstract 2	3,0	7,1	0,3	1013,0	48,2	405,2	2,0	4,4	0,3	666,0	41,6	381,3	16,0	2,3
Abstract 3	2,0	5,4	0,4	692,0	46,1	433,3	5,0	3,4	0,4	505,0	63,1	388,9	8,0	1,1
Abstract 4	1,0	6,3	0,4	927,0	57,9	418,3	6,0	4,4	0,5	545,0	60,6	472,0	9,0	1,3
Final model	4, 3, 3, 2	18.1	0,5	2379,0	61,0	403,1	6, 6, 6, 5	6,7	0,3	961,0	37,0	389,5	26,0	3,7
Total	14,0	82,6	7,3	11157,0	969,4	3766,9	33,0	36,8	3,1	5243,0	440,2	3538,8	132,0	18,9

As a learning support, respondents outlined the potential feedback ability of the tokens in simplifying and clarifying complex processes, helping to learn new information and recall forgotten ones (e.g., about BPMN gateway semantics). However, it is suggested that students should not become over-reliant on such animations and should have other learning support, including teacher explanations and introductions to the fundamentals of PM first.

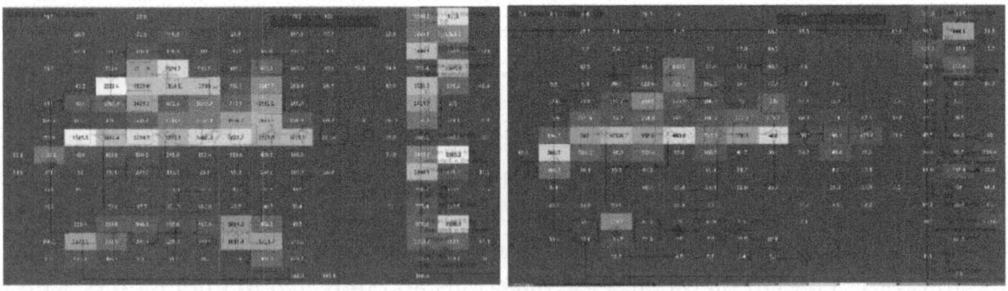

Fig. 2. Comparison of the final static (left) and animated (right) process models overlayed with gridded average dwelling time AOI heatmap analysis (red color indicates a lot of attention, green means limited attention and blue is no attention). (Color figure online)

Overall, we can distill four major suggested reading patterns that are interrelated: (1) tracking the tokens as a guide, (2) using animations to answer the specific questions, (3) using animations to comprehend BPMN patterns truly, and (4) imagining new scenarios, not elaborated in the videos. Respondents reported that by tracking tokens through predetermined scenarios, their attention was directed and enhanced, facilitating a more thorough exploration and generation of a comprehensive representation of the model and process logic. This approach also proved beneficial in answering some questions, for instance, through occurrence (tokens) counting. Not all scenarios showed the relevant information to the answer, thus forcing respondents to think. Comprehension and understanding of BPMN elements (e.g., the gateway splits) and patterns (e.g., deadlock) resulted from the benefits of tokens making BPMN behavior explicit. Imagining non-animated scenarios was often a result of better comprehension of the BPMN logic due to the tokens. Imagining new scenarios implied the mental application of tokens to confirm internal ideas, thus eventually resulting in an answer and suggesting a logical analysis and selective critical engagement with the model based on the animations.

For the visual qualitative analysis, we used typical analyses of fixation scan paths and gridded AOI heatmaps (average dwelling time) of 16 rows by 16 columns. These analyses allowed us to identify the structure (important AOIs) and the sequence of the eye gazes (the connection between AOIs) and identify the four typical patterns discussed above. Given the sheer volume of the data (several participants and several models), we focused on aggregate measures. A single detailed analysis is presented below.

Figure 2 show the heatmap analysis of the final model (in both conditions), which reflects typical occurrences of the eye-gazing behavior. Usually, at the start of the static model viewing, the eye gaze was in the top middle of the screen, immediately shifting to the right of the screen (questions), like in [2]. In the animated condition, we found that participants with each run of the video typically looked at the questions one by one (one trial per question). Typical is the participants' tendency to read models top-down left-to-right like in [16]. We argue that the participants thus initially obtained a mental picture of the model (via ordered top-down, left-right scanning). Still, they didn't scan the model so meticulously in the animated condition. Typically, the parts animated by the tokens in the animated condition had higher attention, while non-animated parts were generally ignored. Participants paid less attention to the questions area. Another common observation is that points of decision and logic (gateways, timer events) attract significantly more attention (dwelling time and fixations) in both conditions. Still, it seems relatively larger in the animated condition than the rest of the BPMN elements. This reflects the 'using animations to comprehend' pattern. Correct answers to the questions often relied on comprehending the logic of these decision points. The participants tend to go back and forth from these special elements to the questions. We thus suggest respondents acquire knowledge of the information in the model from the static condition first and then concentrate the modeler's attention on the animated parts, which are critical to answering the questions (using tokens to comprehend the model).

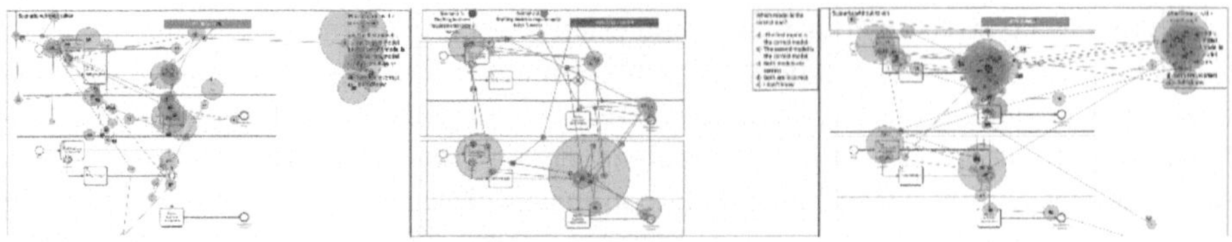

Fig. 3. Participant's scan path in (1) static before animated, (2) animated, (3) static after animated versions in a semantically rich model.

Tracking the tokens is unique to animated conditions. When tracking, the eye-gaze patterns weren't identical but relatively close to (or sometimes in advance of the moving tokens) the tokens' paths. When a token splits (in partial tokens), the participants make a choice. If the split is easy to understand and the partial tokens move one at a time, participants will keep looking at the provided animated tokens. However, if several partial tokens move simultaneously, the participant will try to look at them simultaneously to follow and compare multiple paths.

The use of animations to answer the questions is reflected by the eye-gazes going back and forth between the model and the question area in static and animated conditions, as illustrated in Fig. 3. After reading the question, the participants often looked for the relevant AOI in the process model.

A fourth pattern, as described by the participants, is that imagining new scenarios is not easy to represent via eye-gazing paths. However, Fig. 3 can illustrate the scanning trajectory of a participant as they look at a static model, followed by the animated version of the same model, and subsequently, the automatic reappearance of the static version. Before the animated condition, the eye-gazing behavior was erratic. Then, after watching the animated condition (static after animated), the visual routines resembled the animated token path more. This case can plausibly be interpreted as remembering the animated path or potentially imagining new scenarios, offering a refinement of the mental model. Also, note the abovementioned patterns of tracking the token, answering, and focusing on (comprehension of) two gateways in the 2nd picture (animated condition).

5 Discussion: Results, Limitations, and Future Work

RQ1 is directed at establishing the existence of the difference in comprehension between the static and animated models. It is evident that the way the PM modelers perceive token-animated process models is different in terms of eye-gazing patterns (both quantitatively and qualitatively) and the explanations that the students gave afterward. We find higher scores for most questions in the animated than in the static, with fewer fixations and a lower average dwelling time. Eye-gazing patterns, though similar, were different in the animated and static conditions.

Though expected, we find the findings useful as confirmatory research since there is a lack of research that attempts to explore comprehension of animations in process models via eye-tracking, which can be useful to identify the reasons why animations work. Moreover, we find it remarkable that for a few models, the comprehension decreased, not increased, which requires further investigation.

Theoretically, we suggest that moving tokens, as a method of visualization, are a better external cognition device, as indicated by distributed cognition [10, 12], which is based on the idea that conceptual models are a representation of the internal mental models [7, 11]. Compared to static models, animated process models serve as a scaffolding tool for the cognitive system, which allows for a better understanding of the underlying concepts, rules, and phenomena (e.g., the logic of the process model).

Eye-gazing patterns are concretized in eye-tracking data. In alignment with a study like [16], we propose four major eye-gazing patterns observed when viewing token-animated process models: (1) tracing the token, (2) answering questions, (3) comprehending the models, and (4) imagining new scenarios. Initially, modelers trace the token to understand the model. Comprehending patterns involves dwelling longer on specific BPMN patterns and gateways, typically occurring after an initial tracing. Imagining new scenarios reported by participants involves exploring paths not shown in the model after viewing the token through patterns. Answering is indicated by eye movement between the model and answer area, often considered the final pattern and sometimes seen as confirming the hypothesized answer. The patterns are semi-sequential, starting with tracing the token and ending with answering, comprehending patterns, and imagining scenarios. Tracing and answering is potentially less cognitively demanding since the scenario can show the correct answer, whereas trying to comprehend the pattern and imagining new scenarios can be more cognitively challenging.

Four patterns can work in the educational context, answering RQ2. Tracing can be used to explore and gain an overview of the model and direct a student's attention to specific critical areas in models. Tokens can make BPMN rules and behavior of collaborating elements (patterns) explicit, thus improving the initial identification and comprehension of the BPMN elements and patterns based on descriptions in the course material. Educators can also explore and introduce EMMEs [6], which can serve as a visual example to students of an excellent way to look at a model when using tokens. Tokens can be used in the e-learning to support the independent learning by exploiting the simulation feature of the scenarios [20].

Despite our efforts to keep the study valid and reliable, some limitations and threats to validity still exist. The order of models (static, then animated/static) likely impacted results. Our sample size was small ($n = 7$) and non-heterogenous, which may limit generalizability but is typical for eye-tracking research, which is often used for more qualitative, exploratory purposes [15]. Eye-tracking doesn't show why people look, so we used interviewing. We performed a collaborative analysis to address threats to the validity of typical qualitative research [19].

Our findings may inform future design and instructional approaches for BPMN modeling tools and education, emphasizing the importance of interactive animations that scaffold the viewer's cognitive processes. In future research, the model-modeler interaction strategies can be further developed and decomposed, and EMMEs can be generated based on the interactions with the animations [6]. High-interactive animation tools should be considered [5] and can be used in experiments.

References

1. Dumas, M., La Rosa, M., Mendling, J., Reijers, H.A.: Fundamentals of Business Process Management, 2nd edn. Springer, Heidelberg (2018). https://doi.org/10.1007/978-3-662-56509-4
2. Winter, M., Neumann, H., Pryss, R., Probst, T., Reichert, M.: Defining gaze patterns for process model literacy–exploring visual routines in process models with diverse mappings. Expert Syst. Appl. **213**, 119217 (2023)
3. Moody, D.L., Shanks, G.G.: Improving the quality of data models: empirical validation of a quality management framework. Inf. Syst. **28**, 619–650 (2003)
4. Figl, K.: Comprehension of procedural visual business process models: a literature review. Bus. Inf. Syst. Eng. **59**, 41–67 (2017)
5. Aysolmaz, B., Reijers, H.A.: Animation as a dynamic visualization technique for improving process model comprehension. Inf. Manag. **58**, 103478 (2021)
6. Winter, M., Pryss, R., Probst, T., Reichert, M.: Applying eye movement modeling examples to guide novices' attention in the comprehension of process models. Brain Sci. **11**, 72 (2021)
7. Aysolmaz, B., Reijers, H.A.: Towards an integrated framework for invigorating process models: a research agenda. In: Reichert, M., Reijers, H. (eds.) BPM 2016. Lecture Notes in Business Information Processing, vol. 256, pp. 552–558. Springer, Cham (2016). https://doi.org/10.1007/978-3-319-42887-1_44
8. Emens, R., Vanderfeesten, I., Reijers, H.A.: The dynamic visualization of business process models: a prototype and evaluation. In: Reichert, M., Reijers, H. (eds.) BPM 2016. Lecture Notes in Business Information Processing, vol. 256, pp. 559–570. Springer, Cham (2016). https://doi.org/10.1007/978-3-319-42887-1_45
9. Maslov, I., Poelmans, S.: Facilitating the comprehension of business process models for unexperienced modelers using token-based animations. Inf. Manag. **61**, 103967 (2024)
10. Maslov, I., Poelmans, S.: Advancing the BPMN 2.0 standard with an extended animated notation: a research program for token-based process modeling education. In: Proceedings of BIR (2023)
11. Moody, D.: The physics of notations: toward a scientific basis for constructing visual notations in software engineering. IEEE Trans. Soft Eng. **35**, 756–779 (2009)
12. Bardone, E., Secchi, D.: Distributed cognition: a research agenda for management. In: Current Topics in Management, pp. 183–208 (2018)
13. Maslov, I.: Towards empirically validated process modelling education using a BPMN formalism. In: Guizzardi, R., Ralyté, J., Franch, X. (eds.) RCIS 2022. Lecture Notes in Business Information Processing, vol. 446, pp. 803–810. Springer, Cham (2022). https://doi.org/10.1007/978-3-031-05760-1_58
14. Duggan, G.B., Payne, S.J.: Text skimming: the process and effectiveness of foraging through text under time pressure. J. Exp. Psychol. Appl. **15**, 228 (2009)
15. Wang, W., Chen, T., Indulska, M., Sadiq, S., Weber, B.: Business process and rule integration approaches—an empirical analysis of model understanding. Inf. Syst. **104**, 101901 (2022)
16. Lübke, D., Ahrens, M., Schneider, K.: Influence of diagram layout and scrolling on understandability of BPMN processes: an eye tracking experiment with BPMN diagrams. Inf. Technol. Manag. **22**, 99–131 (2021)
17. Petrusel, R., Mendling, J., Reijers, H.A.: How visual cognition influences process model comprehension. Decis. Support Syst. **96**, 1–16 (2017)
18. Corradini, F., Muzi, C., Re, B., Rossi, L., Tiezzi, F.: Formalising and animating multiple instances in BPMN collaborations. Inf. Syst. **103**, 101459 (2022)

19. Saunders, M., Lewis, P., Thornhill, A.: Research Methods for Business Students. Pearson Education, London (2009)
20. Maslov, I.: Evaluating User Experience (UX) of Students using a Learning Management System Moodle in a Finnish University Through a Holistic UX Model Approach (2020)

Teaching Enterprise Resource Planning Systems and Business Process Management in Tandem Using the Enterprise Online Guide

Hans-Jürgen Scheruhn[1] and Jan Mendling[2,3,4](✉)

[1] Hochschule Harz, Wernigerode, Germany
hscheruhn@hs-harz.de
[2] Humboldt-Universität zu Berlin, Berlin, Germany
[3] Wirtschaftsuniversität Wien, Vienna, Austria
[4] Weizenbaum Institute, Berlin, Germany
jan.mendling@hu-berlin.de

Abstract. Enterprise Resource Planning systems are arguably the most important information systems in almost any larger organization. This makes them an central topic of information systems education. A key challenge for a course on these systems is their sheer complexity with hundreds of transaction types and tens of thousands of database tables. In this paper, we explain how models of the Enterprise Online Guide can be used to facilitate the navigation of this complexity. More specifically, we focus on the relationship between process models and database tables to provide students with an understanding of Enterprise Resource Planning systems. We discuss the effects of this approach for learning key concepts of business process management.

Keywords: Business Process Management · Enterprise Resource Planning Systems · Teaching Information Systems

1 Introduction

Enterprise Resource Planning (ERP) systems are the core information systems of almost any organization in various industries. Vendors of ERP software serve a large market. SAP alone, one of the market leaders, had almost 500,000 customers in 2024. This makes ERP systems a central topic of the curriculum of information systems degree programs [3] and corresponding textbooks [5]. Many more specialized books are offered to practitioners covering various modules of different vendors with the aim to guide adoption and daily usage.

Even though ERP systems are an important topic for information systems education, they often fill a disproportionally smaller coverage in degree programs. A reason appears to be the challenge that lecturers face. "Seeking to teach (...) ERP, there is a significant investment required in understanding

C. Di Ciccio et al. (Eds.): BPM 2024, LNBIP 527, pp. 386–395, 2024.
https://doi.org/10.1007/978-3-031-70445-1_27

the complex melange of technologies, processes and issues involved" [4]. This complexity can be measured in different ways. Modern ERP systems have hundreds of transaction types and tens of thousands of database tables. In larger business-to-consumer companies, a single one of these tables can contain millions of entries for orders per year. This complexity is already difficult for practitioners who are exposed to it on a daily basis. Even more difficult it is to teach students how these systems operate and integrate into the business processes of an organization of realistic size.

The complexity of ERP systems has been the focus of information systems research since the early 1990 s. The proposal of the Architecture of Integrated Information Systems (ARIS) by Scheer was strongly motivated by the idea of using conceptual modeling at the enterprise level [8]. Process models are of central importance in this context. Understanding ERP systems requires an understanding of prominent business processes such as order-to-cash or purchase-to-pay. In turn, understanding business processes equally requires an understanding of ERP systems as their typical information systems infrastructure. This means that teaching ERP systems and business process management is deeply intertwined when the teaching objective is to convey the complexity of realistic business processes together with their corresponding data structures.

In this paper, we address the challenges of teaching ERP systems and corresponding business processes. To this end, we explain how models of the Enterprise Online Guide (EOG) can be used to facilitate the navigation of this complexity.[1] More specifically, we focus on the relationship between process models and database tables to provide students with an understanding of Enterprise Resource Planning systems. We discuss the effects of this approach for learning key concepts of business process management [2].

This paper is structured as follows. Section 2 describes the background of our work with a focus on the Enterprise Online Guide (EOG) developed at Hochschule Harz [10]. Section 3 illustrates how the order-to-cash can be represented using BPMN models that map to various tables of the ERP database schema. Section 4 identifies lessons learnt, before Sect. 5 concludes.

2 Structure of the Enterprise Online Guide

It is a substantial challenge to present the complexity of ERP systems in courses. The Enterprise Online Guide (EOG) was created in order to support the systematic navigation of this complexity by the help of models.

Systematic navigation requires a consistent architecture. The Enterprise Online Guide (EOG) is based on existing Enterprise Architecture Framework and process standards such as TOGAF [11], ARIS [9], APQC [2], and others. EOG uses two structural axes defining a grid of four rows and eight columns. The four rows capture phenomena of the enterprise at different levels of granularity. The eight columns describe different business views of an enterprise towards

[1] Please find previous recordings of a typical EOG presentation at https://dam.sap.com/mac/app/p/assets/saplearning/gxWQRGT?hierarchy=true.

achieving sustainable goals. These eight views include motivation, organization, functions, processes, applications, data, technology, and network. At the top level, for instance, motivation is captured using a business model canvas [6], organization with an organization chart, and process with a process landscape model [2, Ch.2].

The navigation of EOG is loosely inspired by Google Maps, and adopts the term "map" for a model in a cell of the 4 × 8 grid (see Fig. 1). The conceptual challenge in the context of EOG is the definition of distance. Along the hierarchy axis, the distance is naturally defined based on the level of granularity. Along the vertical axis and its columns, proximity is essential for navigation, but more difficult to establish by means of cross-links. Proximity and distance has to be made visible in order to make change traceable. For example, a change of an IT application or its data to be processed often translates to changes of the corresponding business processes and organisational responsibilities. Such impact of change requires navigation mechanisms.

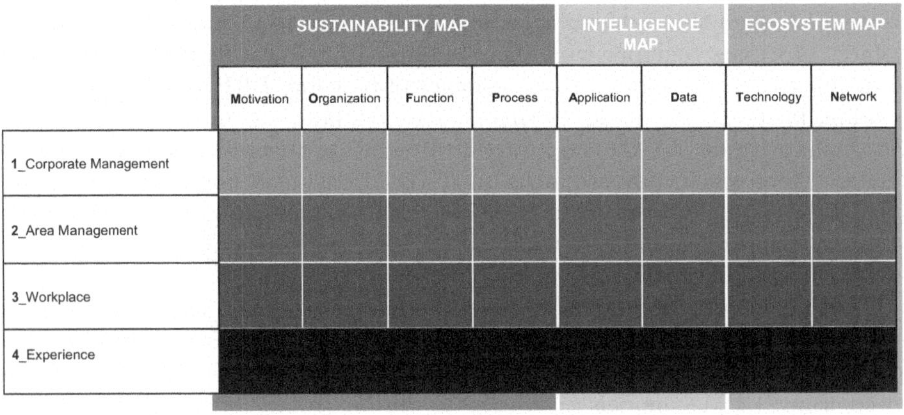

Fig. 1. The Enterprise Online Guide capturing information of an example enterprise.

EOG addresses this problem with a consistent process orientation across all maps. This is achieved through so-called instance links (horizontal) of the process objects and their hierarchy (vertical) in all EOG maps. The basis for this is a shared IT-supported repository in which the same process objects can be used multiple times, thereby enabling consistent horizontal navigation between different maps at the same level of abstraction. In EOG, for example, a BPMN activity from the process map corresponds to an Event-Driven Process Chain (EPC) function [7] from the application map, or to the key activity of a business model canvas from the motivation map at a certain level of abstraction. Furthermore, pools and lanes of a BPMN diagram are hierarchically assigned to the corresponding organisational objects of a process-oriented organisational structure from the organisational map using the corresponding zoom factors. The system organisation units to be considered in ERP systems such as SAP (so-called organisational data) and their multi-level hierarchies are also mapped in a process-oriented manner in EOG and assigned to BPMN diagrams accordingly.

The data flows (typically based on transaction data) depicted in the BPMN diagrams are linked in EOG with ERM models of the data view. The essential referential integrity constraints are represented using the primary keys of the associated master data (e.g. customer number) and the resulting document flow of the transaction data. These are depicted the order-to-cash process as a data model at the lowest level of abstraction. Using the other attributes of the corresponding entities, e.g. Customer Name or Order Net Value, EOG enables horizontal navigation using the input and output fields of the corresponding transactions of the application map at the same level of detail. This provides clarity to the user who is often overwhelmed by the very technical and database schema-like user interface, by minor differences of SAP ERP and its successor SAP S/4HANA, and by the more user-friendly arrangement of the data fields in the alternative GUI of SAP Fiori.

3 Teaching ERP Using Business Process Models

THe EOG has been used by Hochschule Harz for the development of a navigation structure and redocumentation of the Global Bike Group[2] demonstration enterprise and its corresponding SAP system. EOG has been implemented as an extension to Signavio, ARIS, SAP Solution Manager, and LeanIX.

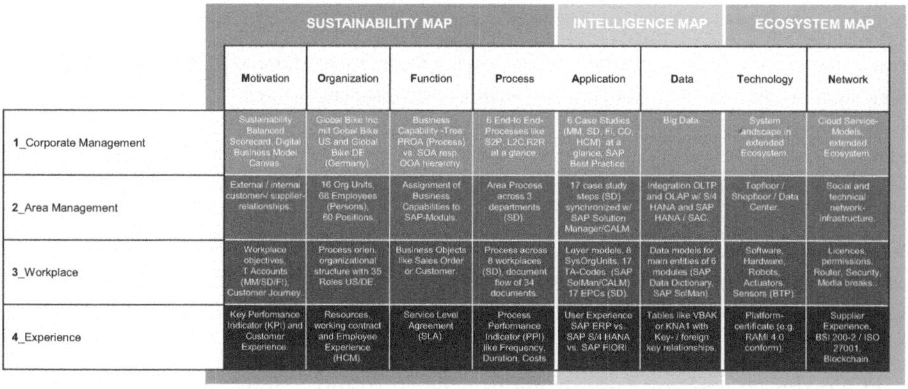

Fig. 2. Enterprise Online Guide capturing information of the Global Bike Group.

The redocumentation of the ERP system and the corresponding enterprise stretches over the 4 × 8 cells of EOG. Figure 2 shows the maps to navigate the

[2] For details on the Global Bike Group installation for SAP, see https://portal.ucc.ovgu.de/sap-model-company-global-bike/.

Global Bike Group demonstration case. For example, a student might want to understand the business of Global Bike first and start navigating in the top-most motivation cell. Here, a business model canvas documents the business model of the company. Its internal structure can be inspected by linking to the top-most organization chart showing two legal entities operating in five different locations. From there, a student can navigate into the details of the Level 3 process of sales and distribution, which describes the order-to-cash process.

Figure 3 shows the order-to-cash process of the Global Bike Group. This process starts by creating customer and contact person entries in the ERP system. Once this data is stored, an inquiry can be created with reference to the customer. For this inquiry, then a quotation is compiled and sent to the customer. A positive response from the customer leads to the creation of a sales order. For this order, the stock status is checked. The delivery is created and a corresponding delivery note sent to the customer. Once the stock status shows the availability of stock, a picking request leads to an outbound delivery. Then, an invoice is created and sent to the customer. Once the payment is received, it is posted and the process completes with a display of the sales order document flow.

Table 1. Connections between Process Activities, SAP Transactions, Entity Typs, and Database Tables.

Actvity	Transaction	Entity Types	Table
Create customer	XD01	Customer Master	KNA1, KNB1
Create contact person	VAP1	Customer Master, Business Partner	BUT000
Create inquiry	VA11	Customer Inquiry	VBAK
Create quotation	VA21	Customer Quot., Customer Quot. Item	VBAK
Create sales order	VA01	Customer Quot. Item, Material Master	VBAK
Check stock status	MMBE	Sales Order, Sales Order Item	MARD, MARC
Display sales order	VA03	Sales Order	VBAK
Create delivery	VL01N	Delivery	LIKP
Change outbound delivery	VL02N	Transport Order	LIKP
Create invoice	VF04	Delivery	VBRK
Display billing document	VF03	Invoice	VBRK
Post incoming payments	F-28	Invoice	BSID, BSAD
Display sales order doc. flow	VA03	Customer Order	VBFA

The order-to-cash process provides a background, against which parts of the database schema can be interpreted. Table 1 gives an overview of connections between database tables, corresponding entities and transactions, as well as process activities. For instance, we observe that the table VBAK stores information on sales orders, customer inquires and quotations. Different transactions of the VA module, specifically VA01, VA03, VA11, and VA21 manipulate these entities.

The connections shown in Table 1 are explicitly conceptually represented in EOG and can be navigated by users. In this way, the create sales order activity includes a link to the corresponding snippet of the Entity-Relationship Diagram [1] of sales orders and related entities. Figure 4 highlights the conceptual

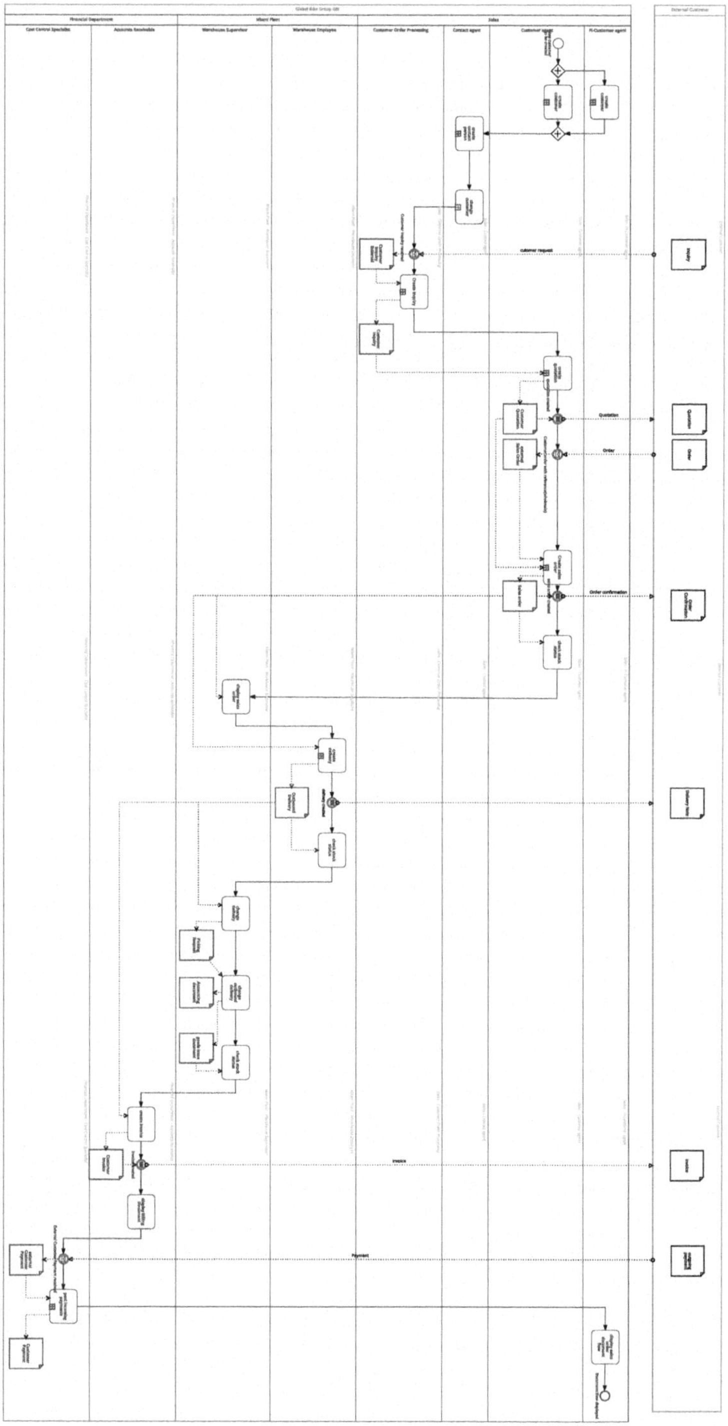

Fig. 3. The order-to-cash process of the Global Bike Group.

Fig. 4. Entity-Relationship Diagram of Sales and Distribution Entities integrated with Financial Entities.

relationships between the entities involved in the business process. Figure 5 shows further details in a separate diagram with the most important attributes, keys, and foreign keys of the four entities Customer Master, Sales Order, Sales Order Item, and Material Master. In this way, we learn that Sales Order Item has both the Sales Order Number and the Material Number as foreign keys and POSNR as a key. For each item, an order quantity has to be specified.

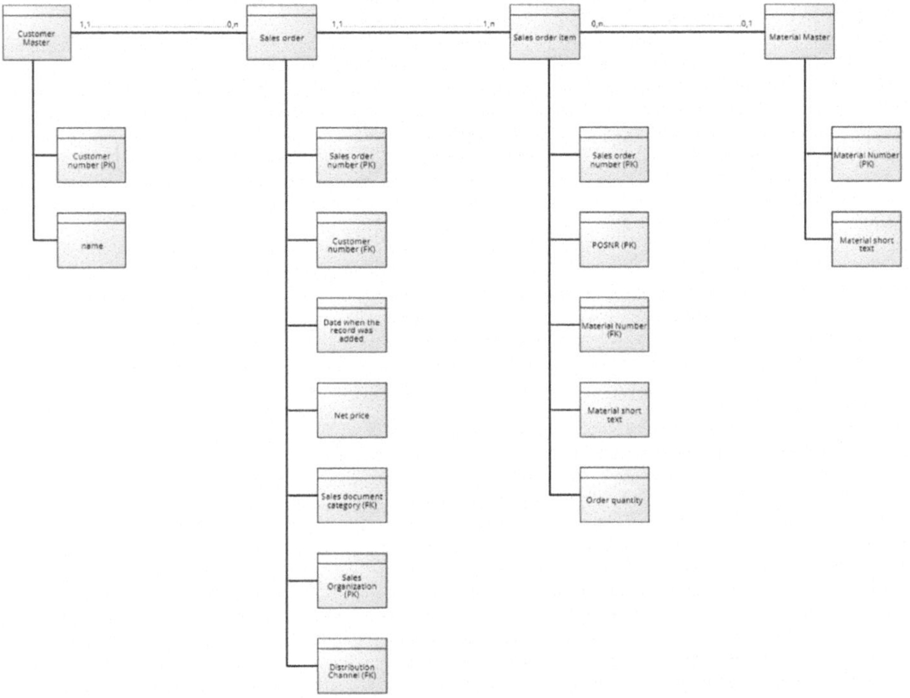

Fig. 5. Entity-Relationship Diagram of Sales Order related Entities.

EOG is accompanied with a series of case studies and exercises. These build on the corporate model Global Bike Inc., which was developed for SAP University Alliance Curricula. The key objective of these exercises is to make the students get to know the corporate models and the model structure as well as model interrelationships. The corresponding exercises are framed in a situation that the student is a new employee of the Global Bike Group. By navigating the EOG models, the student has the tasks to find out, for example,

- who are the internal customers of my organisation and what goals arise from a common Motivation?
- how do I need to set up my Organisation in a process-oriented way to achieve these goals?
- what Functions does my organisation perform for my customers?
- which Processes serve the needs of my customer?
- which company data relates to my processes and which structure do they have (horizontally and vertically)?

All tasks refer to models that are included in EOG. In order to answer the exercise questions, the student has to navigate the models in the right column at the right level of granularity in order to identify the relevant information. In this way, the student is exposed to the different perspectives that can be taken on the enterprise and recognizes the connections between the entities covers by them. This reveals an understanding of the underlying SAP ERP system at different levels of detail, from overview models down to mask models showing the different user interfaces and computer screens.

4 Lessons Learnt

There are some key lessons learnt from developing EOG and using it in teaching. These include the benefits of learning complex processes and corresponding system functionality at different levels of granularity and from different perspectives. Such experience can be further enhanced from a technical angle. To this end, we have used exercises to fetch specific data instances from the database tables of the SAP system. Writing corresponding SQL statements in a programming environment outside of the normal user guidance with an SAP GUI contributes substantially to the understanding of the integration of data and processes. This integrates with EOG and facilitates the basic understanding of an ERP system by an IT consultant.

All previously input and output data, e.g. from an order-to-cash process, can be displayed directly with a select statement for the corresponding tables. For this purpose, EOG offers numerous mapping models from which tables can be selected for the order-to-cash process including entities with their key, foreign key and descriptive attributes. An important learning here is that there is a 1:n relationship between entities such as Customer and the tables implemented in the ERP system such as KNA1 and KNB1. While the referential integrity is represented in the ERM diagrams at the business level, there are also a large number of denormalized tables at the IT concept level. Queries to these tables can, for example, produce inconsistencies in the event of an insert statement to just one of these tables. Therefore this requires more comprehensive handling at the application level (e.g. using object-oriented concepts).

However, reading data is often the first step. If normalized tables are to be joined using key-foreign key relationships on a read-only basis, the corresponding join conditions can be derived from the ERM models mapped in EOG at the subject concept level. Writing such queries with the sense of achievement always amazes the visibly impressed students and leaves them at the end of their first learning journey with a positive feeling of confidence with the concepts they have learnt, in particular regarding the model-based implementation of databases. In more advanced courses, the advanced challenges of a model-based implementation of the entire ERP system can then be discussed by the help of EOG, e.g. with SAP Solution Manager or SAP Cloud ALM. One open challenge for EOG to date is the current lack of referential integrity between all other model objects (not just in the data map) of an enterprise architecture, which are defined in the EOG conventions.

Although EOG is showcased with example entities from the SAP demonstration case, the abstraction of the models comes with conceptual benefits. The abstract representation provides to a certain extent product neutrality, since maybe entities can be found in various companies using different commercial ERP products. EOG users can also choose between different BPM tools at any point (ARIS, Signavio, LeanIX, Adonis, etc.) and learn from concrete examples that EOG has been implemented in a tool-independent manner. User experience this aspect particularly intensively when it comes to creating their own models in the context of the EOG for any company with a tool of their choice.

A considerable challenge is the additional complexity of learning concepts and structures represented in EOG. Students have to invest an initial learning effort in order to ultimately master the complexity of an ERP system more systematically. At the same time, it may be a good option for universities to teach concepts rather than products.

5 Conclusion

Teaching ERP systems and corresponding business processes such as order-to-cash is a challenge for students due their complexity. In this paper, we have explained how models of the Enterprise Online Guide (EOG) can be used to facilitate the navigation of this complexity. We have shown a few examples to illustrate the relationship between process models and database tables, and how EOG can help students to develop an understanding of an ERP systems. We discuss the effects of this approach for learning key concepts of business process management.

Acknowledgments. The research of the authors was supported by the Einstein Foundation Berlin under grant EPP-2019-524, by the German Federal Ministry of Education and Research under grant 16DII133, and by Deutsche Forschungsgemeinschaft under grants 496119880 (VisualMine) and 531115272 (ProImpact).

Disclosure of Interests. The authors have no competing interests to declare that are relevant to the content of this article.

References

1. Chen, P.P.: The entity-relationship model - toward a unified view of data. ACM Trans. Database Syst. **1**(1), 9–36 (1976). https://doi.org/10.1145/320434.320440
2. Dumas, M., Rosa, M.L., Mendling, J., Reijers, H.A.: Fundamentals of Business Process Management, 2nd edn. Springer, Heidelberg (2018). https://doi.org/10.1007/978-3-662-56509-4
3. Gesellschaft für Informatik e. V.: Rahmenempfehlung für die Ausbildung in Wirtschaftsinformatik an Hochschulen (2017)
4. Klaus, H., Rosemann, M., Gable, G.G.: What is erp? Inf. Syst. Front. **2**(2), 141–162 (2000). https://doi.org/10.1023/A:1026543906354
5. Laudon, K., Laudon, J.: Management Information Systems: Managing the Digital Firm, Global Edition. Pearson Education, Boston (2021)
6. Osterwalder, A., Pigneur, Y., Tucci, C.L.: Clarifying business models: origins, present, and future of the concept. Commun. Assoc. Inf. Syst. **16**, 1 (2005). https://doi.org/10.17705/1CAIS.01601
7. Scheer, A.W.: Architecture of integrated information systems foundations of enterprise modelling (1992)
8. Scheer, A., Hars, A.: Extending data modeling to cover the whole enterprise. Commun. ACM **35**(9), 166–172 (1992). https://doi.org/10.1145/130994.131007
9. Scheer, A.-W., Nüttgens, M.: ARIS architecture and reference models for business process management. In: van der Aalst, W., Desel, J., Oberweis, A. (eds.) Business Process Management. LNCS, vol. 1806, pp. 376–389. Springer, Heidelberg (2000). https://doi.org/10.1007/3-540-45594-9_24
10. Scheruhn, H.J., Reiter, C., Bayramli, E.: From process mining to enterprise mining. In: Marx Gomez, J., Elikana Sam, A., Godfrey Nyambo, D. (eds.) ICTA-EMOS 2022, pp. 195–205. Springer, Heidelberg (2022). https://doi.org/10.1007/978-3-031-56576-2_17
11. The Open Group: TOGAF 9 - The Open Group Architecture Framework Version 9 (2009)

Can Business Process Design Space Help Students Generate Innovative Process Redesign Ideas?

Mahendrawathi ER(✉), Angelica Deviana, and Ika Nurkasanah

Institut Teknologi Sepuluh Nopember, Surabaya, Indonesia
mahendra_w@is.its.ac.id, angelicadeviana1@gmail.com,
ika.nurkasanah@its.ac.id

Abstract. The Business Process Management (BPM) community has developed numerous techniques for business process redesign. Business Process Design Space (BPD-Space) offers a method that assists organizations in systematically exploring the design space of processes. This paper presents the result of experiments using the BPD-Space to teach explorative business process redesign. Three free experiments involving students with different backgrounds were conducted. A case study of the Surabaya City Office for Population Administration and Civil Registry (COPACR), which provides population registration and civil registration services, serves as a practical example for the participants to apply the BPD-Space. Results show that participants can apply BPD-Space to generate redesign ideas that vary in number, diversity, and innovativeness. Students with more experience and domain knowledge come up with more creative redesign ideas.

Keywords: Business process redesign · Explorative redesign · Business Process Design Space · BPD-Space · Public sector · Teaching · Case

1 Introduction

Business process redesign is a phase in the Business Process Management (BPM) lifecycle related to making structured and significant changes to the current business process [2]. Many methods, tools, and techniques have been developed to help organizations redesign their business processes. Redesign techniques are well-defined sequences of steps to purposefully transform business processes within and across organizational boundaries [1].

Depending on the ambition (transactional or transformational), nature (analytical or creative) and perspective (inward-looking or outward-looking), the redesign techniques can be visualized as a Redesign Orbit [2]. Re-design techniques that are transactional and analytical in nature depicted in Redesign Orbit are Best Practices of Process Redesign, Lean, and Six Sigma. These techniques enable a process analyst to diagnose problems in the existing process, choose

C. Di Ciccio et al. (Eds.): BPM 2024, LNBIP 527, pp. 396–405, 2024.
https://doi.org/10.1007/978-3-031-70445-1_28

the most suitable approach, and develop redesign ideas to address the identified problems [4]. Such approaches are considered exploitative redesign techniques.

In recent years, as the BPM community aimed to be more outward-looking and opportunity-driven [4], several researchers have advocated for more explorative redesign techniques. These techniques aim to generate innovative and non-obvious redesign ideas inspired by trends beyond organizational boundaries. Explorative redesign techniques aim to be "creative", "outward-looking", and "transformational". One of the explorative redesign techniques is the Business Process Design Space (BPD-Space). The BPD-Space offers a method that assists organizations in systematically exploring the design space of processes.

Exploitative redesign techniques, such as Best Practices of Process Redesign, Lean, and Six Sigma, have been widely researched and extensively taught in BPM courses. However, with the recent trends in digital transformation across various sectors, there is an increasing need to educate students on more explorative techniques. Teaching creative techniques is more challenging due to their subjective nature and heavy reliance on the student's prior knowledge.

This paper addresses the gap in the reported implementation of explorative re-design techniques in research and educational contexts. We present the result of three experiments where students use the BPD-Space to develop redesign ideas for the Surabaya City Office for Population Administration and Civil Registry (COPACR). We aim to answer, "Can students effectively apply the BPD-Space to generate innovative ideas for a real organization?" This paper's main contribution is providing insights and lessons learned for BPM educators in teaching explorative process redesign.

The remainder of this contribution is structured as follows. First, related works in business process redesign, BPD-Space and its implementation are presented in Sect. 2, followed by the description of the experiments. Results from the experiments are presented in Sect. 4. Section 5 discussed the lessons learned from this study that can be useful to other BPM educators. Finally, Sect. 6 provides brief concluding remarks and opportunities for further studies.

2 Related Works

Business Process Management (BPM) is the art and science of overseeing how work is done within an organization to ensure consistent results and to take advantage of improvement opportunities [2]. BPM encompasses six phases in its life cycle: process identification, discovery, analysis, redesign, implementation, and monitoring. Among these phases, the process redesign phase is pivotal to BPM [6].

There are two primary triggers in BPM: those driven by problems and those driven by opportunities. These triggers and their associated actions and value propositions have led to two distinct BPM approaches: exploitative and explorative BPM [3]. Exploitative BPM addresses operational process issues, typically resulting in simplified processes. Conversely, explorative BPM employs an outside-in logic, utilizing environmental scanning capabilities to identify new opportunities in a dynamic business environment [7].

Business Process Design Space (BPD-Space) is a conceptual model of design space for business processes that allows systematic exploration of alternative designs from the design dimensions [5]. BPD-Space consists of 19 business process design dimensions grouped into six layers: customer layer (customer segmentation, customer experience, customer value, and customer channel), product/service layer (scope, component flow, location, and temporality), the business process layer (coordination, triggers, and outcomes), the organizational layer (goals, internal participants, revenue models, and business partners), the information layer (information sources and use of information), and the technology layer (infrastructure and automation).

BPD-Space has been implemented in three companies: financial technology startups, large-scale medical organizations, and medium-sized companies operating transmission systems. Practitioners from all real-world applications found the BPD-Space suitable (applicable) and useful for the generation of redesign alternatives [5]. In more recent experiments, the BPD-Space leads to more redesign ideas, which are more diverse than exploitative techniques [4].

3 Free Experiments Using BPD-Space

To gain insights into how BPD-Space can help students identify innovative process redesign ideas, we conducted experiments involving students in BPM courses and those who volunteered to join workshops. The experiments are designed as free experiments where the surrounding conditions and parameters are not strictly controlled [4]. We particularly tried different settings related to the procedure of introducing the BPD-Space and the format of idea generation.

3.1 Selecting the Case Study

The first step is to select a case study that provides a practical context for the participants to apply the BPD-Space. We selected a case study of the Surabaya City Office for Population Administration and Civil Registry (COPACR), which provides population registration and civil registration services, such as issuing identity cards, child identity cards, family cards, birth and death certificates, etc. This case is selected because participants are likely familiar with the services provided by COPACR, which provide background knowledge for them to propose new redesign ideas.

A description of COPACR's current condition is given to the participant. We provide COPACR business process architecture (available in the supplementary materials) and BPMN models of some of the most important services, including ID cards, family cards, and birth certificate applications. These are typical application-to-approval processes where applicants submit their application and supporting documents. The COPACR staff then verifies the requests and supporting documents, which must be done by accessing the national database system called SIAK. The application is completed after being validated by the section head and signed by the COPACR head.

The large number of requests due to the increase in population in Surabaya, accompanied by changes in conditions due to the COVID-19 pandemic, pressure COPACR to continuously improve its performance in all aspects. Since March 2020, COPACR has innovated by introducing the KLAMPID system to provide easy access to the public during the COVID-19 pandemic. People can make applications independently through the KLAMPID system without having to come in person. KLAMPID allows applicants to upload supporting documents in digital form. Finally, verification and validation processes are also conducted in the system.

3.2 Experiment Preparation

The next step is to prepare materials and rundown for the experiments. BPD-Space Cards consist of an introduction to the concept of design space, an introduction to BPD-Space, a user guide, an explanation of each design dimension (question guide, possible characteristics, and application examples), and supplementary cards. To enhance understanding and assist participants in exploring alternative innovation ideas for business process design, we translated the cards into Indonesian, making them more accessible for the participants. The revenue model dimension was excluded because COPACR is a public service organization that does not generate revenue from its services.

3.3 Conducting the Experiments

We conducted three experiments with different participants and different procedures. The summary of experiments is given in Table 1. The first experiment, the pre-workshop, involved fourth-semester students enrolled in a BPM course. This experiment was part of their BPM course. The 102 students were divided into 35 groups and assigned specific dimensions to work on using BPD-Space before the class. During the class, they were introduced to BPD-Space by one of the inventors. The introduction session is followed by group discussions and presentations of the students' ideas.

The second and third experiments were structured as workshops. In the first workshop, six fourth-year students participated. They were introduced to all the layers of BPD-Space and given an hour to brainstorm redesign ideas, which they then presented and discussed.

Table 1. Summary of Experiments

Date	Background	Number of Participants
19/11/2021	Information Systems Students (Second year)	102 students divided into 35 groups
04/08/2022	Information Systems Students (Fourth year)	6 students
05/13/2022	Information Systems Student (Third year)	8 students

The second workshop, which included eight third-year students, had a modified procedure. First, we updated the materials to include examples of innovation from cases in Indonesia that participants can relate to. Second, the workshop was divided into sessions instead of tackling all layers simultaneously. Each session introduced two layers, and participants applied BPD-Space to these layers for 10 min before discussing their ideas. This procedure was repeated twice. Another workshop is conducted with a representative from Surabaya COPACR, where we discuss whether the proposed ideas can be considered novel/new.

4 Results of Experiments Using BPD-Space

Results from the three experiments are summarized in Table 2. The pre workshop generated the highest number of ideas for several reasons. More participants were involved (35 groups), and each group was assigned one dimension. Each group can generate more than one idea. They are given three days to prepare before the session. The results show that pre-workshops produce 1.14 ideas per group.

Table 2. Description of Innovation Idea Group.

Layer	Dimension	Pre-workshop	Workshop 1	Workshop 2	Total	
Customer	Customer Segmentation	4	0	0	4	20
	Customer Experience	3	0	0	3	
	Customer Value	3	0	6	9	
	Customer Channel	1	0	3	4	
Product/Service	Scope	3	3	0	6	10
	Flow Unit	2	0	0	2	
	Location	2	0	0	2	
	Temporality	0	0	0	0	
Business Process	Coordination	3	0	0	3	16
	Trigger	2	8	0	10	
	Outcome	2	1	0	3	
Organization	Objectives	3	0	0	3	8
	Internal Participants	1	2	1	4	
	Revenue Model	NA	NA	NA	NA	
	Business Partner	1	0	0	1	
Information	Information Source	1	1	3	5	15
	Information Use	1	5	4	10	
Technology	Infrastructure	2	0	0	2	19
	Automation	6	8	3	17	
Total Idea		40	28	20	88	
Number of ideas/groups/participants		1.14	4.67	2.50		

On the contrary, workshop 1 and 2 participants are selected groups of students with limited time to develop their ideas. Workshop 1 participants were the most productive, generating 4.67 ideas per participant. Workshop 2 participants produced 2.5 ideas per participant.

Despite these differences, we will discuss more on how the BPD-Space helps students come up with their ideas, particularly in terms of the dimensions and novelty of their ideas. The pre-workshop participants were assigned specific dimensions to focus on, so, unsurprisingly, they produced ideas across nearly all dimensions except for temporality (note that revenue models are excluded). Participants in Workshops 1 and 2 also did not generate any ideas related to temporality. These results indicate that generating ideas involving temporality is particularly challenging for the students.

4.1 Diversity of Dimensions and Layer of Ideas

The number of ideas that the three experiments produced across different layers in the BPD-Space are summarized in Fig. 1. Participants in the pre-workshops generated many ideas related to the customer (11 ideas), followed by technology (8 ideas), business processes, and product/service (7 ideas for each) layers. They did not come up with many ideas in the information layer.

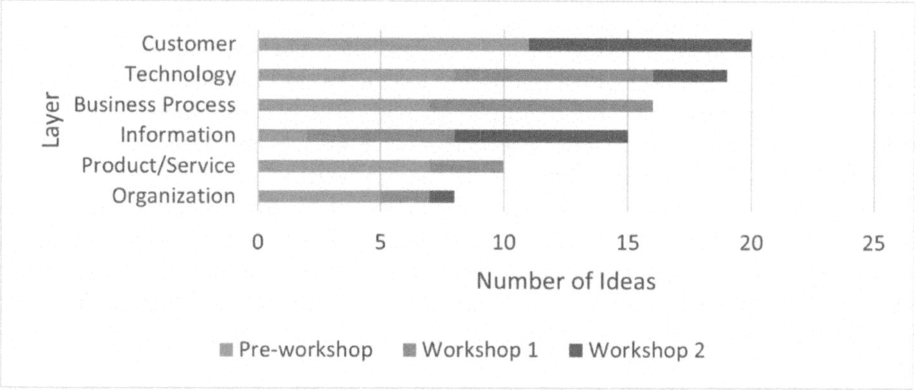

Fig. 1. Participants Ideas Across Different Layers

Workshop 1 participants did not develop any ideas for the customer layer. Instead, they mostly generated ideas about business processes, technology, and information layers. In contrast, Workshop 2 participants came up with many ideas related to the customer layer, none on product/service and business processes, and few on technology. The differences in the layer of BPD-Space innovation appear to be affected by the student's experience. Participants from the pre-workshop are second-year students and probably have less experience regarding the information layer. They can identify many ideas on the customer layer because they put their experience into generating the ideas. The participants of workshops 1 and 2 are more mature students (fourth and third year), so they are probably more exposed to and inspired by the latest technology and information.

4.2 Innovativeness of Ideas

We checked with COPACR representatives to assess the innovativeness of the participants' ideas. For each idea, the COPACR assessed whether it is new, so considered innovative to them or has been done. The innovativeness of the ideas from the three experiments is summarized in Table 3. As seen in Table 3, the pre-workshop participants came up with many ideas, which are not new to COPACR. COPACR has thought about the idea and started or has implemented it. For example, one of the ideas, "Provide an estimate of document processing time once an applicant submits their application", is actually being implemented by COPACR at the time of the validation.

Table 3. Innovativeness of Idea

Experiment	Participant	Dimension	Innovation Idea	Innovativeness	%
Pre-workshop	102 students divided into 35 groups	18	40	11	28%
Workshop 1	6 students	7	28	13	46%
Workshop 2	8 students	6	20	13	65%

Table 4 summarizes the ideas from pre-workshop participants that are innovative. The pre-workshop participants created a lot of innovative ideas related to the customer layer (4), for example, providing a priority lane for different customers, in this case, citizens, a process tracking feature, and live chat text and audio to help applicants throughout the application process.

Participants of Workshop 1 generated a lot of innovative ideas related to technology as summarized in Table 5. They proposed that ID cards or other outcome documents such as birth certificates can be printed digitally, integrating personal data into the ID card and connecting it with an application. A list of all innovative ideas from Workshop 1 can be accessed in the supplementary materials.

Participants from Workshop 2 generated a lot of innovative ideas related to technology and information, as shown in Table 6. Like participants in Workshop 1, Workshop 2 also proposes that COPACR leverage new technology such as Near Field Communication (NFC) into the ID card. Interestingly, Workshop 2 participants came up with many ideas on how COPACR can use the information of the population (average income data, pregnant women) that can be used for various purposes. A list of all innovative ideas from Workshop 2 can be accessed in the supplementary materials.

Table 4. Innovative Ideas from Pre-Workshop

No	Innovation Idea	Dimension
1	Provide priority service lane (elderly, pregnant women, breastfeeding women, women with little children, and people with disability). They can have the 1:1 assistant to help them and the right to reprint their e-ID CARD through the priority lane	Customer Segmentation
2	Provides a process tracking feature that shows the progress of an ID card application. It indicates whether the application is in the queue, currently being processed, and its progress.	Customer Experience
3	Changing the data verification process by the registration officer to occur before printing the application registration through the KLAMPID website	Customer Value
4	Collaboration with the delivery package platform	Business Partner
5	Live chat text and audio to help applicants	Customer Channel

Table 5. Innovative Ideas from Workshop 1

No	Innovation Idea	Dimension
1	ID card is digitally printed.	Automation
2	Outcome documents (such as birth certificates) are digitally printed	Automation
3	Personal data is integrated into one card (QRIS) storing information like name, age, address, date of birth, and marital status	Automation
4	An identity card is connected to an application, functioning like a loyalty card, accessible through a barcode.	Automation
5	The SIAK center detects duplicate data and informs COPACR	Automation

Table 6. Innovative Ideas from Workshop 2

No	Innovation Idea	Dimension
1	Growth data on the population (based on the number of requests for new ID cards) can enhance COPACR's asset system information and increase process effectiveness	Information Use
2	COPACR can gather pertinent information from external organizations, such as average income data specific to each geographic area. This data integration can be facilitated through online mapping platforms, enabling rapid and targeted responses during disaster relief efforts	Information Use
3	Conducting a census of pregnant women to account for the new population	Information Use
4	A national ID card is transformed into an account/portal accessible via barcode scanning or NFC technology embedded in the card.	Automation
5	COPACR implements NFC technology to minimize the reliance on paper photocopies in document management processes	Automation
6	The registration process for ID cards is conducted entirely online, incorporating NFC technology along with Touch ID or fingerprint sensors for enhanced security and convenience	Automation

5 Discussion and Lessons Learned

The aim of this paper is to determine whether students can effectively apply the BPD-Space to generate innovative ideas for the Surabaya COPACR. The goal has been achieved to a certain extent as the participants from the three experiments can apply the BPD-Space and produce some innovative ideas.

However, it must be noted that the experiments' results confirmed that experience is an important aspect of generating innovative ideas. The pre-workshop participants were given three days to prepare for the ideation session and focus on one dimension only, but they did not develop many innovative ideas. The workshop participants are more experienced, so they can generate innovative ideas even in a short period of time. Workshop 1 participants are final year students taking a final project at COPACR, so they have more domain knowledge. They are the most productive, although the innovativeness of their ideas is less than Workshop 2 participants. Workshop 2 participants are third-year students who generate three automation-related ideas, all considered new by COPACR. This warrants that experience and domain knowledge is an important factor that needs to be accounted for if we want to assess the effectiveness of the BPD-Space method in developing innovative redesign ideas.

We noted several lessons learned for other BPM educators wishing to apply the same method in their teaching.

A Perfect Case. In our experiment, we selected a case from the Surabaya City Office for Population Administration and Civil Registry (COPACR). This case suits the experiment well because students are familiar with the organization and its services. Yet, it is sufficiently challenging for them to generate new and innovative ideas. BPM educators can find a similar local case with the same characteristics for their teaching.

Innovativeness in the Eye of Domain Expert. In this study, we can secure access to COPACR representatives who can then judge the innovativeness of the ideas developed by students. However, in other cases, without access to domain experts, educators must find a different mechanism to assess the novelty or innovativeness of ideas proposed by the students.

Appetite for Technology. BPD-Space is developed to embrace opportunities for digital innovation. Our experiments demonstrate that students are indeed able to generate ideas related to the latest technology for the case.

Provide Examples. The BPD-Space already provides a card that explains each dimension and guiding questions. However, we found that more examples from organizations recognizable to the participants (e.g. Indonesian organizations) inspire the participants more. Different countries will also have different cases, so BPM educators should try to find cases to spark the student's ideation process.

Innovation or Modification? While BPD-Space aims to help students develop innovative/novel ideas, our experiments suggest that students require more examples, which can be attributed to the Indonesian culture. Indonesia has an

"Amati Tiru Modifikasi" concept, often associated with entrepreneurial and creative processes. This phrase, translated into "Observe, Imitate, Modify", encourages individuals to observe successful practices or products, imitate them to understand their workings, and then modify or innovate to create something new or improved. This highlights cultural aspects that also need to be considered by other BPM educators. The challenge for educators is how to foster and trigger the students to think beyond what is currently available.

6 Conclusion

This paper presents the result of experiments using the BPD-Space to teach explorative business process redesign. The participants of three experiments can apply BPD-Space to generate redesign ideas that vary in number, diversity, and innovativeness. Younger participants generate more ideas related to customer layers, showing they are more familiar with this layer. We found that experience and domain knowledge allow students to apply BPD-Space better and develop creative redesign ideas. Students with domain knowledge produce more novel ideas.

This study opens up several opportunities for further studies, including 1) replicating the experiment with more controlled parameters, i.e., the same level of participant experience and domain knowledge, 2) comparing the effectiveness of BPD-Space with exploitative re-design techniques as conducted in [4], focusing on the way the students ideate, and applying the experiments with different cases.

References

1. Champy, J., Hammer, M.: Reengineering the Corporation. Bookbytes (2001)
2. Dumas, M., Rosa, L.M., Mendling, J., Reijers, A.H.: Fundamentals of Business Process Management. Springer, Heidelberg (2018). https://doi.org/10.1007/978-3-662-56509-4
3. Grisold, T., Gross, S., Röglinger, M., Stelzl, K., vom Brocke, J.: Exploring explorative BPM - setting the ground for future research. In: Hildebrandt, T., van Dongen, B.F., Röglinger, M., Mendling, J. (eds.) BPM 2019. LNCS, vol. 11675, pp. 23–31. Springer, Cham (2019). https://doi.org/10.1007/978-3-030-26619-6_4
4. Groß, S., et al.: Idea generation in exploitative and explorative business process redesign techniques. Inf. Syst. e-Bus. Manag., 1–29 (2024)
5. Gross, S., Stelzl, K., Grisold, T., Mendling, J., Röglinger, M., vom Brocke, J.: The business process design space for exploring process redesign alternatives. Bus. Process. Manag. J. **27**(8), 25–56 (2021)
6. Kerpedzhiev, G.D., König, U.M., Röglinger, M., Rosemann, M.: An exploration into future business process management capabilities in view of digitalization: results from a Delphi study. Bus. Inf. Syst. Eng. **63**(2), 83–96 (2021)
7. Kohlborn, T., Mueller, O., Poeppelbuss, J., Roeglinger, M.: Interview with Michael Rosemann on ambidextrous business process management. Bus. Process. Manag. J. **20**(4), 634–638 (2014)

Beyond Traditional Learning: The LLM Revolution in BPM Education at University

Marzena Grzesiak[1](✉), Krzysztof Kluza[2,3], Natalia R. Potoczek[4], and Leszek Szała[5]

[1] Gdańsk University of Technology, 11/12 Gabriela Narutowicza Street, 80-233 Gdańsk, Poland
Marzena.Grzesiak@pg.edu.pl

[2] AGH University of Krakow, al. A. Mickiewicza 30, 30-059 Krakow, Poland
kluza@agh.edu.pl

[3] University of Gdańsk, Armii Krajowej 101, Sopot, Poland

[4] Institute of Economics, Polish Academy of Sciences, Nowy Świat 72, 00-330 Warsaw, Poland
npotoczek@inepan.waw.pl

[5] Department of Mathematics, Informatics and Cybernetics, Faculty of Chemical Engineering, University of Chemistry and Technology, Technicka 5, 166 28 Prague 6, Prague, Czech Republic
Leszek.Szala@vscht.cz

Abstract. Large Language Models (LLMs) significantly impact higher education, requiring changes in educational processes, especially in Business Process Management (BPM) practical exercises. The research aims to evaluate the effectiveness of LLMs in BPM education to determine if LLMs can supplement educators. The study involved 33 master's degree students. Students' works were manually evaluated and compared to LLM-generated responses. Results highlight the effectiveness and limitations of LLMs in supporting BPM education, revealing discrepancies between human and AI assessments. Our findings indicate that LLMs like ChatGPT-3.5 and ChatGPT-4o can aid BPM education, but their performance and reliability differ from traditional human grading. The findings underscore LLMs' potential to offer additional perspectives and reduce educators' workload. However, LLMs should be supplementary tools, not replacements for traditional methods. This exploration contributes to understanding the transformative role of LLMs in reshaping educational methodologies. Future research should consider larger samples and a broader range of tasks to validate and extend these findings.

Keywords: BPM · Educators Forum · BPM education · AI · LLM · GPT · ChatGPT · Educational Process

C. Di Ciccio et al. (Eds.): BPM 2024, LNBIP 527, pp. 406–415, 2024.
https://doi.org/10.1007/978-3-031-70445-1_29

1 Introduction

Large Language Models (LLMs) are advanced AI tools capable of generating high-quality texts using Generative Pre-trained Transformer (GPT) networks. As LLMs develop and become more accessible, more and more universities are both struggling and discussing how to incorporate these tools into their curricula. This integration can help teachers and students understand and apply modern technologies in daily academic life, enhancing the educational experience. Although the context of LLMs is extensively discussed in the scholarly literature, there is a gap in understanding how they can be used in teaching Business Process Management (BPM). The core issue is how to adjust exercises to boost students' understanding, ensuring that LLMs help rather than replace their efforts in class. Standardizing BPM teaching is also necessary to ensure consistent, high-quality education across institutions [14].

The research aims to assess the effectiveness of LLMs in BPM education by evaluating LLM-generated responses and those from master's students, exploring the potential of LLMs to grade and provide feedback on BPM-related tasks. This study was conducted in April and May 2024 at one of the Technical Universities in Poland, involving a cohort of 33 master's degree students in a course on process management. Methodologically, the research encompasses a literature review and a survey study using a non-probabilistic sampling technique.

In our approach, the teacher manually evaluated students' works using a specific answer key. LLM-based chats were also utilized to answer the same questions, and their performance was analyzed and compared to the teacher's assessments. The students' works were rated on a scale from no conformance to full conformance with the answer key. Based on the results obtained, we compared the results to provide the analysis of the suitability of LLM to help teachers in the assessment task.

2 LLMs in Educational Processes

Recent publications highlight efforts to implement LLMs in education in various disciplines. According to Stribling et al. [16], ChatGPT performed very well in answering questions requiring a textual answer in biomedical studies. Conversely, GPT-4 performed poorly on questions with figures containing simulated data and those requiring a hand-drawn answer. Other studies on the use of LLM in medical tests showed a great potential of the technology, but also limitations mainly related to the ability to ask questions [1].

Generative AI is also used successfully in the teaching of computer science. Researchers have shown that generative AI tools are capable of solving programming problems, writing extensive blocks of code, and explaining complex code in simple terms [9]. They particularly highlighted its usefulness in improving programming error messages. Studies have shown that students make fewer errors when being trained with ChatGPT 4. Interesting results were provided by studies on using LLM to support students [17]. The studies explored using natural language for problem-solving in learning. The LLM Hint Factory system, offering four levels of hints, showed that hints should be supported by specific code. These results suggest that future work should focus on customizing help responses to better meet students' learning needs. Many researchers' studies also focus on creating support for teachers in assessing and providing feedback for students [11]. The use of LLM to provide feedback significantly improved student work quality. Students who received feedback through ChatGPT 3.5 improved their tasks more than those who did not receive feedback. This shows that LLMs provide timely feedback that benefits students' cognitive and motivational outcomes. Other researchers found LLMs promising for essay grading in language courses [13]. Great hopes lie in creating LLM-based student assistants, which provides opportunities for a better understanding of course content and increases the usefulness of educational materials aimed at using AI technology [3,12]. Undoubtedly, research on the application of AI technology will aim at improving the quality of education but also the competencies to use AI among users, i.e., teachers and students.

2.1 State of the Art in Using LLMs for BPM Tasks

Initial experiments on accomplishing BPM tasks using publicly available models were presented by Grohs et al. in [7]. However, the results are unreliable, as the frequent problem of LLMs are hallucinations. Many problems occurring when querying LLMs for conformance or process improvement questions were noticed in [2]. Although pure LLMs might struggle with answering more complex questions, there are various solutions which might improve the results, such as: using prompt engineering with pre- and post-processing [6], adding to advanced prompt engineering also error handling, code generation techniques, and user feedback [10], providing a variety of business process-related views e.g., process, causal, as input for prompt engineering to achieve better process outcome explanations [5] or using Retrieval Augmented Generation [8].

3 Study Design

Our research aligns with the manifesto to improve BPM education in Poland [15] by integrating emerging IT tools like LLMs into teaching. The tasks for this research are taken from the "Fundamentals of Business Process Management" textbook [4]. The students carrying out the task are students of the Gdańsk University of Technology, Faculty of Management and Economics, second-degree Management, 3- and 4-semester studies. Students of 3-semester studies are graduates of first-degree studies in Engineering Management. All 33 student works were evaluated manually by the teacher according to the answer key to the tasks (included in the book [4]):

- Ex 1.1.1. Who are the actors in this process?
- Ex 1.1.2. Which actors can be considered to be the customer (or customers) in this process?

- Ex 1.1.3. What value does the process deliver to its customer(s)?
- Ex 1.1.4. What are the possible outcomes of this process?
- Ex 1.3. Taking the perspective of the customer, identify at least three performance measures that can be attached to this process.
- Ex 6.1.1. What steps can you extract from this process? Classify these steps into VA.
- Ex 6.1.2. What steps can you extract from this process? Classify these steps into BVA.
- Ex 6.1.3. What steps can you extract from this process? Classify these steps into NVA.

Next, the LLM-based chats were asked to answer these questions and their performance was analyzed. Finally, the LLM-based chats were used for assessing the students' answers and the results were compared with each other and to the results of the teacher's manual check.

The works of students were assessed in the predefined scale:

0 —no conformance (incorrect answer),
0.5 —slight conformance (there are significant elements of the correct answer),
1 —partial conformance (about half of the correct answer),
1.5 —slight discrepancy (most of the answer is correct, only minor errors),
2 —full conformance (fully correct answer).

As LLMs are intended to generate text, we asked chats to assess the answer and provide the results in the above-mentioned scale in the textual (not numerical) way. Then, another prompt was supposed to transform these results into numerical way. In the case of ChatGPT4o, the answers were successfully transformed. In the case of ChatGPT3.5, transforming the answers into points was incorrect in most cases, therefore the numerical points were manually assigned based on the answers if these were provided by the model. In the case of ChatGPT3.5 the answer was taken into account only if the answer was provided for all 8 questions in the above-mentioned scale (in the textual way). If the answer was partial or general (a single answer for the whole set of questions), the result was ignored and a new instance of the chat was used.

4 Results and Discussion

4.1 Students Scores vs Answer Key from the Textbook

Of the 42 people taking the Process Management course, 33 completed the task (21 male and 12 female students). The students' answers were assessed by comparing them with the answers given by the textboox [4]. In the first part of the task (4 questions), entirely consistent with [4], students were to identify the essential elements of the business process. While there were no significant problems with identifying the actors in the process (94% of correct answers or with minor errors) and the customer (70% of correct answers), there were no significant problems with determining the value provided to the customer (only 12% of answers correct or with minor errors) and determining results of the process, some students had problems (21% of answers were correct or with minor errors). The next question differed slightly from the one in the textbook [4]. Students were required to report at least three process measures. The evaluation took into account whether the answers proposed the solutions provided. 33% of students answered correctly or almost correctly. Indicating the steps in the process, especially their classification in the context of the value provided (3 more questions), was also challenging for the students. Steps classified as VA were identified correctly or with minor errors by 58% of students, BVA by 27%, and NVA by 21% of students. In the final scoring, the average score was 9.05 points, and the dominant score was 8. The minimum score is 5.5 points (1 person), and the maximum is 11.5 points (1 person). Figure 1a shows the distribution of point scores.

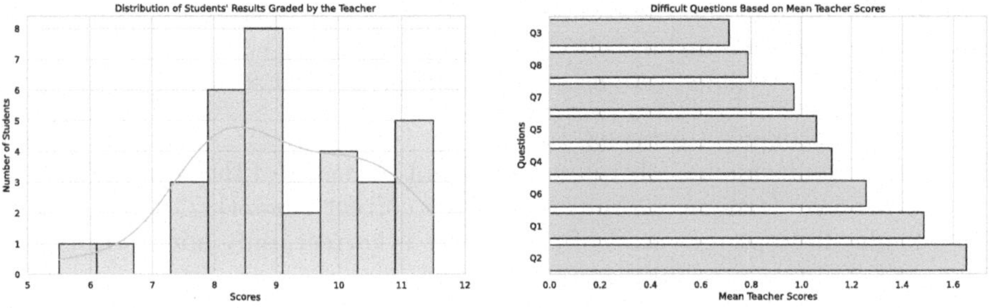

(a) Distribution of grades obtained by students (graded by teacher)

(b) The mean teacher scores for each question (identification of hard questions)

Fig. 1. Comparison of Teacher, ChatGPT3.5 and ChatGPT4o grading

4.2 GPT3.5 vs GPT 4o in Solving the Tasks

Table 1 compares LLM-based ChatGPT3.5 and ChatGPT4o performance across different identification tasks, with and without prompt engineering and self-assessment of the obtained results. The results with and without prompt engineering were analyzed manually, while self-assessment was done by the LLM-based chat itself. Prompt engineering appear to enhance performance in some cases, particularly for identifying value, outcomes, and non-value adding steps[1].

[1] Examples of the effective prompt engineering elements:

- For the task *Value*, the original prompt was:
  `What value does the process deliver to its customer(s)?`
  The improved prompt: `What direct single value does this particular process (only, no further education) generate for its customer?`.
- For the task *Outcomes*, the original prompt was:
  `What are the possible outcomes of this process?`.
  The improved prompt: `Provide different end results of this particular process (process outcomes) so that they reflect the different end states from the candidate's perspective.`

To simplify the comparison, the results are shown as ● (full answer 1.5–2 pts), ◑ (partial answer 0,5-1 pts), and ○ (incorrect answer 0 pts). The best results were obtained with ChatGPT 4o, although its self-assessment was more critical.

Table 1. Comparison of ChatGPT3.5 and ChatGPT4o across identification tasks.

Identification tasks	LLM-based chat					
	ChatGPT 3.5	ChatGPT 3.5 + prompt engineering	ChatGPT 3.5 self-assessment	ChatGPT 4o	ChatGPT 4o + prompt engineering	ChatGPT 4o self-assessment
Actors	●	●	●	●	●	◑
Client	●	●	●	●	●	●
Value	○	●	●	○	●	◑
Outcomes	◑	◑	●	◑	●	◑
Performance Measures	●	●	◑	●	●	◑
Value Adding	◑	◑	●	●	●	●
Business Value Adding	◑	◑	◑	●	●	●
Non-Value Adding	◑	◑	◑	○	●	●

4.3 Grading Student Works by ChatGPT

We have conducted the assessment of student works using ChatGPT-3.5 and ChatGPT-4o. The plot in Fig. 2a compares the overall distribution of scores given by the teacher, ChatGPT-4o, and ChatGPT-3.5. Scores of both chats are more spread out compared to the teacher's scores. Both ChatGPT-4 and ChatGPT-3.5 exhibit more variability in their scoring. The median scores of ChatGPT-4 and ChatGPT-3.5 are slightly lower than the teacher's median score, indicating a tendency towards stricter grading or different interpretation criteria.

The chart in Fig. 2b indicates that both ChatGPT4o and ChatGPT3.5 tend to grade more strictly than the teacher, particularly noticeable in questions Q3, Q4, Q7, and Q8. ChatGPT3.5 generally shows larger differences compared to ChatGPT4o, indicating it may be more conservative or strict in its grading. For ChatGPT4o the differences are more varied but often close to zero, suggesting less deviation from teacher evaluations.

The heatmaps in Fig. 3 display the absolute differences in scores between the teacher and each of the ChatGPT models (GPT-4o and GPT-3.5) for all students. It clearly shows that GPT-4o tends to be more closely aligned with the teacher's scores overall, with fewer and smaller discrepancies.

Questions Q3 and Q7 show the highest absolute differences for both models, indicating significant divergence from the teacher's scores in these questions. In both cases the model tends to underestimate grades compared to the teacher. What is important, both of these questions might be identified as hard, based on the average grades for these questions (Q3: 0.71, Q7: 0.97 – see: Fig. 1b).

Question Q2 shows the smallest differences, suggesting better alignment with the teacher's scores. As there are instances of high differences (e.g., 1.5 or 2.0) for specific students and questions, it indicates that occasionally significant discrepancies between AI and human grading might occur.

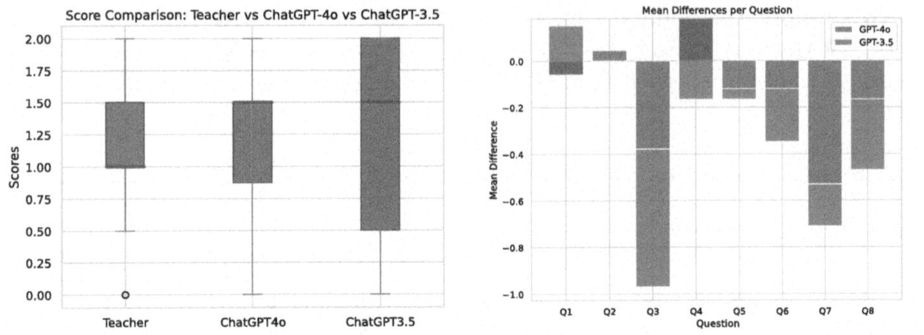

(a) Comparison of scores given by Teacher and Chats (b) Comparison of mean differences per question

Fig. 2. Comparison of Teacher, ChatGPT3.5 and ChatGPT4o grading

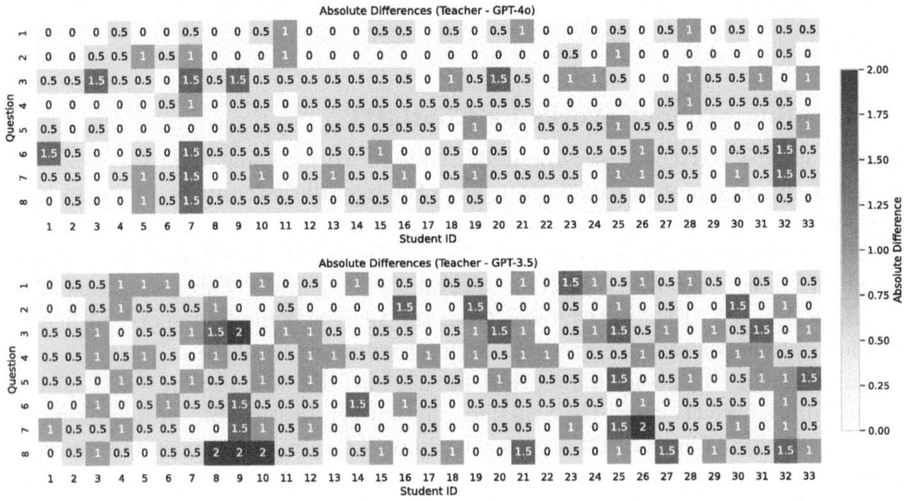

Fig. 3. Absolute differences in the answers of chats and the teacher

4.4 Statistical Evaluation

For the results we obtained, we conducted a statistical evaluation. Cohen's Kappa indices for rounded grades show that ChatGPT-4o has a compliance rate of 0.3097 and ChatPT-3.5 has a compliance rate 0.1344. Cohen's Kappa index is higher for ChatGPT-4o, suggesting that this model should represent teacher evaluations better than ChatGPT-3.5. In order to determine whether there is evidence supporting that the difference is statistically significant, we decided to perform statistical testing.

Applying a t-test requires that the collected data follows a normal distribution. To verify this assumption, the Shapiro-Wilk test was performed. The test

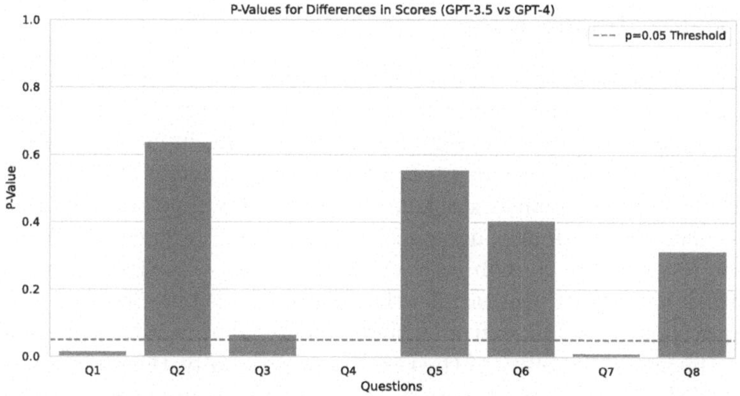

Fig. 4. The Wilcoxon test results for each question

rejected the hypothesis of normality. Therefore the paired samples Wilcoxon test was applied. The significance level of all the tests was set at 0.05.

The paired samples Wilcoxon test between the differences of GPT-3.5 and teacher scores versus the differences of GPT-4 and teacher scores yielded the p-value of 0.422, which do not contradict the hypothesis that the differences of GPT-3.5 and teacher scores have the same distribution as the differences of GPT-4 and teacher scores. This means that both GPT-3.5 and GPT-4 may have a similar level of deviation from the teacher's scores.

Therefore we performed the paired samples Wilcoxon test for each question separately and the results are presented in Fig. 4. Significant differences between GPT-3.5 and GPT-4o in their deviation from the teacher's scores were observed for questions Q1, Q4, and Q7. For the other questions the p-values do not indicate significant differences between the deviations of GPT-3.5 and GPT-4o from the teacher's scores. This suggests that the deviation of GPT-3.5 from the teacher's scores should be significantly different from GPT-4o's deviation for questions Q1, Q4, and Q7. For the other questions, the deviations seem similar.

5 Conclusions

This study explored the integration of LLMs into BPM education, specifically focusing on their potential to enhance and transform educational practices. Our findings suggest that while LLMs like ChatGPT-3.5 and ChatGPT-4o can support BPM education, there are notable differences in their performance and reliability when compared to traditional human grading.

We demonstrated that LLMs could assist in grading and providing feedback on BPM-related tasks, potentially reducing the workload of educators. We highlighted the discrepancies between human and LLM-based assessments, emphasizing the need for careful calibration and prompt engineering to improve the accuracy and reliability of these AI tools. Didactic tasks, regardless of the field, that are based on text analysis can be supported by LLMs.

5.1 Educators Implications

For educators, the findings underscore the importance of integrating LLMs as supplementary tools rather than replacements for traditional teaching methods. LLMs can enhance the learning experience by offering immediate feedback and additional perspectives, but educators should remain aware of their limitations, particularly in handling complex or nuanced tasks. Training students to critically assess AI-generated feedback and encouraging them to engage in higher-order thinking are essential to maximize the benefits of these technologies.

The study has several limitations that should be acknowledged. The sample size was limited to 33 students from a single technical university in Poland, which may affect the generalizability of the results. Additionally, the study focused on a specific set of BPM tasks, and the performance of LLMs might vary with different types of questions or academic disciplines. Future research should consider larger and more diverse samples, as well as a broader range of tasks, to validate and extend the findings.

Our future research will investigate other models and methods of assessment with LLMs. Developing best practices for prompt engineering and AI-human collaboration in educational settings can help optimize the use of LLMs in academia. The analysis of long-term impact of LLM-assisted learning on student performance and engagement will also be crucial.

Acknowledgments. The authors would like to thank Piotr Sliż, the originator of this research, for the idea and encouragement to pursue this research, as well as for hosting Krzysztof Kluza as a scientific intern at the University of Gdańsk.

References

1. Artsi, Y., Sorin, V., Konen, E., Glicksberg, B.S., Nadkarni, G., Klang, E.: Large language models for generating medical examinations: systematic review. BMC Med. Educ. **24**(1), 354 (2024)
2. Berti, A., Schuster, D., van der Aalst, W.M.: Abstractions, scenarios, and prompt definitions for process mining with LLMs: a case study. In: De Weerdt, J., Pufahl, L. (eds.) BPM 2023. LNBIP, vol. 492, pp. 427–439. Springer, Cham (2023). https://doi.org/10.1007/978-3-031-50974-2_32
3. Chen, Y., Jensen, S., Albert, L.J., Gupta, S., Lee, T.: Artificial intelligence (AI) student assistants in the classroom: designing chatbots to support student success. Inf. Syst. Front. **25**(1), 161–182 (2023)
4. Dumas, M., Rosa, M., Mendling, J., Reijers, H.: Fundamentals of Business Process Management. Springer, Heidelberg (2018). https://doi.org/10.1007/978-3-662-56509-4
5. Fahland, D., Fournier, F., Limonad, L., Skarbovsky, I., Swevels, A.J.: How well can large language models explain business processes? arXiv preprint arXiv:2401.12846 (2024)
6. Forell, M., Schüler, S.: Modeling meets large language models. In: Modellierung 2024 Satellite Events. pp. 10–18420. Gesellschaft für Informatik eV (2024)

7. Grohs, M., Abb, L., Elsayed, N., Rehse, J.R.: Large language models can accomplish business process management tasks. In: De Weerdt, J., Pufahl, L. (eds.) BPM 2023. LNBIP, vol. 492, pp. 453–465. Springer, Cham (2023). https://doi.org/10.1007/978-3-031-50974-2_34
8. Jessen, U., Sroka, M., Fahland, D.: Chit-chat or deep talk: prompt engineering for process mining. arXiv preprint arXiv:2307.09909 (2023)
9. Kimmel, B., et al.: Enhancing programming error messages in real time with generative AI. In: Extended Abstracts of the CHI Conference on Human Factors in Computing Systems. pp. 1–7 (2024)
10. Kourani, H., Berti, A., Schuster, D., van der Aalst, W.M.: ProMoAI: process modeling with generative AI. arXiv preprint arXiv:2403.04327 (2024)
11. Meyer, J., et al.: Using LLMs to bring evidence-based feedback into the classroom: AI-generated feedback increases secondary students' text revision, motivation, and positive emotions. Comput. Educ.: Artif. Intell. **6**, 100199 (2024)
12. Neyem, A., González, L.A., Mendoza, M., Alcocer, J.P.S., Centellas, L., Paredes, C.: Towards an AI knowledge assistant for context-aware learning experiences in software capstone project development. IEEE Trans. Learn. Technol. (2024)
13. Pack, A., Barrett, A., Escalante, J.: Large language models and automated essay scoring of English language learner writing: Insights into validity and reliability. Comput. Educ.: Artif. Intell. 100234 (2024)
14. Senkus, P., Berniak-Woźny, J., Gabryelczyk, R., Napieraj, A., Podobińska-Staniec, M., Sliż, P., Szelągowski, M.: Bridging the gap: an evaluation of business process management education and industry expectations-the case of Poland. In: Köpke, J., et al. (eds.) BPM 2023. LNBIP, vol. 491, pp. 250–262. Springer, Cham (2023). https://doi.org/10.1007/978-3-031-43433-4_18
15. Sliż, P., et al.: Business process management education in Poland: A manifesto for academic teaching. e-mentor **2**(104), 9198 (2024)
16. Stribling, D., Xia, Y., Amer, M.K., Graim, K.S., Mulligan, C.J., Renne, R.: The model student: GPT-4 performance on graduate biomedical science exams. Sci. Rep. **14**(1), 5670 (2024)
17. Xiao, R., Hou, X., Stamper, J.: Exploring how multiple levels of GPT-generated programming hints support or disappoint novices. In: Extended Abstracts of the CHI Conference on Human Factors in Computing Systems, pp. 1–10 (2024)

Fostering BPM Learning Through PBL: Development and Application of an Educational Framework

Silvia Inês Dallavalle de Pádua(✉), Marina Lourenção, Sérgio Adriany Santos Moreira, and Luiz Ricardo Brito Ribeiro

University of São Paulo (USP), Ribeirão Preto, São Paulo, Brazil
dallavalle@usp.br

Abstract. The objective of this study is to develop and apply a Business Process Management (BPM) Educational Framework using Project-Based Learning (PBL) with graduate students in administration at a Brazilian institution. The PBL-BPM framework integrates dynamic capabilities and holistic diagnosis for a comprehensive BPM education. The method used was qualitative exploratory research, divided into three phases: (1) development of the PBL-BPM Educational Framework, (2) application of the framework by graduate students in a real project in a cosmetics company, and (3) evaluation of the students' perception of the didactic strategy used and the knowledge acquired. The practical application of the PBL-BPM Educational Framework was carried out in a cosmetics company, providing a unique opportunity to observe the application of theoretical concepts in a real business environment. The results indicated a significant improvement in the students' ability to apply theoretical knowledge in real contexts, develop practical skills, and acquire a holistic understanding of business process management. This study demonstrates that the PBL-BPM Educational Framework is crucial for the development of in-depth BPM skills, generating significant process improvement proposals for the studied company. Besides bridging the gap between theory and practice, this work provides the BPM education community with a flexible and adaptable framework, applicable across various sectors. This study not only better prepares students for the job market but also contributes to the evolution of educational practices in the BPM field.

Keywords: Project-Based Learning · Skill Development · Academic-Industry Collaboration · Educational Framework PBL-BPM · BPM Education

1 Introduction

The dynamic market requires a holistic view of companies [1] focusing on end-to-end processes to deliver customer value [2]. There is a notable gap between BPM knowledge in organizations and academia [3]. We believe that this gap can be reduced by using Project-Based Learning (PBL) for BPM education. PBL encourages lifelong learning and adaptability, essential skills in the rapidly changing job market [4] and also develops

Supplementary Information The online version contains supplementary material available at https://doi.org/10.1007/978-3-031-70445-1_30.

C. Di Ciccio et al. (Eds.): BPM 2024, LNBIP 527, pp. 416–425, 2024.
https://doi.org/10.1007/978-3-031-70445-1_30

essential competencies for sustainable development [5].PBL engages students with real-world problems, enhancing critical thinking and problem-solving skills [4], bridging the gap between theoretical concepts and practical application in BPM education. Integrating PBL in higher education better prepares students for the workforce and improves employability, although more studies are needed to maximize its potential [4].

Therefore, considering that BPM is an important systematic approach to improving an organization's business processes, our study aims to answer the research question: Is PBL an effective method for teaching BPM to graduate students? The objective of this study is to develop and apply a framework for BPM education using PBL with graduate students in administration at a Brazilian higher education institution. Thus, the phenomenon of interest in this work is the relationship between PBL and BPM, highlighting how PBL can be practically implemented to teach BPM to undergraduate students. Insights from students' evaluations were used to measure the framework's effectiveness in bridging academic and industry needs. This research aims to bridge the gap between academic training and industry requirements, offering a replicable model that can contribute to increasing students' readiness for the job market and employability. The structured PBL approach facilitates the education of BPM concepts and develops essential skills such as problem-solving, teamwork, and strategic thinking. This method provides students with a holistic understanding of BPM principles [1] and techniques [6], improving their readiness for the job market.

2 Literature Review

Rapid business changes and increasing complexity require continuous adaptation. The BPM life cycle, with emphasis on strategy, integrates external factors, helping companies identify and respond to external opportunities and threats, thus improving organizational performance and innovation capacity [7], considering the theory of dynamic capabilities [8]. The holistic BPM diagnostic map, encompassing all dimensions of business processes, also emphasizes the importance of an integrated analysis of internal and external aspects [6]. Additionally, understanding the organization through its processes promotes a holistic and detailed view of interactions with each stakeholder [9]. New business models demand new approaches to BPM education [10, 11]. BPM education should be understood in a multidisciplinary manner, as the very essence of this managerial practice suggests a holistic view of the organization as opposed to isolated approaches [12–15]. Problem-solving skills and creativity can be developed through case teaching with BPM projects in companies [16].

Project-Based Learning (PBL), rooted in experiential learning theory, is a well-established teaching approach where working on real-world issues aims to develop students' skills through learning by doing, as opposed to traditional learning such as memorization of facts [17]. PBL has significant implications for how to stimulate meaningful student engagement with external stakeholders, add creativity and innovation, and produce impactful community results [17].

Learning is more efficient when it is actively performed by students and not through traditional methods of passive information reception [13]. Action learning is useful as it combines elements from different methodological categories [18]. Applying Design

Thinking to PBL can help stimulate students' genuine interests in solving problems and increase their dedication to their work, often leading to greater creativity and higher impact of results [6]. Employers value skills developed through PBL, such as teamwork, conflict resolution, informed decision-making, and solving complex problems [5]. However, documented cases of BPM-PBL are scarce in the literature. The growing demand for these skills reveals the gap between academic knowledge and industry needs, highlighting the importance of PBL methods in BPM education to prepare students for the challenges of the business environment.

3 Method

The qualitative exploratory research method is used to provide better insight into the complex phenomenon under investigation [19, 20]. This study aims to develop and apply the PBL-BPM Framework. The methods section is divided into three phases.

3.1 Phase 1 –Development of the PBL-BPM Educational Framework

In the first phase, the framework was developed by aligning BPM concepts with the PBL strategy. This phase starts with the selection and partnership with the company, where dates and availability for meetings, interviews, and presentations are agreed upon, ensuring efficient and productive collaboration. Students receive an introduction to BPM and PBL, establishing theoretical foundations. With this knowledge, they are divided into teams to execute the BPM project stages with the partner company.

Regarding the theoretical concepts of BPM, the framework was proposed based on the BPM life cycle [7], which emphasizes the analysis of the external perspective through dynamic capabilities of the context [1], and on the holistic diagnosis [6] and the understanding of the organization through processes [9]. It incorporates cognitive diagnosis through the Current Reality Tree (CRT) [6, 21] and considers the best practices of Project Management [22] and BPM [23]. Engaging stakeholders in the co-creation process, [17], ensures that proposed solutions meet real needs and promote significant results. These BPM concepts were integrated into the PBL didactic strategy, meaning that after actively seeking knowledge about BPM diagnosis, it was proposed that students apply their knowledge in a real company.

3.2 Phase 2 – Application of the PBL-BPM Educational Framework

This strategy has been applied for 13 years in a 9-week BPM course, with a different company chosen annually for analysis. Each year, a different company is chosen for analysis, with the instructor ensuring partnership availability beforehand. Students then conduct a BPM diagnosis of the organization.

This article presents the application of the PBL-BPM education framework with graduate students for a Brazilian company specialized in developing exclusive topical formulations. Students participated in theoretical classes, seminars, and conducted a BPM project in a real organization. A detailed explanation of the PBL-BPM framework stages can be found in Appendix I. Steps of the project implementation:

1. Analysis and Review of Strategic Guidelines: Identification and review of the company's strategic guidelines, evaluation of mission and vision, definition of strategic objectives, and attributes valued by customers.
2. Customer Experience Analysis: Mapping customer touchpoints, collecting information on customer behavior and feelings, segmenting customers, and building an understanding of customer success.
3. Analysis and Prioritization of Process Improvements: Survey of processes related to stakeholders, construction of the process architecture, evaluation of pains and gains, and identification of improvement priorities.
4. Cognitive Diagnosis: Identification of Undesirable Effects (UEs), construction of the CRT, and identification of root causes of problems.
5. Improvement Proposals: Definition of prioritization criteria, creation of a relationship between improvements and benefits, presentation of improvement suggestions, and their evaluation.
6. Project Closure: Project review, identification of lessons learned, and preparation of the final report with improvement proposals and a list of identified benefits and problems.

3.3 Phase 3 – Evaluation of Students' Perception of the Didactic Strategy and Learning Obtained After Applying the PBL-BPM Educational Framework

The evaluation of the application of the Framework PBL-BPM was conducted through a questionnaire designed to capture students' perceptions of the didactic strategy used and the learning obtained. The questionnaire consisted of 16 statements, each rated on a scale of 1 to 7, where 1 represented total disagreement and 7 total agreement. It was administered to the seven students who participated in the "Alpha Cosmetics Project" within the Business Process Management (BPM) course. The evaluation focused on assessing students' perception of their ability to apply classroom knowledge to real-world contexts, the development of practical skills, and the understanding of BPM concepts and techniques.

4 Results and Discussions

The results were divided into three sections related to the three phases presented in the method.

4.1 Phase 1 – Development of the PBL-BPM Educational Framework

To effectively initiate the PBL project, the following steps were structured to align the theoretical foundations from the literature review with the practical needs of the project. The first part is the Selection and partnership with the company. A partnership is established to set dates and availability for ongoing collaboration through meetings, interviews, and presentations. This ensures that the project timeline is well-coordinated and that the company is fully committed to participating. Following this, an introduction to BPM and PBL is conducted. This foundation is crucial for understanding how

BPM principles will be applied in a real-world context using the PBL methodology. The integration of PBL in higher education [17] improves student engagement and skill development by bridging the gap between theoretical knowledge and practical application. The third step is the formation of project teams. Students are divided into teams, with each team being assigned specific responsibilities and tasks. This division of labor ensures that all aspects of the BPM project are comprehensively covered and that students can collaborate effectively. Team formation aligns with multidisciplinary approaches to BPM education [12, 13], promoting a holistic view of organizational processes. Next, the planning of activities is carried out. Initial meetings with the partner company are organized to start project activities. This step involves setting clear objectives, milestones, and deliverables. Planning and structuring activities are essential components of experiential learning [18], ensuring that students can apply BPM concepts effectively in a real-world scenario.

The project is executed through periodic meetings with company managers and collaborators. During these meetings, students are asked to identify the organization's strategic guidelines, which are fundamental for aligning organizational processes with external and internal demands, allowing for a dynamic and effective response to market changes [7]. Techniques such as SWOT analysis can be used to identify strengths, weaknesses, opportunities, and threats, along with the Business Model Canvas [6, 7].

Students evaluate the customer experience to identify critical touchpoints and improvement areas. Tools like Customer Journey Mapping and Net Promoter Score (NPS) effectively measure customer satisfaction with empathy, an aspect valued in Design Thinking [6]. Subsequently, students are required to conduct an analysis on prioritizing process improvements for the studied company. This analysis is essential for identifying and focusing on areas that offer the most positive impact for the organization [6]. The analysis begins with mapping stakeholder relationships, followed by constructing the company's process architecture, evaluating pains and gains in each process, and prioritizing improvements by identifying high-impact processes [9]. A holistic cognitive diagnosis is necessary to understand the complex interactions between different processes and identify root causes of problems, with CRT being indicated for this diagnosis [6].

The conclusion of the diagnostic project is crucial to ensure that all planned activities are completed and the results communicated to stakeholders. In the final stage, students are asked to deliver a formal presentation and review the lessons learned for continuous improvement [22]. Finally, students present their final BPM diagnostic to the management team with improvement proposals, identified benefits, and problems [23]. Figure 1 illustrates the developed BPM teaching framework.

It is expected that the framework can be used by other BPM educators as an educational strategy that facilitates the application of theoretical concepts in a real case.

4.2 Phase 2 – Application of the PBL-BPM Education Framework

The Education Framework was developed at Alpha Cosmetics, a Brazilian company specialized in developing exclusive topical formulations. Initially, the professor and a group of 7 students attended a lecture given by the company's CEO, who presented the

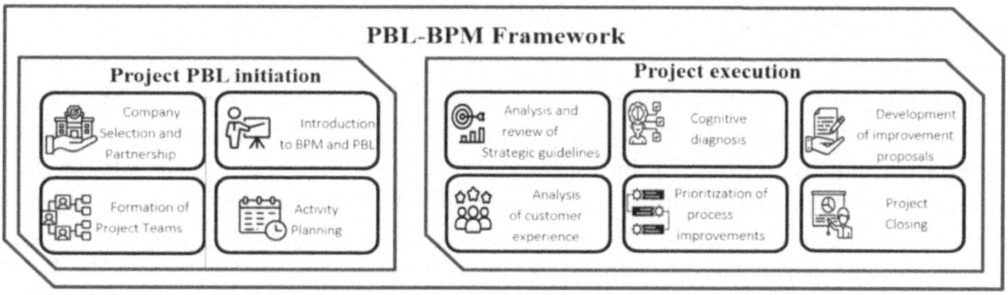

Fig. 1. Framework PBL-BPM. Source: the authors

company's deficiencies and potentialities. During this lecture, the students asked questions about the organizational processes and explained the project stages as described in Appendix II. The PBL-BPM Framework was implemented through six main stages:

1- Analysis and Review of Strategic Guidelines: a) Meeting with the team to discuss the current strategic guidelines. b) Identification of the limitations of the current strategic guidelines. c) Review of strategic guidelines, mission, and vision, d) Definition of strategic objectives. e) Identification of the most valued attributes by customers and evaluation of the company's performance concerning these attributes (Appendix III).
2- Customer Experience Analysis: a) Identification of the customer's touchpoints with the company. b) Collection of information about customer behavior and feelings at each touchpoint. c) Identification of customer segments. d) Survey of customer expectations at each touchpoint. e) Construction of an understanding of customer success. f) Identification of stakeholders involved with the company (Appendix IV).
3- Analysis and Prioritization of Process Improvements: a) Survey of processes related to each stakeholder. b) Construction of the company's process architecture (Appendix V). c) Evaluation of pains and gains in each process. d) Identification of improvement priorities (processes with many pains and many gains) (Appendix VI).
4- Cognitive Diagnosis: a) Identification of Undesirable Effects (UEs). b) Construction of the CRT (Appendix VII). c) Identification of root causes.
5- Improvement Proposals: a) Definition of criteria for prioritizing identified improvements. b) Prioritization of identified improvements according to defined criteria. c) Creation of an improvement-benefits and problems solved relationship. d) Presentation of improvement suggestions. e) Evaluation and correction after the presentation.
6- Project Closure: a) Project review and identification of lessons learned. b) Preparation of the final project report.

4.3 Phase 3 – Evaluation of Students' Perception of the Didactic Strategy and Learning Obtained After Applying the PBL-BPM Education Framework

The results were evaluated through a questionnaire composed of 16 statements regarding the students' perception of the application of the PBL-BPM Framework, rated on a scale of 1 to 7, where 1 represented total disagreement and 7 total agreement. The questionnaire was administered to the seven students who participated in the "Alpha Cosmetics Project" within the Business Process Management (BPM) course and provided relevant insights about the results.

71.4% fully agreed that the project facilitated applying classroom knowledge to real-world contexts, supporting that PBL, rooted in experiential learning, develops skills through learning by doing [17]. Furthermore, 57% stated that the project consolidated their learning and practical skills, 71.4% understood the importance of aligning business processes with strategic objectives, and 42.9% saw how they interconnect to achieve these objectives. The analysis of the students' responses indicated that the PBL-BPM approach was effective in aligning business processes with strategic demands, using techniques such as SWOT analysis to identify and respond to strengths, weaknesses, opportunities, and threats [6, 7].

Students gained knowledge in evaluating business contexts, identifying innovation opportunities (42.9%), using new techniques to analyze processes (57.1%), and understanding stakeholder needs (57.1%). This alignment with dynamic capabilities helps companies adapt to market changes [7]. The evaluation of the customer experience was highlighted as a critical aspect of the project, providing an empathetic and detailed understanding of customer needs and expectations [6]. Additionally, the process architecture creation method (PA) emphasized the importance of a holistic and detailed view of interactions with each stakeholder, facilitating the identification of innovation and continuous improvement opportunities [9]. This reflects the importance of encouraging significant student engagement with external stakeholders, adding creativity and innovation, and producing impactful community results [17].

Moreover, students agreed that identifying the root causes of problems in business processes was fundamental for the engagement of all involved (71.4%). Students reported that the use of the Current Reality Tree (CRT) was crucial for identifying the root causes of problems in business processes, corroborating the importance of holistic cognitive diagnosis [6]. This finding aligns with the assertion that PBL can help increase dedication to their work, leading to greater creativity and more significant impact on results [17].

Regarding the more conceptual aspects of BPM, the results showed that students considered the project an opportunity to become aware of the importance of business process management (57.1%) and feel more confident in identifying problems in these processes and proposing solutions (57.1%). The engagement of stakeholders in the co-creation process was essential to ensure that the proposed solutions were relevant and implementable [17]. This reinforces the need for BPM education to be understood in a multidisciplinary manner, promoting a holistic view of the organization [12–15]. Additionally, the project sparked students' interest in learning more about BPM and its applications in research and organizations (42.9%). This finding supports the notion that BPM education should stimulate students' interest in solving problems and increase their dedication to their work [17].

Although agreement was not as high on this issue, it is still relevant to note that 57.1% of students agreed that the project helped them develop analytical reasoning in structuring problems and proposing solutions. This indicates that the project positively impacted the development of this skill in students, allowing them to approach challenges related to business processes more critically. This reflects the effectiveness of PBL in improving critical thinking and engagement [16, 17]. Additionally, learning is more efficient when students actively engage in the learning process rather than passively receiving information [13], and learning by doing increases students' ability to solve complex problems [18].

Most students (57.1%) strongly agreed that the project helped them understand the importance of analyzing business processes from an end-to-end perspective, considering everything from the need to the realization. This perception demonstrates that the project enabled students to have a broader and more integrated view of processes, understanding the importance of a comprehensive approach to dealing with challenges related to business processes. This aligns with the importance of managing end-to-end processes to achieve organizational objectives [12–15].

These results demonstrate that the application of the PBL-BPM Framework effectively applied the knowledge acquired in the classroom, developing skills and understanding concepts and practices related to BPM. Students reported improved confidence and awareness of the importance of process management, as well as an interest in deepening their knowledge in this area. Appendix VIII presents the questionnaire and a graph with the results. The CEO of Alpha Cosmetics noted that the project was highly beneficial. The initial investigative stage improved understanding of internal processes, providing a realistic view of demands and needs. The final result was a detailed action plan addressing critical issues and supporting company growth.

The results of this application of the PBL-BPM Framework show that the multidisciplinary approach to BPM education, as suggested by various researchers, was also evidenced by students who reported the application of new process analysis techniques, consideration of stakeholders' and customers' needs, and understanding the importance of end-to-end process management [12–15]. These findings also align with the conclusions of Senkus et al., who highlighted the gap between university BPM programs and market demands, emphasizing the need for academic institutions to adapt their curricula to better prepare students for industry demands [3]. He findings also corroborate the assertion that learning by doing is more efficient than traditional methods of passive information reception [13].

5 Final Considerations

Our study addresses the research question: Is PBL an effective method for teaching BPM to graduate students? Results indicate that the PBL-BPM framework is an effective strategy for teaching BPM. Applied with graduate students in a Brazilian cosmetics company, it enabled the implementation of classroom knowledge in real contexts, consolidating learning and practical skills. Student evaluations showed that PBL-BPM Framework helped understand the importance of aligning business processes with strategic objectives and their interconnections. Students highlighted their learning in analyzing the business context, identifying opportunities for innovation, and using new process analysis techniques. The project also increased students' interest in BPM and its applications, raising awareness about the importance of business process management and developing analytical reasoning. The feedback from the CEO of Alpha Cosmetics emphasized the positive impact of the project, providing a realistic view of the organization's demands and a detailed action plan.

In summary, the PBL-BPM Framework effectively applied BPM knowledge and developed essential BPM skills, significantly contributing to student learning and process improvement at Alpha Cosmetics. This PBL-BPM Framework benefits universities by better preparing students for the market and companies by leveraging academic expertise to improve processes. The presented case contributes to BPM education and can serve as a reference for educators, demonstrating that this framework can be effectively applied in companies across various sectors.

Acknowledgments. This study was funded by São Paulo Research Foundation- FAPESP 2021/14363-4 and 2023/08335-3; National Council for Scientific and Technological Development (CNPq) 405977/2023-9.

References

1. Vom Brocke, J., Schmiedel, T., Recker, J., Trkman, P., Mertens, W., Viaene, S.: Ten principles of good business process management. Bus. Process. Manag. J. **20**, 530–548 (2014). https://doi.org/10.1108/BPMJ-06-2013-0074
2. Maddern, H., Smart, P.A., Maull, R.S., Childe, S.: End-to-end process management: implications for theory and practice. Prod. Plan. Control **25**, 1303–1321 (2014). https://doi.org/10.1080/09537287.2013.832821
3. Senkus, P., et al.: Bridging the gap: an evaluation of business process management education and industry expectations – the case of Poland. In: Köpke, J., et al. (eds.) BPM 2023. Lecture Notes in Business Information Processing, vol. 491, pp. 250–262. Springer, Cham (2023). https://doi.org/10.1007/978-3-031-43433-4_18
4. Epler, P., Jacobs, J.: Guide to Integrating Problem-Based Learning Programs in Higher Education Classrooms: Design, Implementation, and Evaluation. IGI Global (2022)
5. Brin, P., Krasnokutska, N., Lombardi, R., Nehme, M.: Project-based learning for responsible education: empirical study of the additional value for students. In: 2021 IEEE International Conference on Modern Electrical and Energy Systems (MEES), pp. 1–5. IEEE, Kremenchuk (2021)
6. Nogueira, C.A., Pádua, S.I.D., Bernardo, R.: A map for the holistic BPM diagnosis. BPMJ **28**, 630–655 (2022). https://doi.org/10.1108/BPMJ-04-2021-0197
7. Bernardo, R., Galina, S.V.R., Pádua, S.I.D.D.: The BPM lifecycle: how to incorporate a view external to the organization through dynamic capability. BPMJ **23**, 155–175 (2017). https://doi.org/10.1108/BPMJ-12-2015-0175
8. Teece, D.J.: Explicating dynamic capabilities: the nature and microfoundations of (sustainable) enterprise performance. Strateg. Manag. J. **28**, 1319–1350 (2007). https://doi.org/10.1002/smj.640
9. Lima Aredes, E., Lourenção, M., Burlton, R., Reijers, H., Dallavale de Pádua, S.: A method for creating process architecture: driving organizational change. In: De Weerdt, J., Pufahl, L. (eds.) BPM 2023. Lecture Notes in Business Information Processing, vol. 492, pp. 539–551. Springer, Cham (2024). https://doi.org/10.1007/978-3-031-50974-2_40
10. Bandara, W., Velmurugan, M., Leemans, S.: Building 'holistic' Business process modelling skills for IS graduates. In: Proceedings of the 41st International Conference on Information Systems (ICIS 2020). Blending the Local and the Global (2020)
11. Levkovskyi, B., Laurim, V., Kayaci, B., Mustroph, H., Utesch, M., Krcmar, H.: Teaching the digital transformation of business processes: design of a simulation game of business process change. In: 2021 IEEE Global Engineering Education Conference (EDUCON), pp. 1262–1271. IEEE, Vienna (2021)

12. Bandara, W., et al.: Business process management education in academia: status, challenges, and recommendations. Commun. Assoc. Inf. Syst. **27**, 743–776 (2010)
13. Chow, W.: Teaching business process management with a flipped-classroom and problem-based learning approach with the use of Apromore and other BPM software in graduate information systems courses. In: 2021 IEEE International Conference on Engineering, Technology & Education (TALE), pp. 1–8. IEEE, Wuhan (2021)
14. Hrabala, M., Opletalovåb, M., Tucekc, D.: Teaching business process management: Improving the process of process modelling course. Istrazivanja i projektovanja za privredu **15**, 113–121 (2017). https://doi.org/10.5937/jaes15-12172
15. Seethamraju, R.: Business process management: a missing link in business education. Bus. Process. Manag. J. **18**, 532–547 (2012). https://doi.org/10.1108/14637151211232696
16. Bandara, W.; Ouyang, C.: Analysing and improving a recruitment process: a teaching case for developing business process improvement capabilities. In: Proceedings of the 16th Pacific Asia Conference on Information Systems (PACIS), pp. 1–13, Vietnam (2012)
17. Cummings, C., Yur-Austin, J.: Design thinking and community impact: a case study of project-based learning in an MBA capstone course. J. Educ. Bus. **97**, 126–132 (2022). https://doi.org/10.1080/08832323.2021.1887795
18. Leyer, M., Moormann, J.: Action learning approach to teaching business engineering methodology. BPMJ **23**, 130–154 (2017). https://doi.org/10.1108/BPMJ-06-2015-0092
19. Malhotra, N.K.: Review of Marketing Research:, vol. 4. Routledge, New York (2017)
20. Eriksson, P., Kovalainen, A.: Qualitative Methods in Business Research. SAGE Publications Ltd., London (2008)
21. De Pádua, S.I.D., Da Costa, J.H.M., Segatto, M., De Souza Júnior, M.A, M., Jabbour, C.J.C, C.: BPM for change management: two process diagnosis techniques. Bus. Process Manag. J. **20**, 247–271 (2014). https://doi.org/10.1108/BPMJ-03-2013-0039
22. Project Management Institute org: A guide to the project management body of knowledge: PMBOK guide. Project Management Institute, Newtown Square, Pennsylvania, USA (2017)
23. Harmon, P.: The scope and evolution of business process management. In: vom Brocke, J., Rosemann, M. (eds.) Handbook on Business Process Management 1. International Handbooks on Information Systems, pp. 37–80. Springer, Berlin, Heidelberg (2015). https://doi.org/10.1007/978-3-642-45100-3_3

Industry Forum

Preface of the Industry Forum

The Industry Forum organized in 2024 as part of the BPM conference in Krakow, Poland has been planned as a platform for bridging the gap between academia and industry, fostering meaningful connections among BPM experts worldwide. The Industry Forum's primary focus was on showing cutting-edge real-world experiences in leveraging BPM for improvement and innovation across sectors, regions and types of economies. It has been a gathering for BPM practitioners to share their rich experiences and viewpoints related to their BPM endeavors as well as academic leaders who are constantly developing BPM and providing new insights through research processes. This mixed audience of the forum ensured a multifaceted approach to BPM. We received eleven submissions, and the top five highest quality papers were selected for presentation and publication. Each submission was reviewed by at least two Program Committee members and meta-reviewed by Industry Forum co-chairs.

The majority of papers presented at the 2024 BPM Industry Forum are about process mining, which reflects the general interest in this topic at the BPM conference. The paper by Nazrina Khurshid, Stephen Palmier, Kerry Buijs and Christopher Klinkmüller discusses how process mining capability was built in a global gaming company. The authors share the insights they gained over 3 years of working in and with a Centre of Excellence for Business Process Management and reflect on the CoE's position and vision. In particular, they share five lessons learned: (1) Masters are made, not born, (2) Learn to mitigate risks, (3) Take the heat out of the value discussion, (4) Enable strategic project selection, and (5) Enable process mining in a broader BPM strategy.

The paper by Piotr Gawrysiak, Olga Kacprzak, Tomasz Romanowski and Kamil Żbikowski also discusses the implementation of process mining, but this time the setting is a large commercial bank in Poland. It presents the bank's process mining journey from its very first beginnings, the various initiatives that were started and the (data quality) issues that the bank experienced. Finally, it presents a holistic approach to use process mining for process optimization, including a discussion on the required skills and knowledge.

The paper by Urszula Jessen, Lucas Schroth and Mario Mühllechner adds AI to the mix, reporting on a case study on the integration between AI and process mining in a manufacturing environment, and how it can help organizations move from data to actionable insights. They present an innovative approach that allows users to interact with the data through ChatGPT, in order to avoid information overload and lack of skills.

The paper by Felix Mannhardt, Ragnhild Halvorsrud, Otas Meironas and Lasse Brurok presents the quest for the comprehensive customer journey at the C2C marketplace. This case study combines traditional qualitative analysis - mystery shopping with data-driven process mining to provide a holistic view of the customer journey in an online marketplace named Bookis. The innovative dual-method approach offers valuable insights by complementing the "outside-in" perspective of mystery shopping with the "inside-out" view of process mining.

Last, in their article "LLM4PM: A case study on using Large Language Models for Process Modeling in Enterprise Organizations", Clara Ziche and Giovanni Apruzzese report on a case study carried out at Hilti Group. In this case study, they develop and test a large language model-based chatbot to support process model creation. They report some valuable and practice-based insights that can inform both BPM research and practice.

We would like to thank the authors and Program Committee members for contributing to this edition of the Industry Forum. We believe that all presentations and publications of this forum have strengthened an important bridge between academia and industry and contributed to the development of this symbiotic relationship. We highly recommend the broader community of BPM researchers, practitioners, and enthusiasts the read these selected papers.

Organization

Program Chairs

Iris Beerepoot	Utrecht University, The Netherlands
Renata Gabryelczyk	University of Warsaw, Poland
Ralf Plattfaut	University of Duisburg-Essen, Germany

Program Committee

Saimir Bala	Humboldt Universität zu Berlin, Germany
Grzegorz Bełz	Wrocław University of Economics and Business, Poland
André Coners	South Westphalia University of Applied Sciences, Germany
Szymon Cyfert	Poznań University of Economics and Business, Poland
Piotr Gawrysiak	WUT, Poland
Urszula Jessen	Technical University Eindhoven, The Netherlands
Christopher Klinkmüller	BPMotion, Australia
Jelmer Jan Koorn	MLC, The Netherlands
Ingo Kregel	University of Applied Sciences and Arts Dortmund, Germany
Benjamin Matthies	Münster University of Applied Sciences, Germany
Jannis Nacke	Enervie Gruppe, Germany
Luise Pufahl	TU Munich, Germany
Malgorzata Pankowska	University of Economics in Katowice, Poland
Stefan Schönig	University of Regensburg, Germany
Piotr Sliż	University of Gdańsk, Poland
Amy Van Looy	Ghent University, Belgium

Building Organizational Process Mining Capability: Experiences at a Global Gaming Company

Nazrina Khurshid[1], Stephen Palmier[1], Kerry Buijs[2], and Christopher Klinkmüller[2](✉)

[1] Aristocrat Leisure Ltd., Sydney, NSW, Australia
{nazrina.khurshid,stephen.palmier}@aristocrat.com
[2] BPMotion, Brisbane, QLD, Australia
{kerry.buijs,christopher.klinkmuller}@bpmotion.com.au

Abstract. For organizations, the promise of process mining is that they can generate additional value from event data that their information systems collect. While case studies and guidelines demonstrate potential benefits and provide support for applying process mining, realizing value at scale requires organizations to overcome challenges that go beyond successfully delivering individual projects. Among others this includes working with low data quality, building data literacy, as well as managing stakeholder expectations and perceptions. In this paper, we present experiences from building organizational process mining capability at a global entertainment and gaming content creation company. We (i) outline the process mining vision of the Center of Excellence for Business Process Management, (ii) provide an overview of specific measures taken over a three year period to deliver on the vision, and (iii) discuss the lessons learned. Our intent is to share valuable insights with researchers and practitioners. We hope to stimulate an exchange of experiences and inspire research to draw further attention to organizational adoption of process mining.

Keywords: Process mining · Business Process Management · Organizational capability · Change Management · Case study

1 Introduction

Process mining offers a set of tools for analyzing event data [1]. For example, *process discovery* [3,14] visualizes the underlying process, *conformance checking* [9] compares the observed to normative behavior, *process enhancement* [19] extends process models with performance information, and *predictive process monitoring* [12,15] forecasts how process instances may unfold during execution.

C. Klinkmüller—The last author of the paper was not involved in any work carried out by the CoE but gave advice and support to shape the content and write the paper.

C. Di Ciccio et al. (Eds.): BPM 2024, LNBIP 527, pp. 431–440, 2024.
https://doi.org/10.1007/978-3-031-70445-1_31

From an organizational perspective, process mining provides an opportunity to generate business value from event data that information systems collect during normal operation [4]. In other words, organizations are interested in using available tools and data to e.g., obtain actionable insights for process improvement, automate controls through conformance checking, or proactive monitoring of operations. To assist organizations in generating value, research has studied aspects of applying process mining, e.g., proposing project methodologies [8], examining analysis patterns [13], and demonstrating practical usage through case studies [2,7,11]. Despite all those insights, there are many technology, people, culture, and data challenges [10,16,17,20] that do not only impact individual projects but the adoption of process mining within organizations altogether.

In this case study, we share our experiences in handling such challenges. In more detail, we provide an overview of the activities of the *Center of Excellence for Business Process Management* (CoE) at a *global gaming content creation company* (the Company). The CoE was established three years ago and since then has organically grown. Its goal is to govern, enable, coordinate, and support *business process management* (BPM) across the Company. Having identified process mining as a critical future capability, the CoE has embarked on a journey to build an enabling process mining infrastructure and to increase organization-wide awareness and application. In this context, we outline specific measures that the CoE has implemented such as the continuous delivery of projects, provisioning of technology, and ongoing stakeholder management. We also summarize lessons learned, highlighting that the Company minimizes the impact of challenges through a range of risk mitigation measures.

Considering that BPM-driven value creation from data and expanding BPM beyond individual processes are key BPM research problems [5], our contributions are twofold. First, we view process mining as an organizational capability, i.e., besides technology we factor in behaviors, processes, competencies, and governance to enable organization-wide usage of process mining. Second, we focus on concrete measures that can contribute to overcoming common process mining challenges. Knowing that many organizations struggle with adopting process mining, our insights are also valuable to practitioners who want to garner ideas for how to approach the roll out of process mining in their organization.

The remainder of the paper is loosely oriented towards guidelines for reporting case studies [18]. In Sect. 2, we introduce the Company's CoE and its vision for process mining. These aspects are important to understand the organizational context, the CoE's actions, and the lessons learned. In Sect. 3, we recap the CoE's process mining journey to date by outlining concrete implementation measures and activities. Next, we present a thematic synthesis [6] that presents the lessons learned from this three year journey in Sect. 4. We conclude the paper with a summary and a discussion of directions for future work in Sect. 5.

2 The CoE and Its Process Mining Vision

The Company is a global entertainment and gaming content creation company powered by technology. Its game and product portfolios span both business-to-business and direct-to-consumer operations. The Company is proud to serve customers, and entertain millions of players worldwide everyday. The Company strives to lead the industry in responsible gameplay and its team of more than 8,500 people are united by its shared values and mission to bring joy to life through the power of play.

In the past, the Company adopted Lean Six Sigma to improve business processes and operations. Yet, improvement teams operated in silos. To enhance the Company's capability to execute transformational projects and programs, the Global Transformation Team decided to establish the Center of Excellence for Business Process Management in 2021. Based on SAP Signavio as an enabling BPM platform, the CoE's task is to govern, enable, coordinate, and support BPM activities including process modeling, process analysis & improvement, performance measurement, and support for strategic decision making.

The Company encourages such initiatives and the Global Transformation Team decided to implement the CoE based on a decentralized model, in which a small core team (comprised of the head of the global transformation team, the CoE manager, and an analyst) organically grows the BPM capability in cooperation with existing transformation and improvement partners. By building strong relationships, the CoE has secured support and collaboration of 15 senior and executive sponsors, 120 practitioners, and 50 functions across the Company.

Since the beginning, the CoE has recognized process mining as a valuable capability. In particular, its vision is to *establish process mining as an organizational capability by driving process mining maturity and adoption across the organization*. Here, maturity concerns the provisioning of adequate process mining technology, know-how, standards, and resources to support data-driven process analysis, improvement, execution, and monitoring. Adoption refers to the number of businesses, regions, functions, and processes in which process mining is used. While the CoE has defined concrete targets for overall BPM maturity and adoption, it did not break them down for process mining specifically. That is because in contrast to other BPM capabilities, process mining is a relatively new topic, requiring creation of broader experience, awareness, and literacy.

3 The CoE's Process Mining Journey

The CoE's process mining journey is embedded in the development of the overall BPM strategy and capability. It is hence important to understand the CoE's BPM implementation approach which is centered around the CoE's decentralized model. That is, acknowledging the need to build and maintain strong stakeholder relationships across the organization, the CoE oriented its BPM implementation plan towards the people-centered implementation framework[1]. As a result, the CoE defined three work streams for building BPM and process mining capability:

1. *Establish:* set up foundations such as standards, technologies, and resources.
2. *Execute:* provide services to support, coordinate, and enable BPM activities.
3. *Engage:* manage stakeholders and organizational change.

[1] More information about the framework is available under https://changesynergy.com.au/change-capability-diagnostic/.

Following, we summarize the measures and activities specifically relevant to process mining. Note that this includes measures that were implemented for general BPM purposes but that process mining activities benefit from nevertheless. Whilst the CoE is yet to fully establish process mining as an organizational capability, it has made excellent progress towards achieving its vision.

Establish
A foundation for process mining is the platform used to extract, transform, and analyze event data. The CoE decided to rely on SAP Signavio's Process Intelligence module for process mining purposes, because it implemented SAP Signavio as the enabling BPM platform (see Sect. 2).

Initially, the CoE members did not have BPM or process mining experience. Hence, the CoE has participated in training courses and leverages internal and business projects as learning opportunities. The CoE also engaged an external BPM advisor who has helped to define and implement BPM governance and initiatives over the last three years. Lastly, external consultants assist the delivery of process mining projects (see 'Execute' section below).

Another cornerstone is the setup of BPM governance mechanisms such as BPM standards, frameworks, and guidelines. Regarding process mining there are two relevant frameworks: an overarching business ownership and a performance measurement framework. The business ownership framework is a result of a collaboration with Data and IT Teams. The intent is to clarify ownership and responsibilities for processes, data, and IT systems. Similarly, the CoE has started the development of an organization-wide performance measurement framework. While both frameworks are not process mining-specific, they are viewed as important drivers for process and data orientation in the Company.

Execute
To drive process mining usage, the CoE conducts process mining projects. The five projects it has been involved in are summarized in Table 1. All projects used process mining for process analysis based on data snapshots and were supported by external consultants. They involved different functions, regions, and IT systems. While initially fully relying on external consultants, the internal CoE took on more and more process mining responsibilities within the projects over time.

Engage
Besides engaging and managing stakeholders within projects and initiatives, the CoE set up two stakeholder forums with the intent to enable and foster ongoing stakeholder collaboration and support. First, there is the steering committee to which the CoE invites senior and executive sponsors. In quarterly meetings, the CoE updates the sponsors on achievements using agreed success measures and presentations by CoE partners and customers. The meetings also give sponsors the opportunity to provide input into the CoE's strategy and plans. Conversely, the CoE uses the steering committee to secure sponsorship support and address potential barriers. Due to its success in enabling and supporting stakeholders, active leadership support for and interest in the CoE has grown over time and the number of committee members has increased from five to 15.

Table 1. Overview of the CoE's projects with process mining focus

#	Business Process In Scope	Region
1	Manufacturing program mgmt. (on-site repairs)	Australia & New Zealand
2	On-site service mgmt. (on-site installation, preventative maintenance, and other scheduled work)	Australia & New Zealand
3	Sales opportunity mgmt. (sales activities from forecasting to contracting)	Australia
4	Source-to-pay (indirect spend activities from vendor selection to payment)	Global
5	Order-to-cash (handling of customer orders from initiation to invoicing)	North America

Second, the CoE established a community of practice that is open to the entire organization and currently comprises 120 people. Through regular meetings and communications, the CoE updates the community on ongoing projects, its services, and BPM standards. Members are encouraged to provide feedback and share their success stories, too. In this regard, the CoE hosts an annual BPM award that rewards exceptional BPM-related projects and practitioners.

4 Lessons Learned

While we summarized initiatives and projects carried out by the CoE to deliver on its process mining vision in Sect. 3, we here discuss how specific activities contributed to or slowed down progress towards the vision by highlighting key lessons learned and outlining supporting observations from the CoE's journey.

Lesson 1: Masters are made, not born! Various challenges make it hard to outright generate value with process mining – but without efforts to apply process mining there will be no progress towards maturity and adoption.

At the beginning of the journey, the CoE's understanding of process mining was largely shaped by case studies and demos which tend to emphasize realized or anticipated benefits and to glance over challenges. Consequently, the CoE wanted to immediately demonstrate the benefits of process mining to create interest within and add value for the organization. To account for a lack of internal skill, it engaged external consultants to deliver the first projects. Despite best efforts, the success of the first projects was mixed. The CoE indeed witnessed how process mining can help business stakeholders to clarify issues and to trigger improvements. Yet, success of individual projects and overall progress towards the process mining vision was and continues to be impeded by many of the known technological, people, culture, and data challenges [10,16,17,20]. For example, low data quality and a lack of internal skills are barriers towards improving process mining maturity. That is, a complex system landscape with data integrity and integration issues limits the ability to derive valuable and actionable insights from data. Similarly, the CoE has to rely on the support of external consultants to deliver projects, requiring it to secure appropriate budget. Adoption is hampered by overall low data and process orientation.

In summary, the CoE recognized that it initially overestimated its ability to generate value and underestimated the efforts required to build necessary process mining capability. However, it also understood that pausing its process mining ambitions would stop progression towards building organizational capability. To increase its understanding of process mining and the challenges, the CoE decided to more actively participate in projects. Combined with training, the knowledge transfer from external consultants put the CoE in a position where it could start to lead projects, more and more relying on consultants for support and advice only. Through project retrospectives the CoE aims to identify pitfalls and improve its approach to process mining.

Lesson 2: Learn to mitigate risks! To drive process mining, organizations need to devise a risk mitigation strategy that helps minimizing the impact of the inevitable cultural, people, data, and technology challenges.

Next, the CoE reviewed the challenges and realized that many of them are outside of its control. For example, the CoE can advise the business to improve data quality, e.g., by eliminating manual data entries, but it cannot actively drive data quality. While the CoE also partners with IT and Data Teams to improve the situation, these teams need to deliver on their own objectives which do not necessarily align with the CoE's needs. Similarly, due to the decentralized operating model and a limited budget, the CoE cannot influence hiring decisions or mandate training that could help to build relevant skills in the business.

In its current position, these challenges are immovable barriers for the CoE. It hence decided to focus on demonstrating value, hoping to create broader support for process mining over time. Considering its limited process mining know-how and resources, it devised a risk mitigation strategy to proactively minimize the impact of the challenges as well as the chance to create stakeholder resistance. Regarding the latter, the CoE believes that in early projects some contextual factors averted the emergence of sustained opposition. This includes the Company's size, the CoE's willingness to continue the journey, its funding model (see Lesson 3) and its broader success story (see Lesson 5). The components of the risk mitigation strategy are discussed in more detail in Lessons 3, 4, and 5.

Lesson 3: Take the heat out of the value discussion! To lower the potential for stakeholder resistance, cover costs for the business and mitigate the impact of challenges through conventional process analysis activities.

When reflecting on the first three projects, the CoE felt that the second project was more successful than the first and third. While the three projects were different in nature (see Table 1), the CoE hypothesized that this is in part because the second project involved process improvement specialists who recently carried out a thorough process analysis and were about to design an improvement plan. For them, process mining did not only provide an opportunity to refine and validate their insights but more importantly to bolster their business case through data storytelling. In fact, the business case was approved and based on the suggested improvements the business generated savings. By contrast, process mining was applied independently of other analysis and improvement activities in the first and third project. While stakeholders were supportive, they were less engaged and had the impression that the data would generate the insights 'on their own'. When the aforementioned process mining challenges limited the value of the insights and the directions for improvement, the conversation tended to focus on limitations and process mining was not perceived useful.

To improve the chances of creating stakeholder support through projects, the CoE decided to follow its observation and embed process mining in conventional process analysis activities. As a next step, it thus applied process mining in a transformation project for which the business requested a process analysis. Here, the CoE primarily relied on interviews, customer surveys, and root cause analysis. Where possible, it also integrated process mining to test hypotheses and for data storytelling. The success of this approach confirmed the CoE's assessment. The stakeholders were engaged in discussions, bought into findings, and initiated concrete improvements. Importantly, they felt that process mining added value and became change champions, self-reliantly promoting the approach in the Company. As a result, the interest in process mining has spiked and business stakeholders for the first time requested a process mining engagement. Note that as a side effect the CoE also hopes to increase data literacy this way. That is because process mining results are organically integrated and used in discussions with business stakeholders and staff supporting the analysis activities.

Another factor that helps to prevent stakeholders from questioning process mining is that the CoE funds process mining activities with its own budget. Considering that costs for tool licenses and support from external consulting can be significant, it is not surprising that business stakeholders would want the value of insights to immediately justify the investment in the project. While further investment could accelerate building of organizational capability, relying on its own budget avoids cost-benefit discussions at the beginning and throughout projects, giving the CoE time to refine its approach and increase awareness.

Lesson 4: Strategically select your projects! To maximize the chance of succeeding, actively identify processes where challenges can be managed and where process mining can bring in a new perspective.

It is crucial for the CoE to select projects that have a high chance to deliver value and to contribute to the vision. Considering its past experiences, the CoE now conducts a more rigorous suitability assessment to identify potential technology, data, culture, and people challenges. Before committing to a project, the

CoE this way aims to anticipate gaps that need to be addressed in other ways, e.g., through conventional process analysis techniques.

In alignment with Lesson 3, the CoE also aims to understand whether process mining can deliver a new, unique perspective for process owners, leadership, or improvement specialists that triggers curiosity and adds value. This includes projects in which process mining can help validating process analysis results and bolstering business cases through data storytelling, as described above. In the fourth and fifth project, the CoE decided to focus on end-to-end processes. Instead of conducting a deep analysis of a single process, the CoE's goal is to, on a high level, connect processes that are otherwise viewed in isolation. Trading data granularity with data integration issues has so far worked well for the CoE. In both projects, this approach has helped to break down silos and has received positive feedback from the involved business and improvement stakeholders.

The CoE also tries to gain visibility into ongoing process improvement initiatives to expand the pool of potential candidates, enabling proactive identification of projects and discussions with process owners. Here, it does not only rely on the stakeholder management forums but more importantly on activities of the broader Transformation Team as well as on ongoing process modeling and analysis activities that it supports throughout the entire organization.

Lesson 5: Embed process mining in a broader BPM strategy! Developing process mining capability as part of a broader BPM strategy can create synergies that can help driving both, process mining maturity and adoption.

The CoE has benefited from embedding process mining in its broader BPM activities and from establishing itself as a trusted and reliable partner in the organization, as confirmed by the growing number of the steering committee members, community of practice members, and service requests. As a result, general support of the steering committee is secured and the pressure to immediately succeed at delivering process mining is lowered. Moreover, the CoE can leverage communication channels and stakeholder relationships that it establishes and maintains independent of process mining projects. Being able to deliver process modeling, analysis, and improvement services enables it to utilize the services as part of its risk mitigation strategy (see Lesson 3) and creates visibility for opportunities to apply process mining (see Lesson 4). Last but not least, having established a clear service offering helps the CoE to avoid potential conflict with other improvement teams and confusion for internal customers.

There are also advantages for driving process mining maturity which the CoE hopes to benefit from in the future. That is, the CoE avails itself of governing BPM standards and especially of developing the business ownership and the performance management frameworks (see Sect. 3). While the business ownership framework will make it easier to identify data, process, and IT owners in the future, the collaboration with Data and IT Teams has already helped to better align process, data, and IT management practices across the organization. Similarly, the performance framework and the top-level leadership support for it are expected to highlight the need to increase process and data orientation.

5 Conclusion

In this case study, we discussed the journey of building organizational process mining capability at a global gaming company. First, we motivated the case study by embedding it in the current academic discourse. We then introduced the Center of Excellence for Business Process Management and outlined its process mining vision. Next, we provided an overview of the CoE's journey over the last three years. Here, we looked at measures and activities related to building the foundations, service delivery, and stakeholder engagement. Lastly, we discussed lessons learned. In summary, the CoE views the implementation of process mining as a journey during which progression towards the vision is ensured by a risk mitigation strategy that addresses common process mining challenges through a variety of organizational measures.

Like any case study, the external and ecological validity of our findings are limited as they rely on the experiences at a single organization with a unique setting over a limited time period. Hence, our lessons might not apply to other organizations with different starting points, operating models, or process mining strategies. Moreover, we took on the roles of both, subjects and observers of the case study. To reduce the impact on the internal validity, e.g., the introduction of a selection bias or the ignorance of confounding variables, and to ensure accurate reporting of insights, we thoroughly discussed all information as a group.

Considering these threats to validity and the challenges for organizations in applying process mining, we want to invite researchers to continue focusing on organizational challenges and, as importantly, on identifying guidelines, measures, and activities that address them. Nevertheless, we believe that our case study provides valuable insights for practitioners and researchers.

References

1. van der Aalst, W.M.P.: Process Mining: Data Science in Action. Springer, Heidelberg (2016). https://doi.org/10.1007/978-3-662-49851-4
2. Andrews, R., Suriadi, S., Wynn, M., ter Hofstede, A.H.M., Rothwell, S.: Improving patient flows at St. Andrew's war memorial hospital's emergency department through process mining. In: vom Brocke, J., Mendling, J. (eds.) Business Process Management Cases. MP, pp. 311–333. Springer, Cham (2018). https://doi.org/10.1007/978-3-319-58307-5_17
3. Augusto, A., Conforti, R., Dumas, M., La Rosa, M., Polyvyanyy, A.: Split miner: automated discovery of accurate and simple business process models from event logs. Knowl. Inf. Syst. **59**, 251–284 (2019)
4. Badakhshan, P., Wurm, B., Grisold, T., Geyer-Klingeberg, J., Mendling, J., vom Brocke, J.: Creating business value with process mining. J. Strateg. Inf. Syst. **31**(4), 101745 (2022)
5. Beerepoot, I., Di Ciccio, C., Reijers, H.A., et al.: The biggest business process management problems to solve before we die. Comput. Ind. **146**, 103837 (2023)
6. Cruzes, D.S., Dybå, T., Runeson, P., Höst, M.: Case studies synthesis: a thematic, cross-case, and narrative synthesis worked example. Empir. Softw. Eng. **20**, 1634–1665 (2015)

7. De Weerdt, J., Schupp, A., Vanderloock, A., Baesens, B.: Process mining for the multi-faceted analysis of business processes - a case study in a financial services organization. Comput. Ind. **64**(1), 57–67 (2013)
8. van Eck, M.L., Lu, X., Leemans, S.J.J., van der Aalst, W.M.P.: PM^2: a process mining project methodology. In: Zdravkovic, J., Kirikova, M., Johannesson, P. (eds.) CAiSE 2015. LNCS, vol. 9097, pp. 297–313. Springer, Cham (2015). https://doi.org/10.1007/978-3-319-19069-3_19
9. García-Bañuelos, L., van Beest, N.R., Dumas, M., Rosa, M.L., Mertens, W.: Complete and interpretable conformance checking of business processes. IEEE Trans. Software Eng. **44**(3), 262–290 (2018)
10. Grisold, T., Mendling, J., Otto, M., vom Brocke, J.: Adoption, use and management of process mining in practice. Bus. Process. Manag. J. **27**, 369–387 (2021)
11. Hobeck, R., Klinkmüller, C., Bandara, H.M.N.D., Weber, I., van der Aalst, W.M.P.: Process mining on blockchain data: a case study of augur. In: Polyvyanyy, A., Wynn, M.T., Van Looy, A., Reichert, M. (eds.) BPM 2021. LNCS, vol. 12875, pp. 306–323. Springer, Cham (2021). https://doi.org/10.1007/978-3-030-85469-0_20
12. Klinkmüller, C., van Beest, N.R.T.P., Weber, I.: Towards reliable predictive process monitoring. In: Mendling, J., Mouratidis, H. (eds.) CAiSE 2018. LNBIP, vol. 317, pp. 163–181. Springer, Cham (2018). https://doi.org/10.1007/978-3-319-92901-9_15
13. Klinkmüller, C., Müller, R., Weber, I.: Mining process mining practices: an exploratory characterization of information needs in process analytics. In: Hildebrandt, T., van Dongen, B.F., Röglinger, M., Mendling, J. (eds.) BPM 2019. LNCS, vol. 11675, pp. 322–337. Springer, Cham (2019). https://doi.org/10.1007/978-3-030-26619-6_21
14. Leemans, S.J.J., Fahland, D., van der Aalst, W.M.P.: Discovering block-structured process models from event logs - a constructive approach. In: Colom, J.-M., Desel, J. (eds.) PETRI NETS 2013. LNCS, vol. 7927, pp. 311–329. Springer, Heidelberg (2013). https://doi.org/10.1007/978-3-642-38697-8_17
15. Maggi, F.M., Di Francescomarino, C., Dumas, M., Ghidini, C.: Predictive monitoring of business processes. In: Jarke, M., et al. (eds.) CAiSE 2014. LNCS, vol. 8484, pp. 457–472. Springer, Cham (2014). https://doi.org/10.1007/978-3-319-07881-6_31
16. Martin, N., et al.: Opportunities and challenges for process mining in organizations: Results of a Delphi study. Bus. Inf. Syst. Eng. **63**, 511–527 (2019)
17. Munoz-Gama, J., Martin, N., Fernandez-Llatas, C., et al.: Process mining for healthcare: characteristics and challenges. J. Biomed. Inform. **127**, 103994 (2022)
18. Runeson, P., Höst, M.: Guidelines for conducting and reporting case study research in software engineering. Empir. Softw. Eng. **14**, 131–164 (2009)
19. Taymouri, F., Rosa, M.L., Dumas, M., Maggi, F.M.: Business process variant analysis: survey and classification. Knowl.-Based Syst. **211**, 106557 (2021)
20. Zerbato, F., Soffer, P., Weber, B.: Process mining practices: evidence from interviews. In: Di Ciccio, C., Dijkman, R., del Río Ortega, A., Rinderle-Ma, S. (eds.) PM 2022. LNCS, vol. 13420, pp. 268–285. Springer, Cham (2022). https://doi.org/10.1007/978-3-031-16103-2_19

Implementing a Process Mining Framework in a Large Commercial Bank Lessons Learned

Piotr Gawrysiak[1,2(✉)], Tomasz Romanowski[2], Olga Kacprzak[3], Kamil Żbikowski[1,2]

[1] Warsaw University of Technology, Warsaw, Poland
[2] Processifier, Warsaw, Poland
piotr.gawrysiak@pw.edu.pl
[3] mBank S.A., Warsaw, Poland
https://www.pw.edu.pl, https://www.processifier.com, https://www.mbank.pl

Abstract. The paper presents an overview of a corporate initiative aiming to implement process mining methodology in one of the largest Polish commercial banks. This undertaking was started several years ago as a small experimental project and resulted in a creation of a robust process mining framework relying on commercial tools. Our paper presents an overview of the above process, describing key developments and summarizing insights that the organization has been able to gather.

Keywords: process mining · task mining · process monitoring · process improvement · process simulation · process-oriented organization · lean management · financial sector

1 Introduction

mBank SA is one of the largest financial institutions in Poland, positioning itself in the market as a high-tech corporation. From the very beginning, the bank has had a relatively small number of traditional branches, first relying on web access and later putting heavy emphasis on mobile channel. This results in a large amount of data generated in the IT systems, which is a valuable asset for a data-driven enterprise. The use of this data for machine learning models is an obvious step. The bank has had solutions based on classifiers and recommendation systems for many years. The adoption of such solutions spans many areas of the organization, including the retail and corporate customer divisions, as well as the security area within anti-fraud models. In recent years, the organization has made a decision, which is also reflected in the bank's strategy, to optimize its business and operational processes based on the process mining approach.

The challenges associated with implementing process mining are largely analogous to those encountered during the adoption of machine learning solutions. These can be divided into two groups: a) technical problems: data availability and quality and b) organizational challenges: integrating new solutions into existing processes and convincing users to adopt the new approach.

C. Di Ciccio et al. (Eds.): BPM 2024, LNBIP 527, pp. 441–450, 2024.
https://doi.org/10.1007/978-3-031-70445-1_32

Looking back, it can be said that the implementation of process mining at mBank is proceeding successfully, but the process is far from complete. In the remainder of the article, we will discuss the difficulties we encountered, how the approach to implementation has evolved over time, and what challenges still lie ahead for the organization.

2 Road to Adoption of Process Mining in mBank

Interest in process mining within the bank dates back to 2017. Initially, there was no structured implementation process or vision. At the outset, the organization lacked even an awareness of process mining as a scientific or engineering discipline.

The initial motivation for implementation was linked to the need for automating the reporting of work volumes by operations staff. This task required the data sources of high frequency and appropriate granularity. The work being done by the operations staff relies on specialized software systems, none of which was specifically designed to generate event log information for the purpose of process mining analysis. However, these systems are usually capable of producing system, or engineering logs, typically designed for debugging purposes and verifying software functionality, without consideration for analytical use.

Despite above problems the gathered data proved to be extremely useful for reporting purposes. Building on the success of this initiative, it was decided that the natural next step would be to attempt a reconstruction of the entire process flow from system logs i.e. building a PM oriented event log.

The first attempt at using system for data modeling process flows involved building a graph database based on the open-source framework Neo4j [5] and the declarative graph query language Cypher [6,7]. Although the initial results seemed promising, the relatively high technical entry barrier (most notably related to the lack of commercial maturity - in 2017 - of above technologies) prevented this solution from being widely adopted within the organization.

Another significant reason for the lack of substantial interest in the new approach to process analysis was undoubtedly the high level of effectiveness of classical process optimization methods such as LEAN. Over time, however, the established optimization indicators became more difficult to achieve, which contributed to renewed attention towards new methods.

Another attempt to implement analytics based on event log data took place in 2019. This approach was more formalized as there was already an awareness within the organization of the existence of a formalized framework for such analyses - *process mining*. This coincided with the first process mining conference, ICPM 2019, in Aachen [8] that the bank representatives attended. This proved crucial for the subsequent efforts, as much information was gained through informal discussions with other conference participants, and important business and research contacts were established. In the same year a commercial process mining software was deployed[1], however the implementation process was not straightforward and required overcoming of several technical obstacles mentioned below.

The implementation of process mining in the bank was one of the first, if not the first, in the region - and definitely in Poland. A natural problem in such a situation is the lack of opportunities to learn from the experiences of other entities that have already undergone similar journey. This necessitates relying on the information obtained from external software providers. Obviously we had some knowledge acquired from in-house work conducted in 2017, however this provided only limited advantage.

Potential data sources had already been identified, which significantly streamlined the proof of concept process. Nevertheless, conducted tests of external solutions revealed a number of data imperfections. It is worth noting that, even at present, there are no commercial solutions available that automatically assess event logs for the detection of so-called event log anti-patterns, i.e., data issues that significantly impact the quality of subsequent analyses.

3 Process Mining Approach

Process mining is a research discipline that sits at the intersection of Business Process Management (BPM) and data science [3]. Its primary aim is to analyze business processes basing on event logs that captures detailed information about the execution of processes within an organization.

It is a relatively young scientific discipline but it has seen rapid development and adoption across various industries in recent years. In 2004, Van der Aalst published a seminal paper [1] that laid the groundwork for process mining by introducing key concepts and techniques. His work highlighted the potential of using event logs to discover, monitor, and improve real business or industrial processes. This marked the beginning of a more structured approach to process mining, leading to the development of various algorithms and tools designed to extract process models from event logs (for a broader overview see e.g. [2]). At the same time information technology advancements played a crucial role in the growth of process mining. The increasing availability of large-scale event data from enterprise information systems, coupled with improvements in data storage and processing capabilities, provided a fertile ground for the practical application of process mining techniques.

As process mining matured, it began to integrate more closely with traditional BPM practices. This integration allowed organizations to use process mining not only for post-hoc analysis but also as a proactive tool for continuous

[1] The first software process mining tool purchased by the bank was a Lana Labs solution (the company was later acquired by Appian [9]). In the long run it proved to be ineffective, mostly due to insufficient analysis capabilities, especially w.r.t. large data volumes and limited customization options. Therefore this solution was phased out and replaced with Processifier process mining software [10], which is being used currently.

process improvement. By aligning process mining insights with BPM initiatives, businesses could achieve more efficient and effective process optimization.

From the perspective of the process mining software commercial market was still in the early stages of its growth in 2017 [4]. Broader adoption of such solutions at that time in the European Union occurred mainly in the German and Dutch markets. This low level of popularity significantly influenced the approach taken in the analysis of processes in the discussed case.

4 Implementation of Process Mining Framework

The initial analyses of processes based on the process mining methodology highlighted the first challenges related to proper data preparation. Given the large scale of operations conducted by a commercial bank, the first significant problem turned out to be the sheer volume of data to be analyzed. The number of records that needed to be processed was in the range of millions. Efficient processing of such data requires appropriate infrastructure and expertise in the field of Big Data. The simplified implementation diagram is shown in Fig. 1.

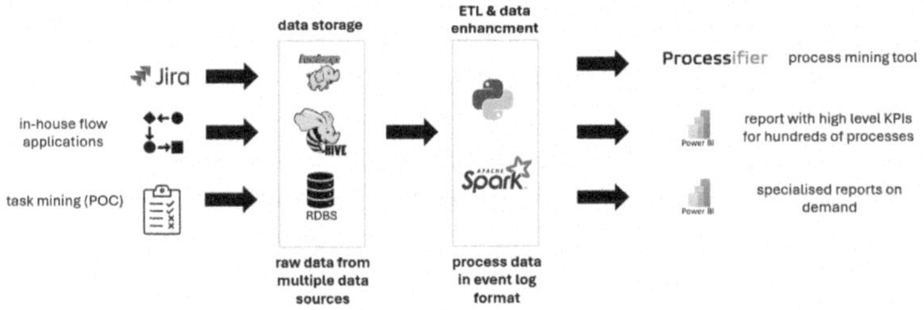

Fig. 1. Implementation of process mining framework in mBank

4.1 Data Quality Challenges

Many companies starting to implement process mining are not aware of the potential data quality issues they may encounter (see Table 1). Direct upload of such data into any commercially available process mining tool proved to be very challenging - the amount of technical events in input data usually resulted in generation of so called *spaghetti processes.*

Another problem that we identified was the occurrence of activities in the logs whose roles changed throughout the process. This means that two different activities - distinct from a business perspective - are marked by the same event in the event log. Failure to detect such data anomalies results in drawing incorrect conclusions about the process flow. The process graph loses clarity and incorrectly indicates rework loops that did not actually occur.

Another problem that was identified was the occurrence of different activities that were assigned the same system time, even though they have been performed at different moments. The reason for this phenomenon was the insufficient precision of timestamps, although it was initially supposed to precise enough. It turned out, that even millisecond time precision may not be sufficient, if the number of events is large enough. Moreover, it was discovered that some systems carried out batch logging of events that occur during very short intervals. Such grouping can result in activities being arranged in a random order, which has serious consequences for the subsequent analysis. The random ordering of activities translates into the generation of additional process variants and extra connections in the graph. Removing events with the same timestamps is not an option in this case, as the occurrence of these entries may be significant from a business interpretation perspective.

Table 1. Presence of data quality issues detected in mBank by classification proposed in [11].

anti-pattern	presence
Form-based Event Capture	x
Inadvertent Time Travel	
Unanchored Event	
Scattered Event	
Elusive Case	
Scattered Case	x
Collateral Events	x
Polluted Label	
Distorted Label	x
Synonymous Labels	x
Homonymous Label	

We developed a way to mitigate this problem by creating a reference sequence of events, which is used when it is not possible to sort events based on timestamps. The reference sequence of events is established in consultation with the business team responsible for the process, allowing for the avoidance of future issues with the proper interpretation of specific sequences in the event log.

As previously mentioned, IT systems used by the bank were not originally designed with future process analysis in mind. The primary focus was on optimal and reliable operation, which influences the way business events are logged. In this case, system logs are structured so that the end of one event directly initiates the next. As a result, it is impossible to isolate the time spent directly working on a task. The duration information available in the log only defines the lead time, which is the time from the completion of one process step to the completion of the next step. When looking for optimization opportunities, it is crucial to be able to divide the process time into its components: idle time and working time.

For newly designed processes, such limitations can be removed by appropriately defining waiting statuses at the IT system level. However, the problem remains for existing processes. Fortunately, since many workflow systems used in the bank are operated via web interfaces, it is possible to collect web browser events indicating user activity on a specific task. By overlaying this information on the original (system) event log, we could better estimate the actual time an employee spends on a given task.

Of course above approach cannot be exploited for processes or process steps that are carried out outside of flow systems, especially in closed source, custom and monolithic applications such as Microsoft Excel, Microsoft Word, etc. The process mining approach might be however ineffective in these cases anyway due to the lack of appropriate data.

4.2 Data Availability Challenges - Task Mining Experiments

Task mining is a technology that focuses on capturing and analyzing detailed activities performed by users on their personal computers/workstations, particularly within various software applications. Unlike process mining, which analyzes event logs from enterprise systems in order to understand process flows, task mining delves into the granular level of individual tasks. This is achieved by recording lowest level user interactions with hardware and software such as keystrokes, mouse clicks, and application usage statistics.

To cover the analysis of those parts of the process that are not directly performed in any flow application, a proof of concept of a task mining tool was conducted in the bank. The recording of work was carried out in a manner previously agreed upon with the employees, which is an important consideration as the task mining tools are potentially able to capture all employee activity - and thus might present significant privacy issues. Additionally, employees were asked not to work on different cases concurrently, but to complete each application[2] fully from start to finish during the recording. For this purpose, the recording period was chosen to coincide with a part of the month characterized by a lower workload.

Despite ensuring these conditions, recordings often contained interwoven tasks, necessitating manual assignment of individual application screens. Such issues cast doubt on the feasibility of using this technique in an automated manner and on a larger scale that would allow for continuous process monitoring company - or at least division - wide. An additional problem, encountered at the data level, was the very high granularity of the collected events, which required filtering and pre-aggregation of the data before it could be used for process mining purposes.

In addition to the challenges of proper data preparation, the issues of security and auditability of the proposed solutions were crucial factors. Since the bank operates in a highly regulated market and processes sensitive customer transactional data, stringent security requirements are imposed on such solutions.

[2] An application refers here to *bank customer application* such as credit application, or a compliant.

Currently, despite the aforementioned challenges, the bank is still in the process of searching for solutions and identifying appropriate areas where task mining could add significant value in the analytical and monitoring layers.

5 Organizational Perspective

As mentioned at the beginning, the second equally important factor is the organization's ability to effectively adopt the new solution. First, we will present observations regarding the challenges we faced during the implementation. In the following section, we will discuss the current approach of the organization to applying process mining within BPM and the types of competencies identified to achieve the best results.

5.1 Assesment of Organizational Adoption of Process Mining

Over the years, the attitude towards the process mining technology in the bank has undergone significant changes. Initially, the prevailing view was that process mining would revolutionize and optimize processes and operational activities across the entire organization. Efforts were made to provide data and metrics for several hundred processes. The expectation was that those responsible for these processes would largely monitor their areas independently. If anything appeared that significantly deviated from the managers' expectations, it would then be subjected to more detailed analysis. In retrospect, it can be assessed that this strategy yielded less than expected results. It is important to understand the reasons behind this.

Firstly, our observations indicate that people often resist the introduction of new technological innovations. When implementing solutions such as process mining, those responsible for processes may feel evaluated and controlled. They do not perceive the new technology as an enhancement of their capabilities in process management. A defensive attitude is more commonly observed.

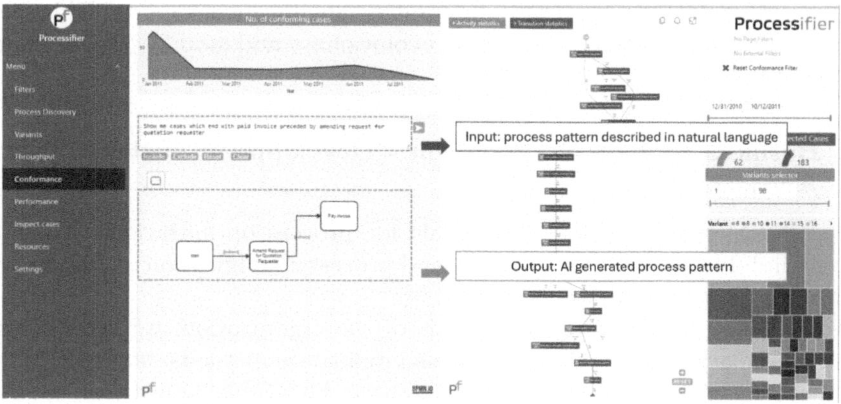

Fig. 2. Implementation of generative AI in process mining tool

Secondly, there is the time required to assimilate a completely new method or tool. Most employees do not have sufficient time to effectively learn a new technique. Ongoing responsibilities naturally take higher priority over trying to apply a something new, the final benefits of which are unknown. Even when a process is appropriately analyzed using the new tools, this new data tends to be abandoned due to the overload of other responsibilities. One remedy for addressing the problem of a relatively high entry barrier is the implementation of generative AI solutions, which facilitate analysis for users with lower technical skills. An example of this is generating a specific process pattern that is later used to filter the analyzed dataset. The user provides a description of the cases of interest, which is then automatically transformed into a simplified BPMN notation (see Fig. 2).

Thirdly, those responsible for processes expect that the presented analyses and KPIs will provide them with specific information on areas that need optimization. This can be challenging when using a generic set of metrics and statistics, especially considering that processes can differ significantly from one another as well as in their level of digitization. Additionally, measuring processes uniformly can lead to controversies regarding the value of certain indicators. As mentioned earlier, people often adopt a defensive stance towards new things, and it takes time to convince them that the prepared indicators are primarily intended to assist them in managing the process. In many cases, the value of the indicator itself is not crucial. More important is observing changes in values, revealing appropriate trends, and drawing the right conclusions from them.

A more effective strategy seems to be project-based, which does not force managers to adopt a specific method but initially focuses on the needs and challenges facing the optimization of a given process. Process mining is thus treated as an additional tool in the analytical teams' repertoire, providing information in areas where it is genuinely needed. For the business side, dedicated reports are prepared using data from flow systems, the usefulness of which is not questioned because they are developed in collaboration with the business side from the outset.

This method also provides space for the gradual introduction of process mining elements, which are naturally integrated into business reports. This allows the new technology to be adopted in an evolutionary and more accessible manner for the end user.

5.2 Business Process Management - Development of the Holistic Approach

An organization usually has clear goals for process optimizations. These are savings (in FTE, processing time etc.) and process simplification and integration resulting in best automation potential.

To achieve the aforementioned goals of the organization, we need to use a holistic approach - based on both process intelligence methods, but also utilizing lean methods and the voice of the process user. The Table 2 illustrates a holistic approach to process optimization.

Table 2. Holistic approach in process optimisation

Reporting the optimization changes	Process Intelligence	Conducting workshops and optimisation projects
Submissions by people working on processes	Process mining analyses	Performed by lean experts
Necessary for implementation by IT	Task mining analyses	Identification if pain points
Qualitative, requiring changes in process procedures	ML models	AS-IS process analysis
		TO-BE process creation

Table 3. Skills and knowledge needed to implement holistic approach

Reporting the optimization changes	Process Intelligence	Conducting workshops and optimization projects
Subject matter experts (SME) involvement	Strong analytical skill set	Process modeling skills
Ecosystem for process changes requests	Tools for process analysis	BPMN notation skills
Clear rules for process changes requests	Tech stack: Python, SQL, Power BI, Hadoop, Spark	Process drawing tools
		Lean and Six sigma approaches
		Project management skills

Accordingly, teams should be formed that are interdisciplinary (see Table 3). Such teams should be held accountable to pre-set goals and KPIs for a given period of time, such as e.g. achieving savings equal to 10% of the FTE of a given area or number of recommendations or optimizations implemented to the process and proposed by the team.

This approach increases the probability of success for the optimization team and the use of advanced quantitative methods. It addresses the real needs of the organization, rather than trying to force the use of new "toys". Adapt methods to optimization needs, never the other way around.

6 Conclusion

This paper presents only brief overview of the ongoing project, which - when it was initially commenced a few years ago - was indented solely as a small scale experiment. However since this time it developed, progressing from a tailor-made process analysis and optimization tools to the self-sufficient framework based on the industrial quality process mining software. The project allowed not only the optimization of bank's operational business, but also influenced other organizational changes, including - but not limited to - introduction of agility inspired approaches to business process management.

The implementation and refinement of process mining solutions have led to several key takeaways. Firstly, data quality issues had to be sorted out before the process analysis could start. This includes solving issues related to missing values, inconsistencies etc. and their influence on modeling of the real process flow. Secondly, it demonstrated the important role of a CoE in driving, standardizing, and further improving process mining activities across the organization. And, last but not least, it demonstrated that a project-focused approach, tailored to the specific needs of each process, usually results in more efficiency than a generic one trying to force all processes to fit one standardized methodology of process mining.

Entire experience of the bank with process mining has also demonstrated how valuable this approach can be for improving business process efficiency and effectiveness. However, one needs also to keep in mind the challenges such an approach can face from data quality, user adoption, and probable resistance to change at the organizational level. Obviously some of the problems will become less important, especially thanks to advances in artificial intelligence methods. In fact exploiting potential of AI, end especially Generative AI, towards automatic identification of event log anomalies and their repair, or for process graph analysis is an area research that we plan to explore in the near future. However one might not forget that the organizational and social aspects are equally important.

References

1. Van Der Aalst, W.: Process mining: a research agenda. Comput. Ind. **53**, 231–244 (2004). https://doi.org/10.1016/j.compind.2003.10.001
2. Van Der Aalst, W.: Data Science in Action, pp. 3–23. Springer, Heidelberg (2016). https://doi.org/10.1007/978-3-662-49851-4
3. The landscape picture of Process Mining. https://www.processmining.org/overview.html. Accessed 25 May 2024
4. Polaris Market Research. https://www.polarismarketresearch.com/industry-analysis/process-mining-software-market. Accessed 25 May 2024
5. Neo4j: Graphs for everyone. https://github.com/neo4j/neo4j. Accessed25 May 2024
6. Graph Query Language Cypher. https://neo4j.com/product/cypher-graph-query-language/. Accessed 25 May 2024
7. Francis, N., et al.: Cypher: An evolving query language for property graphs. In Proceedings of SIGMOD 2001 (2018). https://doi.org/10.1145/3183713.3190657
8. 2019 International Conference on Process Mining (ICPM), IEEE, Aachen, Germany, (2019). https://doi.org/10.1109/ICPM.2019.00031
9. Appian acquires Lana Labs. https://appian.com/about/explore/press-releases/2021/appian-acquires-leading-process-mining-company. Accessed 25 May 2024
10. Processifier Process Mining. https://www.processifier.com. Accessed 25 May 2024
11. Suriadi, S., et al.: Event log imperfection patterns for process mining: towards a systematic approach to cleaning event logs. Inf. Syst. **64**, 132–150 (2017). ISSN 0306-4379

The Quest for the Comprehensive Customer Journey - A Case Study from a C2C Marketplace

Felix Mannhardt[1](✉), Ragnhild Halvorsrud[2], Otas Meironas[2], and Lasse Brurok[3]

[1] Eindhoven University of Technology, Eindhoven, The Netherlands
f.mannhardt@tue.nl
[2] SINTEF Digital, Oslo, Norway
[3] Bookis AS, Tromsø, Norway

Abstract. Bookis is a C2C online marketplace for physical books. It is a product-orientated platform on web and mobile apps that enables customers to buy used books from other customers on the platform. Insights into the customer behavior and how customers experience their customer journey is crucial to improve their marketplace. Service providers, such as Bookis, are accumulating more and more data from their systems that could be leveraged for analysis. Yet, turning such data into comprehensive individual customer journeys is a significant challenge. This difficulty is compounded by several issues among which: (1) converting or connecting raw events to discernible touchpoints, (2) defining the boundaries of journeys in a C2C marketplace, and (3) acquiring and integrating data from external service providers such as delivery services. We provide a case study describing two analysis iterations in an attempt to obtain a comprehensive view on Bookis' main customer journey by combining an "outside in" view (mystery shopping) with an "inside out" view (process mining). We leverage a novel event knowledge graph approach to data integration and conclude that the combination of both viewpoints is important to obtain a comprehensive analysis of the customer journey. In particular, relevant parts of the journey may not be covered by data-driven method, despite the large amounts of event data available.

Keywords: customer journey · mystery shopping · process mining

1 Introduction

Service provisioning has expanded significantly over the past few decades, and today, our society is dominated by services. Private and public service providers are challenged by the fragmentation of service systems, a growing number of customer channels, and customer-to-customer (C2C) interaction through social media [11]. It has become increasingly complex for service providers to assess

C. Di Ciccio et al. (Eds.): BPM 2024, LNBIP 527, pp. 451–461, 2024.
https://doi.org/10.1007/978-3-031-70445-1_33

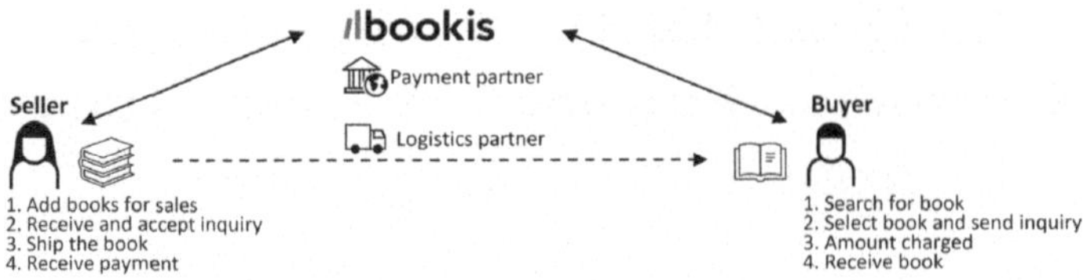

Fig. 1. Overview of the Bookis service system.

service quality. We provide a case study that investigates the services of Bookis, a complex C2C service, through the lens of customer journeys and process mining.

Bookis is a service provider in the circular economy sector that provides a C2C online marketplace for physical books. Shipping and payment are integrated into the buy-and-sell flow. Also, there is an option for both the buyer and seller to donate royalties directly to the author, and authors can sell books directly on the platform. Hosted on Amazon Web Services, Bookis is built natively on mobile apps and websites. Traditionally such a marketplace would only facilitate so that buyers and sellers could find each other. Today, most marketplaces have switched from "unmanaged marketplaces" to "managed marketplaces" resulting in becoming transactional with integrated payment and shipping. Bookis provides such transactional service integrating with external service providers for shipping and payment. Besides Bookis, the service delivery network consists of a payment provider and several logistic partners.

Central to the service design of the product-oriented marketplace is the product page. It allows Bookis to facilitate the service experience for the buyer. For example, among many sellers, one of them will be highlighted as the recommended seller for you as a buyer depending on an algorithm that considers your profile and shopping basket. This can increase the number of books being bought in every transaction and benefits both sellers and buyers by increasing the sales for highlighted sellers and reducing shipping costs for buyers. In summary, Bookis is a prime example of a complex service design in which their own services are integrated with external services providers and the part of the services enacted by the users, seller and buyer, themselves. Figure 1 shows the main steps involved when selling and buying books.

The main objective of our research is to improve the understanding of the *benefits* as well as the *limitations* of traditional customer journey analysis methods and process mining methods along with their *combination*. In principle, traditional observation methods (observation, mystery shopping) provide a comprehensive "outside-in" view on a customer journey using a small sample. In contrast, process mining provides an "inside-out" view, i.e., the source of information are the organizations information systems, and is data-driven so that it can leverage a large sample of all available event data. Thus, the combination of both methods seems to be promising and complementary.

In fact, it can be difficult to holistically capture the user experience with process mining alone [16] since not all aspects of the journey are recorded. Conversely, a small sample observation or mystery shopping study may be biased through the sampling approach and fail to provide the large-scale analysis provided by process mining. It also may fail to connect the external observations to the internal processes of the organization. We are not aware of published studies that systematically combine both approaches, even though a dual view on customer journeys seems crucial to understand all aspects of a customer's journey. In our study, we encountered several challenges when mapping events to the customer journey domain concepts which led us to propose a new data format. We also contribute lessons learned on combining both approaches.

In the remainder of the paper, Sect. 2 describes the problem addressed, provides background on both methods and the case study execution. Then, the results are summarized in Sect. 3, and Sect. 4 concludes with the lessons learned.

2 Case Study

The aim of our case study was to obtain a comprehensive view on the core customer journeys for the Bookis marketplace from a customer perspective. For Bookis the value of such analysis is to avoid blind spots in their understanding of how customers perceive their services. It is also relevant to understand both the nuances of how individual journeys were performed as well as a birds-eye-view on multiple journeys and how multiple journeys connect to each other. Next to the business value, we set out to understand how the available event data on such customer journeys with multiple service providers and multiple customer roles (buyer, seller) can be effectively integrated into a single representation of the customer journeys.

In the Bookis marketplace there are two main journeys (Fig. 1) depending the role of the user acting as seller or buyer. Bookis describes them as follows:

- **Adding books for sale (J1).** With a mobile app, the seller scans the barcode on the book, obtains all the metadata, which prefils the details. The seller needs to only specify the price and condition and can add a book for sale in 15 s.
- **Buying books (J2).** With the mobile app or in a browser, the buyer finds a book or multiple books and sends a request to the seller. The seller receives a push/email/SMS about the book(s) the buyer wants to buy and can then accept or decline the request.

When conducting the analysis, we focused on J2 as it is a more complex journey than J1 involving multiple service providers as well as the buyer and the seller. It is also the main revenue-generating journey.

To comprehensively analyze the buying book journey (J2) and investigate our research objectives, we followed a procedure in which we combined a mix of the customer journey analysis (qualitative and quantitative) as well as process mining (quantitative) that gives rise the a proposed generalized procedure and enabled us to collect the challenges encountered during the study execution and provide the lessons learned reflected on later. In the following Sect. 2.1 and

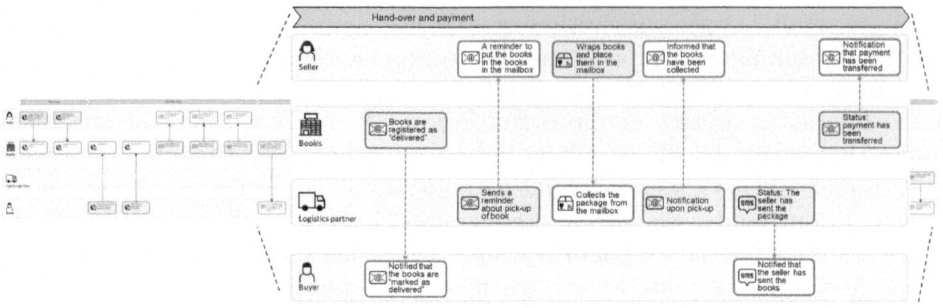

Fig. 2. Zoomed-in segment of the combined journeys for the seller and buyer on Bookis represented as CJML. The entire journey consists of four phases and 27 touchpoints.

Sect. 2.2, we describe our method of switching between an "outside in" and an "inside out" perspective and solutions for some of the challenges. Later in Sect. 3 we summarize the results and the problems that remain open.

2.1 Outside-in Perspective

Customer journeys emerged from marketing and customer-centric thinking and has become a strategic tool to help identify issues within current services, design novel services, and assess customer behaviors and experiences [10]. Journey maps are diagrams that chronologically depict the customer's steps or touchpoints. However, journey maps vary in definition, detail, and abstraction. In this paper we use the Customer Journey Modeling Language (CJML) which defines a journey as a sequence of touchpoint involved for a user to achieve a specific goal in the context of a service process [7]. CJML allows for a precise modeling of individual customer journeys from the viewpoint of the customer ("outside in") and the comparison of these against the service delivery process as planned [13].

We take a look at an example of CJML representing a combined customer journey for the seller and buyer of books using Bookis as service provider. Figure 2 shows the combined journeys using CJML. We highlight a segment of the handover phase to provide an example of the notation. Please note that the CJML diagram in Fig. 2 represent a single actual customer journey which not necessarily complies with the planned ideal journey.

The seller, the service providers, and the buyer appear in separate swim lanes containing their respective touchpoints (nodes). It is assumed that the seller and buyer already have user accounts, and the payment partner is integrated with Bookis. First, the seller, Jane, prepares the books for sale and lists them on the platform. This involves logging into the platform, searching for the appropriate categories, and publishing ads for each book. The buyer, Mike, browses the platform and identifies two books of interest. Upon finding suitable books, he requests to buy them, and payment is reserved on his card. Jane receives notifications and accepts the purchase requests through a link provided via email. Once Jane accepts the requests, Mike is notified by Bookis. Interestingly, the system already "marks the books as delivered" even though the logistics partner did not yet receive them from Jane. This appears as inconsistency in Fig. 2 but corresponds to the actual journey at that time. In due time, the logistics partner reminds Jane to prepare the books for pick-up. Jane wraps the books, marks them with reference codes, and places them in the mailbox for pick-up. The logistics partner collects the books from Jane's mailbox and updates the platform about the pick-up status, notifying Mike that the books are on their way. At this point, the money is withdrawn from Mike's card and transferred to Jane. Mike is kept updated on the expected delivery time and is finally notified when the books are safely delivered to his mailbox.

In the case study, the outside-in perspective was initially explored through service walk-throughs using both PC and mobile app. Following this initial exploration, we applied the mystery shopping method, where researchers took on the roles of seller and buyer, completing the entire transaction process on the C2C platform. In total, five journeys were completed and documented using CJML to cover the use of different end-user devices and the comparison between browser and app usage. A second round of mystery shopping was conducted 9 months later. In this round, two buyer and seller journeys were paired, with two books being purchased at a time. This approach enabled a comprehensive assessment of the customer journey from the outside-in perspectives of both buyer and seller within the same context.

2.2 Inside-Out Perspective

In the digital era, most touchpoints within customer journeys generate numerous events that can be leveraged for process mining [1] as proposed in the literature [2]. Process mining reverse engineers a process by leveraging the recorded events from information systems to infer the as-is process behavior [1]. Several applications of process mining for analysing user journeys have been reported. For instance, web log analysis [14], and service delivery improvements [8]. Bernard and Andritsos [2,3] were among the first to apply process mining for understanding customer experience. Their approach took a customer-centric approach and attempted to consider contextual information, e.g., customer emotions and characteristics into the analysis. In [14], insights into the user behavior are rendered but external factors beyond the web logs that could influence the user behavior are not considered. More importantly, the actual user experience at each of these steps is not considered. Hansson et al. [8] presented a comprehensive mapping of a real-life customer journey using process mining and also performed a qualitative analysis. However, the journey analyzed was very simple on a small sample. Recently Zilker et al. [16], noted that a holistic and customer-centered analysis for process mining would be important for service analysis and recognized such a holistic view of an end-to-end customer journeys with all the steps, various systems and user experience as challenge. In [6] it was also reported that the relationship between atomic events recorded by various systems and a single touchpoint in the journey on the abstraction level of CJML may be complex and multifaceted. Also, event data may need to be raised to

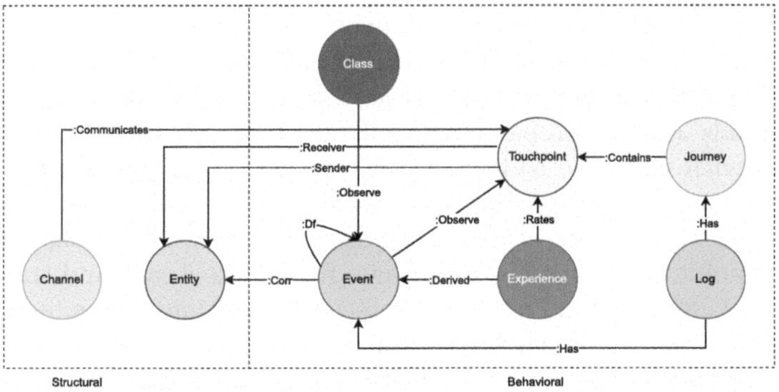

Fig. 3. The extended event knowledge graph schema based on the proposal in [4]. Each of the circles depicts a node type and each edge a relationship type.

a higher abstraction level [15], refer to multiple actors (the service provider's employees, the customer), be related to customer experience, and spread over multiple organizational units or organizations.

For the process mining part of the analysis, we extracted data from Bookis systems supporting the mobile application and website using both database records as well as tracking logs. The multi-faceted nature of the events described by prior work is also recognizable in our case confirming the suggestions of prior work. Events refer to several different actors who can switch roles (buyer, seller) and belong to different organizations. In fact, the data extraction took place after the initial round of mystery shopping, which was crucial to mapping the various technical events to user-facing touchpoints. The data validation with the "outside-in" perspective also led to a repeated extraction complementing the dataset with parcel tracking events from the shipping provider.

To retain the flexibility of investigating the event data from multiple perspectives and to be able to connect the extracted events with CJML domain concepts, we decided to use an Event Knowledge Graph (EKG) [4] representation. An EKG is a recently proposed model that was designed to represent such multi-dimensional or multi-entity event data in a flexible and extensible graph representation. To extend the EKG with CJML domain concepts, we enriched its schema as shown in Fig. 3. An EKG is a graph consisting of nodes and relationships between nodes. In Fig. 3, the possible node types are represented as circles and the possible relationships between nodes of a certain type are represented by labeled edges. The basic EKG schema proposes to use *Entity*, *Event*, and *Class* nodes. Events are of a certain event class and may relate (*:Corr*) to multiple entities and a directly-follows relation (*:Df*) local to each entity can be constructed such that two events are related by *:Df* when they are directly following and relate to the same entity.

We model the different service providers and end users from CJML as entities since this allows us to view the recorded events from their viewpoint. We decided to further extend the basic schema with the other domain concept of CJML.

Table 1. A simplified listing of the buyer journey (J2) touchpoints.

ID	Time	Description	Initiator	Receiver	Channel
T1	Day 1	Searches and selects two books	Buyer	Bookis	web/app
T2	Day 1	Pays for the two books	Buyer	Bookis	web/app
T3	Day 1	Request sent to seller for book 1	Bookis	Buyer	e-mail
T4	Day 1	Request sent to seller for book 2	Bookis	Buyer	e-mail
T5	Day 1	Seller accepts the request for book 1	Bookis	Buyer	e-mail
T6	Day 1	Seller accepts the request for book 2	Bookis	Buyer	e-mail
T7	Day 1	Notification: Book 1 marked as "delivered"	Bookis	Buyer	e-mail
T8	Day 1	Notification: Book 2 marked as "delivered"	Bookis	Buyer	e-mail
T9	Day 3	Seller has sent a package	Logistics	Buyer	SMS
T10	Day 11	Package delivery next day	Logistics	Buyer	e-mail
T11	Day 12	Package is delivered to mailbox	Logistics	Buyer	SMS

Touchpoint nodes can group events that were observed during a touchpoint and relate to communication *Channel*s through which the participating parties (*Entity*) receive and send. Since touchpoints always relate to a single journey, we also included the *journey* as separate node type. For future use, we also included a dedicated node type to capture customer experience. Please note that whereas we validated the proposed extension only with the Bookis case the extensions are based on the generic CJML concepts, which have been shown applicable in several other domains [13] having service processes such at least telecommunication [5] and healthcare [7].

For the case study, we used Neo4J as graph database and developed a Python tool to load the extracted events from a JSON format into the EKG following our propose schema. Based on the EKG, we reviewed several of the actual journeys using the built-in graph visualization of Neo4j Browser as enriched directly-follows graphs. The goal was to understand the challenges in connecting event data to the customer journey as perceived by the customer and to assess which further insights a process mining view on the journeys could provide.

3 Outcomes

We followed both two methods as described in Sect. 2. This led to two main analysis iterations for which we now summarize and reflect on the outcomes.

3.1 Outcome of First Analysis Iteration

Previous studies show a lack of comprehensive overview of the customer journeys [12], particularly during the handover to partners in the service delivery network [5]. The systematic investigation of the outside-in perspective by using CJML and mystery shopping provided new insights for Bookis. By concretizing and visualizing the seller and buyer journeys, along with structured feedback, Bookis gained a more detailed view of the entire customer experience. This process clearly highlighted both barriers and positive experiences.

It became evident that several aspects required immediate attention, particularly substantial back-end changes. These issues were promptly addressed, with a focus on enhancing product stability and streamlining processes. Additionally, the findings corroborated some previously identified problems. The following improvements were initiated:

- Ensure all sellers can use the shipping and payment solutions
- Improve search functionality for relevance and availability.
- Simplify communication via email and SMS
- Review and streamline integrations, such as preventing premature delivery notifications (cf. Sect. 2.1 and T7 and T8 in Table 1).
- Reassess key customer journeys and communication, aiming for clearer and more oncise explanations.
- Ensure listed books are actually available for sale and sellers are active.
- Address various other minor issues to enhance the customer experience.

The initial data collection amounted to a bit more than 25 000 events out of which about a bit more than 3000 events were related to some of the touchpoints identified. Table 1 show the sequence of touchpoints that were identified for one of the mystery shopping journeys. Whereas we could successfully obtain events for T1–T8 from the Bookis system, the analysis revealed blind spots in the data recording. Events from the shipment provider data could not yet be accessed at this stage, which prevented an analysis of the final phase of the journey, i.e., for T9–T11 in Table 1 we were not able to use process mining.

3.2 Outcome of Second Analysis Iteration

The second round of analysis revealed additional areas for improvement. Some issues were relatively easy to address, such as textual changes in notifications. For example, the message sent to buyers stating, "the book is marked as delivered in our systems," required more customer-oriented language. This was particularly important since the buyer was notified that the book would be shipped from the seller two days later. Both the content and timing of this touchpoint needed Bookis' attention. Another issue requiring back-end work was the number of notifications sent when purchasing multiple books. Many touchpoints are sent "per book," causing the total number of touchpoints to grow linearly with the number of books purchased. This overload of touchpoints distracts the buyer from important messages. Additionally, the seller could be notified when the payment is received in their bank account, and when the books have arrived at the buyer's address.

Finally, we revised the data integration to also include events provided by the shipment provider. This yielded a more complete view on the journey as visible in the subgraph shown in Fig. 4. Currently, we are working at aggregating several

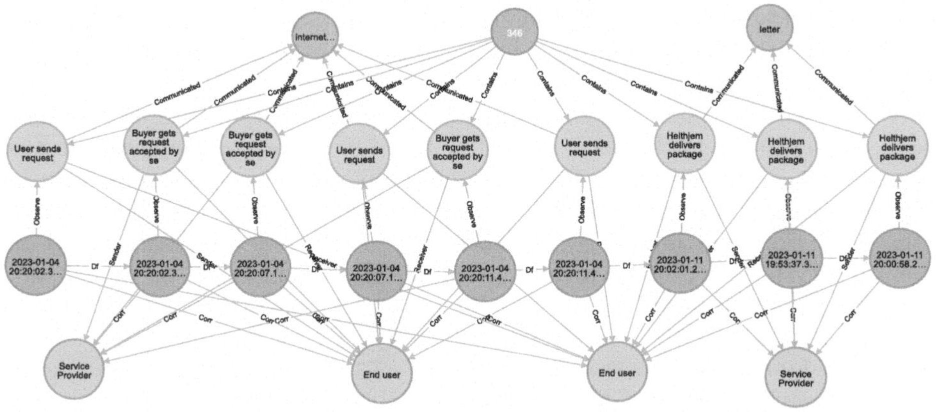

Fig. 4. A subgraph representing a part of a specific journey. The node types are color coded as in the introduced schema. Multiple other journeys that were logged for these particular end users (seller and buyer) are not shown for legibility but are connected to the respective entities.

similar journeys to a multi-entity directly-follows graph providing indications on frequencies and performance. Also, several events that are not directly connected to the touchpoints but may be relevant to the context of the journey, e.g., user navigation in the mobile application, are already added to the EKG but not yet considered for analysis. However, the EKG already provides us with the explicit directionality of the communication through the touchpoints, connecting the buyer, seller, and the service providers through *sender* and *receiver* relations (cf. Fig. 2). We exploited such directionality in a related research topic on using user journey games [9]. The data integration performed enables us to apply this research to the Bookis case in the future.

4 Conclusion

Most process mining initiatives reflect only parts of a full customer journey since many of the relevant details are not captured in event data. Based on industry experience with a C2C marketplace for used books, we described a use study on combining traditional qualitative analysis with data-driven process mining. Several lessons were learned during the case study. Starting with the outside-in approach was needed to estimate even the planned journey since awareness of all the touchpoints generated towards the customer can be lacking. Still, using the events logged from systems supported the successful extraction of touchpoints and allowed to validate them with a larger number of journeys. However, event data several touchpoints of the journeys is lacking. The case study also showed that service providers should plan new service system to allow for effectively capturing user journeys. Capturing the concepts provided by CJML in an graph-based EKG representation was useful to understand the connection of logged events with the user perspective of the service. However, this needs to still be validated in other service domains and for capturing the user experience.

Next steps in the study are to aggregate multiple similar journeys and compare them with the planned journey, investigate if we can leverage a user journey game approach [9] on the data, as well as to integrate user experience into the analysis. The data schema in Fig. 3 already prepares connecting events to user experience about a touchpoint. This would be a prerequisite to enable the analysis, prediction and, ultimately, improvement of the user experience based on event data. However, automatically obtaining high quality data on actual user experience is challenging.

Acknowledgements. This work is part of the *Smart Journey Mining* project, funded by the Research Council of Norway (grant no. 312198).

References

1. Aalst, W.V.D.: Process Mining: Data Science in Action. Springer, Heidelberg (2019)
2. Bernard, G.: Process-Mining-based Customer Journey Analytics. Ph.D. thesis, University of Lausanne (2020)
3. Bernard, G., Andritsos, P.: Cjm-ex: Goal-oriented exploration of customer journey maps using event logs and data analytics. In: BPM (Demos). CEUR Workshop Proceedings, vol. 1920. CEUR-WS.org (2017)
4. Esser, S., Fahland, D.: Multi-dimensional event data in graph databases. J. Data Semant. **10** (2021)
5. Halvorsrud, R., Kvale, K., Følstad, A.: Improving service quality through customer journey analysis. J. Serv. Theory Pract. **26**(6), 840–867 (2016)
6. Halvorsrud, R., Mannhardt, F., Johnsen, E.B., Tarifa, S.L.T.: Smart journey mining for improved service quality. In: SCC, pp. 367–369. IEEE (2021)
7. Halvorsrud, R., Sanchez, O.R., Boletsis, C., Skjuve, M.: Involving users in the development of a modeling language for customer journeys. Softw. Syst. Model. 1–30 (2023)
8. Hansson, M., Angel, K., Mannhardt, F., Kvale, K.: How can a service provider utilize process mining on customer journeys to gain actionable insights for service delivery improvements? In: BPM Industry Forum. CEUR Workshop Proceedings, vol. 3112, pp. 15–26. CEUR-WS.org (2021)
9. Kobialka, P., Mannhardt, F., Tarifa, S.L.T., Johnsen, E.B.: Building user journey games from multi-party event logs. In: Montali, M., Senderovich, A., Weidlich, M. (eds.) ICPM 2022. LNBIP, vol. 468, pp. 71–83. Springer, Cham (2022). https://doi.org/10.1007/978-3-031-27815-0_6
10. Lemon, K.N., Verhoef, P.C.: Understanding customer experience throughout the customer journey. J. Mark. **80**(6), 69–96 (2016)
11. Ostrom, A.L., et al.: Service research priorities: managing and delivering service in turbulent times. J. Serv. Res. **24**(3), 329–353 (2021)
12. Rawson, A., Duncan, E., Jones, C.: The truth about customer experience. Harv. Bus. Rev. **91**(9), 90–98 (2013)
13. SINTEF Digital: Resource site for CJML (2022). https://www.cjml.no
14. Terragni, A., Hassani, M.: Analyzing customer journey with process mining: from discovery to recommendations. In: FiCloud, pp. 224–229. IEEE (2018)

15. van Zelst, S.J., Mannhardt, F., de Leoni, M., Koschmider, A.: Event abstraction in process mining: literature review and taxonomy. Granul. Comput. **6**(3), 719–736 (2020)
16. Zilker, S., Marx, E., Stierle, M., Matzner, M.: Process mining for advanced service analytics - from process efficiency to customer encounter and experience. In: HICSS, pp. 1–10. ScholarSpace (2022)

From Data to Actionable Insights: Utilizing AI and Process Mining in Manufacturing Processes

Urszula Jessen[1,2](✉), Lucas Schroth[2], and Mario Mühllechner[3]

[1] Eindhoven University of Technology, Eindhoven, The Netherlands
u.a.jessen@tue.nl
[2] process science GmbH & Co. KG, Hamburg, Germany
ls@process-science.com
[3] Liebherr Group, Bulle, Germany
Mario.Muehllechner@liebherr.com

Abstract. This paper presents a case study on analyzing process of large manufacturing company specializing in construction machinery. We demonstrate the case of aligning strategic business objectives with process improvement, utilizing advanced data visualization and analytics tools. By integrating domain knowledge, process mining, and machine learning, we analyzed production data to identify inefficiencies and bottlenecks. AI integration, specifically ChatGPT, allowed users to interact with the data through natural language queries, offering detailed insights and identifying critical paths. Our approach addresses data overload and skill gaps in process mining, showing how non-experts can leverage advanced tools for rapid insights.

Keywords: process mining · actionable insights · process optimization · manufacturing processes

1 Introduction

In the contemporary business landscape, companies strive to enhance operations to gain a competitive edge, improve efficiency, and meet customer demands. Central to these efforts is process analysis, involving the systematic examination, modeling, and improvement of organizational processes to identify inefficiencies, highlight areas for enhancement, and ensure regulatory compliance [1].

Despite advancements in commercial software tools, the anticipated value of process mining projects is often not fully realized, leading to user disillusionment [5]. Enhancing the clarity and usability of outputs for non-expert stakeholders, especially within complex industrial environments, remains a significant challenge [15]. This underscores the need for 'analytics translators'-professionals adept at interpreting data to identify business challenges and converting these needs into data-driven inquiries [7].

This case study introduces a novel scoring method to help non-expert users understand inefficiencies in analyzed processes and rapidly identify root causes.

C. Di Ciccio et al. (Eds.): BPM 2024, LNBIP 527, pp. 462–471, 2024.
https://doi.org/10.1007/978-3-031-70445-1_34

The method employs summarization and approximation techniques to generate an overall score for process health, enabling companies to prioritize areas for in-depth analysis and resolve issues. To enhance visual clarity, industrial standards such as the IBCS Standard are utilized, with AI-enhanced support added to simplify root cause identification. This research was conducted with data science specialists from the Liebherr Group.

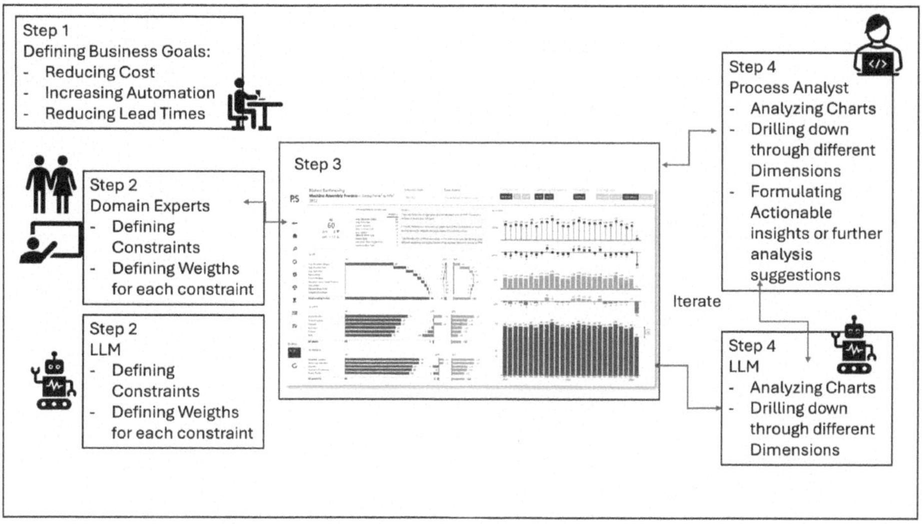

Fig. 1. Process Analysis in Case Study.

The Liebherr Group, a German-Swiss multinational company, is known for its extensive range of construction machinery and high-quality products across various sectors. Founded in 1949, the company includes over 130 companies organized into 11 divisions. With its headquarters in Bulle, Switzerland, Liebherr maintains a decentralized structure, allowing each division to operate autonomously while ensuring consistent quality and innovation [9].

This case study aims to illustrate the analysis and insight generation for production processes using advanced data visualization and analytics tools. The primary goals are:

- To demonstrate the application of analytical tools and standards in creating actionable insights.
- To showcase the integration of machine learning and process mining techniques for operational improvement.
- To explore the benefits of leveraging artificial intelligence, particularly large language models, in identifying critical paths and anomalies in production processes.

A distinctive feature of this case study is the integration of an AI conversational agent, which assists users in delving deeper into the data. Users can query the system for specific insights, and the AI generates detailed explanations, identifies anomalies, and suggests critical paths in the production process. This interactive approach facilitates a deeper understanding of the data and empowers users to make data-driven decisions more efficiently.

2 Problem Statement

Manufacturing organizations in the competitive construction machinery and equipment industry must continuously innovate and optimize their production processes. This case study explores specific challenges such as process mining complexities, data overload, and the skill gap in data analysis.

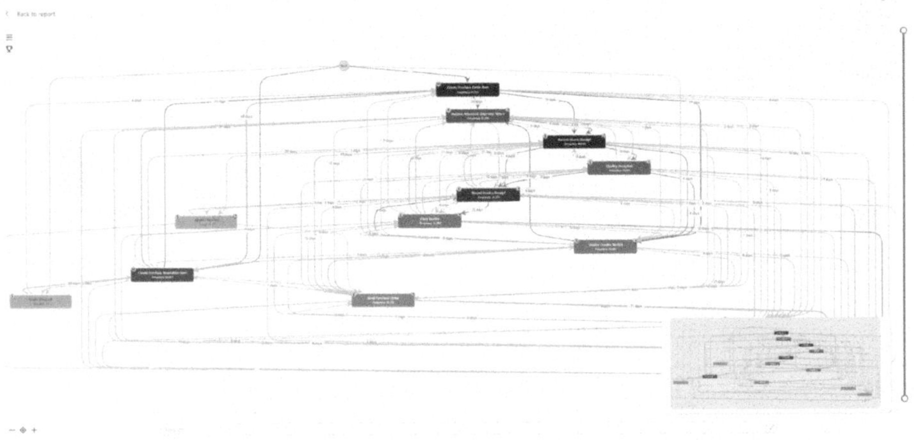

Fig. 2. Process model filtered on main activities and variants.

As the manufacturing industry expands and faces diverse market incentives, enterprises must maintain flexibility, uphold high quality standards, enhance productivity, and ensure sustainability [6]. Performance measurement systems, including Key Performance Indicators (KPIs), are implemented to evaluate and improve operational efficiency and effectiveness [17]. However, the vast volume and variety of data generated present significant challenges in aligning project objectives.

Traditional process mining algorithms often yield overly complex models, making interpretation difficult for non-experts [16]. Additionally, the proliferation of IoT devices and advanced monitoring systems results in data overload, which is challenging to interpret without sophisticated tools and expertise. Figure 2 shows the discovered process model, already filtered for main activities and variants. The project stakeholder and process analysts realised that even such a generalized model was hard to interpret for analytics teams. Therefore based on these findings, the project aimed to:

- **Enable Non-Experts:** Develop tools and methodologies that allow non-process mining experts to uncover anomalies and gain insights from process data and event logs.
- **Leverage AI for Better Insights:** Utilize Large Language Models (LLMs) like ChatGPT to enable complex analyses using natural language queries, bridging the gap between data complexity and user capability.

By enabling non-experts and leveraging cutting-edge technologies, this case study aims to improve efficiency, quality, and decision-making in production environments.

3 Research/Methodological Background

3.1 Related Literature

Process mining research has focused on supporting industry in analyzing real-life processes through the development of various algorithms and methods. Besides Procedural Process Mining Algorithms such as Fuzzy Mining and clustering techniques further methods such as Declarative Process Mining has been developed [13]. Despite these advancements, significant challenges remain unresolved. The Process Mining Manifesto [15] outlines 11 broad challenges, including handling complex event logs and improving the understandability of outputs for non-experts. Numerous studies have identified challenges such as managing intricate business processes [4], securing stakeholder buy-in, and aligning expectations [3] as critical for effective process mining adoption. Similarly, R'Bigui and Cho [12] and Munoz-Gama et al. [10] note that many of these challenges persist. Additionally, Thiede et al. emphasize the need for further research on applying process mining in complex scenarios [14].

3.2 Methodology

The methodology employed in this project was based on the general objectives of the company and involved a structured approach designed to evaluate and optimize process efficiency from multiple perspectives. It began with the creation of a Process Norm, which defines optimal process behavior based on specific business goals. This norm was established through input from domain experts and process analysts, who assigned importance to various process activities to ensure alignment with organizational objectives [2]. Figure 3 shows the results of multiple workshops with process stakeholders, which were integral in developing these norms.

The project also addressed the challenge of collaborating with non-process experts by utilizing the International Business Communication Standards (IBCS) in conjunction with Zebra BI visuals for dashboard creation. IBCS provides a standardized framework for data presentation, ensuring consistency and clarity, which helps process analysts quickly grasp key insights and make

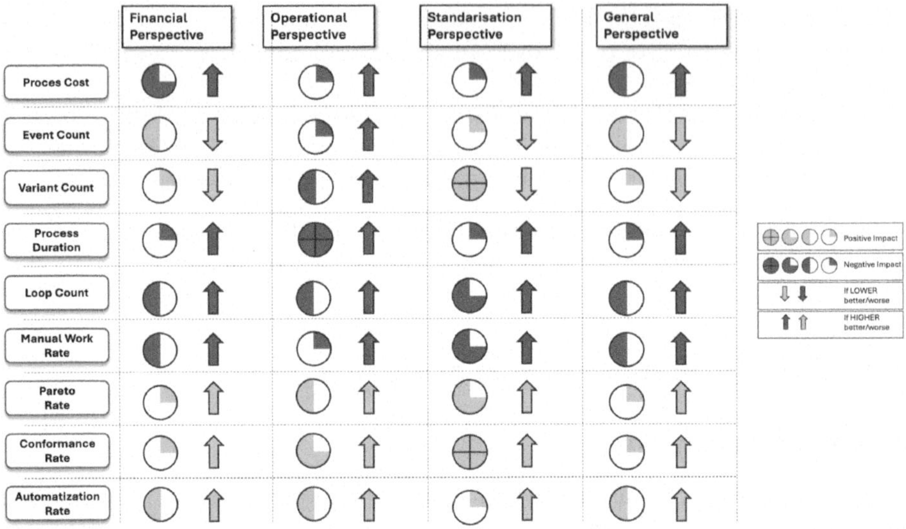

Fig. 3. Scoring Perspectives based on different features.

informed decisions without wading through convoluted data [11]. This standardization unifies symbols, colors, and terms, creating a common language for business communication and making it easier to identify patterns and anomalies in manufacturing data [8].

In addition to establishing the Process Norm and implementing IBCS, the methodology included the use of advanced data analytics and visualization tools. These tools were employed to analyze event logs and other relevant data, enabling the identification of inefficiencies and bottlenecks within the manufacturing processes. The scoring method developed as part of this project involved summarization and approximation techniques to generate an overall score for process health. This score helped prioritize areas for in-depth analysis and identify potential root causes for process inefficiencies.

To ensure the practical application of these insights, process stakeholders were actively involved in investigating identified issues. They engaged in creating hypotheses about root causes and proposing mitigation strategies. This collaborative approach ensured that the solutions were grounded in real-world contexts and aligned with the company's strategic goals.

In summary, this methodology integrated expert input, standardized data presentation, advanced analytics, and active stakeholder engagement to optimize manufacturing processes and facilitate continuous improvement.

4 Case Study and Lessons Learned

To implement the methodology within the context of production processes, the following steps were undertaken (see Fig. 1):

1. **Identifying Business Goals:** Goals were categorized into financial, operational, standardization, and general process management perspectives. This step involved collaboration with stakeholders, particularly from management, to align the business objectives of the analysis with the company's strategic goals.
2. **Establishing the Process Norm:** Optimal process behavior was defined, recognizing different goals for various stakeholders. This step was executed with domain experts who have expertise in different aspects of the analyzed process.
3. **Analyzing Process Data:** Advanced process mining algorithms were utilized to analyze event logs and process data.
4. **Detecting Anomalies:** Deviations from the Process Norm were identified to uncover inefficiencies and areas for improvement. This step was executed both with and without support of LLM-Models. As a result, the main areas for improvement were identified and described.

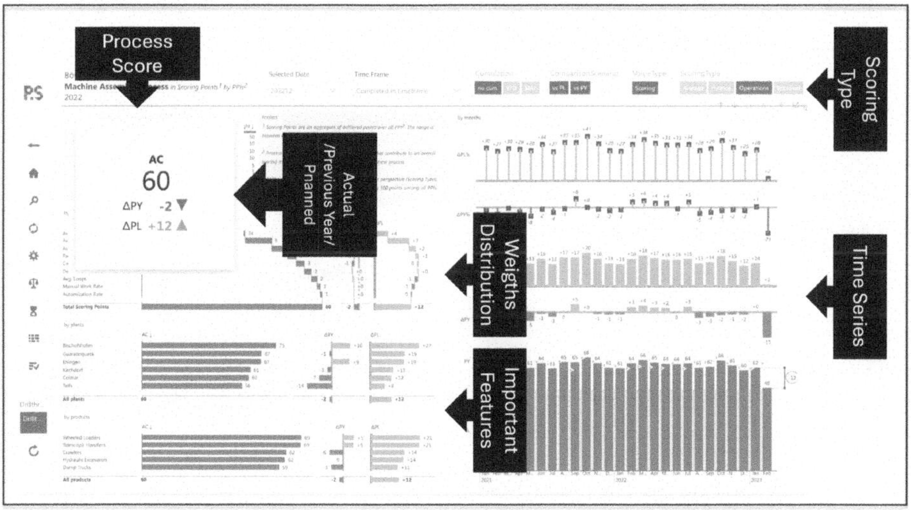

Fig. 4. Scoring Perspectives based on different features.

Figure 3 depicts an example of the defined Process Norm. The Process Norm describes various features of the process such as Process Cost, Variant Count, or Automation Rate and defines the Process Norm for different perspectives. For example, in the Financial Perspective, the impact of Process Cost is very high, and with the increase in cost, the health of the process for this perspective decreases. Conversely, in the Operational Perspective, Process Duration plays a significant role.

Based on the Process Norm, the process analyst can structure the analysis and add the most important dimensions to the dashboards. Figure 4 depicts the most important parts of the dashboard. One of the challenges that process analysts faced with non-process experts is the difficulty of establishing a clear message about whether the process is performing well. The clear process score, with actual value and comparison against the previous year and planned value, established common ground for discussing the processes from different points of view. The weights distribution shows the process analyst the factors impacting each perspective the most. Scoring type allows for changing the perspective depending on the business goal that should be observed. Important Features and Time Series Data enable further dicing and slicing of data to look for hidden root causes of problems.

Figure 4 shows the process from an operational perspective. The main information shown in the left corner is the overall process health, which in this case is 60%. The analyst can further analyze data, observing problems with operational processes in different plants and by particular products (Fig. 5).

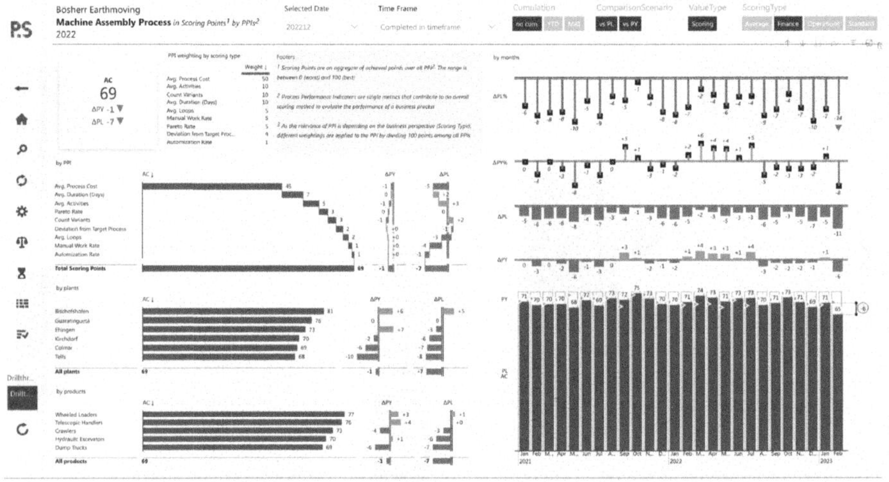

Fig. 5. Process Dashboard viewed from a financial perspective.

Similarly, the analyst can observe the process from a financial perspective. The financial goal focuses on cost reduction and profitability enhancement, while the quality goal aims at minimizing defects and improving customer satisfaction. The operational perspective prioritizes maximizing efficiency and throughput while ensuring just-in-time production. The business process management perspective concentrates on standardizing, streamlining, and automating processes to improve overall organizational performance.

To support users in finding intricate patterns in data, Large Language Model (LLM) support was added. Users can interact with the data through natural language queries, making complex analyses more accessible and intuitive. This approach democratizes data analytics, enabling non-experts to uncover insights and facilitating a more collaborative and interactive process of optimization. Figure 6 depicts an example of user-LLM conversation. Users are able not only to ask questions in natural language but also query the data with the use of questions and commands. By enabling users to query data using natural language, AI technologies bridge the gap between data complexity and user capability. This interaction allows for:

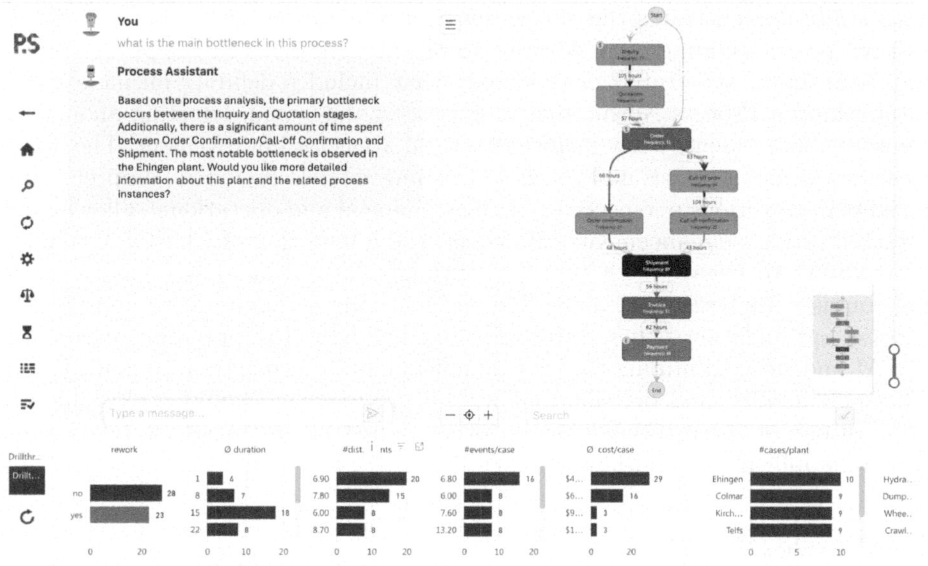

Fig. 6. Process Dashboard with AI-driven query support.

- **Enhanced Accessibility:** Non-experts can engage with data without needing deep technical knowledge.
- **Rapid Insights:** Users can quickly obtain detailed explanations, identify anomalies, and understand critical paths in the production process.
- **Collaborative Analysis:** Facilitates a more interactive approach to data analysis, fostering collaboration among stakeholders.

By combining the process scoring methodology with multi-perspective process mining, a holistic analysis of the production processes was achieved. This approach considers multiple perspectives and identifies potential areas for improvement across the entire process flow.

4.1 Lesson Learned

This project demonstrates that clearly stating expected results and focusing on the most important aspects, using Pareto principles to avoid getting lost in details, can result in successful process mining project implementation. The integration of non-process-mining experts, who understand the process from different perspectives, is essential and can determine the success or failure of such projects.

5 Conclusion

This study demonstrates the effectiveness of integrating advanced data visualization, process mining, and AI tools to optimize manufacturing processes. By implementing a structured methodology that includes defining business goals, establishing a Process Norm, analyzing process data, and detecting anomalies, we were able to identify key inefficiencies and areas for improvement. The use of clear, transparent scores and machine learning models provided comprehensive insights from various perspectives, such as financial and operational, allowing for targeted process enhancements. Moreover, the integration of ChatGPT enabled non-experts to engage with the data through natural language queries, making complex analyses more accessible and fostering a collaborative approach to process optimization. These advancements highlight the potential of combining AI and process mining to drive significant improvements in manufacturing efficiency, quality, and decision-making. Future research should explore further applications of these technologies in other industrial settings to validate their broader applicability and impact.

References

1. Wil van der Aalst. Process mining: data science in action. *(No Title)* (2016)
2. Adesola, S., Baines, T.: Developing and evaluating a methodology for business process improvement. Bus. Process. Manag. J. **11**(1), 37–46 (2005)
3. Dakic, D., Sladojevic, S., Lolic, T., Stefanovic, D.: Process mining possibilities and challenges: a case study. In 2019 IEEE 17th International Symposium on Intelligent Systems and Informatics (sisy), pp. 000161–000166. IEEE (2019)
4. Eggert, M., Dyong, J.: Applying process mining in small and medium sized it enterprises–challenges and guidelines, pp. 125–142 (2022)
5. Emamjome, F., Andrews, R., ter Hofstede, A.H.M.: A case study lens on process mining in practice. In: Panetto, H., Debruyne, C., Hepp, M., Lewis, D., Ardagna, C.A., Meersman, R. (eds.) OTM 2019. LNCS, vol. 11877, pp. 127–145. Springer, Cham (2019). https://doi.org/10.1007/978-3-030-33246-4_8
6. Franceschini, F., Galetto, M., Maisano, D.: Management by Measurement: Designing Key Indicators and Performance Measurement Systems. Springer, Heidelberg (2007)
7. Herring, L., Mayhew, H., Midha, A., Puri, A.: How to train someone to translate business problems into analytics questions. Harvard Business Review Digital Articles (2019). Accessed 11 Feb 2019
8. Hichert, R., Faisst, J.: Notation standards in business communication and their practical benefits. *HICHERT® IBCS WHITE PAPER. Hichert Partner*, pp. 15 (2014)

9. Liebherr Group. Liebherr group official website (2024). Accessed 31 May 2024
10. Munoz-Gama, J., et al.: Process mining for healthcare: characteristics and challenges. J. Biomed. Inform. **127**, 103994 (2022)
11. Nogueira, R.M.: Improving data visualization reports for a financial services company (2022)
12. Hind R'bigui and Chiwoon Cho: The state-of-the-art of business process mining challenges. Int. J. Bus. Process. Integr. Manag. **8**(4), 285–303 (2017)
13. Stefanini, A., Aloini, D., Benevento, E., Dulmin, R., Mininno, V.: A process mining methodology for modeling unstructured processes. Knowl. Process. Manag. **27**(4), 294–310 (2020)
14. Thiede, M., Fuerstenau, D., Bezerra Barquet, A.P.: How is process mining technology used by organizations? a systematic literature review of empirical studies, vol. 24, pp. 900–922. Emerald Publishing Limited (2018)
15. van der Aalst, W., et al.: Process mining manifesto. In: Daniel, F., Barkaoui, K., Dustdar, S. (eds.) BPM 2011. LNBIP, vol. 99, pp. 169–194. Springer, Heidelberg (2012). https://doi.org/10.1007/978-3-642-28108-2_19
16. Van Der Aalst, W.M.P.: Process mining: discovering and improving spaghetti and lasagna processes. In: 2011 IEEE Symposium on Computational Intelligence and Data Mining (CIDM), pp. 1–7. IEEE (2011)
17. Zhu, L., Johnsson, C., Mejvik, J., Varisco, M., Schiraldi, M.: Key performance indicators for manufacturing operations management in the process industry. In: 2017 IEEE International Conference on Industrial Engineering and Engineering Management (IEEM), pp. 969–973. IEEE (2017)

LLM4PM: A Case Study on Using Large Language Models for Process Modeling in Enterprise Organizations

Clara Ziche and Giovanni Apruzzese(✉)

University of Liechtenstein, Vaduz, Liechtenstein
giovanni.apruzzese@uni.li

Abstract. We investigate the potential of using Large Language Models (LLM) to support process model creation in organizational contexts. Specifically, we carry out a case study wherein we develop and test an LLM-based chatbot, `PRODIGY` (PROcess moDellIng Guidance for You), in a multinational company, the Hilti Group. We are particularly interested in understanding how LLM can aid (human) modellers in creating process flow diagrams. To this purpose, we first conduct a preliminary user study (n=10) with professional process modellers from Hilti, inquiring for various pain-points they encounter in their daily routines. Then, we use their responses to design and implement `PRODIGY`. Finally, we evaluate `PRODIGY` by letting our user study's participants use `PRODIGY`, and then ask for their opinion on the pros and cons of `PRODIGY`. We coalesce our results in actionable takeaways. Through our research, we showcase the first practical application of LLM for process modelling in the real world, shedding light on how industries can leverage LLM to enhance their Business Process Management activities.

1 Introduction

Organizations perform business processes to deliver value-adding outcomes to their customers. Hence, Business Process Management (BPM) capabilities, such as *process modeling*, are a pivotal task in modern enterprises [3]. However, despite decades of efforts [7], process modeling still remains a costly activity due to, e.g., the difficulty of providing clear, up-to-date and easy-to-retrieve *documentation* [3] to those tasked to carry out such activities—the process modelers.

Inspired by recent developments in artificial intelligence (AI), such as large language models (LLM), researchers have proposed various techniques that can facilitate BPM-related tasks (e.g., [6]). Indeed, LLM can elaborate large collections of documents. Hence, by receiving an input from a given user, LLM can quickly produce an output that *(i)* accounts for existing documentation, while simultaneously *(ii)* answering the request of the user—i.e., a human. Yet, we found no evidence of *practical* applications of LLM for BPM in real contexts and, in particular, for process modeling. Hence, there is a need to investigate the effectiveness of such automation in industry [15]. Here, we tackle this challenge and showcase how a large enterprise, **Hilti**, can benefit from a LLM-powered chatbot—which we developed ad-hoc for Hilti—for BPM.

C. Di Ciccio et al. (Eds.): BPM 2024, LNBIP 527, pp. 472–483, 2024.
https://doi.org/10.1007/978-3-031-70445-1_35

CONTRIBUTIONS. We present a (the first) real-world case study showcasing the application of LLM for Process Modeling in operational contexts. Specifically, to advance the state of the art on BPM, we:

- describe the *problems faced by the considered organization*, Hilti, providing evidence of the necessities of modern enterprises (Sect. 2);
- carry out a *requirement analysis by conducting interviews* with Hilti's employees, shedding light on the pain-points of professional process modelers (Sect. 3.1);
- use our interviews as a scaffold to *develop an original LLM-based chatbot*, PRODIGY (Fig. 1), designed to support Hilti's process modelers (Sect. 3.2);
- evaluate the ability of PRODIGY to *generate practical value* by *(i)* having Hilti's employees use PRODIGY and *(ii)* collecting and analysing their feedback (Sect. 3.3).

Our results (Sect. 4) show that PRODIGY is generally well-received, and identify room for improvement. We also derive lessons learned that future work can use to drive practical deployment of LLM-based technologies (Sect. 5).

Industrial Secret and Ethics: In this paper, we describe some elements pertaining to the internal processes of Hilti: we have been granted permission to share such information. Furthermore, we carry out our user studies ethically: our institutions are aware of this research, our participants have been informed of their rights and of the purpose of their contributions, and we have their consent to post them.

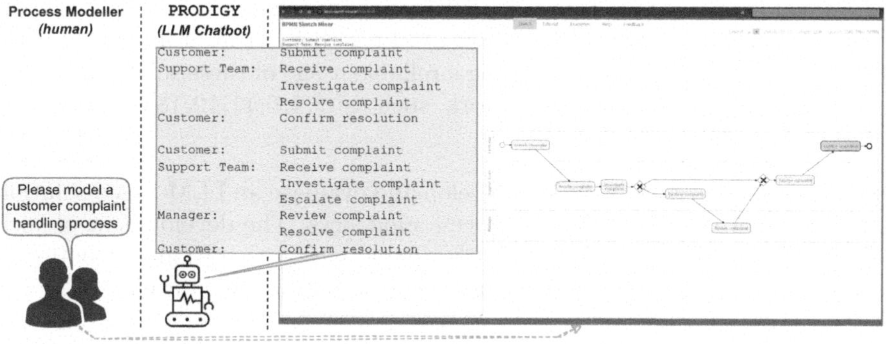

Fig. 1. An exemplary usage of PRODIGY – Our LLM-powered chatbot can fulfill various BPM-related tasks. Its most appreciated functionality is generating an output for a complete process model (if copy-pasted into the open-source tool BPMN Sketch Miner [10])

2 Organizational Context and Problem Statement

Our case organization, Hilti Group, is a multinational company that was founded in 1941 in Schaan, Liechtenstein. It is a world market leader in fastening and demolition technology for construction professionals and provides tools, technologies, software and services to the global construction industry. In 2023, Hilti's workforce consists of about 33.000 employees in more than 120 countries, making it a highly diverse, distributed organization that operates in complex and competitive markets all over the world. The size, complexity and business model of Hilti make it an ideal use case for testing the capabilities of LLM for process modeling: BPM is essential to ensure cooperation and consistent outcomes within Hilti's ecosystem; furthermore, it is crucial for Hilti to optimize customer-facing processes. Hence, a smooth process modelling is pivotal for Hilti.

Challenge. Hilti has an extensive and heterogeneous documentation landscape which adds to the intrinsically complex nature of process modeling. Hilti's employees spend abundant time searching through such documentation (our interviews revealed an average of ~40 min of search *before* modelling a process). Hence, to improve the productivity of their process modellers, and to actively explore innovative technologies, Hilti is interested in novel solutions that facilitate the routines of their employees.

Technological Gap. LLM-based solutions are common (in 2024). However, existing techniques cannot be applied to Hilti's use case. This is because of the *confidential nature* of Hilti's documents: publicly available models (e.g., ChatGPT) should not be able to access Hilti's data to shape their responses; furthermore, even *interacting* with certain LLM (or their APIs) from within Hilti's networks triggers warnings, preventing a reliable usage of these solutions—which are leveraged also by renown prior work, such as [2,4,6,11,12,18].

> **OUR GOAL.** We seek to design, develop and evaluate an LLM-based solution that facilitates the job of Hilti's process modellers. The development of such a solution should be driven by Hilti's distinctive organizational's context—including its employee's viewpoint, and its existing documentation.

3 Research and Methods

Inspired by Peffers et al. [14], we followed a Design Science Research (DSR) process consisting of four phases depicted in Fig. 2.[1] DSR is appropriate given

[1] **Background:** DSR is a methodology that focuses on creating and evaluating artifacts to solve complex problems. Such procedure is rooted on the coming together of people, organizations and technology, with the ultimate intention of "extending the boundaries of human and organizational capabilities" [8].

our goal of examining LLMs for process modelling in organizations, as DSR emphasizes the creation of innovative solutions (in our case, PRODIGY) while also considering the context in which these solutions will be applied.

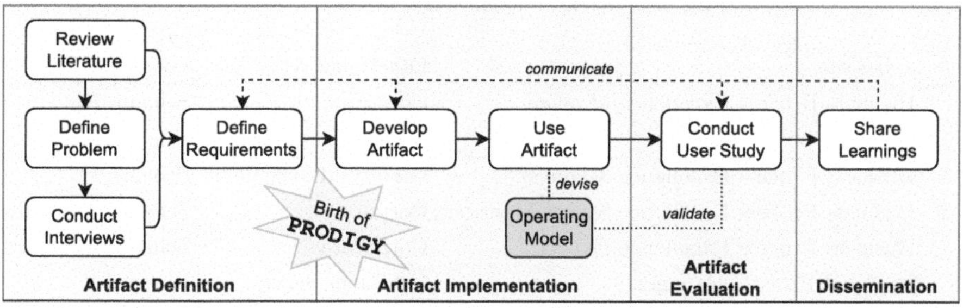

Fig. 2. Method. We rely on design science research to design, develop, and deploy our artifact. During the implementation of PRODIGY, we also devise an "operating model" (which we validate in the evaluation of PRODIGY) through which we explain how PRODIGY should be used in real organizations.

3.1 Artifact Definition

As a preliminary step, we carried out a systematic literature review [20] which we used as a foundation to investigate the state of the art and define the scope of our project (see Sect. 2). Then, we carried out structured interviews [5] meant to identify pain-points and desiderata by professional process modelers[2] working for Hilti. We found an agreement with 10 employees, summarised in Table 1. The complete questionnaire is provided in our repository [1]. Among the most relevant questions, we ask: "what *challenges* do you experience when modelling processes?", "how helpful is *existing documentation* when you model processes?" and "what would you like to see in a new AI artifact that supports process modelling?"; we also provide a list of functionalities for the AI artifact and ask to rate them on a 1–5 scale, as well as potential concerns. Finally, we inquire about the time spent looking for, and reviewing, existing documentation.

[2] This is in stark contrast with a closely related work that does not carry out any user study [12], thereby preventing to fully capture the organizational context.

Table 1. Overview of process modellers. Our participants pertain to various geographical locations of Hilti, and have diverse backgrounds. *[Demographics]* Each participant has 5–25 years of experience in BPM, and they are within 26–60 years of age. Seven hold a MSc. degree. The male:female ratio is 6:4. They all have "above average" or "advanced" computer knowledge, and all have a basic understanding of LLM. Five perform process modelling activities at least weekly.

#	Job Title	Functional Area	Location
1	Business Process Excellence Manager	Corporate	Schaan, FL
2	Global Process Manager	Communications	Schaan, FL
3	Business Process Excellence Manager	Quality Management	Schaan, FL
4	Business Process Excellence Senior Manager	Corporate	Schaan, FL
5	Business Process Excellence Expert	Corporate	Schaan, FL
6	Regional Process Manager	Customer Service	Plano, US
7	Regional Process Manager	Logistics	Kaufering, GER
8	Business Process Excellence Lead	Corporate	Schaan, FL
9	Global Process Manager	Repair	Schaan, FL
10	Global Process Manager	Repair	Schaan, FL

3.2 Artifact Implementation

We use the results of our interviews alongside those of our investigation of the state of the art to define the requirements of our technical artifact, i.e., the LLM-based chatbot `PRODIGY`. To develop `PRODIGY`, we rely on Botpress, a platform to build custom AI chatbots powered by GPT-based LLMs; for our prototype version of `PRODIGY`, we used GPT-3.5 Turbo, which we found provided satisfactory performance while also requiring less resources to generate an output.

A crucial aspect of `PRODIGY` is its reliance on the BPMN Sketch Miner tool [10]. The syntax for this tool is entirely text-based, human-readable and light in terms of token consumption, making it appropriate for our case study. Therefore, we use few-shot prompting to teach `PRODIGY` to provide an output that matches the format expected by BPMN Sketch Miner. This output serves as the input for the model generation and transformation pipeline of BPMN Sketch Miner [9]. Such a design choice enables users of `PRODIGY` to directly paste the AI outputs into the online tool and get their model visualized (see Fig. 1).

Furthermore, we have leveraged retrieval-augmented generation (RAG) [13] to embed Hilti's documentation into `PRODIGY`. Such documentation included: process descriptions from Hilti's internal documentation repository (anonymised); and information about Hilti's process management, and how to model processes at Hilti (taken verbatim from the learning platform for Hilti's process modellers). These procedures enabled us to instill some knowledge about Hilti's processes in `PRODIGY`—a functionality that was heavily endorsed by our interviewees.

Demonstration: We recorded a video [1] showing the functionalities of `PRODIGY`.

3.3 Artifact Evaluation

We conducted a user study with our artifact and process modellers. Our aim was to answer evaluative questions on the quality of PRODIGY for Hilti.

First, the process modellers tested all functionalities of PRODIGY by creating custom prompts, with the intention of simulating their routine tasks. Their inputs and the corresponding AI-generated outputs are fully observable in our repository [1]. Then, we carried out semi-structured interviews during which the participants answered 28 questions. Among these, we ask to give an 1–5 rating to the statement "Using PRODIGY would make it easier for me to do process modeling tasks." The complete questionnaire is provided in our repository [1].

3.4 Dissemination and Communication of the Results

To conclude our DSR process, we formalized our learnings and made them accessible to interested parties. We documented our observations, analyzed our findings, identified lessons learned, stated limitations, and recommended directions for future work. We shared our learnings within Hilti Group and the wider BPM community in academia and practice—some companies reached out to us and expressed their interest about the development process of PRODIGY.

4 Key Findings and Lessons Learned

We first summarise the major results of our user studies, and then outline our proposed "operating model" for our developed LLM-based chatbot, PRODIGY.

Confidentiality Statement: To protect the privacy of the participants to our user studies, we cannot reveal the full transcript of their interviews. However, we are able to answer questions about their generic viewpoint on some specific issues.

4.1 Preliminary Interviews: What Do Hilti Process Modellers Want?

These open interviews lasted for 60 min, and the results shed light on the pain-points and desiderata of our participants. We found that, before modelling a process, 60% spend between 15–60 min *searching* for documentation; and also 60% spend between 5–60 min to *review* such documentation. As a matter of fact, 90% state that it is "extremely important" that an LLM-based chatbot has access to Hilti's documentation; however, we also found that, on a 1–10 rating (low to high) scale, the average usefullness of current Hilti's documentation is 6.7—indicating helpfulness, but with huge margins for improvement. Nonetheless, with respect to AI-related concerns, some stated that "humans may misinterpret the AI's outputs" or "AI may negatively impact collaboration with colleagues" or even about accountability ("the mindset that [the machine] does everything and we no longer have to worry about it is dangerous").

TAKEAWAYS. After analysing all our responses, we identified two design objectives which we used as basis to develop our LLM-based chatbot, PRODIGY.

- The chatbot should *support process modellers* in creating BPMN models. In doing so, the chatbot should hint at the larger picture, i.e., emphasize and guide in purpose, usage, and value creation of the resulting models.
- The chatbot should be able to access and utilize existing documentation, and hence *be aware of organizational specifics*. Such knowledge should drive the formulation of the output, which will be tailored to the organization.

The name PRODIGY stands for "PROcess moDellIng Guidance for You".

4.2 Evaluation: What Do Hilti Process Modellers Say About PRODIGY?

After letting our process modellers use PRODIGY, we collected their feedback via 90-min long semi-structured interviews; one participant to the preliminary interviews did not provide feedback since they were not available, so we obtained responses from nine employees. **The general opinion was positive**. Five of our participants asserted that they would use PRODIGY on a daily or weekly basis (i.e., whenever they have to carry out process-modeling duties). Moreover, six participants asserted that PRODIGY *would speed-up* their tasks (three remained neutral), and eight believe that PRODIGY *makes their tasks easier* (one remained neutral). Finally, we report in Fig. 3 the participants' perception on the functionalities we integrated in PRODIGY, showing great appreciation.

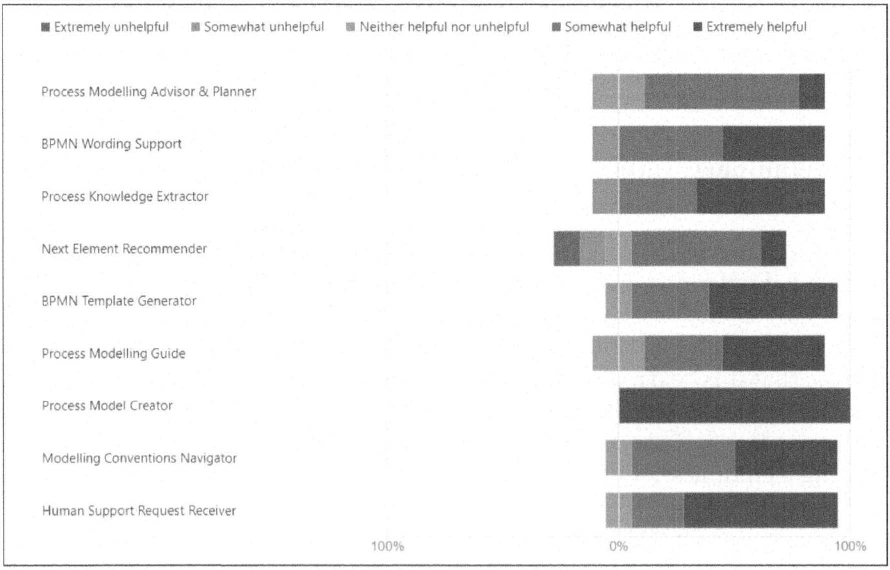

Fig. 3. Helpfulness of PRODIGY's functionalities. According to our participants, most of our implemented features are helpful—especially for *creating process models* and *supporting human requests*.

4.3 Operating Model: How Should PRODIGY Be Used in Practice?

During our implementation, we devised an *operating model* that describes how PRODIGY should be leveraged by real organizations; we have further refined our model (shown in Fig. 4) after receiving the feedback by our interviewees.

At a high-level, our model emphasizes continuous improvement through regular evaluations, feedback, and updates. To this end, our model delineates the interaction between a *Governance Team* (i.e., the set of employees within a company that oversee the development and maintenance of PRODIGY) and a *Process Modeller* (i.e., the end-users of PRODIGY). These two actors work collaboratively to ensure that the system performs well over time. For instance, the Process Modeller should be familiar with existing documentation and with the specific process, and scrutinize the response of PRODIGY accordingly; they should also be willing to provide feedback (collected in a dedicated repository) and receive guidance from the Governance Team—who must, in turn, define clear performance indicators for PRODIGY and periodically review the performance of prodigy (e.g., by analysing logs [19]) and apply updates if needed; as well as ensure that existing documentation is properly embedded in PRODIGY (in a timely manner).

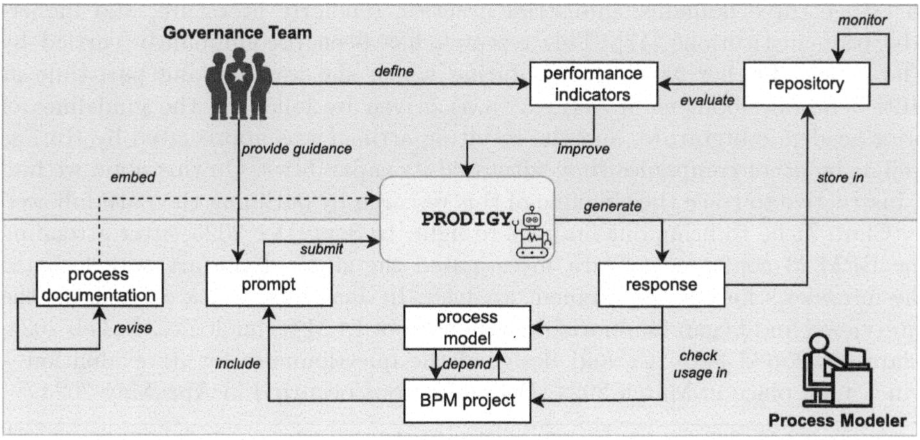

Fig. 4. Operating model of PRODIGY – We visualize the interactions between the *Governance Team* (e.g., developers and managers) of a given organization with the *Process Modeller* (i.e., the end-users of PRODIGY) that ensure a smooth operation of PRODIGY for real-world deployments.

5 Significance and Relevance in Research and Practice

Besides our key findings we underscore three orthogonal aspects of our research.

The Perspective of Process Modellers in Organizations. We coalesce the responses—not pertaining to AI—of our preliminary interviews, and derive an *original framework* representing dynamics of process modellers' issues at Hilti. This is instructive because, during our literature analysis, we found some works mentioning pitfalls of process modeling (e.g., [16]) but without accounting for context. Our framework (displayed in Fig. 5, and described in the caption of Fig. 5) attempts to rectify this shortcoming, providing guidance for future work.

Evaluating LLM-/AI-Based Solutions. Upon further analysing the results of our evaluation interview, we have found that the reception of PRODIGY by our process modellers was highly dependant on their expectations and overall attitude towards AI and IT innovation. Indeed, some participants had "lower expectations" and provided prompts that were "more aligned" to the expected input of PRODIGY—and these participants rated PRODIGY more positively. In contrast, participants who were expecting that PRODIGY would "do their work for them" by issuing a single (and typically poorly phrased and/or unclear) prompt were more skeptical of PRODIGY's helpfulness. These results underscore the importance of *(i)* accounting for each end-user's expectations while evaluating the performance of operational AI-based solutions; as well as *(ii)* educating end-users on the potential (and limitations) of AI-based tools.

The Role Played by Higher Education. With this paper, we (also) seek to bridge three domains: industrial practice, scholarly literature, and higher-education institutions [17]. This research has been predominantly carried by Clara Ziche for her MSc. thesis, during which she was working part-time at Hilti. The development of PRODIGY was driven by following the guidelines of prior academic literature, and the resulting artifact was appreciated by Hilti as well as by other companies that witnessed its capabilities. On this note, we find it instructive to trace the timeline of this research by outlining the path followed by Clara Ziche to bring our findings to light. In Sept–Dec 2023, after attending the BPM'23 conference, Clara investigated the state of the art and designed the interviews for the requirement analysis. In Jan 2024, Clara carried out the interviews, and began familiarizing with current LLM technologies. In Feb 2024, Clara developed PRODIGY and designed the questionnaire for its evaluation—which took place in March 2024. Disseminations occurred in Apr–May 2024.

Technical feasibility. After having collected the input from experts, the development of PRODIGY took only two weeks *from a MSc. student in Information Systems.*

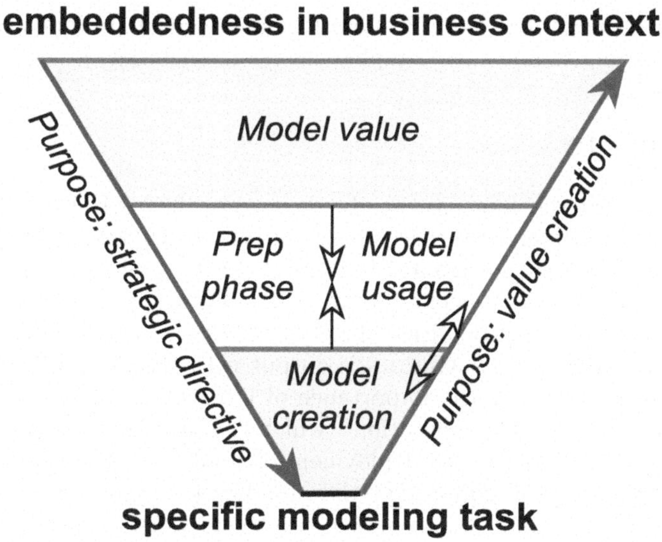

Fig. 5. Issues of process modellers. Our framework has three levels with two-way transitions between each level—each having its own set of issues. *Model value→Model creation:* the organization must communicate a clear strategic directive for process model creation, balancing the cost/benefit of model creation and usage [issue: during the "Prep phase", process modellers find existing communication to lack clarity, leading to time waste and unproductive discussions among various stakeholders]. *Model creation→Model value:* While creating the model, process modelers should have a clear vision of "who and how" is going to use the model (which is what leads to the model becoming valuable) [issue: lack of clarity and/or poor documentation may lead to process models representing "standalone exercises" which do not bring any value to the company.]

6 Discussion: Scope and Limitations

We showcased an exemplary application of an LLM-based chatbot that can assist process modellers in a large organization, Hilti. In doing so, we have carried out a twofold user study with 10 employees of Hilti, and developed an original artifact, PRODIGY. Our research has a number of limitations. For instance, we do not claim that our findings can apply to other organizations—irrespective of their similarity to Hilti. Moreover, we cannot claim that even our own findings can apply to the entirety of Hilti: The participants of our user study are mostly based in Liechtenstein, and therefore cover the global headquarters perspective rather than regional and local perspectives. Furthermore, PRODIGY uses GPT-3.5 Turbo (which is not privacy-compliant), and it relies on BPMN Sketch Miner: if such a tool is taken down, the output of PRODIGY may lose its immediate usefulness. Finally, even our own participants have pointed out some shortcomings of PRODIGY, such as a poor "knowledge" of Hilti's documentation. Such a result, however, was expected: the documents that PRODIGY has access to (with RAG) are just a drop in the deluge of files and logs included in Hilti's databases (and we, as researchers, do not have complete access to such data).

7 Conclusions

We have presented the first case study showcasing how LLM can be used for process modeling in large enterprises—specifically, Hilti Group. We follow DSR guidelines and develop an original LLM-based chatbot, PRODIGY, which we test with professional process modellers from Hilti. Our findings revealed that end-users appreciate the functionalities of PRODIGY. However, concerns were raised about the poor alignment of PRODIGY's output with Hilti's specifications. Such a shortcoming underscores the importance of integrating LLM-based solutions with the organization's documentation—which is a task outside the responsibilities of process modellers. Hence, deployment of similar solutions in real contexts should be done with the support of the organization's governance team: it is unrealistic to expect that "off the shelf" solutions work properly to drive the process modeling routines of complex and large organizations (see Fig. 6).

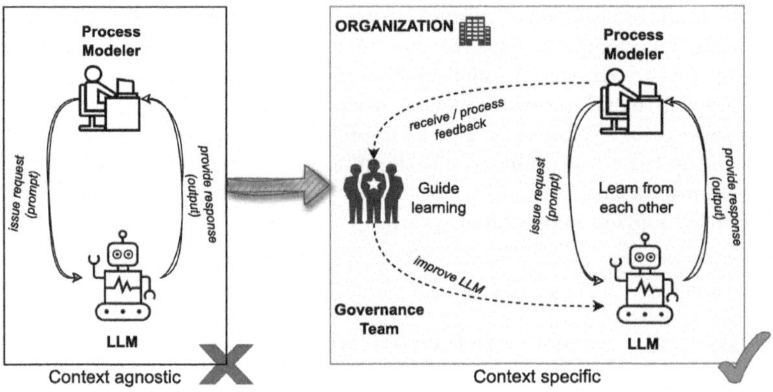

Fig. 6. Takeaway. LLMs are hardly usable for process modeling in a context-agnostic setting [left]. Deployment of LLM in organizations for process modeling should follow a context-specific approach, in which the governance team ensures that LLM and end-users "learn from each other" [right].

Acknowledgements. We thank Hilti for enabling and funding this research; and the participants to our user study for their contributions and feedback.

References

1. Our repository. https://github.com/Nouronihar/BPM24_LLM4PM
2. Bellan, P., Dragoni, M., Ghidini, C.: Extracting business process entities and relations from text using pre-trained language models and in-context learning. In: Almeida, J.P.A., Karastoyanova, D., Guizzardi, G., Montali, M., Maggi, F.M., Fonseca, C.M. (eds.) EDOC 2022. LNCS, vol. 13585, pp. 182–199. Springer, Cham (2022). https://doi.org/10.1007/978-3-031-17604-3_11
3. Dumas, M., La Rosa, M., Mendling, J., Reijers, H.A., et al.: Fundamentals of Business Process Management. Springer, Heidelberg (2018)
4. Fill, H.G., Fettke, P., Köpke, J.: Conceptual modeling and large language models: impressions from first experiments with chatgpt. EMISAJ (2023)
5. Franch, X., Palomares, C., Quer, C., Chatzipetrou, P., Gorschek, T.: The state-of-practice in requirements specification: an extended interview study at 12 companies. Requirements Eng. **28**, 377–409 (2023)
6. nce Grohs, M., Abb, L., Elsayed, N., Rehse, J.R.: Large language models can accomplish business process management tasks. In: International Conference BPM (2023)
7. Herbst, J., Karagiannis, D.: An inductive approach to the acquisition and adaptation of workflow models. In: IJCAI (1999)
8. Hevner, A.R., March, S.T., Park, J., Ram, S.: Design science in information systems research. MIS Q. (2004)
9. Ivanchikj, A., Serbout, S., Pautasso, C.: From text to visual BPMN process models: Design and evaluation. In: ACM/IEEE MoDELS (2020)
10. Ivanchikj, A., Serbout, S., Pautasso, C.: Live process modeling with the BPMN sketch miner. Softw. Syst. Model. **21**, 1877–1906 (2022)
11. Klievtsova, N., Benzin, J.V., Kampik, T., Mangler, J., Rinderle-Ma, S.: Conversational process modelling: state of the art, applications, and implications in practice. In: International Conference on Business Process Management (2023)
12. Kourani, H., Berti, A., Schuster, D., van der Aalst, W.M.: Process modeling with large language models. arXiv:2403.07541 (2024)
13. Lewis, P., et al.: Retrieval-augmented generation for knowledge-intensive nlp tasks. In: Advances in Neural Information Processing Systems (2020)
14. Peffers, K., Tuunanen, T., Rothenberger, M.A., Chatterjee, S.: A design science research methodology for information systems research. JMIS **24**, 45–77 (2007)
15. Plattfaut, R., Borghoff, V., Godefroid, M., Koch, J., Trampler, M., Coners, A.: The critical success factors for robotic process automation. Comput. Ind. **138**, 103646 (2022)
16. Rosemann, M.: Potential pitfalls of process modeling: part a. BPM J. (2006)
17. Senkus, P., et al.: Bridging the gap: an evaluation of business process management education and industry expectations–the case of Poland. In: International Conference on Business Process Management (2023)
18. Sola, D., van der Aa, H., Meilicke, C., Stuckenschmidt, H.: Activity recommendation for business process modeling with pre-trained language models. In: European Semantic Web Conference (2023)
19. Stein Dani, V., et al.: Towards understanding the role of the human in event log extraction. In: Marrella, A., Weber, B. (eds.) BPM 2021. LNBIP, vol. 436, pp. 86–98. Springer, Cham (2022). https://doi.org/10.1007/978-3-030-94343-1_7
20. Vom Brocke, J., Simons, A., Riemer, K., Niehaves, B., Plattfaut, R., Cleven, A.: Standing on the shoulders of giants: challenges and recommendations of literature search in information systems research. CAIS (2015)

Author Index

C. Di Ciccio et al. (Eds.): BPM 2024, LNBIP 527, pp. 485–486, 2024.
https://doi.org/10.1007/978-3-031-70445-1

The manufacturer's authorised representative in the EU is Springer Nature Customer Service Centre GmbH, Europaplatz 3, 69115 Heidelberg, Germany. If you have any concerns regarding our products, please contact ProductSafety@springernature.com

Printed and bound by CPI Group (UK) Ltd, Croydon, CR0 4YY
02/10/2026
02227135-0006